# Sustainable Environmental Waste Management Strategies

**Series Editors**

Alok Prasad Das, Department of Life Sciences, Rama Devi Women's University, Bhubaneswar, Odisha, India

Eric van Hullebusch, Équipe ACE, Institut de Physique du Globe de Paris, Paris Cedex 05, France

Sustainable waste management is an absolute imperative in a world that is already dealing with the effects of climate change. This book series seeks to publish an extensive collection of scientific books targeting researchers, scientists, professionals, academics, students, and anyone else involved in the wide, diverse field of sustainable waste management. The series accepts monographs, edited volumes, textbooks, and conference proceedings. Topics covered include but are not limited to environmental pollution and remediation, sources of environmental waste, environmental and human health, sustainable waste management, circular economy, nature-inspired environmental solutions, pollution prevention, and energy-efficient technologies for waste management. The series puts a particular emphasis on developing countries/ the majority world but accepts proposals globally as long as the topic is in scope.

Alok Prasad Das · Selcan Karakuş
Editors

# Synergizing Sustainability for Integrated Waste Management

## Artificial Intelligence, Economic Stability and Energy Recovery

Springer

*Editors*
Alok Prasad Das
Department of Life Sciences
Rama Devi Women's University
Bhubaneswar, Odisha, India

Selcan Karakuş
Department of Chemistry
Faculty of Engineering
Istanbul University-Cerrahpaşa
Istanbul, Türkiye

ISSN 3005-1622 ISSN 3005-1630 (electronic)
Sustainable Environmental Waste Management Strategies
ISBN 978-3-032-03717-6 ISBN 978-3-032-03718-3 (eBook)
https://doi.org/10.1007/978-3-032-03718-3



This Springer imprint is published by the registered company Springer Nature Switzerland AG
The registered company address is: Gewerbestrasse 11, 6330 Cham, Switzerland

# Contents

**Overview of Sustainable Waste Management** . . . . . . . . . . . . . . . . . . . . . . . . 1
Şakir Altınsoy and Nuray Bekoz Üllen

**Relationship Between Artificial Intelligence and Carbon Footprint** . . . . . 17
Gulcicek Dere

**A Multidimensional Approach to E-Waste with Challenges and Strategies for Effective Sustainable Waste Management** . . . . . . . . . . . 35
Elif Tüzün, Meryem Demir Güldal, Tarık Küçükdeniz, and Canan Hazal Akarsu

**Waste Management and Its Impact on Environment and Health: Addressing Emerging Waste Challenges in a Post-pandemic World** . . . . . 53
Sweta Sinha, Rahul Singh, Sudhir Lima, Mayank Srivastava, Nitya Garg, and Sona Kumari

**Management of Plastic Wastes: History, Current Applications, and Future Perspectives of Recycling, Upcycling, and Reclaiming Technologies** . . . . . . . . . . . . . . . . . . . . . . . . . . . . . . . . . . . . . . . . . . . . . . 69
Mine Begum Alanalp, Mustafa Fatih Ergin, and Ali Durmus

**Innovative Technologies for Heavy Metal Management** . . . . . . . . . . . . . . . 99
Murat Torun and Nursel Pekel Bayramgil

**Cutting-Edge Technologies for Heavy Metal Waste Management** . . . . . . . 125
S. Charulatha and K. S. Thangamani

**Innovative Strategies for Sustainable Waste Disposal** . . . . . . . . . . . . . . . . . 151
Kiran Bala, Renuka Harit, Arushi Dogra, Raj Kumar, and Alok Prasad Das

**Waste Management's Effects on the Environment and Health** . . . . . . . . . . 163
Yasemin Yilmazoğlu and Emre Yilmazoğlu

**Integrated Waste Management: Importance, Legislation, and Research Strategies Towards Sustainability** .......... 191
V. P. Sharma, Pragya Sharma, and Alok Prasad Das

**Leveraging AI for Intelligent Waste Management in Smart Cities** ...... 205
Ritah Nabawagga, Jeanritah Roenah Birungi, Shikta Singh, and Satya Ranjan Dash

**Waste-to-Energy Conversion** .......... 227
Selcan Karakuş and Alok Prasad Das

**Nanotechnology Sustainability and Nano-waste Concerns** .......... 245
Gizem Karabulut Sevk and Nuray Beköz Üllen

**Sustainable Environmental Solutions Through Nanotechnology: Opportunities, Challenges, and Nanowaste Management** .......... 265
Noor Alam Chowdhary and L. S. Songachan

**Application of Nanotechnology for Sustainable Agricultural Waste Management and Energy Recycling** .......... 297
Soumya Hota, Sudeshna Dey, and Alok Prasad Das

**Recent Advances in Valorizing Spent Coffee Grounds for Biodiesel Production with Sustainability Considerations** .......... 323
Merve Ceylan, Ibrahim Metin Hasdemir, Ekin Selvi, and Dilek Özmen

**Economic Potential of Recyclable Wastes in Municipal Solid Waste: A Case Study of the Marmara Region** .......... 345
Mirac Nur Ciner, Serdar Aydın, Emine Elmaslar Özbaş, H. Kurtulus Ozcan, Sinan Güneysu, and Atakan Öngen

**Waste Water Treatment Using the Constructed Wetland: Recent and Future Prospects** .......... 365
Vivek Raj, Santosh Kumar Karn, Rakesh Kumar, Aman Prakash, Kanchan Kumari, Alok Prasad Das, and Neeraj Choudhary

# Overview of Sustainable Waste Management

**Şakir Altınsoy and Nuray Beköz Üllen**

## 1 Introduction

In general terms, waste management consists of the cycle of collecting, transporting, processing, recycling, purifying, and disposing of waste materials and managing, monitoring, and regulating them (Mishra et al., 2023). It focuses on the management of waste materials generated by human activities and reducing the negative effects of waste on human health and natural life (Dey et al., 2023). Waste materials can consist of solid, liquid, gas, or radioactive substances. Activities related to waste management may differ in domestic and industrial wastes depending on the development status of countries (Das et al., 2019; Elsaid & Aghezzaf, 2015). Waste management is necessary for a sustainable and sensitive world. Waste management addresses not only the disposal of garbage or recycling of landfill waste but also what needs to be done to avoid creating waste (Priyadarsini & Das, 2024, 2025). Sustainability should develop important strategies such as reducing waste and moving waste collection to an economical place. There are potential environmental risks associated with waste from industrial establishments. Environmentally friendly waste management principles need to be adopted in a sectoral sense. Previously, the economic system adopted the make, use, and throw away model for most areas. This was the case in automotive, textile, electrical-electronics, white goods, and many other areas. However, today this model has given way to sustainable processes. Waste management has changed from a linear economy to a circular economy. Sustainable and renewable

Ş. Altınsoy
Department of Biomedical Engineering, Faculty of Engineering and Architecture, Istanbul Yeni Yuzyil University, Istanbul, Turkey

N. Beköz Üllen (✉)
Department of Metallurgical and Materials Engineering, Faculty of Engineering, Istanbul University-Cerrahpaşa, İstanbul, Turkey
e-mail: nbekoz@iuc.edu.tr



A. P. Das and S. Karakuş (eds.), *Synergizing Sustainability for Integrated Waste Management*, Sustainable Environmental Waste Management Strategies,
https://doi.org/10.1007/978-3-032-03718-3_1

energy-generated processes have replaced the use of raw materials (Castaldi, 2014; Elsaid & Aghezzaf, 2015; Yeheyis et al., 2013).

Sustainable waste management is one of the important elements of environmental protection and responsible resource use today. Sustainable waste management aims to support recycling, reduce waste, and recover energy from waste. For this purpose, it is necessary to reduce waste sources, implement effective and efficient programs, generate energy from waste, and support the use of products obtained from waste (Das et al., 2024; Mishra et al., 2019). When technological, institutional, and financial factors are taken into consideration, the path to sustainable waste management is quite long. Waste management approaches should provide options with more effective results in terms of sustainability (Al-Maaded et al., 2012; Chen et al., 2010; Das et al., 2019). This section addresses the strategic perspectives of sustainable waste management. These are generally to identify existing innovations and resources, to analyze the waste management system in the international arena, to define and evaluate the waste management life cycle, and to present effective approaches for sustainable recycling of solid waste. The geographical location and economic situation of countries are also very important in sustainable waste management.

Waste management involves a decentralized process that is linked to the national economy. Advanced waste monitoring technologies used by developed countries have accelerated their development in recent years (Das & Ghosh, 2022; Mohanty et al., 2018). The landfill technique is the most widely adopted method of waste disposal globally due to its easy applicability. Due to its high cost, most waste cannot be recycled. Modern landfill technologies have been developed in developed countries that provide optimum use of waste with minimum pollution risk. Since these areas are constructed with many architectural improvements, they are generally environmentally safer than commonly used landfills (Al-Maaded et al., 2012; Elsaid & Aghezzaf, 2015). These areas typically consist of four layers. The lower layer consists of a dense clay liner covered with high-density polyethylene sheets that restrict the leakage of toxic waste. The second layer contains suitable drainage channels that direct the flow of incoming toxic liquids from the landfill to a treatment plant. The third layer collects the gases formed in the storage area and then uses them for cleaning and different usage processes. The last layer at the top forms the largest reservoir to accommodate solid waste (Saha, 2023). The reason why such an application cannot be applied in underdeveloped countries is the lack of specialized labor. However, deterioration in waste treatment infrastructure is often due to countries' economic stagnation and low environmental awareness. Internationally, waste management has become a vital problem for most countries in the world. With increasing industrialization, this problem has also grown. The main obstacles to overcoming this problem are limited resources, inadequacy of technology to be applied for treatment and disposal, and ineffective implementation of relevant legislation. European countries in particular have made progress in sustainable waste management. They have established standards for sustainable waste management with eco-cities based on environmental protection (Das et al., 2019; Kiddee et al., 2013; Moustakas et al., 2020).

This chapter presents various strategies for sustainable waste management: (i) identifying waste management technology, strategic innovations, and waste monitoring tools, (ii) recognizing countries' waste management strategies, (iii) conducting waste management life cycle assessment by identifying other modeling processes, and (iv) identifying applicable approaches for sustainable recycling and utilization of waste (Das et al., 2019; Elsaid & Aghezzaf, 2015). The geographical location and economic conditions of countries play a significant role in determining waste characteristics. Countries with strong economies have defined a number of economic and LCA models for the selection and performance evaluation of appropriate waste management algorithms. Various innovative solutions are being demonstrated in many countries for the implementation of smart and sustainable waste management plans. Soil, air, and water pollution pose a constant risk to sustainable development. Therefore, waste management is directly related to sustainable development (Pires et al., 2019). This chapter clearly states that traditional waste recycling processes are not as adequate as they should be. Various sustainable waste management strategies are mentioned to solve the problems encountered in waste disposal. A society can achieve sustainable waste management goals with the participation of all stakeholders in the production and consumption areas. In this chapter, sustainable solution suggestions for waste management will also be discussed.

## 2 Sustainable Waste Management

The extent to which sustainable waste management will be successful varies in terms of the economy, wealth, standard of living, culture, policy, and administrative systems of countries. A strong economy and advanced technology provide an opportunity for the sustainable development of industrialized countries (Mehta et al., 2023). Effective use of material resources, development of elements that will reduce waste production, and implementation of sustainable development projects in economic, social, and environmental terms are required. Sustainable waste management can be achieved with the correct planning of processes. These are listed below (Das et al., 2019; Elsaid & Aghezzaf, 2015; Wilson et al., 2013):

- Reducing the amount of waste produced in all areas
- Encouraging the use of recovered or recycled materials
- Supporting recycling areas and technological facilities
- Ensuring that legal regulations for recycling are made and put into practice

To develop sustainable waste management, it is crucial to identify the types of waste that require sustainable management. These can be considered as natural waste, industrial waste, medical waste, hazardous radioactive waste, construction waste, and e-waste. Among these wastes, hazardous waste poses a serious threat to natural life and living beings. In order to manage hazardous waste within the scope of sustainable waste management, it is necessary to comply with strict legal regulations and use approved carriers and technologically based disposal facilities. Careful and effective management of clinical and medical waste is mandatory. Following country-specific coding and disposal rules are the standards required for effective management. Sustainable waste management aims to reduce waste and improve recycling. Waste management solutions and the implementation of effective waste management strategies are also necessary to achieve the goals (Wilson et al., 2013; Yeheyis et al., 2013).

## 3 Principles of Sustainable Waste Management

The waste management hierarchy is the basic application in sustainable waste management that offers different waste management options that reflect relative sustainability (Ghosh & Das, 2017; Pradhan et al., 2023; Sahu et al., 2024a). The 3Rs (reduce, reuse, and recycle) mentioned in the previous section form the infrastructure of the waste management hierarchy. The "prevention process," which is at the top of the waste management hierarchy, aims to minimize possible risks by reducing waste production. Reducing unnecessary consumption in every area of life, not using single-use products, and reducing the amount of packaging in which products are stored will significantly reduce waste production. Reuse is the best medicine for sustainability and offers opportunities for a sustainable waste process. Recycling saves energy, reduces the demand for new products, and helps to create creative ideas for creating new products. In sustainable waste management, waste should be addressed with the concept of the waste hierarchy (Kiddee et al., 2013; Theron & Dowden, 2017). Environmentally, the options for the disposal of waste are limited. For this, as little waste as possible should be produced; therefore, it is at the top of the waste prevention hierarchical pyramid. It is followed by reuse and recovery processes. Landfills and incineration techniques are the least accepted options due to high carbon emissions and are at the bottom of the hierarchy. Sustainability of waste management is possible if the waste materials have the least negative impact during processing and disposal. The mentioned impacts can be considered as environmental, social, and economic impacts. Targets to be adopted for a sustainable waste management model (Das et al., 2019; Elsaid & Aghezzaf, 2015; Lakiotia et al., 2017) include:

- Creating climate-friendly technologies and energy sources
- Creating targets to reduce waste production in every area
- Extending the usability of products over long periods
- Adjusting the use of resources to optimum levels
- Developing processes that protect the balance of the ecological system
- Not preferring to create landfill waste and use the incineration process
- Ensuring that waste is monitored and reported by effective institutions
- Ensuring that suppliers have sustainable standards

# 4 The Hierarchy of Sustainable Waste Management

The waste hierarchy or waste management hierarchy leads to sustainable processes that minimize waste production and maximize efficiency with effectively used resources. It is the approach and general outline of sustainable waste management. It creates solutions to find answers to the question of "How to deal with waste?" This framework focuses on waste management practices that prioritize environmental sustainability and resource conservation. Waste hierarchy processes can be used in many different applications and remain valid (Castaldi, 2014; Elsaid & Aghezzaf, 2015; Lee et al., 2020). All processes described in the waste hierarchy given in Fig. 1 provide environmental and economic benefits within themselves.

## *4.1 Waste Prevention/Reduction*

At the top of the Sustainable Waste Management Hierarchy is the process of reducing waste. In general, it means reducing the amount of waste to a minimum. Using less packaging at every stage of consumption, avoiding single-use products, and preferring to use reusable products in every area of our lives can be possible. A significant amount of general domestic and municipal waste consists of domestic packaging waste. Preventive activities begin at the design and production stage of products (Chen et al., 2010; Phillips et al., 2002). Consumers support the process by ensuring that products are used for a long time in order to prevent waste. The more waste that can be collected and then processed and disposed of, the more recycled raw materials will be released. This formation reduces costs and reduces environmental impact. Waste prevention/reduction, the highest level of the hierarchy, is a prevention activity also called source reduction or simply reduction. This prevention activity

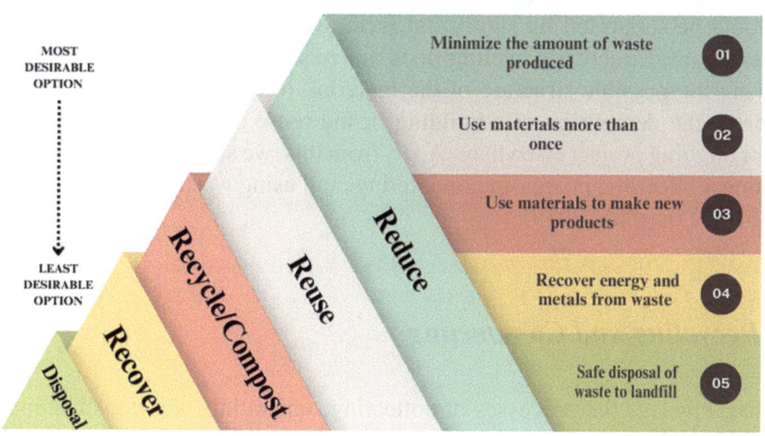


**Fig. 1** Schematic representation of waste management hierarchy processes

aims to reduce waste formation at its source. Generally accepted prevention activities may include the following (Elsaid & Aghezzaf, 2015; Phillips et al., 2002; Zorpas, 2016):

- Designing products that reduce resource use at the design stage
- Reducing the packaging ratio of products
- Increasing the durability of products to increase their lifespan
- Encouraging conscious and responsible processes in consumption models

The basic aim of all of these is to prevent waste generation by providing resource optimization. It aims for a process that is achieved through preventive activities without the need for corrective action.

### *4.2 Materials Reuse*

This stage, which is the second level of the hierarchy, is the reusability of products without undergoing any significant change or requiring any processing. The process that must occur for waste products to be reused is defined at this stage. Reusing a product for similar purposes or a different purpose is the next choice. The status of the product used for reuse should be reviewed, and it should be evaluated whether it meets the necessary standards. Studies have shown that 30% of the products that enter the recycling process are still usable, while this rate is 25% for electronic waste. What is important at this stage is the demand for products produced from waste products, that is, encouraging reuse. To achieve this, it is necessary to donate products that are seen as waste but usable, repair unusable products, give different forms to old products, and create new designs (Cucchiella et al., 2017; Saha, 2023).

With this stage, the life of the products is extended, and the demand for new resources decreases. It also seriously affects the processes of selling and purchasing products. Manufacturers need to work on the reuse of waste products. This situation should be supported by significant incentives and developed with training by an expert team. Effective recycling does not require additional time or energy by preserving the possible structure of the material, for example, the on-site reuse of waste from the demolition of a building or the reuse of remaining materials for a future or ongoing project elsewhere. Apart from this, we should consider recycling as separating, collecting, processing, marketing, and using waste materials (Cucchiella et al., 2017; Das et al., 2019).

### *4.3 Recycling and Composting*

Recycling involves the processes of collecting, separating, and transforming waste into a new product, for example, the collection of paper waste in educational institutions, government offices, and many offices and its reuse for paper production.

Composting is a process that uses organic waste, although it is similar. An example is the composting of vegetable and fruit waste for the purpose of obtaining fertilizer. The recycling and composting process requires less energy and cost than obtaining new products. The economic viability of this stage, which is the third step of the hierarchy, interacts with processes that provide quality waste streams, the storage and shipping status in the waste total process, and the market price of recycled materials. For example, construction waste should be recycled close to where it is removed because it is heavy (Das et al., 2019; Lakiotia et al., 2017). Research shows that even in areas where recycling is possible, approximately 60% of recyclable products are thrown away. Materials such as paper, glass, plastic, and metal are generally recyclable, meaning they can be collected, processed, and transformed into new products. Packaging products in particular need to be easily recyclable. However, even in this area, it has been found that almost 40% cannot be easily recycled. This process, which plays an important role in sustainable waste management, is an important step in the hierarchy as it saves new resources, reduces energy consumption and emissions, and reduces waste in landfills or incinerators (Kiddee et al., 2013).

## 4.4 *Energy Recovery*

This level of the hierarchy is detailed in converting non-recyclable waste into electricity, heat, or fuel. There are many ways to obtain energy from waste products in today's technology. In waste-to-energy facilities, waste management processes such as collecting, burning, gasifying, and pyrolyzing waste, as well as landfill gas recovery, are included. The energy obtained from waste incineration can recover some energy from the waste stream. In the first stage, this energy is much less than that obtained by recycling, composting, reusing, or reducing the waste produced. If recycling is not possible, the waste should be evaluated in one of the other recovery options rather than in the landfill. Energy recovery has a lower environmental impact than recycling or reuse. However, it is still a better way than sending waste directly to landfills. For example, burning non-recyclable materials reduces the consumption of fossil fuels needed to generate power. Thus, combustion reduces the amount of waste sent to landfills in addition to providing energy. This stage of the hierarchy is also important for renewable energy production by converting organic waste into electricity, heat, or fuel through technologies such as waste-to-energy plants and anaerobic digestion (Castaldi, 2014; Das et al., 2019; Lakiotia et al., 2017).

## 4.5 *Treatment*

In cases where the reuse and recycling of waste is not possible or cannot be recovered for energy production, treatment is the case. This stage may include processes such as chemical treatment, bioremediation, or sterilization to reduce the volume or hazardous nature of the waste. There is a clear position that incineration is not environmentally friendly due to the emissions released into the air. However, it is the only permitted disposal option for many flammable, volatile, toxic, and infectious hazardous wastes. This is especially true for medical waste due to the potential health risks. Apart from this, hazardous materials such as lead-containing and asbestos-containing products, mercury-containing equipment, and batteries must receive appropriate and necessary treatment in terms of health and safety (Castaldi, 2014; Das et al., 2019; Elsaid & Aghezzaf, 2015; Lakiotia et al., 2017).

### *4.6 Landfill Disposal*

This stage, which constitutes the last step in the waste management hierarchy, is the disposal of waste. This stage, which is the least preferred in the waste hierarchy, is usually the process that sends waste to landfills or incineration facilities without energy recovery. The main goal should be to reduce the dependency on landfills and incineration by removing waste from this final destination. Technological landfills and management practices try to reduce environmental damage. The waste hierarchy is a guide for the life cycle of the waste product. Wastes that cannot be reused, recycled, or converted into energy should be disposed of. As urbanization increases, usable land areas decrease, and the garbage rate increases. This situation is a threat to environmental protection, resource conservation, and human well-being. In order to increase high recycling or reuse rates, a waste management plan should be in place and should be implemented carefully. The waste sustainability index should also be developed (Elsaid & Aghezzaf, 2015; Zorpas, 2016).

## 5 Circular Economy for Sustainable Waste

The concept of circular economy aims to increase the lifespan of products and maximize their value before responsibly disposing of them. At the heart of sustainable waste management, it aims to reuse waste products and create new materials from recycled resources (Bala et al., 2024; Tripathy et al., 2024). A circular economy in waste management is an economic system that includes more processes to find solutions to problems related to sustainable waste management practices. The traditional linear economy is an alternative process that adopts the take-make-dispose approach. It is based on the principles of reducing, reusing, and recycling resources and materials by creating a closed-loop system to achieve its goals (Elsaid & Aghezzaf, 2015; Saha, 2023; Zorpas, 2016). In order to use natural resources effectively and reduce the negative impacts on the life cycle and the environment, a transition has been made toward a circular economy in sustainable waste management. Strategies based on a circular economy in many industrial areas can provide new sources of income by reducing waste and greenhouse gas emissions and improving product quality.

Circular economy in waste management is a restorative and regenerative economic system by design. The basic components on which this concept is based are (i) eliminating waste and pollution, (ii) extending the life of products and materials, and (iii) renewing natural systems (Saha, 2023).

Circular economy strategies for sustainable waste management include several approaches: (i) optimizing production processes, reducing losses during storage and transportation, and consequently reducing waste production and (ii) reusing waste materials as input materials in other industries or products. The circular economy approach enables more sustainable waste management practices in all industrial areas, preventing environmental damage and providing economic benefits (Castaldi, 2014; Das et al., 2019; Elsaid & Aghezzaf, 2015; Saha, 2023). The concepts underlying the circular economy, reducing waste, protecting resources and creating economic efficiency, are presented in Fig. 2.

More powerful and effective circular economy strategies should be developed for sustainable waste management. These strategies may require costly processes to implement at the beginning, but economic investment opportunities can be created in the long term. More information should be acquired about the economic benefits obtained through circular economy strategies. The audience that should be actively included in the process is the decisive element, the consumers. For a sustainable waste management and circular economy cycle, there should be solid cooperation between companies, governments, and civil society organizations. As industrial organizations continue to focus on sustainability and environmental responsibility, it is assumed that circular economy strategies will become increasingly important (Saha, 2023).

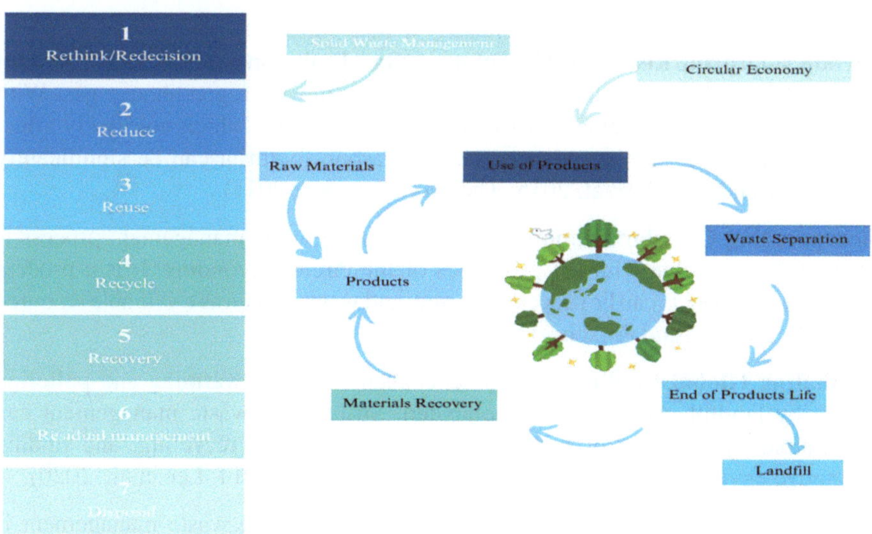


**Fig. 2** Concepts forming the basis of the circular economy

## 6 Significance of Sustainable Waste Management

The importance of sustainable waste management is given below in bullet points:

**Energy Recovery**: Energy production from waste is a vital component of sustainable waste management, reduces the amount of waste sent to landfills, and also produces renewable energy, which is accepted as more sustainable energy (Das et al., 2019; Moustakas et al., 2020).

**Environmental Approach**: The main problems in waste management are inadequate waste storage areas and inappropriate disposal processes. When these processes are not effectively controlled, they destroy vital areas such as water and soil and endanger the living spaces of living beings. The protection of the ecological system is the starting point of sustainable waste management (Das et al., 2019; Okedu et al., 2022).

**Protection of Resources**: Recycling and reuse of waste products can support the protection of resources. The implementation of these periods not only reduces the demand for new raw materials but also preserves energy and reduces carbon emissions (Castaldi, 2014; Elsaid & Aghezzaf, 2015).

**Controlling Global Warming**: Sustainable waste management plays a key role in both reducing global warming and carbon emissions. Both landfills and burning or burying waste are huge sources of greenhouse gases. We can control global warming by reducing landfill waste, expanding recycling, and activating the energy conversion cycle of waste with technological developments (Elsaid & Aghezzaf, 2015; Phillips et al., 2002).

**Carbon Footprint/Reducing Greenhouse Gas Emissions**: Areas where landfill waste is located are the source of methane, a very powerful greenhouse gas. With sustainable waste management processes, these emissions, which negatively affect the global balance, can be reduced, and the carbon footprint can be significantly reduced (Elsaid & Aghezzaf, 2015; Theron & Dowden, 2017).

**Incorporating Circular Economy**: The sustainable waste management process is inherently compatible with the principles of the circular economy. It is a process where waste products can be recycled, reused, and waste is seen as a valuable resource (Castaldi, 2014; Elsaid & Aghezzaf, 2015).

**Combating Global Challenges**: Problems such as global warming and pollution caused by landfill waste must be combated. Sustainable waste management can be possible by reducing or controlling waste, supporting recycling, and finding alternative environmental solutions to landfills (Castaldi, 2014; Lee et al., 2020).

**Protection of Natural Resources**: The aim of sustainable waste management is to protect our valuable and existing natural resources. Many materials, especially materials whose basic materials are plastic, paper, and metal, enter the recycling process. This application reduces the use of natural resources, and thus the ecosystem is protected (Yeheyis et al., 2013).

**Supporting Sustainable Product Management**: Sustainable waste management addresses the entire life cycle from production to use and disposal of the product. Preferring green products and applying green design techniques in production reduces waste generation and increases the efficient use of natural resources (Elsaid & Aghezzaf, 2015; Zorpas, 2016).

**Support for Recycling in Industry**: Adopting recycling processes in every area of the industry is essential for strong, sustainable waste management. It is a process that supports the growth of the recycling industry and is quite advantageous in the long term. It contributes to a circular economy, the economy is revitalized, the need for new materials is reduced, and a more sustainable area is created (Cucchiella et al., 2017; Yeheyis et al., 2013).

## 7 Benefits of Sustainable Waste Management

It would be better to list the benefits of sustainable waste management to better understand them. Some of these benefits (Cucchiella et al., 2017; Elsaid & Aghezzaf, 2015; Gunnarsson & Fält, 2020; Ludwig et al., 2012; Okedu et al., 2022) include:

- The greatest benefit of sustainable waste management is focused on the green environment.
- It supports the fight against global warming process by reducing greenhouse gas emissions.
- It reduces energy consumption and provides energy savings.
- It protects natural resources and is one of the pioneers in the fight against climate change.
- It eliminates both visual and bad odors by controlling and reducing waste areas.
- It provides the creation of markets for recycling.
- It helps the formation of more conscious and green industrial institutions.
- It saves labor, time, and resources.

The efforts of sustainable waste management are aimed at reducing pollution, which is a common problem for all living things, and protecting natural resources (Das & Ghosh, 2018; Ray et al., 2024; Sahu et al., 2024b). In addition to its environmental and economic benefits, it also provides social benefits that encourage all humanity to act together. Some of the economic benefits of buying recycled products are given in Table 1. Sustainable waste management will find alternatives by preventing waste from going to landfills or incineration facilities. In general, it tends to manage the process by educating people on this issue and encouraging responsible recycling. Forest areas will be damaged less with recycling, and energy and water consumption will be significantly reduced. These processes help reduce environmental stress and ensure more efficient use of resources. In terms of economics,

**Table 1** Some benefits of buying recycled products

| Area of influence | Benefits |
|---|---|
| Local and general governments | • Savings are achieved by using less water and energy in the production of recycled products<br>• Long-term profit is possible by reusing resources<br>• Strong stance of the administrations toward the society is ensured |
| Sector and industrial organizations | • The increasing demand for recycled products contributes to the growth and development of the existing market<br>• Increasing the product range of the local and general sector<br>• Contributing to the green and circular economy |
| Society and environment | • New sectors and opportunities opened by recycling provide job opportunities<br>• Safer use of natural resources<br>• Less water and energy consumption<br>• Reduction of greenhouse gas emissions that will reduce air and water pollution |

sustainable waste management can provide significant economic savings. The social benefit to be obtained from sustainable waste management is more conscious and improved public health. Green approaches and quality air and water consumption make life more livable and protect the health of all living things. In addition to all these, sustainable waste management also provides job opportunities in areas such as waste collection, separation, and processing (Deviatkin et al., 2023; Esmaeilian et al., 2018; Vargas & de Souza Campos, 2020).

## 8 Challenges in Sustainable Waste Management

Insufficient environmental awareness, unconscious social attitudes, and loss of motivation have prevented the acceptance of new technologies that will improve sustainable waste management and approaches toward innovation. The challenges facing the development of sustainable waste management are as follows: high costs in waste collection, uncontrolled waste monitoring process, inadequate budget allocation, inadequate equipment and lack of qualified personnel, inadequate training and public awareness on waste management, inadequate technological processes and developments, lack of laws and regulations on waste management, visual problems experienced in waste collection areas, etc. Overcoming these challenges is possible with innovative solutions and investment in infrastructure. This process requires joint efforts between governments, industries, and individuals (Morrissey & Browne, 2004; Phillips et al., 2002). Dependence on traditional waste disposal methods is also one of the main challenges in sustainable waste management. There is a tendency toward landfilling and incineration for waste disposal to a large extent. This process, which leads to environmental pollution and resource depletion, can be eliminated with investment and infrastructure development. The ease of use and low cost of disposable products often hinder efforts to reduce waste at source. Education, awareness campaigns, and incentives may be an option to reduce waste production. Reusing waste products is a way to reduce waste and conserve resources. Lack of infrastructure for waste collection and processing and limited consumer availability of recycled materials make reuse efforts difficult. Creating new markets for recycled products and incentives to raise awareness among businesses about using recycled products are potential solutions. The lack of a complete waste management system is a significant challenge. The development and implementation of advanced recycling and waste treatment technologies require significant investment and expertise. The adoption of new waste management technologies can be hindered by technological limitations. Overcoming these limitations requires research and development efforts as well as collaboration between industry and government organizations (Brunner & Rechberger, 2015; Gunnarsson & Fält, 2020; Okedu et al., 2022).

## 9 Conclusion

Sustainable waste management needs to be addressed for the future of our world. Waste management has become a prerequisite for healthy, safe, and sustainable living spaces. Waste should be considered as a resource rather than a problem. The process can be carried out from many perspectives, from implementing the waste hierarchies discussed in this section to waste-to-energy conversion and zero-waste targeting. Sustainable waste management, which strives for a greener future, requires innovation, patience, dedication, awareness, and discipline. Unbalanced living conditions have caused an imbalance in the ecosystem through the production, consumption, and unnecessary use of natural resources. Today, it is understood that the most sustainable way to manage waste is recycling and composting. One of the aims of this section is to address and understand the paths leading to sustainable waste management. Scientific and industrial studies on the waste cycle are still insufficient, and, therefore, every step taken is important. The concept of sustainable waste management develops a methodology that encompasses all living things on earth. With this methodology, the decision-making process, strategic development, evaluation, and implementation are achieved. The success of the waste management system depends on technical elements such as collection, disposal, and recycling, as well as public and social management processes. In developed countries, there are integrated waste management facilities. The use of these facilities, which facilitate waste management, should be expanded in low-income countries. Solutions for sustainable waste management should be developed locally and adapted to local needs and conditions. Waste management services should be at a cost that can be afforded locally. Research shows that sustainable waste and resource management applies to developing countries and that more stable progress is carried out reasonably. Technological advances have made it possible to solve environmental problems in more economical and trouble-free ways. Developments in technology, depending on the status, type, amount, and characteristics of the waste, will create more economical and sustainable waste management systems.

For environmental sustainability in waste management, it is essential to create a waste management system that reduces the effects of energy consumption and air, soil, and water pollution. For economic sustainability in waste management, the financial picture in the waste management system must be accepted by all segments of society. For a sustainable system in environmental and economic terms, all waste management infrastructures such as storage areas, incineration facilities, waste types, and amounts must be addressed. Alternatives with minimum environmental impact and low cost for low-pollution recycling and energy recovery are the reasons for the emergence of different treatment systems. Pyrolysis, gasification, the combination of the two, and thermal treatment options of composting systems are increasingly gaining interest as economical and green alternative methods for waste recycling. To make sustainable waste management effective, all types of waste such as industrial, agricultural, and construction waste must be treated in order to be included in economic life. Although sustainable waste management is seen as a technical issue, it is understood that it is a political and economic issue in terms of the functioning of the process.

## References

Al-Maaded, M., Madi, N. K., Kahraman, R., Hodzic, A., & Ozerkan, N. G. (2012). An overview of solid waste management and plastic recycling in Qatar. *Journal of Polymers and the Environment, 20*(1), 186–194. https://doi.org/10.1007/s10924-011-0332-2

Bala, K., Anshul, Verma, S., Dogra, A., Das, A. P., Kumar, R., & Nandini. (2024). Sustainable management for urban plastic waste generation. *Sustainable Management of Urban Plastic Waste Through Circular Economic Approaches*, 211–236. https://doi.org/10.1201/9781003465447-13

Brunner, P. H., & Rechberger, H. (2015). Waste to energy–key element for sustainable waste management. *Waste Management, 37*, 3–12.

Castaldi, M. J. (2014). Perspectives on sustainable waste management. *Annual Review of Chemical and Biomolecular Engineering, 5*(1), 547–562. https://doi.org/10.1146/annurev-chembioeng-060713-040306

Chen, X., Geng, Y., & Fujita, T. (2010). An overview of municipal solid waste management in China. *Waste Management, 30*(4), 716–724.

Cucchiella, F., D'Adamo, I., & Gastaldi, M. (2017). Sustainable waste management: Waste to energy plant as an alternative to landfill. *Energy Conversion and Management, 131*, 18–31.

Das, A. P., & Ghosh, S. (2018). Bioleaching of Manganese from mining waste materials. *Materials Today: Proceedings, 5*(1), 2381–2390. https://doi.org/10.1016/J.MATPR.2017.11.459

Das, A., & Ghosh, S. (2022). Role of microorganisms in extenuation of mining and industrial wastes. *Geomicrobiology Journal, 39*, 173–175. https://doi.org/10.1080/01490451.2022.2038953

Das, S., Lee, S.-H., Kumar, P., Kim, K.-H., Lee, S. S., & Bhattacharya, S. S. (2019). Solid waste management: Scope and the challenge of sustainability. *Journal of Cleaner Production, 228*, 658–678.

Das, A., Ray, S. K., Mohanty, M., Mohanty, J., Dey, S., & Das, A. P. (2024). Ecotoxicity of microplastic wastes and their sustainable management: A review. *Environmental Chemistry and Ecotoxicology, 6*, 144–152. https://doi.org/10.1016/J.ENCECO.2024.05.003

Deviatkin, I., Horttanainen, M., & Havukainen, J. (2023). Sustainability of waste management systems: final disposal. In: Içinde S. O. Idowu, R. Schmidpeter, N. Capaldi, L. Zu, M. Del Baldo, & R. Abreu (Eds.), *Encyclopedia of Sustainable Management* (ss. 3356–3365). Springer International Publishing. https://doi.org/10.1007/978-3-031-25984-5_413

Dey, S., Das, A., Mallick, K., Sahu, A., & Das, A. P. (2023). Environmental petroleum waste: Pollution, toxicity, sustainable remediation, pp. 159–175. https://doi.org/10.1007/978-3-031-48220-5_7

Elsaid, S., & Aghezzaf, E.-H. (2015). A framework for sustainable waste management: Challenges and opportunities. *Management Research Review, 38*(10), 1086–1097. https://doi.org/10.1108/MRR-11-2014-0264

Esmaeilian, B., Wang, B., Lewis, K., Duarte, F., Ratti, C., & Behdad, S. (2018). The future of waste management in smart and sustainable cities: A review and concept paper. *Waste Management, 81*, 177–195.

Ghosh, S., & Das, A. P. (2017). Bioleaching of manganese from mining waste residues using Acinetobacter sp. *Geology, Ecology, and Landscapes, 1*(2), 77–83. https://doi.org/10.1080/24749508.2017.1332847

Gunnarsson, J., & Fält, E. (2020). *Sustainable Waste Management: Identifying and Analysing the Present and Future Waste Situation in Linga Linga*. https://www.diva-portal.org/smash/record.jsf?pid=diva2:1461685

Kiddee, P., Naidu, R., & Wong, M. H. (2013). Electronic waste management approaches: An overview. *Waste Management, 33*(5), 1237–1250.

Lakiotia, E. N., Moustakas, K. G., Komilisc, D. P., Domopouloua, A. E., & Karayannisa, V. G. (2017). Sustainable solid waste management: Socio-economic considerations. *Chemical Engineering, 56*. https://www.academia.edu/download/114626948/111.pdf

Lee, R. P., Meyer, B., Huang, Q., & Voss, R. (2020). Sustainable waste management for zero waste cities in China: Potential, challenges and opportunities. *Clean Energy, 4*(3), 169–201.

Ludwig, C., Hellweg, S., & Stucki, S. (2012). *Municipal solid waste management: Strategies and technologies for sustainable solutions*. Springer Science & Business Media.

Mehta, D., Paliwal, D., & Tege, S. (2023). Sustainable waste management: An approach towards sustainability. *International Journal of Emerging Technologies and Innovative Research, 5*(9), 101–104.

Mishra, S., Ghosh, S., van Hullebusch, E. D., Singh, S., & Das, A. P. (2023). A critical review on the recovery of base and critical elements from electronic waste-contaminated streams using microbial biotechnology. *Applied Biochemistry and Biotechnology, 2023*, 1–30. https://doi.org/10.1007/S12010-023-04440-X

Mishra, S., Rout, P. K., & Das, A. P. (2019). Solar photovoltaic panels as next generation waste: A review. *Biointerface Research in Applied Chemistry, 9*(6).

Mohanty, S., Ghosh, S., Bal, B., & Das, A. P. (2018). A review of biotechnology processes applied for manganese recovery from wastes. *Reviews in Environmental Science and Bio/Technology 2018 17*(4), 791–811. https://doi.org/10.1007/S11157-018-9482-1

Morrissey, A. J., & Browne, J. (2004). Waste management models and their application to sustainable waste management. *Waste Management, 24*(3), 297–308.

Moustakas, K., Rehan, M., Loizidou, M., Nizami, A.-S., & Naqvi, M. (2020). Energy and resource recovery through integrated sustainable waste management. İçinde *Applied Energy* (C. 261, s. 114372). Elsevier. https://www.sciencedirect.com/science/article/pii/S0306261919320598

Okedu, K. E., Barghash, H. F., & Al Nadabi, H. A. (2022). Sustainable waste management strategies for effective energy utilization in Oman: A review. *Frontiers in Bioengineering and Biotechnology, 10*, Article 825728.

Phillips, P. S., Clarkson, P., Barnes, N. J., & Adams, J. (2002). A UK county sustainable waste management programme. *International Journal of Environment and Sustainable Development, 1*(1), 2. https://doi.org/10.1504/IJESD.2002.000714

Pires, A., Martinho, G., Rodrigues, S., & Gomes, M. I. (2019). Sustainable solid waste collection and management. *Springer International Publishing.* https://doi.org/10.1007/978-3-319-93200-2

Pradhan, G., Tripathy, B., Ram, D. K., Digal, A. K., & Das, A. P. (2023). Bauxite mining waste pollution and its sustainable management through bioremediation. *Geomicrobiology Journal*, 1–10. https://doi.org/10.1080/01490451.2023.2235353

Priyadarsini, S., & Das, A. P. (2024). Analytical and structural characterization of waste lithium-ion batteries for their effective recycling strategy. *Environmental Chemistry and Ecotoxicology.* https://doi.org/10.1016/j.enceco.2024.12.004

Priyadarsini, S., & Das, A. P. (2025). Lithium bioleaching: A review on microbial-assisted sustainable technology for lithium bio-circularity. *Journal of Water Process Engineering, 69*, Article 106744. https://doi.org/10.1016/j.jwpe.2024.106744

Ray, S. K., Dey, S., Das, A. P., & John, A. (2024). Microplastic waste management through greener technologies : A review. *Sustainable Management of Urban Plastic Waste Through Circular Economic Approaches*, 237–261. https://doi.org/10.1201/9781003465447-14

Saha, A. (2023). Circular economy strategies for sustainable waste management in the food industry. *Journal of Recycling Economy & Sustainability Policy, 2*(2), 1–16.

Sahu, A., Dey, S., Mallick, K., Priyadarsini, S., Behera, I. D., & Das, A. P. (2024a). A review on advanced technology for sustainable management of synthetic microplastic waste. *Sustainable Management of Urban Plastic Waste Through Circular Economic Approaches*, 262–279. https://doi.org/10.1201/9781003465447-15

Theron, C., & Dowden, M. (2017). *Strategic sustainable procurement: Law and best practice for the public and private sectors.* Routledge. https://www.taylorfrancis.com/books/mono/https://doi.org/10.4324/9781351274487/strategic-sustainable-procurement-colleen-theron-malcolm-dowden

Tripathy, B., Dey, S., Dubey, N. K., Lo, W.-C., & Das, A. P. (2024). An omics-based approach for managing the risk of COVID-19-generated wastes, pp. 51–66. https://doi.org/10.1007/978-3-031-50840-0_4

Vargas, D. B., & de Souza Campos, L. M. (2020). Waste management in higher education institutions: A state-of-the-art overview. In *Proceedings of the 5th NA International Conference on Industrial Engineering and Operations Management Detroit, Michigan, USA*, pp. 75–86. https://www.ieomsociety.org/detroit2020/papers/16.pdf

Wilson, D. C., Velis, C. A., & Rodic, L. (2013). Integrated sustainable waste management in developing countries. *Proceedings of the Institution of Civil Engineers—Waste and Resource Management, 166*(2), 52–68. https://doi.org/10.1680/warm.12.00005

Yeheyis, M., Hewage, K., Alam, M. S., Eskicioglu, C., & Sadiq, R. (2013). An overview of construction and demolition waste management in Canada: A lifecycle analysis approach to sustainability. *Clean Technologies and Environmental Policy, 15*(1), 81–91. https://doi.org/10.1007/s10098-012-0481-6

Zorpas, A. A. (2016). Sustainable waste management through end-of-waste criteria development. *Environmental Science and Pollution Research, 23*(8), 7376–7389.

# Relationship Between Artificial Intelligence and Carbon Footprint

**Gulcicek Dere**

## 1 Introduction

Sustainability is now a deciding element in the corporate world; it is not an alternative (Bala et al., 2024; Singh Jadaun et al., 2022). In many industries, it is now a major force behind company strategy, innovation, and regulatory compliance. In terms of sustainability, solutions like carbon capture and the circular economy are becoming more popular. As a component of long-term decarbonization routes, they are being widely embraced. Compared to many other industries, AI energy usage is contributing to a fast-increasing carbon footprint. This sharp rise necessitates quick, efficiency-focused mitigating measures. The advancement of AI technology significantly impacts carbon emissions and energy usage. Because of these effects, responsible AI development that is in line with sustainability objectives is required. Large language models and Deep Learning algorithm training in particular result in substantial energy consumption and notable carbon emissions (Digital Sustainability, 2024; Mills, 2023; Sustainable Organizations, 2020). Reducing emissions can be greatly aided by optimizing such models. It takes a tremendous amount of data and processing power to train and run these models. Both the financial and environmental expenses are increased by this need.

Thousands of high-performance graphics processors (GPU/TPU) must be used for prolonged periods to train Deep Learning models with billions of parameters (e.g., GPT-3, 175 billion parameters) (Grigorev, 2021; OECD, 2018). Research on energy efficiency is expanding as a result of this hardware need. Carbon emissions and energy use are thus rising quickly. International requests for green AI techniques have been sparked by this trend. Technology companies like Google, Microsoft, and Amazon have data centers that hold hundreds of servers and use a lot of power since

G. Dere (✉)
Bahcesehir University, Istanbul, Turkey
e-mail: gulcicekdere@yahoo.com

 
A. P. Das and S. Karakuş (eds.), *Synergizing Sustainability for Integrated Waste Management*, Sustainable Environmental Waste Management Strategies,
https://doi.org/10.1007/978-3-032-03718-3_2

they run around the clock. Their activities account for a significant amount of the environmental impact of the digital economy. Out of all AI applications, chatbots and large language models use the most energy (Patterson et al., 2021a, 2021b). They are now the main topic of discussion in the field of sustainable AI.

A single ChatGPT query, for instance, uses ten times as much power as the typical Google search query. The significant energy cost of sophisticated language models is demonstrated by this ratio. During operation, the OpenAI-created GPT-3 model generates 502 tons of carbon dioxide and 1,287 MWh of energy. These numbers demonstrate the magnitude of emissions linked to cutting-edge models. In the United States, it produces 120 kWh of electricity and 110 kWh of carbon dioxide. These values are often used in carbon footprint benchmarking studies. Google's AlphaGo Zero model produces 96 tons of $CO_2$ during its 40 h of operation. Despite its short runtime, the environmental impact is notable. A popular chatbot service, ChatGPT, produces 260,930 kg of $CO_2$ during its operation. This emission level is comparable to thousands of transatlantic flights. This indicates a significant outcome for the company, as it currently uses 18.3 TWh of electricity, which is a substantial increase compared to the 2% TWh of electricity used in all services in Atlanta (DeepMind, 2018; Silver et al., 2018; Vanderbauwhede, 2024). The scale of energy use has sparked debates around AI accountability.

## *1.1 Operational Energy Footprint of AI Technologies*

The development and deployment of chatbot systems necessitate strategic planning to enhance energy efficiency and minimize carbon emissions. Frameworks for energy-conscious model development are being widely explored. As AI becomes more integrated into various industries, concerns regarding its environmental footprint have gained prominence. Sustainability assessments are now a vital component of AI implementation strategies.

In particular, chatbot technologies are increasingly scrutinized for their role in global energy consumption and carbon emissions (International Energy Agency, 2023). This scrutiny is prompting industry-wide reflection on responsible AI usage. On the other hand, several striking examples highlight the immense energy consumption of AI in human-level tasks. Such comparisons are key to contextualizing the ecological cost of advanced computation. For instance, AlphaGo, an AI-powered system that defeated the world champion in the game of Go, required approximately 1 megawatt of power during the match. This energy cost illustrates the gap between machine and biological efficiency. In contrast, the human brain operates on just 20 watts—indicating that AlphaGo uses 500,000 times more energy than a human brain to achieve victory (Moore, 1965). The disparity reinforces calls for energy-aware AI architectures. This stark contrast underscores the inefficiency of AI systems when compared to biological intelligence, raising concerns about scalability and sustainability in AI applications (Schwartz, 2024). It also pushes for bio-inspired computing models that mimic natural efficiency.

## 1.2 Efficiency Strategies and Industry Response

In response to these challenges, various strategies are being developed to reduce the environmental impact of AI and foster a more sustainable digital ecosystem. The aim is to achieve high performance without compromising ecological balance. Companies that are at the forefront of technology are actively working toward the goal of making data centers more energy-efficient, with a strong emphasis on investments in renewable energy (DeepSeek, 2024). Through the utilization of AI intelligence, they are managing resource allocation and reducing operational inefficiencies.

In addition, AI-driven services, in particular search engines, are likely to raise the demand for computational electricity by five to ten times for each query. This will result in significant extra stress on the energy supply, considering the billions of searches that are performed every day (Hölzle, 2016). This trend underscores the need for algorithmic innovation and hardware efficiency to control future demand. This exponential increase in energy demand necessitates robust regulatory frameworks and innovations in AI efficiency to balance technological advancement with environmental responsibility. Interdisciplinary collaboration between governments, academia, and industry is becoming essential.

## 1.3 The Demand for Energy and the Development of Hardware

It is anticipated that the energy consumption of data centers will considerably increase, reaching 1.8% of the total power use worldwide by the year 2030 (Koomey et al., 2016). This is due to the rapid growth of digital technology and applications of AI. The importance of establishing environmentally friendly data infrastructure on a worldwide scale is brought to light by this scenario. Furthermore, according to the findings of some research, the need for computing power brought about by AI and cryptocurrency mining might boost the energy consumption of data centers to approximately 8% of the total world electricity consumption by the same year (Luccioni, 2024). A rise of this magnitude, if not controlled properly, has the potential to dramatically undermine climate objectives. As our dependence on high-performance computing continues to grow, we are beginning to have concerns over the sustainability and efficiency of our digital infrastructure. The use of eco-centric design ideas in the next generation of supercomputing systems is becoming more popular. One of the primary contributors to this ever-increasing need for energy is the ongoing development of information processing power. According to Moore's Law, the number of transistors on an integrated circuit is estimated to double about every 18 months. This leads to fast advances in computing power without large increases in production costs (Nakagawa, 2024).

On the other hand, this exponential development in capacity has not always been the goal of environmental sustainability. Even though this approach has made it possible for AI skills to expand exponentially, it has also resulted in a significant rise in the amount of computing energy that is required, which has contributed to environmental concerns (PolicyWatch, 2025). To establish a separation between performance and power consumption, new paradigms in chip design are required. As AI models get more complicated and need a bigger processing capacity, businesses and researchers need to investigate energy-efficient solutions to reduce the negative effect that these models have on the environment. Quantization, knowledge distillation, and dynamic computing are some of the methods that are included in this investigation.

## *1.4 Moore's Law and the Transition to Green AI*

According to the ChatGPT 4o model, the transition will take place as previously described by the year 2030 (Rong et al., 2016). This is in agreement with Moore's Law, which states that the expansion will take place. However, if architectural changes are not emphasized, this estimate indicates that there will be an increase in the need for energy. As a consequence of this, there has been an increase in the area of AI research that focuses on efficiency. The objective is to get performance that is comparable to or even better than the model while using a much lower amount of energy. Through the development of algorithms and models that accomplish the same goal but use less processing power, the goal is to achieve this. Because of this, specific lightweight designs and alternate transformers have been developed as a result. Figure 1 represents Moore's Law: Transistor Count Over Time.

In the field of Deep Learning, several methods, including model pruning, network reduction, and efficient attention mechanisms, have been implemented to reduce the amount of computing effort that is required (The Guardian, 2016). When it comes to

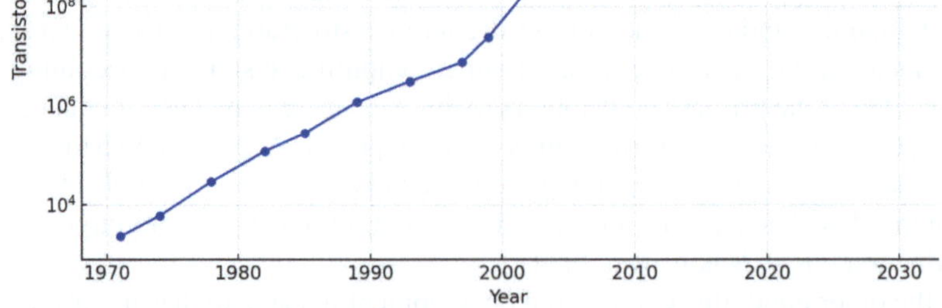


**Fig. 1** Moore's law: transistor count over time

constructing AI infrastructures that are both scalable and sustainable, these strategies are essential. The goal of these technologies is to reduce carbon emissions by reducing the size of models and, therefore, the amount of energy that is used (Vincent, 2016). Smaller versions not only decrease the amount of energy that is required, but they also cut the price of hardware and make access more equal.

In addition, a system known as "Green AI" is being used, which evaluates energy efficiency in conjunction with the results of research on AI (Vinuesa et al., 2020). This paradigm change redefines success measures in AI by emphasizing environmental effects. Within this context, several research teams have started releasing the amount of energy that AI models use as well as their carbon footprint throughout the publication process. This is done to raise more awareness about the importance of sustainability in the development of AI. In the field of AI research, transparency in energy measurements is becoming an ethical requirement (UNEP, 2023; Lindner et al., 2019). As a consequence of this, carbon labeling methods are gaining hold in both the academic world and the business world, which encourages AI developers to favor energy-efficient models. A change in market expectations is brought about as a result of this step, which encourages innovation in low-carbon computing.

## *1.5 Optimization of Artificial Intelligence in Data Centers*

While this is going on, the DeepMind team at Google is working hard to improve the energy efficiency of its data centers by using AI-driven systems that continually evaluate sensor data (ISO, 2019). In the realm of intelligent infrastructure management, this real-time analysis has established itself as a standard. It has been shown that AI can optimize operational efficiency while minimizing environmental effects (Desislavov et al., 2023; Dhar, 2020). This method has successfully decreased energy usage at Google's facilities by forty percent when it was implemented. There is a compelling argument to be made for the use of AI not just in goods but also in the maintenance of their infrastructure.

As a result of these improvements, there is an urgent need for sustainable AI solutions, as the demand for computing resources continues to increase (European Commission, 2021). In the absence of such modifications, the ecological cost of AI could be greater than the advantages it provides to society.

# 2 Regulatory Mechanisms

## *2.1 International Policy Frameworks and Strategic Guidelines*

To reduce the carbon effect of AI, intergovernmental cooperation between states and companies is necessary. Collaboration on a global scale ensures that efforts are coordinated and that standards are established across boundaries. There is a voluntary commitment being made by enterprises in the technology sector to reduce their emissions. Frequently, these commitments are included in the ESG (environmental, social, and governance) initiatives of corporations. Corporate entities such as Amazon, Google, and Microsoft, for example, have committed to achieving carbon neutrality or negative carbon emissions by certain dates and have increased their investments in renewable energy (European Investment Bank, 2022). These kinds of agreements affect the whole ecosystem of technology, driving partners and suppliers toward more sustainable practices.

On the other hand, organizations such as the United Nations and the European Union are working toward the establishment of consensus by distributing guiding principles about environmentally friendly technology and digital sustainability. To foster international collaboration on digital climate responsibility and bridge regulatory gaps, some efforts have been launched. In its report for the year 2023, the United Nations Environment Program (UNEP) presents a plan to increase awareness of the relationship between AI intelligence and carbon emissions on a worldwide scale, while simultaneously cultivating an understanding of sustainable AI (Masanet et al., 2020). The guideline provided by the United Nations Environment Program establishes a core policy framework for the use of AI globally. As a result, increasing the efficiency with which AI uses power may result in significant climatic advantages. This roadmap delineates five major proposals for governments and corporations:

1. Creating standardized methods to assess the environmental consequences of artificial intelligence
2. Regulations mandating that organizations disclose the environmental consequences of their AI-based products and services
3. Companies repurposing resources such as water and hardware and enhancing the energy efficiency of their AI algorithms
4. Encouraging carbon offsetting and promoting the use of renewable energy in the data center
5. The integration of general environmental legislation with AI policies

The World Wide Fund for Nature (WWF) and the World Input–Output Database (WIOD) are among the organizations that provide detailed statistics on greenhouse gas emissions on an annual basis. Their databases provide assistance for decision-making that is evidence-based and for planning over the long term for sustainability.

These reports provide emission information from a variety of industries, such as the energy sector, commercial industry, agricultural sector, and waste sector. As a result of this data, practitioners of AI may better connect their technology with emission reduction aims. The results of the measurements suggest that there has been a large rise in the carbon footprint, and the degree to which it contributes to the overall footprint is increasing. This development is a reflection of the critical necessity for all businesses to embrace AI techniques that are focused on energy efficiency. The electrical industry is responsible for the largest amount of the carbon footprint (Williams, 2004). As a result, increasing the efficiency with which AI uses power may result in significant climatic advantages.

## 2.2 *Certification Standards and Monitoring Practices*

For international organizations that are entrusted with evaluating, reporting, and regulating carbon emissions, certificates are provided under the ISO 14064, EUROCERT, ISO 14067, and QSI CERT standards. Certifications like this provide credibility to environmental responsibility and give established criteria for measuring performance. Their primary emphasis is on sustainability and performance management, and they provide information on carbon emissions as well as strategies for reducing emissions. Through the establishment of quantitative objectives, they make it possible to measure the progress made in green technology developments. Providing strategic analysis of the sustainability initiatives used by corporations is what TSE Global does. The assessments it provides assist in aligning the ambitions of corporations with sustainability standards that are recognized worldwide.

Auditing organizations such as TUV Rheinland, EUROCERT, and DQS assist businesses and projects throughout the process of evaluating the environmental impacts of their operations, ensuring that they comply with international standards, and developing strategies via the use of reporting and verification. Checkpoints for compliance are provided by these audits, which also help to assure a long-term commitment to environmental objectives. In the area of monitoring, reporting, and verifying carbon footprints and greenhouse gas emissions, some companies provide consultancy services, auditing services, and certification services. These organizations operate domestically and globally. Because of their knowledge, the implementation of climate policy is sped up in both the public and commercial sectors. Bureau Veritas, Sigmacert Global, Quality Services International, and AURA International Certification are some of the organizations that assist businesses and projects in the management and certification of their carbon footprint and greenhouse gas emissions by international standards (Hook et al., 2020). When it comes to investor confidence, their certification methods often affect investor confidence, promote openness, and create trust.

Organizations such as AFNOR Group, IQ & Partners, TSE GLOBAL, Quick-Carbon, and QSI CERT also support businesses and projects in managing and verifying their carbon footprint and greenhouse gas emissions by international standards. These institutions are important stakeholders in the process of adopting policies that reduce carbon emissions throughout the sector. A standard has been set for the carbon footprint of the product, which ensures that the quantification of greenhouse gas emissions and reductions connected with the product, as well as its manufacturing, shipping, and disposal processes, is accurate. Product-level footprinting raises awareness among consumers and encourages them to buy environmentally friendly products.

A standard for measuring the carbon footprint of products is ISO 14067. It assists businesses and projects in the process of measuring and certifying their carbon footprint and emissions of greenhouse gases by international regulations. The environmental product declarations (EPDs) are built on this standard, which serves as their foundation. The following are examples of websites that provide statistical data and research on AI energy consumption and carbon footprint: some investors have disclosed that Google's use of AI in 2021 amounted to ten to fifteen percent of the entire amount of power used.

The data in question bring to light the disproportionate amount of energy that is used by the most advanced AI models. In the context of climate change, CarbonSmart explores the possible effects and threats that AI may have. They provide policymakers and business executives with risk evaluations that are based on hypothetical scenarios. Robot Dreams discusses the role of AI intelligence in the future of the world, while the Green Energy Journal examines the impact of AI systems on energy consumption and carbon emissions. These kinds of publications encourage critical conversation on the ecological responsibilities of AI (Gomes et al., 2019; BrainBox AI, 2024).

## 3 Global Consequences

### *3.1 International Regulatory and Financial Response*

Officials from the governments of the United States of America and the European Union are now working together to help reduce the environmental effects caused by the use of energy on a worldwide scale. These endeavors are geared toward reaching a global agreement on environmentally responsible AI techniques.

The European Parliament has made the sustainability of energy consumption a priority, while at the same time taking specific steps to mitigate the impacts of energy use on the environment. Digital sustainability is increasingly being included in conventional energy policy frameworks via the implementation of policy directives.

The "Environmental Impact of Artificial Intelligence Act" of 2023 has as its primary purpose the reduction of carbon emissions and the cutback of energy consumption across a variety of industries and geographical areas. This legislation is one of the first legal measures to govern the ecological footprint that AI leaves behind. Improving global energy strategy and reducing environmental hazards are the goals of these policies, which are intended to be implemented. They act as models for other countries that are interested in addressing emissions connected to AI. The principles that should be followed to ensure that AI is environmentally sustainable are defined in this recommendation paper; however, these concepts are not legally obligatory. Standardizing worldwide compliance is made more difficult by the fact that this lack of enforcement already exists. Because these principles are not legally obligatory, the environmental sustainability component is usually ignored (The Guardian, 2024). More stringent legal requirements are required to transform suggestions into concrete effects.

On a worldwide scale, sector-based carbon footprint data can be found in databases such as the International Energy Agency (IEA), World Resources Institute (WRI), Climate Action Initiative (CAIT), United Nations Framework Convention on Climate Change (UNFCCC), and World Bank—Carbon Footprint Data (Buscheck et al., 2023). To facilitate evidence-based decision-making and policy assessment, several repositories are available. Both historical and current information about energy-related emissions from a variety of industries and nations are included in the database. The ability to evaluate progress and anticipate future trends requires such temporal insights to be of crucial importance. When it comes to investments in carbon-intensive industries, which emphasize sustainability and attempts to reduce carbon footprints, there is a possibility that the value of these investments may decrease because of the global scale. As a result, investors are put under pressure to change their portfolios toward more ecologically responsible investments. In light of this, the financial industry is also making the move to low-carbon investments; environmental, social, and governance funds; and green bonds (Nguyen & Associates, 2023). This change is a reflection of a more comprehensive linkage between environmental stewardship and organizational financial success.

## 3.2 *Environmental Burden of Digital Infrastructure*

The significant amount of energy required by large-scale data centers contributes to the ecological footprint that this business leaves behind. These establishments have come to represent the energy issues that are being faced in this era of digital technology. In the same way that it is the case in other industrial sectors, the environmental viability of electronic circuits and vehicles is dependent on the efficient management of waste. In addition to carbon emissions, the management of electronic waste is becoming an increasingly important challenge.

The transfer of technological companies to renewable energy sources is becoming an increasingly important need (Hao, 2020). As the technology market becomes more competitive, sustainability credentials have emerged as a significant differentiation.

The remote work and online services that have grown more widespread in our lives and that we have experienced in several locations during the pandemic are also contributing factors that make the digital carbon footprint even more significant (Strubell et al., 2019).

Consequently, investments in environmentally responsible infrastructure are very necessary to accompany digital change. By reducing the amount of energy that is used, AI plays a significant part in reducing the carbon footprint (Wu et al., 2022). Due to its ability to forecast and adapt, it offers a tremendous potential for environmentally friendly innovation.

## *3.3 Role of AI in Energy Optimization and Decarbonization*

Intelligent energy management systems can decrease needless usage by analyzing the energy consumption of buildings (Calvert, 2024). In the fight against climate change, smart grids and intelligent infrastructure are becoming more important instruments. There are a great number of industries that generate a significant amount of energy consumption, especially those that operate on a considerable scale. Among the industries that contribute the most carbon to the atmosphere are manufacturing, transportation, and computers. These processes require a significant amount of energy, notably in the production and transmission of electricity. This portion of the energy consumption is very significant. For this reason, it is necessary to modernize energy distribution networks to achieve greater efficiency. A further factor that leads to increased carbon emissions is the usage of businesses that need a significant amount of energy. Immediately beneficial effects on the environment will result from lowering the energy intensity of these industries. For instance, the energy output of a large-scale facility is higher than that of a smaller facility, which indicates that the former uses more energy than the latter. The effects of scaling operations need to be counterbalanced by compensatory mechanisms. Because all activities lead to an increase in energy consumption, the rise in carbon emissions is also a consequence of the use of the Internet (Heikkilä, 2023). At this point, the patterns of digital consumption are comparable to those of conventional industrial activity. The reduction of carbon emissions and the reduction of waste by ten to twenty percent are both attainable via the optimization of waste management systems. Increasingly, AI models are being used to assess inefficiencies in waste management systems. In addition, it is feasible to reduce the amount of energy that is used and the amount of carbon emissions that are produced in facilities by refining energy systems. In this context, "automating HVAC systems" and "integrating sensor-based feedback loops" are both included.

On the other hand, by capitalizing on the environmental consequences of a variety of technologies, it is possible to accomplish the optimum waste management techniques, energy efficiency, and new energy consumption. Decarbonization of the whole system is made possible by the synergy of these developments. There is the potential for this to result in cost savings, as well as long-term advantages for the economy and the environment. Benefits of this nature illustrate the economic feasibility of environmentally responsible innovation.

# 4 Carbon Footprint Reduction Initiatives

## *4.1 Emerging Technologies for Sustainability and Energy Optimization*

To reduce carbon emissions and improve energy efficiency, Carbon Capture and Storage (CCS) Technologies are essential. Renewable energy solutions, including solar panels, wind turbines, hydrogen energy, electric vehicles (EVs), smart grids, IoT sensors, and digital twins, optimize energy use and lower emissions across sectors. These innovations contribute to sustainability by enhancing energy efficiency and reducing carbon footprints. Additionally, Software-Based Solutions play a crucial role in minimizing emissions without requiring physical infrastructure. Technologies such as Cloud Computing, Virtualization, Carbon Accounting Software, Blockchain-based Carbon Credit Systems, and Digital Education platforms help reduce energy consumption and support environmental awareness. By integrating these solutions, industries can achieve long-term economic and environmental benefits while fostering a sustainable future (Hiller, 2024).

The study conducted by Nguyen and associates assesses the most effective contributions of carbon capture and storage (CCS) technologies and renewable energy systems to attain carbon neutrality by 2050. The investigation concentrates on the advancement of novel energy and carbon capture and storage (CCS) technologies to decrease carbon emissions by 55% and 67%, respectively, by 2050. Furthermore, it investigates the impact of individual technologies on carbon emissions and energy efficiency. The study also investigates the potential of digital technology to reduce carbon emissions and the impact of digital technologies on carbon emissions. The study also investigates the potential of smart technologies to reduce energy consumption and carbon emissions, with a particular emphasis on the use of BrainBox AI, which in the 45 Broadway area of Manhattan, results in a 15% reduction in energy consumption when compared to HVAC. The significance of incorporating these technologies into energy management strategies is underscored by the research (Rahman-Jones, 2024).

## *4.2 AI-Driven Reductions in Energy Use and Carbon Emissions*

In 2021, the GPT-3 model, which was developed by OpenAI, was able to decrease energy consumption by 1287 MWh and $CO_2$ emissions by 502 kg during its operation. This is equivalent to the annual carbon emissions of 110 vehicles and the electricity consumption of 120 households in the United States (Anny, 2024). Similarly, during its 40-day training period, Google's AlphaGo Zero reduced $CO_2$ emissions by 96 kg, highlighting the potential for energy-efficient AI training (Columbia

University, 2024). Furthermore, ChatGPT, a widely used AI service, has contributed to an emission reduction of approximately 260,930 kg, which is equivalent to 260 transatlantic flights between New York and London (Zhang & Peters, 2023). These advancements emphasize the role of AI-driven optimizations in reducing carbon footprints and improving energy efficiency across industries (Han & Zhou, 2023).

In the past few years, deep learning systems have stood out because they can deal with unstructured data like text, images, and sounds. CNN, RNN/LSTM models for Time Series, and Transformer-based designs like BERT/GPT have made big steps forward in many areas, including natural language processing, classification, and prediction.

## 5 Artificial Intelligence Applications in Waste Management

Artificial intelligence (AI) has significantly transformed traditional waste management systems by improving classification accuracy, forecasting capabilities, and operational efficiency. Table 1 provides an overview of specific AI models employed across various waste management domains.

ResNet-50, an image processing-based classification model, comes in first place in Table 1. It lowers the need for human help on the recycling line by automatically identifying the types of trash. Second place goes to the reinforcement learning algorithm DQN, which figures out the best tracks for trash trucks. This also cuts down on carbon pollution. In third place, important parts of electronic waste are sorted by using natural language processing (BERT) and visual recognition (YOLOv5)

**Table 1** Specific AI models applied in waste management domains

| Application area | AI model used | Performance/observed impact |
|---|---|---|
| Waste sorting | ResNet-50 (CNN) | Plastic and paper wastes were classified with ~ 95% accuracy, enhancing sorting speed and precision |
| Collection route optimization | Deep Q-Network (DQN) | Enabled route planning that reduced travel distance and fuel usage by over 20% |
| E-waste recovery | YOLOv5 + BERT | Facilitated identification of reusable e-waste components with 90% + precision |
| Compost quality forecasting | Random Forest Regressor | Achieved 88% accuracy in classifying organic compost types using chemical data |
| Carbon emission prediction | LSTM + Attention Mechanism | Improved CO2 prediction accuracy by 25%, aiding emission control strategies |

together. In fourth place, the random forest algorithm was used to guess the quality of composts made from biological garbage based on their chemical makeup. Fifth place goes to LSTM models that use time series data to correctly predict carbon emissions from trash processing systems.

# 6 Conclusion and Discussion

## *6.1 Summary of Findings and Present Challenges*

The energy intensity of current applications and the associated carbon emissions are a significant concern in terms of environmental sustainability, despite the potential for artificial intelligence technologies to make positive contributions in areas such as energy efficiency and environmental monitoring. Consequently, it is crucial to implement strategies that optimize energy efficiency during the development and implementation of AI systems. This demand may be further exacerbated by the rapid increase in duties resulting from artificial intelligence. A projection indicates that energy efficiency may only increase by fourfold under the present trend, even though computing processing demands are anticipated to increase by 50-fold by 2028 (Dockrill, 2024). This situation suggests that the carbon footprint of data centers operated by major technology companies could increase exponentially in the future years if effective measures are not taken.

AI algorithms optimize energy consumption by minimizing resource consumption and reducing waste.

Large corporations, including Amazon, Microsoft, and Google, implement renewable energy strategies to enhance their competitiveness (World Economic Forum, 2023). AI also assists in the optimization of carbon management and storage systems by reducing greenhouse gas emissions.

The ecological footprint of AI research and implementations is also expanding in comparison to traditional information and communication technologies (ICT). The widespread adoption of AI has the potential to reduce the carbon footprint of the digital sector to that of industries such as transportation and manufacturing, as the emissions of data centers exceed those of aviation (Kaur & Patel, 2022). It has been reported that the number of data centers worldwide has increased from approximately 500,000 in 2012 to an extraordinary 8 million by 2022, a figure that has been influenced by AI growth (Jain & Müller, 2023). This trend suggests that the negative impact of AI on climate change will swiftly escalate if energy sources are not converted to renewable energy sources, and efficiency is not enhanced.

## *6.2 Strategic Directions and Future Opportunities*

To mitigate the environmental impact of AI, the following strategies can be implemented:

- Optimizing hardware and algorithms to train AI models with minimal energy consumption.
- Prioritizing renewable energy sources for data centers operating AI systems (Hiller, 2024).
- Consistent monitoring and reporting of the energy consumption and carbon emissions associated with AI applications.

The development and implementation of smart grid systems are also crucial for sustainable development and energy efficiency. In the future, utilizing smart grid technologies has the potential to substantially decrease greenhouse gas emissions and energy consumption.

Large AI models, like GPT and BERT, use a lot of energy while they are being trained, which indirectly adds to carbon emissions. Two hundred eighty-four tons of $CO_2$ can be made by training just one NLP model. Long-term, real-time systems also leave behind an enormous carbon impact. Different approaches, such as TinyML, compact models, and green AI, are better for the environment.

The design of extremely energy-efficient systems is heavily influenced by technological breakthroughs. AI technologies, in particular, enable not only the digitization of commercial operations but also their environmental sustainability (OECD, 2019). Nowadays, reducing the carbon footprint is central to many technologies developed to improve the efficiency of energy-intensive operations.

AI algorithms are one of the most recent approaches developed in this regard. These algorithms optimize themselves in response to environmental factors. Processing capacity may be adjusted by these algorithms using real-time energy sources and carbon intensity data, which can lead to emissions reduction and energy savings (Goldman Sachs Research, 2024). Furthermore, the initial model architectures that incorporate carbon awareness have been developed; these structures are assessed for their environmental impact in addition to their accuracy (Bilancioni, 2024).

The utilization of AI-powered World Observatories is an additional noteworthy application area. These systems have a wide range of applications, including the identification of illicit harvesting in wooded areas and the creation of carbon emission maps. They give policymakers and environmental monitoring organizations real-time data support. They provide environmental monitoring groups and legislators with real-time data help. They help environmental monitoring agencies and legislators with real-time data (CEVA, Inc., 2016).

From a hardware perspective, AI-created adaptive and low-power CPUs are amazing. These hardware elements improve energy efficiency and lower the total carbon footprint of the system using self-reorganization in line with the load of the application (Polytechnique Insights, 2024) and issue sustainability credits (Lindner &

Arto, 2019). Micro-environmental modeling is a data-driven approach to infrastructure development that analyzes energy use and carbon consumption in rural or disadvantaged regions at the local level (Anny, 2024).

Lastly, Digital Twin Worlds, which enable the simulation of the entire world, hold significant potential for the comprehensive monitoring of carbon emissions and the simulation of future scenarios (Zhang & Peters, 2023).

## References

Anny, D. (2024). *Energy-Efficient AI Architectures for Sustainable Computing*. ResearchGate.

Bala, K., Anshul, Verma, S., Dogra, A., Das, A. P., Kumar, R., & Nandini. (2024). Sustainable management for urban plastic waste generation. *Sustainable Management of Urban Plastic Waste Through Circular Economic Approaches*, 211–236. https://doi.org/10.1201/9781003465447-13

Bilancioni, M. (2024). Energy efficiency: AI vs. the human brain. Medium.

BrainBox AI. (2024). Making buildings more efficient. Time.

Buscheck, T. A., Sun, Y., & Hao, Y. (2023). S-plume geothermal (CPG): Power net generation from 3D fluvial aquifers. *International Journal of Greenhouse Gas Control, 127*, Article 103828.

Calvert, B. (2024). AI already uses as much energy as a small country. It's only the beginning.

CEVA, Inc. (2016). Another look at AlphaGo vs. Lee Sedol: The power angle. CEVA's Official Blog.

Columbia University. (2024). *Landscape Assessment of AI for Climate and Nature*. Columbia University Press.

DeepMind. (2018). DeepMind, meet Android. DeepMind Blog. Retrieved from https://deepmind.com/blog/article/deepmind-meet-android

DeepSeek. (2024). DeepSeek says it built its chatbot cheaply. What does that mean for AI's energy needs and the climate? Associated Press.

Desislavov, R., Martínez-Plumed, F., & Hernández-Orallo, J. (2023). Trends in AI inference energy consumption: Beyond the performance-vs-parameter laws of Deep Learning. *Sustainable Computing: Informatics and Systems, 28*, Article 100461.

Dhar, P. (2020). The carbon impact of Artificial Intelligence. *Nature Machine Intelligence, 2*(8), 423–425.

Digital Sustainability. (2024). In Palgrave studies in digital business & enabling technologies. https://doi.org/10.1007/978-3-031-61749-2

Dockrill, P. (2024). *How AI Is Making Buildings More Energy-Efficient*. TIME Magazine. https://time.com/ai-buildings-energy

European Commission. (2021). Proposal for a Regulation laying down harmonized rules on Artificial Intelligence (Artificial Intelligence Act) and amending certain Union legislative acts.

European Investment Bank. (2022). EIB Group sustainability report 2021. Retrieved from https://www.eib.org/en/publications/eib-group-sustainability-report-2021

Goldman Sachs Research. (2024). AI is poised to drive a 160% increase in data center power demand. Goldman Sachs.

Gomes, C., Dietterich, T., Barrett, C., Conrad, J., & Dilkina, B. (2019). Computational sustainability: Computing for a better world and a sustainable future. Communications of the ACM.

Grigorev, A. (2021). Machine learning bookcamp: Build a portfolio of real-life projects. Simon and Schuster.

Han, W., & Zhou, Y. (2023). AI-Driven digital twins for enhancing indoor environmental quality. *Buildings*, MDPI.

Hao, K. (2020). Training a single AI model can emit as much carbon as five cars in their lifetimes. MIT Technology Review.
Heikkilä, M. (2023). AI's carbon footprint is bigger than you think.
Hiller, J. (2024). Tech industry wants to lock up nuclear power for AI. *The Wall Street Journal.* 11687654321
Hölzle, U. (2016). 100% renewable is just the beginning. Google Blog.
Hook, A., Sovacool, B. K., Sorrell, S., & Court, V. (2020). A systematic review of the energy and climate impacts of teleworking. *Environmental Research Letters*
International Energy Agency. (2023). Data Centres and Data Transmission Networks. Retrieved from https://www.iea.org/reports/data-centres-and-data-transmission-networks
ISO. (2019). ISO 14064-1:2019 Greenhouse gases—Part 1: Specification with guidance at the organization level for quantification and reporting of greenhouse gas emissions and removals. International Organization for Standardization.
Jain, A., & Müller, S. (2023). Digital twin of the earth and global emissions monitoring systems. *AI for Earth, 6*(1), 45–60.
Kaur, R., & Patel, D. (2022). Micro-environmental energy models enhanced by AI. *Journal of Environmental Modeling, 18*(2), 121–138.
Koomey, J., Masanet, E., Horner, N., Azevedo, I., & Lintner, W. (2016). United States data center energy usage report. Lawrence Berkeley National Laboratory.
Lindner, T.D., Arto, S., Román, I., Rueda-Cantuche, M.V., Velázquez Afonso, J.M., Amores, A., Neuwahl, A.F. (2019). World input-output database environmental accounts update 2000-2016 Corsatea, F., JRC Technical Report.
Luccioni, S. (2024). Sasha Luccioni: Saving the planet, one AI tool at a time. TIME. Retrieved from https://time.com/7012753/sasha-luccioni/
Masanet, E., Shehabi, A., Lei, N., Smith, S., & Koomey, J. (2020). Recalibrating global data center energy-use estimates. *Science, 367*(6481), 984–986. https://doi.org/10.1126/science.aba3758
Mills, G. B. (2023). International hydrographic survey standards. *The International Hydrographic Review, 29*(1), 154–157. https://doi.org/10.58440/ihr-29-a17
Moore, G. E. (1965). Cramming more components onto integrated circuits. *Electronics, 38*(8), 114–117.
Nakagawa, M. (2024). Melanie Nakagawa: Leading Microsoft's sustainability efforts. TIME. Retrieved from https://time.com/7172427/melanie-nakagawa/
Nguyen, T. H., & Associates. (2023). Assessing the contributions of carbon capture and storage (CCS) technologies and renewable energy systems to achieve carbon neutrality by 2050. *Journal of Sustainable Energy Research, 15*(4), 123–145
Oecd. (2019). Artificial Intelligence in society. OECD Publishing.
Oecd. (2018). OECD science, technology and innovation outlook 2018 adapting to technological and societal disruption. OECD Publishing.
Patterson, D., Gonzalez, J., Le, Q., Liang, C., Munguia, L. M., Rothchild, D., & Dean, J. (2021a). Carbon emissions and large neural network training. *Proceedings of the Conference on Machine Learning and AI, 1*(3), 245–261.
Patterson, D., Gonzalez, J., Le, Q., Liang, C., Munguia, L., Rothchild, D., So, D., & Dean, J. (2021). Carbon emissions and large neural network training. ArXiv Preprint, arXiv:2104.10350
PolicyWatch. (2025). The UK says AI will super-charge the economy. But will it scupper net zero? Reuters.
Polytechnique Insights. (2024). Generative AI: energy consumption soars. Polytechnique Insights.
Rahman-Jones, I. (2024). AI means Google's greenhouse gas emissions up 48% in 5 years.
Rong, H., Zhang, H., Xiao, S., Li, C., & Hu, C. (2016). Optimizing energy consumption for data centers. *Renewable and Sustainable Energy Reviews, 58*, 674–691. https://doi.org/10.1016/j.rser.2015.12.283
Schwartz, E. (2024). The exponential rise of AI computing power and its energy implications. *Journal of Emerging Technologies, 12*(3), 45–60. https://doi.org/10.1007/s12345-024-09876-x

Silver, D., Hubert, T., Schrittwieser, J., Antonoglou, I., Lai, M., Guez, A., Lanctot, M., Sifre, L., Kumaran, D., & Graepel, T. (2018). A general reinforcement learning algorithm that masters chess, shogi, and Go through self-play. *Science, 362*(6419), 1140–1144. https://doi.org/10.1126/science.aar6404

Singh Jadaun, J., Bansal, S., Sonthalia, A., Rai, A. K., & Singh, S. P. (2022). Biodegradation of plastics for sustainable environment. *Bioresource Technology, 347*, Article 126697. https://doi.org/10.1016/J.BIORTECH.2022.126697

Strubell, E., Ganesh, A., & McCallum, A. (2019). Energy and policy considerations for deep learning in NLP. In *Proceedings of the 57th Annual Meeting of the Association for Computational Linguistics*, pp. 3645–3650.

Sustainable Organizations—models, applications, and new perspectives. (2020). In IntechOpen eBooks.

The Guardian. (2016). DeepMind AI reduces Google's data center cooling bill by 40%. The Guardian—Technology

The Guardian. (2024). Concerned about your data use? Here is the carbon footprint of an average day of emails, WhatsApps, and more.

UNEP. (2023). AI and the environment: Assessing Artificial Intelligence's impact on global carbon emissions. United Nations Environment Programme. Retrieved from https://wesr.unep.org

Vanderbauwhede, W. (2024). Estimating the increase in emissions caused by AI-augmented search. ArXiv Preprint. arXiv:2407.16894

Vincent, J. (2016). Google uses DeepMind AI to cut data center energy bills. The Verge. Retrieved from https://www.theverge.com/2016/7/21/12246210/google-deepmind-ai-data-center-cooling

Vinuesa, R., Azizpour, H., Leite, I., Balaam, M., Dignum, V., Domisch, S., Felländer, A., Langhans, S. D., Tegmark, M., & Fuso Nerini, F. (2020). The role of Artificial Intelligence in achieving the sustainable development goals. *Nature Communications, 11*(1), 233. https://doi.org/10.1038/s41467-019-14108-y

Williams, E. (2004). Energy intensity of computer manufacturing: Hybrid assessment combining process and economic input-output methods. *Environmental Science & Technology, 38*(22), 6166–6174.

World Economic Forum. (2023). *Pairing AI and Digital Twin Technology to Cut Emissions*. https://www.weforum.org/ai-digital-twins

Wu, J., Zhang, J., & Wang, W. (2022). Sustainable AI: Environmental implications, challenges, and opportunities. *Artificial Intelligence Review, 55*, 1–23. https://doi.org/10.1007/s10462-021-10072-3

Zhang, L., & Peters, M. (2023). Artificial Intelligence in net-zero carbon emissions for sustainable buildings. *Buildings*, MDPI.

Silver, D., Hubert, T., Schrittwieser, J., Antonoglou, I., Lai, M., Guez, A., Lanctot, M., Sifre, L., Kumaran, D., Graepel, T., [illegible] (2018). A general reinforcement learning algorithm that masters chess, shogi, and Go through self-play. Science, 362(6419), 1140–1144. [illegible] science.aar6404

[illegible] plastics for sustainable environment. [illegible] https://doi.org/[illegible]

Strubell, E., Ganesh, A., & McCallum, A. (2019). Energy and policy considerations for deep learning in NLP. [illegible] Computational Linguistics (pp. 3645–3650).

[illegible]

[illegible]

[illegible]

[illegible]

[illegible]

[illegible]

[illegible]

# A Multidimensional Approach to E-Waste with Challenges and Strategies for Effective Sustainable Waste Management

**Elif Tüzün, Meryem Demir Güldal, Tarık Küçükdeniz, and Canan Hazal Akarsu**

## 1 Introduction

Waste management issues have become increasingly complex in recent decades. Waste management is a public service that should be available to all residents. This helps the system to work better, which improves the quality of life (Dey et al., 2023; Mishra et al., 2023). Relevant to the environmental management system is the management of household waste. Increasing waste volumes and social environmental awareness are major drivers for environmental managers to achieve sustainable waste management (Das et al., 2024; Mishra et al., 2019). The waste management system includes collecting, transporting, pretreating, processing, and final disposal of residues. Mismanagement of this waste can have a significant impact on the environment and public health and can also have a negative impact on the overall recovery process. In the future, sustainable management of resources and waste will require a more systemic approach. This will address the root causes of the problems. The dangers of domestic waste management to human well-being and the environment remain a concern for cities and semi-cities in developing and developed countries. Waste is inevitable as daily human activities generate it through discarded materials

E. Tüzün (✉)
Department of Chemistry, Faculty of Engineering, Istanbul University-Cerrahpaşa, İstanbul, Türkiye
e-mail: elif.tuzun@iuc.edu.tr

M. Demir Güldal
Department of Materials and Materials Processing Technologies, Vocational Technical Science School of Higher Education, Istanbul University-Cerrahpasa, Istanbul, Türkiye

T. Küçükdeniz · C. H. Akarsu
Department of Industrial Engineering, Faculty of Engineering, Istanbul University-Cerrahpaşa, İstanbul, Türkiye

 
A. P. Das and S. Karakuş (eds.), *Synergizing Sustainability for Integrated Waste Management*, Sustainable Environmental Waste Management Strategies,
https://doi.org/10.1007/978-3-032-03718-3_3

from domestic, agricultural, industrial, commercial, medical and natural, and man-made disasters. Increased consumption and production generate a significant amount of waste. People want more goods and services and so a lot of waste goes into the environment (Das & Ghosh, 2022; Mohanty et al., 2018).

Waste usually has a negative connotation: people think of rubbish, garbage, or perhaps even dangerous or toxic material. Waste is something you want to get rid of, and you are prepared to pay a fee for the service. Apart from household rubbish, many substances and objects are waste, particularly in the process and manufacturing industries (Mishra & Das, 2023; Tripathy et al., 2024). However, a substance is only waste if it is perceived or labelled as such. Waste is a subjective concept, or rather a qualification of a particular substance or object that does not disappear after disposal. The qualification can change: what is considered waste today may be a resource in the future. A substance or object is waste if it is not used to its full potential. It is possible to create a waste hierarchy by considering waste management principles (Mishra & Das, 2021; Mishra et al., 2024). The waste hierarchy is formed by evaluating how the waste is treated and which method is more appropriate according to the environmental damage and sustainability approach. According to this hierarchy, there are different preferred methods of waste management. When these are ranked from most preferred to least preferred, the ranking is as follows (Mishra et al., 2022a, 2022b):

1. Prevention.
2. Reduction.
3. Reuse.
4. Recycling and composting: It is the separation and collection of those parts of municipal solid waste that can be used in production and their reuse in production as raw materials. Composting is the biochemical decomposition of organic waste by micro-organisms.
5. Energy recovery (thermal conversion, incineration, gasification, biomethane energy):

It is waste incineration with energy recovery of waste. Incineration produces steam energy and electricity. At the same time, the volume of waste is reduced to one tenth.

6. Disposal: Landfill is the sine qua non of solid waste management.

Wild is completely different from landfill. Hazardous waste is not placed in these storage areas. Gas and water leakage control systems are in place, and the landfill floor is impermeable. Modern sanitary landfills aim to minimize the negative effects of open dumping.

Everyone deserves to live in a clean environment. Everyone deserves access to sustainable sanitation and waste management.

Integrated Sustainable Waste Management (ISWM) Model is a model that allows complex and multidimensional systems to be studied in an integrated manner (Guerrero et al., 2013).

## 2 Solid Waste Management

The terms "integrated" and "sustainable" are used to describe solid waste management systems in both developed and developing countries. By the mid-2000s, "integrated" was standard. It had been adopted by researchers and reflected in the names of academic research centers, like 3R: Residual Resources Research, a PhD school; Resource and waste management at DTU, Denmark; the Integrated Waste Management Centre at Cranfield University, UK; Integrated Waste Systems at The Open University, UK; the CSIR Centre for Integrated Waste Management, South Africa; and the Center for Integrated Waste Management at the University at Buffalo, USA (Wilson et al., 2013). In the literature, the terms "integrated waste management" and "integrated solid waste management" are technical and generally contain process information. Diaz et al. argued for a holistic and systems approach to "integrated and sustainable solid waste management (SWM)," addressing all three aspects of sustainability (Wilson et al., 2013). Local authorities are responsible for solid waste management, but this is one of the poorest services. The systems are outdated and inefficient, resulting in poor coverage and marginalization of the poor.

Waste is strewn everywhere, leading to unhygienic living conditions. Municipal laws do not have adequate provisions to deal with the problem of solid waste management. Rapid urbanization has made the situation critical. The urban population has increased fivefold in the last six decades, with 285.35 million people living in urban areas (Asnani, n.d.).

Solid waste management is a basic service provided by local authorities to keep urban centers clean, but it is often poorly delivered. Systems are outdated and inefficient, cover a small percentage of the population and marginalize the poor (Asnani, n.d.). Managing municipal solid waste is a major global issue. The high consumption rates and population growth are leading to high levels of municipal solid waste. This waste must be disposed of or recycled effectively to prevent environmental harm. Municipal solid waste is the mix of discarded solid waste from urban and rural populations. Nanda and Berruti said that municipal solid waste originates in the solid waste of municipalities collected from households, offices, institutions, and businesses, varying globally in composition and classification (Nanda & Berruti, 2021). It consists of biodegradable and nonbiodegradable fractions from organic and inorganic materials.

Typically, municipal solid waste is made up of:

- Plastic/rubber
- Kitchen waste
- Glass
- Yard waste
- Paper/cardboard
- Metal
- Miscellaneous trash
- Electronic waste
- Inert material

Istanbul is one of the largest cities in the world. Activity map of solid waste management in Istanbul is as follows:

- Waste incineration and energy generation plant
- Biomethanization plant
- Garbage leachate treatment plants
- Compost and recovery facilities
- Waste-derived fuel preparation plant
- Urban cleaning services
- Solid waste transfer stations
- Landfill sites
- Electricity generation plants from landfill gas
- Waste disposal
- Collection and disposal of medical waste

According to the new law no. 5216, which came into force in 2004, the duties and responsibilities of metropolitan municipalities in relation to solid waste management are as follows:

- To ensure the protection of the environment in accordance with the principle of sustainable development
- To provide services in relation to industrial and medical waste
- To establish, have established, operate, or have operated the necessary facilities for this purpose
- To draw up or have drawn up the metropolitan solid waste management plan
- Building materials, scrap metal storage and sales areas, excavated earth, rubble, construction materials, scrap metal storage and sales areas, excavated soil, rubble, sand and gravel storage areas, wood and coal sales and storage areas, to determine the locations, to ensure that the transport of these will not cause environmental pollution to take measures
- Except for the collection of solid waste at the source and transport to the transfer station, recycling, storage, and disposal of solid waste, excavation to perform the services related to the provision of services, and to establish facilities for this purpose, to establish, operate, or have operated
- Collection and treatment of waste from marine vessels and to provide the necessary arrangements (Kavut, 2023)

With the increase in population, the increase in needs and the resulting increase in consumed products are becoming a serious environmental pollution because consumption in every sector means that waste is created as a result (Kumar et al., 2023; Mishra & Das, 2022). The fact that these wastes consume our natural resources is a serious environmental problem. The decrease in our natural resources poses a serious threat to all living things. Therefore, this problem should be studied in every field, and a solution should be produced. Disposing of these wastes or, more importantly, recycling them and contributing again is a social duty for all branches of science. It is important for all countries to determine their strategies in many areas in order to prevent environmental pollution and to contribute to the reduction of the consumption of our natural resources by these wastes. Sustainability has been defined as environmental sustainability by trying to find solutions to environmental pollution, ecological sustainability by making efforts to protect our natural resources and prevent them from decreasing, and economic sustainability in terms of correctly evaluating the costs and contributing to the whole society. The issue of waste is a subject that requires many disciplines to work together and should be evaluated accordingly (Bala et al., 2024; Mishra et al., 2022a, 2022b).

The wastes from every product we consume are encountered as wastes from every sector. If we do not understand what needs to be done in this regard, the amount of wastes generated and the extent of the damage they cause to the environment will become very dangerous for both humans and animals (Das & Ghosh, 2018; Ray et al., 2024). In order to eliminate the wastes generated in cities in a harmless manner and even to return them to nature if possible, cooperation should be made in all areas, especially economy, health, and environment. Since it concerns all disciplines, every discipline in the scientific world can eliminate this danger by working together on this issue.

First of all, the type, amount, and environmental damage of solid waste should be determined in this regard. Then, the organization and gathering of the wastes, their transportation and storage under appropriate conditions, disposal and recycling, and solid waste management and sustainability are remarkable issues for the whole world.

According to recent studies on waste, waste originating from electronic devices, industrial waste, waste resulting from the development of the construction sector, etc., especially multilayered plastics, poses a problem for all cities and municipalities.

In this study, we will discuss the problems in this regard and the positive and negative effects of the concept of sustainability in eliminating these problems and which sectors are affected. We will examine the solution suggestions and applications that emerge when which disciplines come together in this regard.

## 3 Waste Management Applications

### a. Plastic Waste

Multilayered plastics have become a serious environmental problem in Indonesia, as in many countries, since it is not possible to break down naturally. The fact that these materials are very difficult to recycle is also another factor (Das et al., 2025; Singh Jadaun et al., 2022). Today, these wastes need to be kept under control and managed. For this purpose, an effective and versatile method is used in Indonesia to analyze the effects of the environment, economy, social life and infrastructure for the management of these wastes (Dey et al., 2024; Priyadarshini et al., 2024). The multi-dimensional sustainability analysis method called multidimensional sustainability analysis (MSA) is a good analysis method for Indonesia due to its holistic perspective for the difficulties and precautions in the management of its wastes. Their main purpose is to be able to prioritize the issues to be improved and to produce solutions by increasing activity in recycling. The study conducted by Anwar and his friends is one of the first studies to use technology in recycling and to increase the responsibilities of producers (Anwar et al., 2024). In this regard, the lack of a systematic strategy for waste management is a major problem. Especially wastes that are covered with plastic paper and metal materials in order to have longer shelf life products and then become waste materials are increasingly difficult to separate and recycle (Anwar et al., 2023). The difficulty of separating waste from garbage and the fact that it has little value and attempting to destroy it by burning it are the main causes of environmental pollution. If the serious effects of this waste accumulation continue, it is estimated that the production of plastics will double by 2040. They should take urgent measures in the political arena and make strategic plans in this regard (Anwar et al., 2023). Although many positive measures were taken before, such as not using single-use plastics and collecting waste, they were not successful in practice. Different working methods have also been tried, but high-cost recycling technologies do not solve the problem (Zhao et al., 2021).

In conclusion, for sustainability in recycling:

1. Classification and storage of waste
2. Good use of technology
3. Support of the state
4. Increasing awareness of people
5. Cooperation in many areas
6. Correct planning in every area

It is of critical importance in its subjects (Anwar et al., 2025).

b. **Demolition Waste**

In the construction sector, as in every field, a large amount of waste is generated during demolition. Reducing waste from construction demolitions, which constitute an important part of global waste, is an important step in sustainability. It is aimed to find a solution to reduce construction-demolition waste (C&DW) or develop a new strategy for its reuse. The materials resulting from these demolitions consist of a wide variety of products that require different processes. The aim of the study is to evaluate these material differences and provide solutions to the difficulties that arise. For this purpose, they examined all the research in the last 25 years, analyzed the studies conducted on this subject, and tried to facilitate planning on this subject. As a result of the examinations, they revealed that recycling issues related to plastics, metals, and wood materials were researched less than concrete materials. Although applications such as building-information-modeling (BIM) and life-cycle-assessment (LCA) provide convenience in the digital field, they cause alternative searches due to their high costs. The political approaches of governments in this regard can be significantly advanced. While European countries stand out in successful studies on recycling, developing countries are also struggling with the weaknesses of their practices. This study helps to make new regulations on recycling by examining the contributions and solution suggestions of different disciplines and to better analyze the progress in the academic field. In the future, it is planned for the construction sector to minimize the environmental impact of the wastes resulting from this destruction, to increase sustainability in the urban process and to contribute to the circular economy. The importance of waste coordination studies together with the artificial intelligence support of technological developments should not be ignore (Hasibuan et al., 2025).

c. **Metal Waste**

Nowadays, electronic devices and home appliances that make our lives easier and cars are the tools that make consumers dependent in every field (Priyadarsini & Das, 2024, 2025). The consumption rates of these many electrical devices and vehicles that have become a necessity continue to increase. However, with the development of technology, the introduction of better versions of these electronic devices to the market at any moment leaves us with the disposal or recycling of scrap devices. In the electronics and automotive sectors, they consume a lot of materials, mostly metal, in order to improve these products. As the demands of consumers increase, the number of these metal wastes also increases. For example, we can say that a good car consists of many and different metal parts and compounds. Considering the many different compounds used for the assembly methods for these vehicles, recycling is quite difficult. The recycling processes of materials that have reached the end of their life that we do not use can be provided by metallurgical methods (Bal et al., 2018; Ghosh et al., 2016). In addition to these materials, there are also plastic parts in vehicles. Plastics are also included in the recycling according to their quality. Metal-containing wastes remain as inseparable residues in the first mechanical transformation stage. In this content, there are metals that cannot be separated by physical means as well as organic wastes. Bachér et al., in their study, tested the distribution of metals and achieved the removal of organic materials with high carbon yield (Bachér et al., 2025).

e. **Agricultural Waste**

In recent studies, the concept of sustainability is also very important for environmental awareness in recycling, which supports food safety in agriculture. Mistakes made in the management of agricultural waste and inadequate processing of agricultural lands cause problems in air pollution. In order to eliminate these problems, waste management should be developed. They tried to provide a comprehensive solution with a model to eliminate the deficiencies in this area. The model, which was developed by using the method called multi-criteria decision-making (MCDM), tried to contribute to waste transformation by analyzing the information of the geographical system (Ma et al., 2025). While creating this model, they developed this model by dividing the planning into three groups.

These are the following:

(1) Making arrangements for environmental quality in urban areas
(2) Planning the location of facilities and reducing investment costs
(3) Optimizing and allocating resource use in this regard

For these groups, the regulation of agricultural waste, the increase in the number of facilities, the development of the importance of distances, and strategies in waste collection have been carried out under optimum conditions, and progress has been made in the field of sustainability. As a result of the planning, agricultural waste has been provided to be beneficial in the environment, energy, and economic areas.

e. **E-Waste**

With the increase in electronic waste (e-waste) along with technological developments, these wastes that pollute the environment should concern all environmentally conscious individuals. Singh et al. are looking for effective recycling methods for the management of these wastes. Sustainability in terms of product is a basic concept. Their aim is to develop a model with a proposition that examines the search for perfect products in production and the repair of defective ones. For this purpose, they tried to produce a model that would reduce costs by combining both environmental and mathematical engineering disciplines using the optimization technique.

## 4 A Multidimensional Approach to E-Waste: Integrating Circular Economy, Sustainability Metrics, and AI Solutions

Electronic waste (e-waste) consists of discarded electrical or electronic devices, including computers, smartphones, televisions, and household appliances. E-waste management has become essential because this waste stream grows at an unprecedented rate, as statistics show 57.4 million metric tons of e-waste were produced in 2021, and projections indicate this number will rise significantly by 2030 (Farjana et al., 2023). The rapid technological progress with reduced product lifespans and expanding global consumer behavior patterns serve as primary factors behind this trend. The current situation demonstrates why proper e-waste management strategies must be implemented to minimize environmental damage and health risks from improper disposal.

E-waste management faces major challenges because its hazardous materials such as heavy metals and toxic substances create the main problems. Environmental damage from improper handling leads to soil and water pollution while creating major health risks for humans. Sustainable e-waste management requires implementation because machine learning-based approaches offer effective solutions (Farjana et al., 2023). Sustainability science integrated with data analytics enables innovative solutions for e-waste recycling and management that recover resources while minimizing environmental damage.

The purpose of this literature review is to present a systematic overview of the current situation regarding e-waste production and its management practices and sustainability effects worldwide. This review aims to identify new approaches, particularly those based on machine learning technologies, to determine the current state of knowledge and to propose directions for future research. The analysis will include an extensive evaluation of the regulatory and economic frameworks that affect e-waste management.

a. **The Current Landscape of E-Waste Management**

The generation patterns of e-waste show substantial variations across different regions because developed countries produce more e-waste due to their elevated electronic product consumption (Andeobu et al., 2021). The production of e-waste in Asia will reach 24.4 million tons by 2022 because of the extensive use of consumer electronics, including smartphones and laptops (Farjana et al., 2023). The economic and technological disparities between Sub-Saharan Africa and other regions produce unique e-waste management challenges because of insufficient waste management systems (Djampou, 2023).

The worldwide regulatory framework for e-waste shows significant differences that directly affect how e-waste should be managed. The European Union's Waste Electrical and Electronic Equipment (WEEE) directive serves as an example of strict regulations that support recycling practices and correct waste disposal. Developing nations face inadequate enforcement of their regulatory frameworks, which leads to dangerous informal disposal methods (Abalansa et al., 2021). The current management approaches that include landfilling together with incineration demonstrate multiple severe drawbacks. The widespread e-waste disposal methods waste valuable recovery possibilities from electronic waste while creating major environmental pollution that harms human health (Farjana et al., 2023; Fatima, 2023).

The e-waste management industry presents both major obstacles and potential advantages from an economic standpoint. The high costs of sophisticated recycling technologies create a barrier to the recovery of substantial value from recycling programs. The improper disposal of e-waste results in severe environmental and social problems that harm both community health and environmental quality. Studies have proven definitive links between hazardous substance exposure during e-waste recycling operations, which leads to adverse health effects for workers and local residents.

b. **Machine Learning and Advanced Analytics for E-Waste Management**

Machine learning applications in e-waste management create innovative solutions that improve both operational efficiency and decision-making capabilities. Machine learning algorithms show promise in e-waste characterization according to recent studies, which help identify materials for recycling and recovery purposes (Farjana et al., 2023; Khanal et al., 2024). The application of predictive modeling enables better e-waste generation forecasting through the analysis of historical data and current market trends, which leads to improved strategic planning and resource management (Farjana et al., 2023).

The application of machine learning-based optimization techniques enhances e-waste logistics operations from the collection to the transportation and processing stages. The efficiency of resource extraction methods improves through these techniques, which leads to higher recovery rates and smaller environmental impact

(Leena & Kirti, 2022). Feature importance analysis has become essential for identifying crucial variables affecting e-waste management practices, which helps direct attention toward the most impactful factors.

Clustering approaches based on unsupervised learning have been employed to optimize e-waste classification for customized recycling procedures according to Ononiwu et al. (2024). Visual analytics tools enable stakeholders to understand e-waste data patterns and relationships and recovery possibilities, which leads to more achievable and actionable data-driven decisions (Ngambo et al., 2023).

c. **Circular Economy Integration for E-Waste Management**

The base of sustainable e-waste management is based on circular economy principles, which include extending the product lifecycle and reducing waste through innovative design and recycling approaches. The current literature emphasizes different product lifecycle extension strategies and stresses the importance of disassembly and recyclability design to align manufacturing processes with sustainability goals (Farjana et al., 2023). Circular economy business models are becoming more popular, showing a shift from traditional ownership models to leasing and sharing systems with the goal of reducing electronic consumption (Muhammad et al., 2024).

Extended producer responsibility (EPR) is another important component of circular economy frameworks that makes manufacturers responsible for the entire lifecycle of their products, from production to end-of-life disposal (Nandan et al., 2023). Through the formalization of consumer awareness of the environmental impacts of e-waste, initiatives can encourage more responsible disposal practices (Andeobu et al., 2021).

Research conducted by Khanal et al. demonstrates that environmental awareness directly influences how consumers intend to recycle their e-waste. The integration of digital technologies into circular economy practices enables better e-waste flow monitoring and enhances the transparency of the recycling process, which builds trust and accountability (Sofian et al., 2023).

The circular economy approach to e-waste represents a basic shift from the linear "take-make-dispose" model to a regenerative system that focuses on resource conservation. This transformation includes several key strategies. The design for circularity requires manufacturers to develop durable products that can be easily repaired and disassembled for recycling purposes. The right to repair movement demonstrates this approach by pushing for laws that force manufacturers to enable independent product repairs.

Product-as-a-Service (PaaS) arrangements as innovative business models drive manufacturers to focus on durability instead of obsolescence. The ownership of devices remains with companies under these systems, which customers access through leasing or subscription payments. Performance-based contracts together with sharing platforms and refurbishment operations implement circular principles to maximize product utilization and extend useful lifespans.

EPR implementation methods vary across countries because different regions adopt either collective industry responsibility or individual producer accountability. Successful systems generally include clear performance metrics, transparent reporting, and financial mechanisms, such as advanced disposal fees or deposit-refund systems that incorporate environmental costs into product prices.

Digital technologies have greatly helped circular practices. Blockchain systems create secure product "passports" which record material composition and optimal end-of-life processing. IoT applications track device performance, which allows for preventative maintenance. AI and machine learning improve recycling processes by using advanced sorting and predictive analytics. Digital marketplaces for used electronics and repair services have built strong secondary markets that help extend product lifespans.

System effectiveness depends on continuous consumer engagement. Educational campaigns enhance public understanding of the environmental impacts and recycling options, and incentive systems encourage sustainable behaviors. The social dynamics of community initiatives, including repair cafés and collection events, help establish new norms about electronic product use and disposal. The combination of these approaches enables the development of circular e-waste management systems that convert environmental issues into sustainable resource recovery opportunities.

d. **Integrating Sustainability Metrics and Assessment Methods**

E-waste management evolution demonstrates why sustainability metrics and assessment methods need to be integrated into decision-making frameworks. Multiple sustainability assessment frameworks have proven effective for evaluating e-waste management alternatives because they assess environmental, economic, and social impacts (Fatima, 2023; López-Olvera et al., 2025). Life Cycle Assessment (LCA) stands out as a leading method in research for measuring the complete environmental impacts that electronic products create from extraction through their entire lifecycle (Avis, 2022).

The evaluation of e-waste management strategies heavily depends on multi-criteria decision-making approaches. The frameworks enable the integration of multiple stakeholder viewpoints and evaluation criteria, which results in decisions that properly represent e-waste management solution complexities (Djampou, 2023). The implementation of measurement frameworks that track e-waste sustainability indicators enables better practices and policies through quantifiable performance benchmarks (Sutawong et al., 2020).

The United Nations Sustainable Development Goals (SDGs) need to be integrated with e-waste management to establish sustainable practices that resolve environmental and socioeconomic problems. The improper disposal of e-waste creates health hazards for humans because toxic substances in e-waste lead to multiple health problems according to Shahabuddin et al. (2023). E-waste recycling and recovery operations create employment opportunities and establish protected work environments, which advance SDG 8 (Decent Work and Economic Growth) objectives

(Anuardo et al., 2023). The circular economy framework under SDG 12 (Responsible Consumption and Production) uses strategies to minimize e-waste production while maximizing resource recovery, which supports responsible production practices and environmental protection goals (Trummer, 2022). The sustainable management of e-waste fulfills two essential targets of the United Nations Sustainable Development Goals by protecting both SDG 6 (Clean Water and Sanitation) and SDG 15 (Life on Land) through its ability to stop soil and water pollution (Schröder et al., 2020). The achievement of SDG 17 requires partnerships because joint work between governments, businesses, and civil society enables the creation of successful e-waste management policies that improve both resource distribution equity and technological access fairness (Anuardo et al., 2023). Several countries have implemented e-waste recycling programs that combine community health benefits with economic development, thus illustrating how these SDGs function together in real-world scenarios (Anuardo et al., 2023). The strategic implementation of SDGs within e-waste management presents a necessary opportunity to develop a sustainable environment that promotes equity and health.

e. **Successful E-Waste Management Case Studies**

Different regions have achieved successful e-waste management through effective strategies that prove sustainable methods work when adapted to local conditions.

The European region considers Switzerland's e-waste management system as the standard for developed economies. The Swiss SENS/Swico system requires an advanced recycling fee for all electronic purchases to support complete collection and processing operations. The system reaches outstanding e-waste collection performance by recovering more than 75% of the total e-waste produced, which surpasses typical global collection standards. The Swiss system provides easy e-waste disposal through its wide range of collection facilities located at retail stores and public areas and specialized recycling facilities. The system's achievement of high success rates depends on its transparent financial system and its strong regulatory framework, which maintains electronics lifecycle compliance.

The Indian organization Saahas, based in Bengaluru, operates a community-based collection system in urban areas as an equally impressive yet contrasting example. Saahas works with informal sector waste pickers to establish a hybrid model that protects their working safety while preserving their economic benefits from informal recycling operations. The organization delivers training programs, protective equipment, and appropriate recycling technologies to its members. Through this approach, Saahas has managed to keep 80 tons of e-waste from landfills each month while securing better financial stability for workers who were previously marginalized. The Saahas model shows that formal systems can integrate existing informal sector operations to create socially inclusive solutions.

The global takeback program of Dell serves as an example of corporate leadership in e-waste management because it has recovered more than 2.1 billion pounds of used electronics since 2007. Dell operates a unique closed-loop recycling system that transforms plastic waste from old computers into new products. The initiative reduces virgin plastic production by thousands of tons while simultaneously establishing market demand for recycled materials. The company achieves its success through its easy collection system, free return shipping, and clear disclosure of material recovery statistics. The manufacturer-led initiative at Dell demonstrates how supply chain innovation spreads throughout the industry while shaping standard industry practices.

The developing world context shows Rwanda's national e-waste strategy as an example of a successful public–private partnership. The Rwandan government built a centralized e-waste recycling facility in Kigali through partnerships with international organizations and private sector companies. The facility has trained staff who evaluate devices for potential refurbishment before performing disassembly on items that cannot be restored. The program receives support from complete regulations and tax benefits for companies that participate in takeback schemes for e-waste collection. Rwanda's method establishes structured employment while minimizing unsafe disposal methods, which proves that middle-income nations can establish efficient systems even when resources are limited.

E-waste management in South Korea has been revolutionized by technology-enabled collection and tracking systems that use a sophisticated Radio Frequency Identification (RFID) system to monitor e-waste flows throughout the recycling chain. The deposit-refund system, which is administered through a digital platform, provides economic incentives to consumers, while real-time tracking guarantees that all collected materials are delivered to certified processors. The technology-forward approach has enabled South Korea to achieve a 95% recovery rate for major appliances and has provided valuable data for continuous system optimization. The Korean model shows how digital technologies can improve transparency and efficiency in e-waste management systems.

Multiple case studies show that e-waste management success depends on several key elements, including convenient collection infrastructure, appropriate financial mechanisms, stakeholder collaboration, clear regulatory frameworks, and technological innovation. The examples showed that e-waste management needs context-specific solutions that must consider local economic conditions, waste management practices, and cultural factors. The examples show how tracking technologies in developed economies and community-based approaches in emerging markets provide adaptable solutions to manage growing global e-waste while generating environmental, economic, and social advantages.

f. **Challenges and Barriers**

Multiple major obstacles remain in the way of effective e-waste management practices and policy implementation despite recent advancements. The current recycling methodologies demonstrate poor material recovery efficiency, which creates a significant technical barrier (Uhunamure et al., 2021). The high expenses of sophisticated recycling technologies create economic barriers that prevent sufficient funding for e-waste management sustainability (Abalansa et al., 2021).

The scientific knowledge about the long-term health and environmental impacts of e-waste is still limited, which hampers the improvement of management practices (Khanal et al., 2024). The lack of detailed information on e-waste flows hinders policy-making and resource management, thus weakening regulatory structures (Andeobu et al., 2021).

The improper disposal of e-waste is also hindered by social factors such as deeply ingrained disposal habits within communities and a lack of public understanding of the consequences of improper disposal methods (Ononiwu et al., 2024). Furthermore, there is still a problem of uneven policy implementation and enforcement across the world, particularly in developing countries where informal e-waste recycling is the norm (Achinas, 2020). These complex challenges can only be addressed through coordinated efforts across sectors, including technological advancement, regulatory improvement, and community participation.

## 5 Research Gaps and Future Directions

The ongoing research on e-waste management shows promise, but numerous knowledge gaps exist that future studies can effectively explore. The current lack of knowledge about how technological advancements affect regulatory systems to enhance e-waste management stands out as a significant research need. Research should explore how artificial intelligence and Internet of Things (IoT) applications could affect e-waste management systems according to Khanal et al. (2024).

Future research should explore the potential of combining machine learning methods with life cycle assessment (LCA) and material flow analysis to develop advanced models for e-waste generation patterns and recycling efficiencies (Nandan et al., 2023).

The field would gain significantly from interdisciplinary research that combines environmental science, social sciences, and data analytics to gain a better understanding of stakeholder behaviors and regulatory impacts (Dirja & Dwandaru, 2024). A thorough analysis of the socioeconomic factors that influence consumer behavior regarding e-waste disposal would provide valuable information for policymakers to promote sustainable practices (Farjana et al., 2023).

## 6 Conclusion

The literature demonstrates that sustainable e-waste management strategies must be developed to handle the complex problems resulting from increasing electronic waste generation. The research review demonstrated that machine learning techniques offer substantial potential to boost e-waste management approaches. The circular economy framework stands as a fundamental concept that supports sustainability by developing innovative consumption patterns and disposal methods.

The identified research gaps need to be bridged to advance e-waste management. Further research into the convergence of technology, regulatory frameworks, and social behavior patterns is necessary for the development of effective solutions to mitigate the environmental and health impacts of e-waste. Sustainable e-waste management has the potential to transform the future of the environment and society, and this can only be achieved through the collaborative efforts of industries, academic institutions, and policy sectors.

## References


Abalansa, S., El Mahrad, B., Icely, J., & Newton, A. (2021). Electronic waste, an environmental problem exported to developing countries: The GOOD, the BAD and the UGLY. *Sustainability, 13*(9), 5302. https://doi.org/10.3390/su13095302

Achinas, S. (2020). An overview of the technological applicability of plasma gasification process. In *Contemporary Environmental Issues and Challenges in Era of Climate Change* (pp. 261–275). Springer Singapore. https://doi.org/10.1007/978-981-32-9595-7_15

Andeobu, L., Wibowo, S., & Grandhi, S. (2021). A Systematic review of e-waste generation and environmental management of Asia Pacific Countries. *International Journal of Environmental Research and Public Health, 18*(17), 9051. https://doi.org/10.3390/ijerph18179051

Anuardo, R. G., Espuny, M., Costa, A. C. F., Espuny, A. L. G., Kazançoğlu, Y., Kandsamy, J., & de Oliveira, O. J. (2023). Transforming e-waste into opportunities: driving organizational actions to achieve sustainable development goals. *Sustainability, 15*(19), 14150. https://doi.org/10.3390/su151914150

Anwar, M. A., Sasongko, N. A., Suprihatin, & Najib, M. (2023). Sustainable plastic packaging waste management strategy based on a circular economy. *IOP Conference Series: Earth and Environmental Science, 1267*(1), 012062. https://doi.org/10.1088/1755-1315/1267/1/012062

Anwar, M. A., Suprihatin, S., Sasongko, N. A., Najib, M., Pranoto, B., Firmansyah, I., & Soekotjo, E. S. (2025). Sustainable waste management strategies for multilayer plastic in Indonesia. *Cleaner and Responsible Consumption, 16*, Article 100254. https://doi.org/10.1016/j.clrc.2025.100254

Anwar, M. A., Suprihatin, Sasongko, N. A., Najib, M., & Pranoto, B. (2024). Challenges and prospects of multilayer plastic waste management in several countries: A systematic literature review. *Case Studies in Chemical and Environmental Engineering, 10*, 100911. https://doi.org/10.1016/j.cscee.2024.100911

Asnani, P. U. (n.d.). *Solid waste management.*

Avis, W. (2022). *Technical Aspects of e-Waste Management*. https://doi.org/10.19088/K4D.2022.051

Bachér, J., Laatikainen-Luntama, J., Rintala, L., & Horttanainen, M. (2025). The distribution of valuable metals in gasification of metal-containing residues from mechanical recycling of end-of-life vehicles and electronic waste. *Journal of Environmental Management, 373*, Article 123526. https://doi.org/10.1016/j.jenvman.2024.123526

Bal, B., Ghosh, S., & Das, A. P. (2018). Microbial recovery and recycling of manganese waste and their future application: A review. *Geomicrobiological Journal, 36*(1), 85–96. https://doi.org/10.1080/01490451.2018.1497731

Bala, K., Anshul, Verma, S., Dogra, A., Das, A. P., Kumar, R., & Nandini. (2024). Sustainable management for urban plastic waste generation. *Sustainable Management of Urban Plastic Waste Through Circular Economic Approaches*, 211–236. https://doi.org/10.1201/9781003465447-13

Das, A., Dey, S., Gacem, A., Yadav, K. K., Nath, H., Ram, D. K., & Das, A. P. (2025). Revolutionizing solid hydrocarbon sourcing: Plastic's journey from waste to upcycled treasure. *Environmental Science and Engineering, Part F218*, 331–351. https://doi.org/10.1007/978-3-031-81615-4_14

Das, A., & Ghosh, S. (2022). Role of microorganisms in extenuation of mining and industrial wastes. *Geomicrobiology Journal, 39*, 173–175. https://doi.org/10.1080/01490451.2022.2038953

Das, A. P., & Ghosh, S. (2018). Bioleaching of Manganese from mining waste materials. *Materials Today: Proceedings, 5*(1), 2381–2390. https://doi.org/10.1016/J.MATPR.2017.11.459

Das, A., Ray, S. K., Mohanty, M., Mohanty, J., Dey, S., & Das, A. P. (2024). Ecotoxicity of microplastic wastes and their sustainable management: A review. *Environmental Chemistry and Ecotoxicology, 6*, 144–152. https://doi.org/10.1016/J.ENCECO.2024.05.003

Dey, S., Das, A., Mallick, K., Sahu, A., & Das, A. P. (2023). *Environmental petroleum waste: Pollution, toxicity, sustainable remediation*, pp. 159–175. https://doi.org/10.1007/978-3-031-48220-5_7

Dey, S., Dash, N., Ray, S. K., Priyadarsini, S., Chhotaray, C., & Das, A. P. (2024). Occurrence and quantification of tire wear plastics pollutants from highway City Dust of Bhubaneswar City of Odisha. *Sustainable Microbial Technology for Synthetic and Cellulosic Microfiber Bioremediation*, 77–89. https://doi.org/10.1007/978-3-031-62600-5_5

Dirja, E. Z. A., & Dwandaru, W. S. B. (2024). Utilizing household organic waste as a precursor for carbon nanomaterial production: challenges and outlook for sustainable waste management. *IOP Conference Series: Earth and Environmental Science, 1425*(1), Article 012018. https://doi.org/10.1088/1755-1315/1425/1/012018

Djampou, A. (2023). *2050 Electronic and Electrical Waste Outlook in West Asia.* United Nations Environment Programme. https://doi.org/10.59117/20.500.11822/42147

Farjana, M., Fahad, A. B., Alam, S. E., & Islam, Md. M. (2023). An IoT- and cloud-based e-waste management system for resource reclamation with a data-driven decision-making process. *IoT, 4*(3), 202–220. https://doi.org/10.3390/iot4030011

Fatima, M. and A. G. (2023). Understanding strategic environmental management response to the sustainability challenges: a case of electronic waste management. *Pakistan Social Sciences Review, 7*(IV). https://doi.org/10.35484/pssr.2023(7-IV)37

Ghosh, S., Mohanty, S., Akcil, A., Sukla, L. B., & Das, A. P. (2016). A greener approach for resource recycling: Manganese bioleaching. *Chemosphere, 154*, 628–639. https://doi.org/10.1016/J.CHEMOSPHERE.2016.04.028

Guerrero, L. A., Maas, G., & Hogland, W. (2013). Solid waste management challenges for cities in developing countries. *Waste Management, 33*(1), 220–232. https://doi.org/10.1016/j.wasman.2012.09.008

Hasibuan, G. C. R., Al Fath, M. T., Yusof, N., Dewi, R. A., Syafridon, G. G. A., Jaya, I., Anas, M. R., & Syahrizal. (2025). Integrating circular economy into construction and demolition waste management: a bibliometric review of sustainable engineering practices in the built environment. *Case Studies in Chemical and Environmental Engineering*, 101159. https://doi.org/10.1016/j.cscee.2025.101159

Kavut, S. (2023). *İklim değişikliği açısından katı atık yönetimi: Doğu Anadolu Bölgesi belediyeleri üzerine inceleme* [Public administration]. İnönü Üniversitesi.

Khanal, A., Sherpa, P., Chataut, P., Khanal, A., & Suja, G. (2024). Transforming e-waste management: challenges and opportunities. *International Research Journal of Multidisciplinary Technovation*, 108–115. https://doi.org/10.54392/irjmt2429

Kumar, M., Pandit, S., Patel, V., Khanna, N., Nag, M., Lahiri, D., Ray, R. R., Das, A., & Das, D. (2023). Maximization of energy recovery from starch processing wastewater by thermophilic dark fermentation coupled with microbial fuel cell technology. *Geomicrobiology Journal*, 1–12. https://doi.org/10.1080/01490451.2023.2209555

Leena N., M., & Kirti J., M. (2022). Impact of biomedical waste on environment and human health. *International Journal of Zoological Investigations, 08*(02), 484–489. https://doi.org/10.33745/ijzi.2022.v08i02.060

López-Olvera, S., Olivo-Escudero, J. C., Hernández-Marroquín, M., López-Olvera, F. R., Vallejo-Rivera, Y. L., Rodríguez-Gómez, M. A., & Rodríguez-Herrero, P. H. (2025). Towards fiscal sustainable practices in Mexico: The quality of municipal tax regulation and tiered tariffs improve income from solid waste management services. *Waste Management Bulletin, 3*(1), 243–254. https://doi.org/10.1016/j.wmb.2025.01.009

Ma, C., Zhang, Y., Yi, S., & Lu, Y. (2025). Optimizing urban agricultural waste planning and management to enhance sustainability: Strategies for three types of cities. *Sustainable Cities and Society, 120*, Article 106168. https://doi.org/10.1016/j.scs.2025.106168

Mishra, S., & Das, A. P. (2021). Current treatment technologies for removal of microplastic and microfiber pollutants from wastewater. *Wastewater Treatment: Cutting-Edge Molecular Tools, Techniques and Applied Aspects*, 237–251. https://doi.org/10.1016/B978-0-12-821881-5.00011-8

Mishra, S., & Das, A. P. (2022). Treatment of the wastewater polluted with synthetic microfiber released from washing machine. In B. B. Das, H. Hettiarachchi, P. K. Sahu, & S. Nanda (Eds.), *Lecture Notes in Civil Engineering* (Vol. 207, pp. 109–117). Springer Science and Business Media Deutschland GmbH. https://doi.org/10.1007/978-981-16-7509-6_9/COVER

Mishra, S., & Das, A. P. (2023). Microbial remediation of synthetic microfiber contaminated wastewater. In M. P. Shah (Ed.), *Microbial Technologies in Industrial Wastewater Treatment* (pp. 337–350). Springer, Singapore. https://doi.org/10.1007/978-981-99-2435-6_17

Mishra, S., Dash, D., & Das, A. P. (2022a). Detection, characterization and possible biofragmentation of synthetic microfibers released from domestic laundering wastewater as an emerging source of marine pollution. *Marine Pollution Bulletin, 185*, Article 114254. https://doi.org/10.1016/J.MARPOLBUL.2022.114254

Mishra, S., Dash, D., & Das, A. P. (2024). Isolation and molecular characterization of polyester-degrading bacterial strains from municipal wastewater and their application in microplastic degradation in contaminated wastewater. *Geomicrobiology Journal*. https://doi.org/10.1080/01490451.2024.2376101

Mishra, S., Ghosh, S., van Hullebusch, E. D., Singh, S., & Das, A. P. (2023). A critical review on the recovery of base and critical elements from electronic waste-contaminated streams using microbial biotechnology. *Applied Biochemistry and Biotechnology, 2023*, 1–30. https://doi.org/10.1007/S12010-023-04440-X

Mishra, S., Rout, P. K., & Das, A. P. (2019). Solar photovoltaic panels as next generation waste: a review. *Biointerface Research in Applied Chemistry, 9*(6).

Mishra, S., Singh, R. P., Rout, P. K., & Das, A. P. (2022). Membrane bioreactor (MBR) as an advanced wastewater treatment technology for removal of synthetic microplastics. *Development in Wastewater Treatment Research and Processes*, 45–60. https://doi.org/10.1016/B978-0-323-85583-9.00022-3

Mohanty, S., Ghosh, S., Bal, B., & Das, A. P. (2018). A review of biotechnology processes applied for manganese recovery from wastes. *Reviews in Environmental Science and Bio/Technology 2018, 17*(4), 791–811. https://doi.org/10.1007/S11157-018-9482-1

Muhammad, A., Muhammad Hussein, M. Z. S., Zulfakar, M. H., & Sundram, V. P. K. (2024). The study of environmentally responsible behaviour in household electronic waste reverse logistics: A systematic review. *Malaysian Journal of Consumer and Family Economics, 32*(1), 515–549. https://doi.org/10.60016/majcafe.v32.20

Nanda, S., & Berruti, F. (2021). Municipal solid waste management and landfilling technologies: A review. *Environmental Chemistry Letters, 19*(2), 1433–1456. https://doi.org/10.1007/s10311-020-01100-y

Nandan, A., Suresh, A., Saole, P., Jeevanasai, S., Chandrasekaran, R., Meili, L., Wan Azelee, N., & Selvasembian, R. (2023). An integrated approach for electronic waste management—overview of sources of generation, toxicological effects, assessment, governance, and mitigation approaches. *Sustainability, 15*(24), 16946. https://doi.org/10.3390/su152416946

Ngambo, G., Hanna, E. G., Gannon, J., Marcus, H., Lomazzi, M., & Azari, R. (2023). A scoping review on e-cigarette environmental impacts. *Tobacco Prevention & Cessation, 9*, 1–8. https://doi.org/10.18332/tpc/172079

Ononiwu, N., Ako, P., Anyaoha, C., Ikwuagwu, C., & Jacobs, I. (2024). Sustainable considerations in additive manufacturing processes: A review. *Journal of Mechanical Engineering and Sciences*, 9853–9871. https://doi.org/10.15282/jmes.18.1.2024.5.0780

Priyadarshini, B., Santhosh Kumar, M., & Das, A. P. (2024). Microbial degradation of synthetic plastics released from COVID-19 generated single use disposable face masks. *Environmental Science and Engineering*, *Part F2827*, 1–16. https://doi.org/10.1007/978-3-031-50840-0_1

Priyadarsini, S., & Das, A. P. (2024). Analytical and structural characterization of waste lithium-ion batteries for their effective recycling strategy. *Environmental Chemistry and Ecotoxicology*. https://doi.org/10.1016/j.enceco.2024.12.004

Priyadarsini, S., & Das, A. P. (2025). Lithium bioleaching: A review on microbial-assisted sustainable technology for lithium bio-circularity. *Journal of Water Process Engineering, 69*, Article 106744. https://doi.org/10.1016/j.jwpe.2024.106744

Ray, S. K., Dey, S., Das, A. P., & John, A. (2024). Microplastic waste management through greener technologies : A review. *Sustainable Management of Urban Plastic Waste Through Circular Economic Approaches*, 237–261. https://doi.org/10.1201/9781003465447-14

Schröder, P., Lemille, A., & Desmond, P. (2020). Making the circular economy work for human development. *Resources, Conservation and Recycling, 156*, Article 104686. https://doi.org/10.1016/j.resconrec.2020.104686

Shahabuddin, M., Uddin, M. N., Chowdhury, J. I., Ahmed, S. F., Uddin, M. N., Mofijur, M., & Uddin, M. A. (2023). A review of the recent development, challenges, and opportunities of electronic waste (e-waste). *International Journal of Environmental Science and Technology, 20*(4), 4513–4520. https://doi.org/10.1007/s13762-022-04274-w

Singh Jadaun, J., Bansal, S., Sonthalia, A., Rai, A. K., & Singh, S. P. (2022). Biodegradation of plastics for sustainable environment. *Bioresource Technology, 347*, Article 126697. https://doi.org/10.1016/J.BIORTECH.2022.126697

Sofian, D. D., Hanafiah, M. M., Woon, K. S., & Hassan, F. M. (2023). Characteristics of consumer towards development of sustainable e-waste management in Malaysia. *IOP Conference Series: Earth and Environmental Science, 1167*(1), Article 012043. https://doi.org/10.1088/1755-1315/1167/1/012043

Sutawong, P., Wattanachaıyıngcharoen, D., Itsubo, N., & Suvannarat, K. (2020). The Risks To Human Health And Environmental Impacts From Community E-Waste Separation. *Geographia Technica*, 159–168. https://doi.org/10.21163/GT_2020.151.33

Tripathy, B., Dey, S., Dubey, N. K., Lo, W.-C., & Das, A. P. (2024). *An Omics-Based Approach for Managing the Risk of COVID-19-Generated Wastes,* pp. 51–66. https://doi.org/10.1007/978-3-031-50840-0_4

Trummer, I. (2022). CodexDB. *Proceedings of the VLDB Endowment, 15*(11), 2921–2928. https://doi.org/10.14778/3551793.3551841

Uhunamure, S. E., Nethengwe, N. S., Shale, K., Mudau, V., & Mokgoebo, M. (2021). Appraisal of households' knowledge and perception towards e-waste management in Limpopo Province South Africa. *Recycling, 6*(2), 39. https://doi.org/10.3390/recycling6020039

Wilson, D. C., Velis, C. A., & Rodic, L. (2013). Integrated sustainable waste management in developing countries. *Proceedings of the Institution of Civil Engineers—Waste and Resource Management, 166*(2), 52–68. https://doi.org/10.1680/warm.12.00005

Zhao, J., Faqiri, H., Ahmad, Z., Emam, W., Yusuf, M., & Sharawy, A. M. (2021). The Lomax-Claim model: Bivariate extension and applications to financial data. *Complexity*, *2021*(1). https://doi.org/10.1155/2021/9993611

# Waste Management and Its Impact on Environment and Health: Addressing Emerging Waste Challenges in a Post-pandemic World

**Sweta Sinha, Rahul Singh, Sudhir Lima, Mayank Srivastava, Nitya Garg, and Sona Kumari**

## 1 Introduction

The COVID-19 pandemic has brought unprecedented changes to nearly every aspect of human society. As the world grappled with the health crisis, one of the less-discussed yet critically important challenges that emerged was the management of waste, particularly biomedical and plastic waste (Singh, 2021). The rapid spread of the virus and the urgent need to protect public health led to a massive increase in the production and disposal of single-use medical supplies and plastics (Smith et al., 2023). Products such as face masks, gloves, gowns, and various forms of packaging became ubiquitous, forming an integral part of daily life in hospitals, public spaces, and households alike (Mohanty et al., 2024; Priyadarshini et al., 2024a; A. Sahu et al., 2024). However, the sudden and extensive reliance on these materials has exposed significant shortcomings in the existing waste management infrastructure.

Before the pandemic, waste management systems in many regions were already under pressure due to rapid urbanization, inadequate funding, and limited technological integration. The crisis exacerbated these issues by introducing an additional layer of complexity to waste segregation, treatment, and disposal (Priyadarshini & Das, 2024). Traditional methods of handling waste, which were designed for relatively stable waste streams, were quickly overwhelmed by the surge in hazardous and non-hazardous waste. As a consequence, many countries witnessed a spike in

S. Sinha (✉) · R. Singh · S. Lima · M. Srivastava
Amity University Jharkhand, Ranchi, Jharkhand, India
e-mail: sweta.sinha2203@gmail.com

N. Garg
Sarla Birla University, Ranchi, Jharkhand, India

S. Kumari
RTC Institute of Technology, Ranchi, Jharkhand, India

 
A. P. Das and S. Karakuş (eds.), *Synergizing Sustainability for Integrated Waste Management*, Sustainable Environmental Waste Management Strategies,
https://doi.org/10.1007/978-3-032-03718-3_4

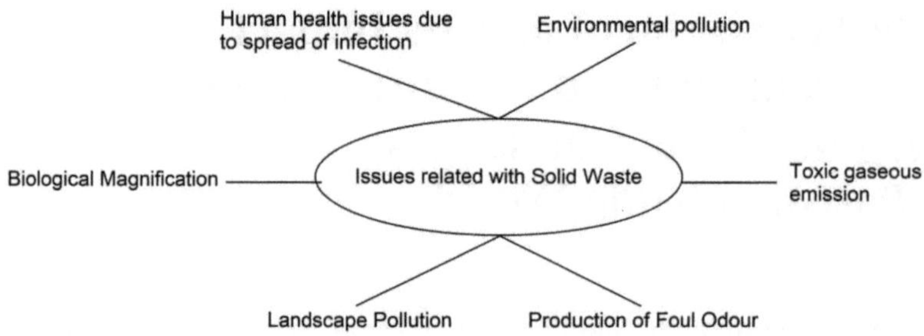


**Fig. 1** Issues related with solid waste generation during and post pandemic

environmental pollution, which in turn has raised public health concerns that extend well beyond the immediate impact of the virus itself (Salaria et al., 2024; Kumar et al., 2023).

The objective of this review is to analyze the post-pandemic waste challenges from an environmental and public health perspective. In doing so, it examines the nature and volume of waste generated during the COVID-19 crisis, discusses its environmental fate, and evaluates the associated health risks (Fig. 1). Furthermore, the paper explores emerging strategies and technologies that promise to improve waste management practices and reduce the adverse impacts of waste on ecosystems and human populations. By adopting a comprehensive approach, the review seeks to contribute to the dialogue on sustainable waste management in a rapidly changing global landscape.

## 2 The Rise of Medical and Plastic Waste in the Post-Pandemic Era

### *2.1 Increased Medical Waste Generation*

The outbreak of COVID-19 precipitated a dramatic shift in waste generation patterns across the globe. In many ways, the pandemic served as a catalyst for increased consumption of disposable products, especially in the medical and healthcare sectors (Luhar et al., 2022). Hospitals and clinics, in an effort to prevent the spread of the virus, began to use vast quantities of personal protective equipment (PPE) on a daily basis (Eke & Eke, 2021). This included not only masks, gloves, and gowns but also items such as face shields and protective covers. The urgency of the situation led to a temporary relaxation of regulations that previously limited the use of single-use plastics, allowing for the rapid production and distribution of these essential items (Salaria et al., 2024; Kumar et al., 2023).

As healthcare facilities prioritized the immediate need to protect patients and staff, the disposal of used medical supplies became a pressing concern. Many regions lacked the necessary infrastructure to process the increased volume of hazardous biomedical waste, leading to improper handling and disposal practices (Smith et al., 2023; Khatun et al., 2023a, 2023b; Sadasivuni et al., 2019). This surge in waste generation was not limited to hospitals; it also extended to households, where individuals began using and disposing of PPE in large quantities. The combined effect was an exponential increase in the total volume of waste that was often contaminated and hazardous, yet not always treated as such.

### 2.2 *Escalation of Plastic Waste*

The increase in plastic waste is another significant concern in the post-pandemic era. With a heightened emphasis on hygiene and safety, single-use plastics became the material of choice for a wide range of products (Bala et al., 2024; Das et al., 2025; Singh Jadaun et al., 2022). Packaging for food, medicines, and consumer goods saw a notable rise, driven by consumer demand for products that minimize the risk of contamination. Although plastic has been a convenient and inexpensive material for decades, its durability and resistance to biodegradation now pose a major environmental challenge (Dey et al., 2024; John & Das, 2024; Priyadarshini et al., 2024b). When disposed of improperly, these plastics fragment over time into microplastics—tiny particles that infiltrate soil, water bodies, and even the atmosphere (Belioka & Achilias, 2024; Sinha, 2024; Sinha et al., 2024; Dayal et al., 2024).

The rise in medical and plastic waste during the pandemic underscores a broader systemic issue: the lack of a resilient, scalable, and environmentally sound waste management infrastructure (Acharya et al., 2024). In many parts of the world, waste management systems were not designed to handle the sudden influx of contaminated materials. As a result, the pandemic has revealed the urgent need for innovative waste management strategies that can adapt to crisis situations while also safeguarding public health and the environment.

## 3 Environmental Impact of Improper Waste Management

The environmental implications of the increased waste generation during the COVID-19 pandemic are both profound and far-reaching. One of the most visible impacts is the proliferation of plastic pollution in natural ecosystems. When medical and plastic waste is not managed properly, it often finds its way into landfills, water bodies, and even remote natural habitats. In these environments, plastics undergo degradation processes such as photo-degradation, thermal degradation, and chemical degradation, albeit overextended periods (Key, 2023; Shanker & Rani, 2024). The gradual breakdown of these materials results in the formation of microplastics, which have now been detected in some of the most pristine environments on Earth.

### 3.1 Microplastic Contamination

Microplastics, due to their minute size, can be ingested by a wide range of organisms, from plankton to larger marine animals (Mishra et al., 2021; Sahu & Das, 2024a, 2024b). This ingestion leads to bioaccumulation, where toxic substances concentrated in the plastics move up the food chain, eventually impacting human health (Ghosh et al., 2023; Mishra et al., 2022; Mohanty et al., 2024). Studies have shown that microplastics can serve as vectors for toxic chemicals, including persistent organic pollutants (POPs), heavy metals, and endocrine-disrupting compounds (Belioka & Achilias, 2024; Sinha, 2024; Sinha et al., 2024). As these substances enter the human food chain, there is growing concern about their long-term health impacts, which may include hormonal imbalances, developmental disorders, and even cancer (Dayal et al., 2024).

### 3.2 Air Pollution and Greenhouse Gas Emissions

Another significant environmental impact of improper waste management is the contribution to air pollution (GodvinSharmila et al., 2023). In many regions, the incineration of medical waste is a common practice. While incineration can reduce the volume of waste, it often releases harmful pollutants into the atmosphere. Toxic gases, such as dioxins and furans, are known by-products of incineration processes, especially when plastics are burned at high temperatures without adequate emission controls (Kanhai et al., 2021; Verma et al., 2016). These pollutants not only affect air quality but also have the potential to cause respiratory problems, cardiovascular diseases, and other chronic health issues in exposed populations.

Landfill sites, particularly those that are poorly managed, also pose a serious threat to the environment. When plastic and biomedical waste are deposited in landfills, there is a significant risk of leachate formation: a liquid that can seep through the waste and contaminate soil and groundwater (Godvin et al., 2023). The leachate often contains a complex mixture of hazardous chemicals that can persist in the environment for decades, affecting agricultural productivity and water quality. The cumulative impact of these processes is a gradual deterioration of environmental quality, with implications for both ecosystems and human communities (Chand et al., 2022).

### 3.3 Ecosystem Disruption

The environmental degradation resulting from improper waste management is further compounded by the disruption of ecosystems. Wildlife, from marine species to terrestrial animals, is frequently impacted by the presence of waste in their habitats. Discarded PPE items such as masks and gloves, along with other plastic debris, can entangle animals or be mistaken for food, leading to injury or death. The loss of biodiversity and disruption of natural habitats undermine ecosystem services that are vital for environmental sustainability, including water purification, nutrient cycling, and carbon sequestration.

# 4 Health Implications of Poor Waste Management

The health implications of inadequate waste management extend beyond the immediate risk of infection during a pandemic. When hazardous waste is not properly segregated and disposed of, it creates multiple avenues for pathogen transmission (Priyadarsini & Das, 2024, 2025). Healthcare facilities that generate large volumes of biomedical waste must adhere to strict protocols to prevent the spread of infectious diseases (Fig. 2) (Reinhardt, 2018). However, in many cases, these protocols are either poorly implemented or altogether absent, particularly in regions where resources are limited. As a result, sanitation workers, waste handlers, and even local communities are at risk of exposure to pathogens that can cause a range of illnesses, from mild infections to life-threatening diseases.

## 4.1 *Spread of Infectious Diseases*

One of the key concerns is the potential for waste-related disease outbreaks. Contaminated waste disposal sites can serve as breeding grounds for vectors such as rodents, insects, and other animals that can transmit diseases to humans. The improper handling of waste not only increases the risk of infection for those directly involved in waste management but also poses a broader public health threat. Communities living in the vicinity of waste disposal sites are exposed to the risks associated with

| Category | Type of Waste |
|---|---|
| Category I (Yellow) | Microbiology or Biotechnology and other Clinical laboratory waste |
| Category II (Red) | Wastes generated from disposable items such as tubing, bottles, etc. |
| Category III (White) | Sharp wastes including metals, needles, syringes, etc. |
| Category IV (Blue) | Broken or discarded and contaminated glass, etc. |

**Fig. 2** Standardized color-coded segregation system for biomedical waste management for proper handling and disposal of different waste types in healthcare sectors

contaminated air, water, and soil, creating an environment where infectious diseases can thrive (Kumar et al., 2024).

In addition to the spread of infectious diseases, poor waste management can also result in chronic exposure to toxic substances. The degradation of plastics and other materials releases chemical additives and pollutants that can have serious health consequences over time. For example, substances like phthalates, bisphenol A (BPA), and heavy metals are commonly found in plastic products. When these chemicals leach into the environment, they may contaminate water supplies and agricultural products (Talreja et al., 2025; Woodruff, 2024). Chronic exposure to such toxins has been associated with a range of health issues, including endocrine disruption, reproductive disorders, and an increased risk of certain cancers.

## 4.2 *Toxic Chemical Exposure*

The health impacts of waste mismanagement are not limited to chemical exposure and pathogen transmission. The mental health and social well-being of communities can also be adversely affected by the presence of waste in their environment. The sight of overflowing waste, polluted water bodies, and contaminated public spaces can create a pervasive sense of neglect and environmental injustice. In many urban areas, residents living near poorly managed waste disposal sites report feelings of anxiety, stress, and reduced quality of life. Such psychosocial stressors, when combined with the physical health risks, create a multifaceted public health challenge that demands urgent attention (Taylor et al., 1991).

## 4.3 *Water and Food Contamination*

Furthermore, the occupational hazards associated with waste management cannot be overlooked. Sanitation workers and waste collectors often operate in environments that expose them to a host of risks (Mishra et al., 2024; Mishra & Das, 2023). Inadequate protective equipment, lack of proper training, and unsafe working conditions contribute to a higher incidence of work-related injuries and illnesses among these workers (Abdalla et al., 2018). During the COVID-19 pandemic, the situation became even more dire as these individuals were exposed to potentially contaminated materials on a daily basis (Weldearegay et al., 2024). Ensuring the safety and well-being of waste management personnel is a critical component of any effective waste management strategy.

## 5 Emerging Waste Management Strategies

In response to the mounting environmental and health challenges posed by the surge in medical and plastic waste, researchers, policymakers, and industry leaders are exploring innovative waste management strategies. Figure 3 presents an integrated framework that outlines the key components of effective waste management in the post-pandemic era.

As illustrated in Fig. 3, a holistic framework is emerging that integrates several key components designed to manage waste more effectively. The figure outlines a continuum of strategies that begins with sustainable waste segregation and disposal. It

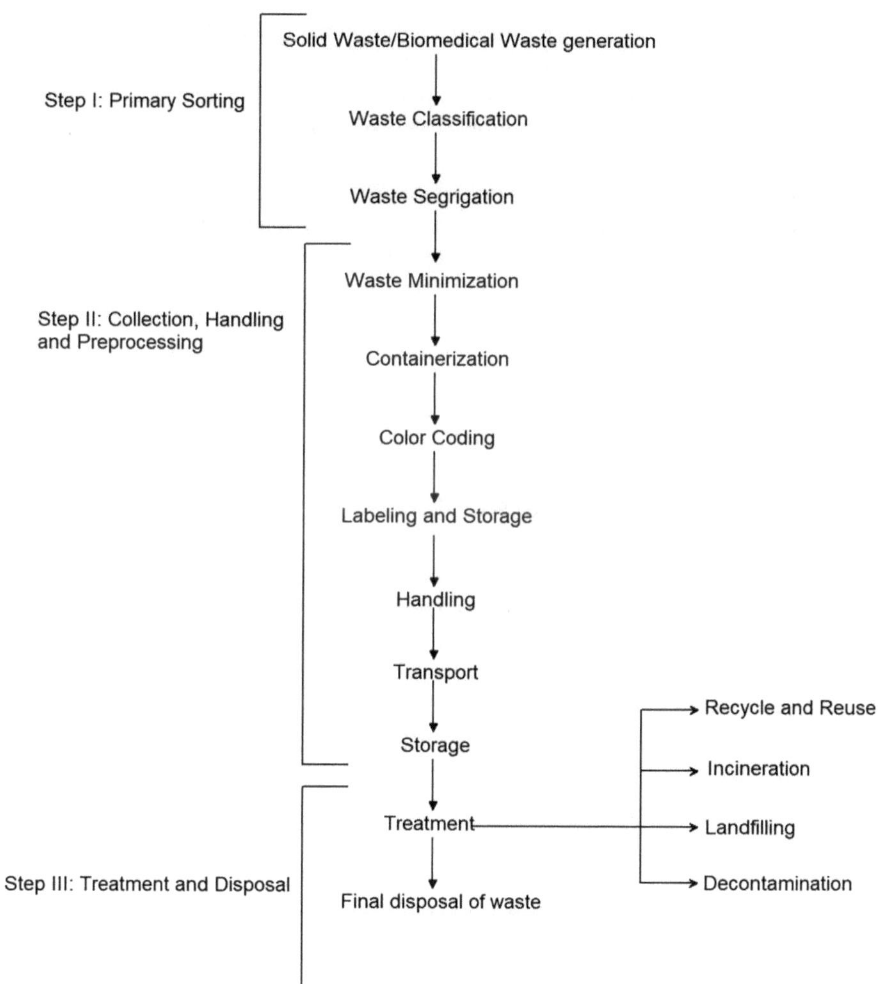


**Fig. 3** Steps involved in emerging waste management strategies

demonstrates how improved segregation at the source through the use of color-coded systems and clearly labelled disposal containers can prevent cross-contamination and ensure that hazardous materials are efficiently isolated from non-hazardous waste. This initial step is crucial for minimizing the environmental and health risks associated with contaminated waste streams.

Following segregation, Fig. 3 emphasizes the role of technological advancements in recycling. Modern recycling technologies, such as chemical recycling and pyrolysis, are highlighted as methods capable of breaking down plastics into their constituent monomers. These processes enable the recycling of waste into new products, thereby closing the loop in the production cycle. The figure also references emerging biotechnological approaches, including enzymatic degradation, which may provide innovative solutions for converting nonbiodegradable plastics into environmentally benign substances. Traditional incineration, while effective in reducing waste volume, often leads to the emission of toxic pollutants (Quina et al., 2008). Alternative methods, such as autoclaving, plasma gasification, and microwave sterilization, are presented as safer options for disinfecting and processing contaminated waste (Kaushal et al., 2024). These techniques help preserve the integrity of recyclable materials and reduce the risk of environmental contamination.

Figure 3 demonstrates how a shift from the conventional "take, make, dispose" model to a circular framework can drive the sustainable management of waste. By emphasizing product reuse, repair, and recycling and by holding manufacturers accountable for the entire lifecycle of their products through extended producer responsibility (EPR), the circular economy model offers a robust pathway to reducing the overall waste footprint.

## *5.1 Sustainable Waste Segregation and Disposal*

Effective segregation at the source is fundamental to minimizing contamination and ensuring that hazardous materials are isolated from non-hazardous waste. In many countries, efforts are underway to educate the public and healthcare workers about proper waste separation techniques, emphasizing the need for clear labelling and designated disposal containers. The adoption of color-coded systems, as shown in Fig. 2, and standardized protocols has shown promise in streamlining waste collection processes, reducing the risk of cross-contamination, and enhancing overall waste management efficiency.

## *5.2 Advancements in Recycling Technologies*

Technological innovations are also playing a significant role in the evolution of waste management systems. Advanced recycling technologies, such as chemical recycling and pyrolysis, offer viable methods for transforming plastic waste into valuable resources (Chen & Hu, 2024). These technologies break down polymers into their constituent monomers, which can then be used to manufacture new plastic products, effectively closing the loop in the production cycle. In addition, researchers are exploring the use of enzymatic degradation and other biotechnological approaches to accelerate the breakdown of plastics in the environment. By harnessing the power of microorganisms and engineered enzymes, it may be possible to convert non-biodegradable plastics into environmentally benign substances, thereby reducing the long-term ecological footprint of plastic waste (Sinha, 2025; Kumari, et al., 2025).

### *5.3 Disinfection and Safe Disposal of Medical Waste*

Another promising area is the development of non-incineration treatment methods for biomedical waste. Traditional incineration, while effective in reducing waste volume, often results in the emission of toxic pollutants that can harm both human health and the environment. In contrast, alternative methods such as autoclaving, plasma gasification, and microwave sterilization provide safer ways to treat contaminated waste. Autoclaving, for example, uses high-pressure steam to disinfect waste, rendering it safe for further processing or disposal. These methods not only reduce the risk of airborne pollutants but also preserve the integrity of recyclable materials, allowing for their subsequent reprocessing and reuse.

### *5.4 Circular Economy Approaches*

The concept of a circular economy has gained traction as a holistic approach to waste management. In a circular economy, the emphasis is placed on designing products and systems that minimize waste generation by promoting reuse, repair, and recycling. This approach challenges the traditional linear model of "take, make, dispose" and instead advocates for a sustainable cycle in which materials are continuously repurposed (Duc & Dinh, 2024). Policies that encourage extended producer responsibility (EPR) have been instrumental in this regard, as they require manufacturers to take accountability for the entire lifecycle of their products, including post-consumer waste management. By incentivizing companies to invest in recyclable materials and sustainable packaging designs, the circular economy model offers a pathway toward reducing the overall volume of waste and mitigating its environmental impact (Sinha, 2023).

In addition to these technological and systemic innovations, there is a growing recognition of the importance of public awareness and community engagement. Educating citizens about the environmental and health consequences of improper waste disposal is critical to fostering behavioral change. Public campaigns that emphasize the benefits of responsible waste management from improved local environmental quality to enhanced public health outcomes can help to build a culture of sustainability. In this context, community-based initiatives and local government programs play an essential role in bridging the gap between policy and practice, ensuring that waste management strategies are implemented effectively at the grassroots level (Ruiz Bravo, 2023).

## 6 Policy and Regulatory Interventions

The complexities introduced by the COVID-19 pandemic have underscored the necessity for robust policy and regulatory frameworks in the realm of waste management. Governments worldwide are now faced with the challenge of revising existing waste management regulations to better accommodate the increased volume and hazardous nature of post-pandemic waste (Ganguly & Chakraborty, 2024). Effective policies must address not only the technical aspects of waste treatment but also the broader socioeconomic and environmental dimensions of waste management.

Regulatory interventions in this field have taken many forms. Some governments have enacted stricter guidelines for hospitals and healthcare facilities to ensure that biomedical waste is properly segregated, treated, and disposed (Singh et al., 2024). These guidelines often require facilities to adhere to detailed protocols that specify the handling of different waste streams, from infectious materials to chemical residues. In addition, policies aimed at curbing the use of single-use plastics have gained momentum, with some jurisdictions introducing bans or levies on plastic products that are deemed non-essential. Such measures are designed to reduce the overall production of plastic waste and encourage the adoption of alternative materials that are more sustainable and biodegradable.

The enforcement of these regulations, however, remains a significant challenge. In many developing countries, limited resources and weak institutional frameworks hamper the effective implementation of waste management policies. The lack of comprehensive monitoring and accountability mechanisms often results in poor compliance, with hazardous waste being improperly disposed of in open dumps or landfills. To address these issues, international cooperation and support from global organizations have become increasingly important. Collaborative initiatives that provide technical assistance, capacity-building programs, and financial resources can help to strengthen local waste management infrastructures and ensure that regulatory measures are properly enforced (Sinha & Hamdan, 2023).

In parallel with regulatory reforms, there has been a growing emphasis on integrating waste management with broader environmental and public health policies. Governments are increasingly recognizing that waste management is not an isolated issue but is interconnected with challenges such as climate change, resource scarcity, and social inequality. For instance, waste-to-energy technologies, which convert waste into usable energy, not only help to reduce the volume of waste but also contribute to renewable energy goals. Similarly, policies that promote the circular economy can drive innovation in product design, foster new business models, and create employment opportunities in recycling and waste management sectors (Aiguobarueghian et al., 2024).

The development of comprehensive policies is also critical for addressing the transboundary nature of waste pollution. Plastic waste, once released into the environment, does not respect national boundaries. International agreements and collaborations are therefore essential to tackle the global dimensions of waste management. Frameworks such as the Basel Convention, which governs the transboundary movements of hazardous waste, provide a foundation for international efforts to reduce the environmental impact of waste (Mihai et al., 2022). Strengthening these agreements and ensuring their effective implementation can help to mitigate the long-term consequences of post-pandemic waste pollution on a global scale.

## 7 Future Directions and Recommendations

Looking ahead, the challenge of managing waste in a post-pandemic world calls for a multidisciplinary approach that combines technological innovation, robust policy frameworks, and active community participation (Khorram-Manesh et al., 2024). Researchers, policymakers, and industry leaders must collaborate to develop sustainable solutions that not only address the immediate issues but also lay the groundwork for a more resilient and environmentally friendly waste management system.

One key area of future research involves the development of advanced materials and technologies that reduce the reliance on single-use plastics and other non-biodegradable substances. Innovations in biodegradable polymers and green chemistry hold promise for creating products that maintain the necessary functionality while minimizing environmental harm (Ncube et al., 2023). Additionally, further exploration of enzymatic and microbial degradation methods may offer novel pathways for breaking down persistent plastics in natural settings, thereby reducing the formation of microplastics and their associated risks.

Another important direction for future work is the enhancement of waste management infrastructure at the local level. Investments in state-of-the-art waste treatment facilities, including advanced recycling centers and non-incineration treatment plants, are crucial to managing waste safely and efficiently. Integrating digital technologies, such as smart sensors, real-time monitoring systems, and data analytics, into waste management practices can optimize collection routes, improve segregation practices, and provide timely insights into waste flows. These innovations not only improve operational efficiency but also help to reduce the environmental footprint of waste management activities (Nagarale, 2024).

From a policy perspective, there is an urgent need for governments to align waste management strategies with broader sustainability goals. This includes fostering partnerships between public agencies, private enterprises, and civil society organizations to create a unified approach to waste reduction. Policies that incentivize sustainable practices, such as tax breaks for companies that adopt circular economy principles or grants for community recycling projects, can catalyze change (Nagarale, 2024).

Furthermore, comprehensive public awareness campaigns are essential to inform citizens about the importance of responsible waste disposal and the long-term benefits of sustainable waste management.

In addition to technological and policy innovations, the human dimension of waste management must not be overlooked. Training programs and capacity-building initiatives for waste management workers are critical to ensuring that those on the frontlines are equipped with the knowledge and resources necessary to perform their duties safely. Efforts to improve the working conditions of sanitation workers by providing them with appropriate protective equipment, fair wages, and safe working environments are also essential components of a sustainable waste management strategy.

International cooperation will undoubtedly play a significant role in shaping the future of waste management. As countries continue to face the repercussions of the pandemic, there is a growing recognition of the need for cross-border collaboration to address environmental challenges that do not adhere to national boundaries. By sharing best practices, technological innovations, and financial resources, the global community can work together to build more resilient waste management systems that benefit both human populations and the natural world.

## 8 Conclusion

The COVID-19 pandemic has exposed critical weaknesses in global waste management systems, highlighting the challenges posed by increasing medical and plastic waste. The surge in hazardous waste has intensified environmental pollution, microplastic contamination, and public health risks. If left unaddressed, these issues will continue to threaten ecosystems and human well-being. This review emphasizes the necessity of a comprehensive approach to waste management, incorporating technological advancements, policy enhancements, and public participation. Strengthening waste segregation, adopting non-incineration treatment technologies, and integrating circular economy principles are key to minimizing waste-related hazards. Improved regulations and infrastructure development can ensure sustainable waste management while reducing ecological damage. Collaboration among governments, industries, and communities is essential to implementing long-term waste solutions. Encouraging the use of biodegradable alternatives, enhancing recycling facilities, and enforcing stricter environmental policies will be crucial in preventing future waste crises. The pandemic has provided an opportunity to transition from conventional waste disposal methods to more sustainable resource management strategies. By prioritizing innovation, sustainability, and regulatory reforms, society can transform this crisis into an opportunity for a healthier planet and safer communities. The measures taken today will determine the future of environmental resilience, making it imperative to commit to responsible waste management for generations to come.

## References


Abdalla, S., Apramian, S. S., Cantley, L. F., & Cullen, M. R. (2018). *Occupation and risk for injuries*.

Acharya, B., Behera, A., Deshmukh, K., & Moharana, S. (2024). Plastic waste management during and post COVID-19 pandemic: Challenges and strategies. In *Plastic waste management: methods and applications* (pp. 173–199).

Aiguobarueghian, I., Adanma, U. M., Ogunbiyi, E. O., & Solomon, N. O. (2024). Waste management and circular economy: A review of sustainable practices and economic benefits. *World Journal of Advanced Research and Reviews, 22*(2), 1708–1719.

Bala, K., Anshul, Verma, S., Dogra, A., Das, A. P., Kumar, R., & Nandini. (2024). Sustainable management for urban plastic waste generation. In *Sustainable management of urban plastic waste through circular economic approaches* (pp. 211–236). https://doi.org/10.1201/9781003465447-13

Belioka, M. P., & Achilias, D. S. (2024). How plastic waste management affects the accumulation of microplastics in waters: A review for transport mechanisms and routes of microplastics in aquatic environments and a timeline for their fate and occurrence (past, present, and future). *Water Emerging Contaminants & Nanoplastics, 3*(3), N-A.

Chand, S., Pradhan, B., Chand, S., & Naik, S. K. (2022). Solid waste and landfill leachate: A transient source of emerging microbes and legacy contaminants for groundwater pollution. In *Contaminants of emerging concerns and reigning removal technologies* (pp. 123–149). CRC Press.

Chen, S., & Hu, Y. H. (2024). Advancements and future directions in waste plastics recycling: From mechanical methods to innovative chemical processes. *Chemical Engineering Journal*, 152727.

Das, A., Dey, S., Gacem, A., Yadav, K. K., Nath, H., Ram, D. K., & Das, A. P. (2025). Revolutionizing solid hydrocarbon sourcing: Plastic's journey from waste to upcycled treasure. *Environmental Science and Engineering, Part F218*, 331–351. https://doi.org/10.1007/978-3-031-81615-4_14

Dayal, L., Raj, D., Kumari, P., & Sinha, S. (2024). Abundance of microplastics in marine and freshwater ecosystem and its impact on biotic and abiotic components. *Water, Air, & Soil Pollution, 235*(7), 416.

Dey, S., Dash, N., Ray, S. K., Priyadarsini, S., Chhotaray, C., & Das, A. P. (2024). Occurrence and quantification of tire wear plastics pollutants from highway city dust of Bhubaneswar City of Odisha. In *Sustainable microbial technology for synthetic and cellulosic microfiber bioremediation* (pp. 77–89). https://doi.org/10.1007/978-3-031-62600-5_5

Duc, K. N., & Dinh, N. C. (2024). Circular economy. *Navigating the circular age of a sustainable digital revolution* (Vol. 1).

Eke, U. A., & Eke, A. C. (2021). Personal protective equipment in the siege of respiratory viral pandemics: Strides made and next steps. *Expert Review of Respiratory Medicine, 15*(4), 441–452.

Ganguly, R. K., & Chakraborty, S. K. (2024). Plastic waste management during and post Covid19 pandemic: Challenges and strategies towards circular economy. *Heliyon, 10*(4).

Ghosh, S., Sinha, J. K., Ghosh, S., Vashisth, K., Han, S., & Bhaskar, R. (2023). Microplastics as an emerging threat to the global environment and human health. *Sustainability, 15*(14), 10821. https://doi.org/10.3390/SU151410821

GodvinSharmila, V., Shanmugavel, S. P., Tyagi, V. K., & Banu, J. R. (2023). Microplastics as emergent contaminants in landfill leachate: Source, potential impact and remediation technologies. *Journal of Environmental Management, 343*, Article 118240.

John, A., & Das, A. P. (2024). Government strategies for achieving circular economy for plastic waste management in Southeast Asia. In *Sustainable management of urban plastic waste through circular economic approaches* (pp. 72–82). https://doi.org/10.1201/9781003465447-6/GOVERNMENT-STRATEGIES-ACHIEVING-CIRCULAR-ECONOMY-PLASTIC-WASTE-MANAGEMENT-SOUTHEAST-ASIA-AKBAR-JOHN-ALOK-PRASAD-DAS

Kanhai, G., Fobil, J. N., Nartey, B. A., Spadaro, J. V., & Mudu, P. (2021). Urban municipal solid waste management: Modeling air pollution scenarios and health impacts in the case of Accra, Ghana. *Waste Management, 123*, 15–22.

Kaushal, R., Rohit, & Dhaka, A. K. (2024). A comprehensive review of the application of plasma gasification technology in circumventing the medical waste in a post-COVID-19 scenario. *Biomass Conversion and Biorefinery, 14*(2), 1427–1442.

Key, S. (2023). *Investigation of factors affecting the breakdown of plastics in the environment* (Doctoral dissertation, University of Leicester).

Khatun, R., Singh, S., Dubey, N. K., & Prasad Das, A. (2023a). A review on marine-based phytochemicals and their application in biomedical research. In *Recent frontiers of phytochemicals: Applications in food, pharmacy, cosmetics, and biotechnology* (pp. 383–395). https://doi.org/10.1016/B978-0-443-19143-5.00013-X

Khatun, R., Singh, S., Dubey, N. K., & Prasad Das, A. (2023b). A review on marine-based phytochemicals and their application in biomedical research. In *Recent frontiers of phytochemicals: Applications in food, pharmacy, cosmetics, and biotechnology* (pp. 383–395). Elsevier. https://doi.org/10.1016/B978-0-443-19143-5.00013-X

Khorram-Manesh, A., Burkle, F. M., Jr., & Goniewicz, K. (2024). Pandemics: Past, present, and future: Multitasking challenges in need of cross-disciplinary, transdisciplinary, and multidisciplinary collaborative solutions. *Osong Public Health and Research Perspectives, 15*(4), 267.

Kumar, A., Thakur, A. K., Gaurav, G. K., Klemeš, J. J., Sandhwar, V. K., Pant, K. K., & Kumar, R. (2023). A critical review on sustainable hazardous waste management strategies: A step towards a circular economy. *Environmental Science and Pollution Research, 30*(48), 105030–105055.

Kumari, P., Sinha, S., Raj, D., Kumar, P., & Gupta, U. (2025). The potential of fungi in the bioremediation of xenobiotics: a mycological approach to water treatment. *Toxicological & Environmental Chemistry, 107*(6), 1208–1245

Kumar, G., Vyas, S., Sharma, S. N., & Dehalwar, K. (2024). Challenges of environmental health in waste management for peri-urban areas. In *Solid waste management: Advances and trends to tackle the SDGs* (pp. 149–168). Springer Nature Switzerland.

Luhar, I., Luhar, S., & Abdullah, M. M. A. B. (2022). Challenges and impacts of COVID-19 pandemic on global waste management systems: A review. *Journal of Composites Science, 6*(9), 271.

Mihai, F. C., Gnoni, M. G., Meidiana, C., Schneider, P., Ezeah, C., & Elia, V. (2022). A global outlook on the implementation of the basel convention and the transboundary movement of e-waste. In *Paradigm shift in E-waste management* (pp. 49–75). CRC Press.

Mishra, S., & Das, A. P. (2023). Microbial remediation of synthetic microfiber contaminated wastewater. In M. P. Shah (Ed.), *Microbial technologies in industrial wastewater treatment* (pp. 337–350). Springer. https://doi.org/10.1007/978-981-99-2435-6_17

Mishra, S., Dash, D., & Das, A. P. (2024). Isolation and molecular characterization of polyester-degrading bacterial strains from municipal wastewater and their application in microplastic degradation in contaminated wastewater. *Geomicrobiology Journal*. https://doi.org/10.1080/01490451.2024.2376101

Mishra, S., Singh, R. P., Rout, P. K., & Das, A. P. (2022). Membrane bioreactor (MBR) as an advanced wastewater treatment technology for removal of synthetic microplastics. In *Development in wastewater treatment research and processes* (pp. 45–60). https://doi.org/10.1016/B978-0-323-85583-9.00022-3

Mishra, S., Swain, S., Sahoo, M., Mishra, S., & Das, A. P. (2021). Microbial colonization and degradation of microplastics in aquatic ecosystem: A review. *Geomicrobiology Journal*. https://doi.org/10.1080/01490451.2021.1983670

Mohanty, M., Mohanty, J., Dey, S., Dutta, K., Shah, M. P., & Das, A. P. (2024a). The face mask: A tale from protection to pollution and demanding sustainable solution. *Emerging Contaminants, 10*(2), Article 100298. https://doi.org/10.1016/J.EMCON.2023.100298

Mohanty, M., Tripathy, B., Koteswara Reddy, G., & Das, A. P. (2024). Greener technologies for establishing restraint over microplastic pollution. In *Sustainable microbial technology for synthetic and cellulosic microfiber bioremediation* (pp. 239–258). https://doi.org/10.1007/978-3-031-62600-5_13

Nagarale, S. B. (2024). Management of biomedical waste in India: A holistic approach towards environmental protection, human rights, and judicial oversight with reference to the pandemic. *Christ University Law Journal, 13*(2), 123–153.

Ncube, A., Mtetwa, S., Bukhari, M., Fiorentino, G., & Passaro, R. (2023). Circular economy and green chemistry: The need for radical innovative approaches in the design for new products. *Energies, 16*(4), 1752.

Priyadarshini, B., & Das, A. P. (2024). The adverse impact of synthetic microplastics released from disposable face masks pollutants on aquatic microbiota, aquatic species and humans: A review. *Environmental Science and Engineering, Part F2827*, 67–81. https://doi.org/10.1007/978-3-031-50840-0_5

Priyadarshini, B., Santhosh Kumar, M., & Das, A. P. (2024). Microbial degradation of synthetic plastics released from COVID-19 generated single use disposable face masks. *Environmental Science and Engineering, Part F2827*, 1–16. https://doi.org/10.1007/978-3-031-50840-0_1

Priyadarsini, S., & Das, A. P. (2024). Analytical and structural characterization of waste lithium-ion batteries for their effective recycling strategy. In *Environmental chemistry and ecotoxicology*. https://doi.org/10.1016/j.enceco.2024.12.004

Priyadarsini, S., & Das, A. P. (2025). Lithium bioleaching: A review on microbial-assisted sustainable technology for lithium bio-circularity. *Journal of Water Process Engineering, 69*, Article 106744. https://doi.org/10.1016/j.jwpe.2024.106744

Quina, M. J., Bordado, J. C., & Quinta-Ferreira, R. M. (2008). Treatment and use of air pollution control residues from MSW incineration: An overview. *Waste Management, 28*(11), 2097–2121.

Reinhardt, P. A. (2018). *Infectious and medical waste management*. CRC Press.

Ruiz Bravo, D. (2023). *Shaping social practices and behaviors to implement community-based organic waste management strategies* (Master's thesis, University of Twente).

Sadasivuni, K. K., Rattan, S., Waseem, S., Brahme, S. K., Kondawar, S. B., Ghosh, S., Das, A. P., Chakraborty, P. K., Adhikari, J., Saha, P., & Mazumdar, P. (2019). Silver nanoparticles and its polymer nanocomposites—Synthesis, optimization, biomedical usage, and its various applications. *Lecture Notes in Bioengineering* (pp. 331–373). https://doi.org/10.1007/978-3-030-04741-2_11

Sahu, A., Mallick, K., & Das, A. P. (2024). *Environmental impact of single-use synthetic face mask and its recycling: A sustainable approach* (pp. 197–212). https://doi.org/10.1007/978-3-031-50840-0_11

Sahu, S., & Das, A. P. (2024a). *Microplastics degradation mechanism by aquatic plastiomes: A review* (pp. 213–226). https://doi.org/10.1007/978-3-031-50840-0_12

Sahu, S., & Das, A. P. (2024). Microplastics degradation mechanism by aquatic plastiomes: A review. *Environmental Science and Engineering, Part F2827*, 213–226. https://doi.org/10.1007/978-3-031-50840-0_12

Salaria, N., Kumar, A., Singh, S., Rajput, A., Shukla, M., & Tripathi, S. K. (2024). Biotreatability of industrial hazardous wastes. In *Zero waste management technologies* (pp. 227–249). Springer Nature Switzerland.

Shanker, U., & Rani, M. (n.d.). Plastic degradation and its environmental implications. In *Materials research foundations* (Vol. 99).

Sinha, S. (2025). An overview of phytoremediation for the treatment of wastewater. *Recent Advances in Bioremediation and Phytoremediation*, 273–2842

Sinha, S. (2023). Synthesis of biopolymer based superabsorbent: An eco-friendly approach towards future sustainability. *In Bio-based superabsorbents: recent trends, types, applications and recycling* (pp. 29–49). Singapore: Springer Nature Singapore

Singh Jadaun, J., Bansal, S., Sonthalia, A., Rai, A. K., & Singh, S. P. (2022). Biodegradation of plastics for sustainable environment. *Bioresource Technology, 347*, Article 126697. https://doi.org/10.1016/J.BIORTECH.2022.126697

Singh, H., YT, K., Mishra, A. K., Singh, M., Mohanto, S., Ghumra, S., Thangadurai, D. (2024). Harnessing the foundation of biomedical waste management for fostering public health: Strategies and policies for a clean and safer environment. *Discover Applied Sciences, 6*(3), 89.

Singh, R. (2021). *Solid waste management in Eswatini: Challenges and opportunities.*

Sinha, S. (2024). Food security challenges and microplastics: A comprehensive review. In *Microplastics in African and Asian environments: The influencers, challenges, and solutions* (pp. 361–371).

Sinha, S., & Hamdan, A. (2023). E-plastic waste—A sustainable construction material: An Indian perspective. In *Handbook of sustainable materials: Modelling, characterization, and optimization* (pp. 213–232). CRC Press.

Sinha, S., Singh, S., & Dey, K. P. (2024). Microplastics: A significant threat to freshwater ecosystems. In *Sustainable microbial technology for synthetic and cellulosic microfiber bioremediation* (pp. 91–105). Springer Nature Switzerland.

Talreja, N., Hegde, C., Kumar, E. M., & Chavali, M. (2025). Emerging environmental contaminants: Sources, consequences, and future challenges. In *Green technologies for industrial contaminants* (pp. 119–149).

Woodruff, T. J. (2024). Health effects of fossil fuel–derived endocrine disruptors. *New England Journal of Medicine, 390*(10), 922–933.

# Management of Plastic Wastes: History, Current Applications, and Future Perspectives of Recycling, Upcycling, and Reclaiming Technologies

**Mine Begum Alanalp, Mustafa Fatih Ergin, and Ali Durmus**

## 1 Introduction

Since the 1950s, production and consumption of plastics have readily increased, with global production reaching over 370 million metric tons in recent years. This trend is expected to continue, with projections estimating it could reach over 1 billion metric tons by 2030. The highest plastic consumption is found in developed regions like North America, Europe, and parts of Asia. However, emerging economies are rapidly increasing their consumption related to growing industrialization and urbanization in these countries. Overall, while plastics play a crucial role in modern society, their increasing consumption raises important questions about sustainability and environmental health. Despite all the benefits, plastics have a major downside—pollution. Mismanagement of plastic wastes leads to ocean contamination, wildlife harm, and microplastic pollution in nature (Mishra et al., 2021; Sahu & Das, 2024a, 2024b). The rapid increase in plastic production and consumption has led to significant environmental challenges such as pollution, wildlife harm, and waste management issues. An estimated 11 million metric tons of plastic enter the oceans each year. There is a growing push toward a circular economy model, emphasizing the importance of reducing, reusing, and recycling plastics to minimize environmental impact (Ghosh et al., 2023; Mishra et al., 2022; Mohanty et al., 2024a).

Managing plastic wastes effectively involves a combination of various strategies aimed at reducing, reusing, recycling, and disposing of plastics in an environmentally

M. B. Alanalp
Department of Chemical Engineering, Faculty of Engineering and Natural Sciences, Istanbul Health and Technology University, 34445 Beyoglu, Istanbul, Turkey

M. F. Ergin · A. Durmus (✉)
Engineering Faculty, Department of Chemical Engineering, Istanbul University-Cerrahpasa, 34320 Avcılar, Istanbul, Turkey
e-mail: durmus@iuc.edu.tr

 
A. P. Das and S. Karakuş (eds.), *Synergizing Sustainability for Integrated Waste Management*, Sustainable Environmental Waste Management Strategies,
https://doi.org/10.1007/978-3-032-03718-3_5

responsible manner (Mohanty et al., 2024a, 2024b; Ray et al., 2024). Modern societies can significantly reduce plastic waste and its environmental impact, promoting a more sustainable future by integrating these strategies (Das & Bal, 2025; Das et al., 2023).

Plastic materials can be classified in several ways, depending on several criteria such as their structure, properties, and applications (Priyadarshini & Das, 2024; Tripathy et al., 2024). The most commonly used and basic criterion is the definition of plastics as thermoplastics and thermosets regarding their structural and thermal properties (Das et al., 2025; Priyadarshini et al., 2024). Commodity plastics, including some types of thermoplastics such as polyethylene (PE), polypropylene (PP), polyvinyl chloride (PVC), polystyrene (PS), polyamides (also called nylons), thermoplastic polyesters (e.g., polyethylene terephthalate [PET] and polybutylene terephthalate [PBT]), and some thermoset elastomers such as rubbers and thermoplastic elastomers, have been extensively used in modern society (Mallick et al., 2023; Swain et al., 2025). It is known that approximately half of the plastics used in daily life are thermoplastic packaging materials. Thus, management of thermoplastics and some elastomer wastes has become a major industrial problem.

This chapter briefly summarizes and discusses the current applications of recycling, upcycling, and reclaiming technologies of thermoplastic wastes. Future perspectives and projections will also be highlighted to manage plastic wastes effectively and establish sustainable development in the world.

## 2 Management of Thermoplastic Wastes

The management of thermoplastic waste has become a crucial aspect of sustainable environmental policies, driven by the increasing production and consumption of plastic materials. Thermoplastics, including polyethylene terephthalate (PET), polypropylene (PP), and polyvinyl chloride (PVC), are widely employed across diverse industries such as packaging, automotive, and construction. Implementing effective recycling strategies for these materials not only aids in mitigating environmental pollution but also diminishes dependence on virgin raw materials, thereby contributing to the advancement of a circular economy.

The recycling of thermoplastics can be classified into three primary processes: (i) mechanical recycling, (ii) chemical recycling, and (iii) energy recovery. The choice of an appropriate recycling method is contingent upon multiple factors, including the type of plastic waste, the degree of contamination, and the intended end-use applications. A structured approach to plastic waste management aligns with the 3R principle (Reduce, Reuse, and Recycle), wherein recycling represents the most widely implemented strategy for processing thermoplastics such as PET, high-density polyethylene (HDPE), low-density polyethylene (LDPE), and PP (Abukasim et al., 2020).

Mechanical (or physical) recycling entails the reprocessing of plastic waste through advanced techniques such as twin-screw extrusion, a method demonstrated to improve the mechanical properties of recycled thermoplastics (Singh & Singh, 2022). Nevertheless, successive recycling cycles may result in thermal degradation and a reduction in polymer stability, although overall trends in thermal properties tend to remain relatively consistent (Ignatyev et al., 2014).

Chemical recycling, encompassing processes such as pyrolysis and depolymerization, provides an alternative strategy by decomposing plastics into their monomeric or oligomeric components or converting them into valuable feedstocks. Depolymerization can be achieved in several ways depending on the polymer type and reaction conditions. The main types of depolymerization reactions are (i) thermal depolymerization, (ii) catalytic or thermocatalytic depolymerization, (iii) hydrolytic depolymerization, (iv) enzymatic depolymerization, (v) electrochemical depolymerization, (vi) photodepolymerization, and (vii) solvochemical depolymerization or solvolysis. The latest one can also be defined as alcoholysis, glycolysis, aminolysis, etc. depending on the chemical species used in the process.

Pyrolysis, in particular, has attracted considerable interest as a feasible approach for converting plastic waste into fuel oil and other petrochemical derivatives (Eze et al., 2021). This technique not only facilitates the recovery of valuable materials but also aids in mitigating reliance on fossil fuels (see Fig. 1).

Energy recovery, a critical component of plastic waste management, entails the conversion of nonrecyclable plastic waste into energy through combustion processes. Although this approach contributes to waste minimization and energy production, it simultaneously raises concerns related to emissions and long-term sustainability.

Although mechanical and chemical recycling present viable pathways for thermoplastic waste management, their environmental and economic ramifications must be critically evaluated to ensure long-term sustainability. From an environmental standpoint, mechanical recycling typically exhibits a lower carbon footprint due to its reduced energy consumption and the absence of chemical reagents (Mohanty et al., 2024b). However, the progressive degradation of polymer properties over successive

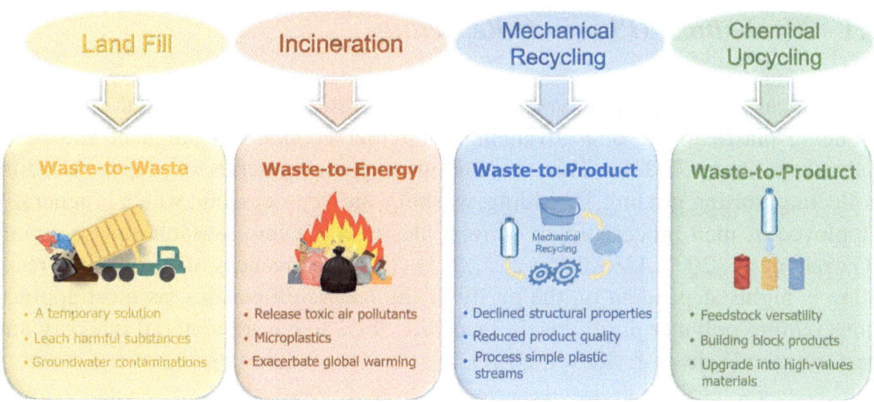


**Fig. 1** Comparison of plastic recycling methods (Chen et al., 2021)

recycling cycles may constrain its long-term viability. Conversely, chemical recycling facilitates the recovery of high-quality raw materials; nevertheless, its substantial energy requirements and potential emissions introduce sustainability challenges. Furthermore, certain chemical recycling processes generate hazardous by-products, necessitating stringent handling measures and regulatory oversight (Fig. 1).

From an economic perspective, mechanical recycling is generally regarded as a more cost-effective approach due to its lower operational expenditures and relatively straightforward processing techniques. Nevertheless, the volatility in the market value of recycled plastics, coupled with constrained demand for specific secondary materials, poses challenges to profitability. Although chemical recycling yields higher-quality outputs, it necessitates significant capital investment and sophisticated infrastructure, thereby limiting its feasibility in certain regions. Conversely, energy recovery methods offer an immediate disposal solution for nonrecyclable plastics; however, they remain contentious due to concerns regarding greenhouse gas emissions and long-term sustainability.

Enhancing the efficiency of plastic waste management necessitates a comprehensive approach that integrates mechanical and chemical recycling with progressive policy frameworks and cutting-edge technological advancements. Innovations in sorting methodologies, development of biodegradable polymers, and refinement of chemical catalysts hold the potential to further optimize recycling processes and mitigate environmental repercussions.

In summary, the synergistic integration of mechanical and chemical recycling techniques, in conjunction with responsible waste management practices, is imperative for mitigating the environmental impact of thermoplastic waste. Future advancements in recycling technologies, coupled with robust policy frameworks, will be instrumental in fostering a sustainable and closed loop plastic waste management system.

## *2.1 Mechanical (Physical) Recycling*

Mechanical recycling of polymer materials represents a fundamental strategy in reducing plastic waste and advancing a circular economy (Lamtai et al., 2023; Schyns & Shaver, 2021). This process encompasses a series of steps, including collecting, sorting/grading, shredding, washing, and reprocessing, which is generally employed as melt processing, to convert plastic waste into reusable raw materials (Garraín et al., 2007; Uzosike et al., 2023). The effectiveness of mechanical recycling is highly dependent on the purity of the recovered plastics, as most applications require polymer purity levels exceeding 99% to ensure material integrity and performance (Schyns & Shaver, 2021).

### 2.1.1 Sorting and Separation

The initial stage of the mechanical recycling process involves the systematic sorting and classification of plastic waste, a critical step in preserving the purity and mechanical properties of the recycled materials (see Fig. 2). Various advanced technologies, including near-infrared (NIR) spectroscopy, optical color recognition, and density-based separation, are employed to enhance sorting efficiency. NIR spectroscopy, in particular, facilitates the precise identification of polymer types based on their spectral absorption characteristics, thereby optimizing the classification process.

Additional sorting techniques, such as X-ray analysis, electrostatic separation, and hydrocyclone-based density sorting, provide complementary advantages depending on the chemical composition and physical properties of the plastics being processed (Hopewell et al., 2009; Ragaert et al., 2017; Ruj et al., 2015). While manual sorting remains a labor-intensive approach, it is still utilized in small-scale recycling facilities where high accuracy is prioritized over automation. Conversely, large-scale industrial operations increasingly adopt advanced recognition systems, including fluorescent marking technologies and artificial intelligence (AI)-driven automated sorting, to improve efficiency and minimize contamination risks (Sayed).

**Fig. 2** Efficient mechanical separation and sorting process for plastic recycling

However, the most important challenges in these sorting, identification, and classification steps are the similarity in chemical and physical properties of some thermoplastics—for example, polyolefins (e.g., polyethylenes and polypropylene)—and the transportation speed of solid materials on conveyor belts. Thus, visual identification and manual separation methods are still applied in small- and/or medium-scale recycling plants.

Once plastics are sorted, they undergo a size reduction process through shredding, followed by further separation techniques such as flotation, air classification, and thermal-based sorting. These additional refining steps play a crucial role in eliminating residual impurities and optimizing particle size for downstream processing (Hong & Chen, 2017; Hopewell et al., 2009).

### 2.1.2 Cleaning and Drying

Following the sorting stage, plastic wastes undergo an intensive cleaning process to remove contaminants such as adhesives, labels, organic residues, and other impurities that could compromise the quality of the final recycled material. Effective cleaning is essential to ensure that the recycled plastics maintain their mechanical integrity and meet regulatory standards for reuse in various applications.

The washing process typically involves a combination of detergents, high-pressure water treatments, and, in some cases, alkaline solutions such as sodium hydroxide (NaOH), which has been demonstrated to enhance decontamination efficiency (Schyns & Shaver, 2021). Due to the widespread use of polyolefins (e.g., polyethylenes and polypropylene) in food packaging, thorough and effective cleaning and decontamination are particularly critical to comply with food safety regulations. For instance, studies indicate that washing PET flakes with NaOH at elevated temperatures significantly improves purity levels, thereby increasing their suitability for high-value recycling applications (Picuno et al., 2020).

Once washed, plastic materials must be adequately dried to prevent residual moisture from affecting subsequent processing stages. Various drying techniques, including hot air drying and vacuum drying, are employed depending on the polymer type and intended recycling process. Proper drying not only enhances the mechanical performance of the recycled plastic but also minimizes the risk of hydrolytic degradation during reprocessing.

### 2.1.3 Shredding and Reprocessing

Washed and dried plastic wastes are subjected to mechanical size reduction operations, including crushing and grinding through shredding, a critical step in preparing materials for further processing. Shredding transforms plastic waste into flakes or pellets, which serve as raw materials for new plastic products. This process is particularly beneficial for thermoplastics, as it facilitates their reuse in various applications, including extrusion and injection molding (El Haggar, 2010).

The efficiency of the shredding process depends on several factors, including polymer type, contamination levels, and the intended application of the recycled material. However, it can be inferred that the most important parameter in shredding and grinding is the size of flakes as the process output. Smaller flake sizes increase the surface area, improving homogeneity in subsequent melt processing operations. However, excessive fragmentation may negatively impact sorting efficiency, particularly for heterogeneous plastic waste streams. Research indicates that HDPE shreds with an average size of 0.85 mm exhibit superior mechanical properties when blended with calcium carbonate additives, highlighting the importance of optimizing shred size for specific applications (Awan et al., 2021).

Following shredding, the processed flakes or pellets are directly utilized in melt compounding either to manufacture recycled pellets or sometimes recycled end products such as packaging films, boxes, pipes, sheets, profiles, etc. These steps further enhance material uniformity, enabling better mechanical performance in final applications. Advances in reprocessing techniques, including compatibilization strategies and additive incorporation, have been shown to improve the quality and durability of recycled plastics, thereby expanding their potential applications in high-performance sectors (Ragaert et al., 2017). Figure 2 describes the efficient mechanical separation and sorting process for plastic recycling.

### 2.1.4 Advantages and Limitations of Mechanical Recycling

Mechanical recycling presents a cost-effective and energy-efficient alternative to chemical recycling, as it eliminates the need for high-temperature treatments or complex chemical processes. This method is particularly advantageous for commonly used thermoplastics such as polyethylene terephthalate (PET), polyethylenes (LDPE, linear low-density polyethylene [LLDPE], HDPE, etc.), and polypropylene (PP), which can be repeatedly reprocessed with relatively minimal energy input (Schyns & Shaver, 2021). Furthermore, mechanical recycling supports the principles of a circular economy by reducing reliance on production of virgin (primer) plastics which are directly dependent on crude oil and petrochemical feedstocks and lowering overall carbon emissions (Lamtai et al., 2023).

Despite these benefits, mechanical recycling faces several critical limitations. One of the most significant challenges is polymer degradation, as repeated recycling cycles lead to a decline in mechanical strength, molecular weight, and overall material performance. This phenomenon is particularly evident in polyamides and thermoplastic polyesters (e.g., PET and PBT) where thermal and oxidative degradation reduce polymer integrity over successive processing steps (Hopewell et al., 2009). Fortunately, various types of chain extenders have been commercialized to increase the molecular weight and improve the physical properties of polyamides and polyesters via coupling reactions between functional groups of recycled polymer chains and chain extender compounds. Chain extenders are organic or oligomeric compounds and/or copolymers having functional groups (like amine, epoxy, or isocyanate groups). The structural features of chain extenders are based on their thermal stability in melt processing conditions, up to 300 °C, and chemical reactivity due to unique chemical structures and functional groups. Additionally, contamination with incompatible polymers or additives can severely impact the quality of the final recycled product, necessitating highly efficient sorting and purification technologies (Schyns & Shaver, 2021).

To address these limitations, recent research has focused on polymer compatibilization strategies and advanced purification techniques. The incorporation of compatibilizers has been shown to enhance the mechanical properties of polymer blends, allowing for more effective recycling of mixed plastic waste streams (Ragaert et al., 2017). Additionally, innovations in automated sorting technologies, such as AI-driven recognition and robotic sorting systems, hold promise for improving material purity and recyclability (Sayed et al., 2023).

Future advancements in material science and recycling infrastructure will be crucial in overcoming these challenges. By integrating improved sorting methods, additive-assisted stabilization techniques, and policy-driven incentives, mechanical recycling can be further optimized to support a more sustainable and economically viable plastic waste management system (Browne et al., 2011).

## 2.2 *Chemical Recycling*

Chemical recycling represents an advanced strategy for converting plastic waste into high-value chemicals and fuels, surpassing the capabilities of traditional mechanical recycling and energy recovery. Unlike mechanical processes, which primarily involve physical reprocessing, chemical recycling enables the breakdown of polymer structures into their fundamental monomers or synthetic feedstocks, allowing for the regeneration of virgin-quality materials (Garcia & Robertson, 2017).

This approach is particularly effective for polyethylene terephthalate (PET), polyurethane (PUR), and polyamides, which can undergo well-established depolymerization and repolymerization processes. These polymers are highly valued due to their high hydrocarbon content and structural resemblance to petroleum-derived feedstocks, making them ideal candidates for chemical recycling.

A distinguishing feature of plastic waste—particularly polyolefins—compared to biomass-derived materials is its low oxygen content, which enhances carbon efficiency and yield potential. This characteristic renders chemical recycling an attractive strategy for both resource recovery and waste reduction.

This section examines key chemical recycling technologies, including gasification, pyrolysis, fluid catalytic cracking (FCC), and hydrocracking, with a particular emphasis on monomer and feedstock recycling. These technologies are increasingly recognized as crucial solutions for enhancing resource sustainability and reducing the accumulation of plastic waste (Al-Salem et al., 2009, 2010).

### 2.2.1 Chemolysis

Chemolysis, a method of chemical recycling, has garnered increasing interest due to its alignment with sustainable development principles and its applicability in food-grade polymer recycling. This process enables the depolymerization of plastics into high-purity monomers, which can be repolymerized into virgin-quality materials. Consequently, chemolysis facilitates the production of value-added products suitable for diverse industrial applications, particularly in sectors requiring stringent material purity, such as food packaging and medical industries (Garcia et al., 2021).

Despite its advantages, chemolysis presents economic challenges, as chemically recycled polymers are often more expensive than their virgin counterparts. This cost disparity arises from several factors such as high feedstock prices, substantial capital investment requirements, and the necessity for large-scale operations to achieve profitability. Studies have indicated that a PET chemolysis facility must process at least 15,000 tons of plastic waste annually to remain economically viable (George & Kurian, 2014). Therefore, the scalability and economic feasibility of chemolysis remain critical challenges in integrating this technology into the circular economy framework.

### 2.2.2 Solvolysis in Plastic Waste Recycling

Solvolysis is a promising technique for recycling of plastic wastes, involving the breakdown of polymer chains into their monomeric or oligomeric units. This process occurs when excess of a proper solvent, which also acts as reactant, is present in the reaction medium and interacts with the polymer chains through nucleophilic substitution. A variety of solvents, including water, methanol, polyols, amines, and ammonia, have been explored for solvolysis, typically under elevated temperature conditions or with the aid of homogeneous or heterogeneous catalyst that accelerates the depolymerization rate.

Besides the conventional depolymerization catalysts, new green materials or physical methods have been suggested to facilitate depolymerization reactions, in recent years. An interesting study conducted by Liu et al. (2018) focused on the application of ionic liquids (ILs) as catalysts to improve the alcoholysis of polycarbonate (PC) waste at 120 °C. This research demonstrated that methanol was the most effective solvent compared to ethanol and propanol. They reported that larger alcohol molecules decreased the depolymerization efficiency. The IL [HDBU][LAc] was found to be particularly effective, achieving near-total conversion of PC to bisphenol A (BPA) with an impressive yield of approximately 99% after only 1 h of reaction. Notably, the catalyst showed excellent reusability, maintaining its catalytic activity even after multiple reaction cycles. They also proposed a reaction mechanism involving the dissolution of PC chains and the formation of active intermediates when the [LAc] anion from the IL interacted with methanol molecules. This interaction enabled the breakdown of ester bonds in PC, leading to the production of soluble oligomers, which were further depolymerized into BPA and dimethyl carbonate (DMC).

In addition to studies on PC, a two-step alkaline hydrolysis method has been successfully applied to PET waste, operating under mild conditions, with a reaction temperature of 80 °C and atmospheric pressure (Ügdüler et al., 2020). This process was optimized to achieve high monomer yields, converting PET pellets into terephthalic acid (TPA) and ethylene glycol (EG) with a yield of 95 wt% in a 5 wt% of NaOH aqueous–ethanol solution within 20 min. The reaction kinetics were found to follow a diffusion model. The process was also applied to monolayer or multilayer postconsumer PET materials, such as bottles, trays, and films. It was observed that smaller and thinner materials underwent hydrolysis more efficiently than thicker ones. For example, multilayer PET trays showed a significantly lower hydrolysis yield compared to monolayer trays. A detailed process flow for the treatment of multilayer colored PET trays was proposed, incorporating steps such as conditioning, milling, alkaline hydrolysis, filtration, and acidification to remove impurities such as carbon black and other unhydrolyzed polyolefins. The resulting products were purified TPA and EG monomers. A life cycle assessment (LCA) of this alkaline hydrolysis method revealed that, when optimized for feedstock-to-solvent ratio, this process has a significantly lower carbon footprint compared to conventional incineration methods.

### 2.2.3 Fluid Catalytic Cracking (FCC)

The thermal degradation of plastic wastes often yields a very broad range of hydrocarbon distribution regarding both carbon number and type of compounds (e.g., aliphatic, aromatic, cyclic, etc.) producing highly different products. In contrast, catalytic degradation has emerged as a promising alternative, offering greater selectivity and control over the final product composition. Research indicates that catalytic cracking promotes the formation of valuable aromatic and naphthenic compounds, which serve as key intermediates in fuel and chemical feedstock applications (Buekens & Huang, 1998; Li et al., 2014). Additionally, catalytic processes enhance gasoline yield by shifting the product spectrum toward lower carbon-number hydrocarbons, thereby improving fuel quality.

Catalytic cracking is generally classified into two main types based on the reaction phase:

1. Liquid-phase catalytic cracking: In this process, catalyst interacts directly with the molten polymer, enabling efficient molecular fragmentation.
2. Vapor-phase catalytic cracking: A part of thermal degradation products in vapor phase react with the active sites on the surfaces and/or pores of the catalyst, and a gas–solid interaction yields a more controlled conversion and product distribution.

Commonly used solid catalysts include silica-alumina, ZSM-5, and zeolites, which facilitate hydrocarbon fragmentation and promote the formation of desirable end products (Buekens & Huang, 1998). Unlike pyrolysis, catalytic cracking operates at lower temperatures (300–350 °C), leading to reduced energy consumption and enhanced cost-effectiveness.

Despite its advantages, catalytic cracking faces several technical challenges, primarily related to:

- **Catalyst deactivation**: Plastic waste contains contaminants such as chlorine and nitrogen compounds (from plastic additives), which can rapidly deactivate cracking catalysts, diminishing their efficiency over time.
- **Inorganic impurities**: Heavy metals and mineral residues can block the catalyst's porous structure, leading to irreversible deactivation and necessitating feedstock pretreatment (Miskolczi et al., 2009).
- **Product quality**: While catalytic cracking optimizes gasoline and diesel yields, some commercial processes produce fuels with lower octane ratings compared to conventional gasoline. These fuels require further refining, increasing processing costs and limiting economic feasibility (Pinto et al., 1999).

Beyond catalyst-related challenges, the lack of optimized reactor technology remains a major barrier to large-scale implementation. To enhance process stability and industrial scalability, research efforts should focus on:

- Developing more durable catalysts with higher resistance to deactivation.
- Innovating advanced reactor designs to improve heat and mass transfer efficiency.
- Integrating catalyst regeneration techniques to enhance longevity and reduce operational costs.
- Mitigating contaminants through pretreatment processes and feedstock purification strategies.

Addressing these limitations will be essential for establishing fluid catalytic cracking as a viable and sustainable solution for plastic waste valorization within the circular economy framework.

### 2.2.4 Hydrocracking

Hydrocracking differs from traditional catalytic cracking by introducing hydrogen into the process. This catalytic reaction operates under high hydrogen pressure (typically around 70 atm) and at temperatures ranging from 375 to 400 °C. Commonly used catalysts include nickel–sulfur (Ni/S) and nickel–molybdenum–sulfur (NiMo/S) supported catalysts (Ding et al., 1997). The hydrocracking process effectively converts plastic waste into valuable fuels and chemicals, as the introduction of hydrogen facilitates the breakdown of plastic polymers into smaller hydrocarbons.

During hydrocracking, raw plastic waste—often containing various inorganic impurities—is first liquefied. Non-distillable substances are removed, and low-temperature pyrolysis is used to eliminate contaminants before hydrogen is introduced. The presence of hydrogen significantly improves fuel quality by increasing the hydrogen-to-carbon (H/C) ratio and reducing the aromatic content, which is a common issue in many other recycling methods. The process predominantly yields paraffins, which are high-quality fuel components.

A key advantage of hydrocracking is its ability to process mixed plastic waste (solid plastic waste [SPW]) containing various polymer types and additives. Additionally, it effectively manages heteroatoms such as sulfur and nitrogen, preventing the formation of toxic by-products like dioxins—a frequent concern in other plastic degradation processes. This makes hydrocracking a safer and more environmentally friendly option for plastic waste recycling.

However, hydrocracking has notable challenges, particularly its high energy and cost requirements. The continuous hydrogen supply is a significant expense, with hydrogen production via electricity costing approximately €2500 per ton. Additionally, the high-pressure conditions and specialized equipment required for the process increase both investment and operational costs, limiting its large-scale adoption. While hydrocracking holds great potential for producing high-quality naphtha and efficiently processing mixed plastic waste, addressing these economic and technical challenges is crucial for its commercial feasibility.

The integrated hydropyrolysis and hydroconversion (IH2) process is a catalytic thermochemical technology designed to convert organic materials into hydrocarbon fuels. This process can handle a wide range of feedstocks, including cellulose fractions, wood, agricultural waste, and municipal solid waste (Marker et al., 2012). Developed by the Gas Technology Institute, IH2 operates in three key stages: (i) hydropyrolysis, (ii) hydroconversion, and (iii) reforming.

In the hydropyrolysis stage, a fluidized bed reactor containing a catalyst operates at 400–500 °C and pressures between 15 and 35 atm. During this stage, deoxygenation reactions remove oxygen atoms from the gas phase, converting them into water ($H_2O$), carbon monoxide (CO), and carbon dioxide ($CO_2$). These reactions release heat, which helps offset the endothermic pyrolysis process. The resulting hydrogen-enriched gas then moves to the hydroconversion reactor, where oxygen content is reduced from approximately 2.7% to below 1% (Marker et al., 2014).

The hydroconversion reactor separates the product into two main phases:

- A liquid phase with low oxygen content, which consists of high-value hydrocarbons
- A water phase, where remaining oxygenated compounds are removed

Meanwhile, the remaining gas phase—containing light hydrocarbons and hydrogen—is directed to a vapor reformer, which produces additional hydrogen for the process. This integrated hydrogen recycling is a key feature of IH2, reducing the need for an external hydrogen supply and improving overall efficiency.

A key benefit of IH2 is its ability to directly produce liquid hydrocarbons from plastic waste without requiring an external hydrogen supply, making the process economically attractive. Furthermore, the IH2 method eliminates the need to handle highly acidic pyrolysis oils, thus reducing the demand for specialized reactors and storage systems.

The IH2 process has strong economic potential due to its high efficiency, simplified design, and integrated hydrogen recycling system. All key components have been commercialized, improving its scalability and reducing investment risks. While initially designed for biomass-to-liquid fuel applications, IH2 has also been successfully tested with solid plastic waste (SPW). However, additional research is needed to fully optimize its use in plastic waste recycling (Ragaert et al.,.).

The KDV process, developed by Alphakat GmbH, is an advanced technology designed to convert biomass and plastic waste into liquid fuels. This process effectively removes oxygen atoms from the feedstock, enabling the production of fuels compatible with conventional internal combustion engines. The KDV technology allows for the production of diesel, kerosene, and oil from a diverse range of organic and mineral-based materials. Compared to pyrolysis, the KDV process operates under milder reaction conditions, making it a more energy-efficient alternative (Fakhrai & Saadatfar, 2015). Industrial-scale KDV plants typically have processing capacities ranging from 150 to 5000 L/h.

The process begins with feedstock preparation, where waste materials are reduced to particles smaller than 3 mm, with a moisture content of approximately 5%. The prepared feedstock is then mixed with a catalyst, lime, and carrier oil before being heated to 250 °C and introduced into a turbine reactor. Inside the reactor, centrifugal force facilitates the separation of liquid and solid materials, yielding a hydrocarbon-rich product. The reaction temperature remains within the range of 250–320 °C, promoting the selective formation of diesel fuel. Additionally, within the reactor environment, plastic components undergo neutralization and dehalogenation, reducing harmful contaminants (Scheirs & Kaminsky, 2006).

With an energy efficiency of approximately 70%, the fuels produced by the KDV process exhibit properties similar to automotive diesel. However, the sulfur content and cetane number of the produced fuels do not fully meet European standards, raising concerns regarding potential emissions (Labeckas & Slavinskas, 2013). Despite these limitations, the KDV process holds significant potential for the production of synthetic fuels from waste materials. Further research into chemical optimization and process engineering is crucial to enhance fuel quality and overall process efficiency.

The toxicity of gases produced during the thermal decomposition of plastics is a significant concern. Research conducted by the National Bureau of Standards has explored the toxicity of effluents generated from the decomposition of various plastics, including polyurethane foams, PVC, and acrylonitrile–butadiene–styrene (ABS) (Levin & Materials, 1987).

For polyurethane foams, the thermal decomposition process releases toxic gases such as carbon monoxide (CO) and hydrogen cyanide (HCN), with flaming combustion being particularly hazardous. Decomposition of PVC predominantly produces hydrogen chloride (HCl), benzene, and unsaturated hydrocarbons. HCl, a major irritant, and CO, an asphyxiant, contribute significantly to the toxicity of PVC combustion products.

The lethal concentration for different materials, represented as LC50 values (lethal concentration for 50% of a population), varies from 0.045 to 57 mg/L. PVC exhibits moderate toxicity relative to other plastics, largely due to the substantial amounts of HCl produced during its degradation.

In comparison, catalytic cracking offers a potential solution to reduce the toxicity of PVC decomposition. By lowering reaction temperatures and enhancing product selectivity, catalytic cracking can decrease the production of chlorinated hydrocarbons, thereby mitigating the toxicity of the process.

Similarly, the degradation products of acrylonitrile–butadiene–styrene (ABS) are found to be as toxic as those from other common polymers, reinforcing the need for improved methods to manage the toxicity of gases released during plastic degradation.

## *2.3 Energy Recovery from Thermoplastics*

Pyrolysis is a thermochemical recycling method that decomposes complex plastic waste into valuable hydrocarbons, monomers, and fuel-grade products under oxygen-free conditions. This technique is particularly effective for processing plastics that are challenging to recycle mechanically, such as mixed polyethylene (PE), polypropylene (PP), polystyrene (PS), multilayer packaging, and polyurethane-based construction waste.

Multilayer packaging represents a significant challenge for conventional mechanical recycling due to its heterogeneous composition, which often includes PE, polyethylene terephthalate (PET), ethylene vinyl alcohol (EVOH) (used for oxygen barrier properties), and metallized films. Given the difficulty of separating these layers efficiently, pyrolysis offers a viable alternative, as it can handle mixed and contaminated plastic waste more effectively than sorting-dependent mechanical recycling (Vermeulen et al., 2011).

The pyrolysis process is conducted at temperatures around 500 °C in an inert (oxygen-free) atmosphere, where polymer chains undergo thermal cracking, yielding gaseous, liquid, and solid fractions. The distribution of these products varies depending on the polymer composition and reaction conditions (Angyal et al., 2007). Key polymer behaviors during pyrolysis include:

- PE and PP undergo random thermal degradation, generating a broad range of hydrocarbon products.
- PTFE (polytetrafluoroethylene) and PMMA (polymethyl methacrylate) undergo monomer recovery, with PMMA achieving up to 98% monomer yield (Garforth et al., 2004).

Due to its ability to convert mixed plastics into high-value hydrocarbons, pyrolysis has emerged as a key solution in the circular economy for plastic waste valorization. However, process optimization, economic feasibility, and contamination challenges must be addressed to facilitate large-scale implementation.

The economic feasibility of pyrolysis is strongly dependent on operational scale, making it a cost-effective solution for processing high-volume polymers such as polyethylene (PE), polypropylene (PP), polystyrene (PS), and polyvinyl chloride (PVC). Together, these plastics constitute approximately 80% of plastic production in Europe, highlighting their significance in waste management strategies. However, the recycling of less common plastics at industrial scales remains a major challenge, as limited availability and heterogeneous composition increase processing costs and technical complexity.

One of the most critical technical hurdles in pyrolysis is the presence of polyvinyl chloride (PVC) in mixed plastic waste streams. During thermal degradation, PVC releases chlorine-containing compounds, which pose significant risks to equipment integrity and environmental safety. Chlorine emissions can lead to:

- Corrosion of reactor components, reducing equipment lifespan and increasing maintenance costs
- Formation of hazardous by-products, such as hydrochloric acid (HCl) and dioxins, which require additional treatment

To mitigate these challenges, low-temperature pyrolysis (~300 °C) has been proposed as a strategy for chlorine removal, with studies indicating that up to 98% of chlorine can be extracted through careful pretreatment and sorting processes (Okuwaki, 2004). However, this approach also generates secondary waste streams, necessitating further treatment to ensure environmental compliance.

Beyond chlorine-related complications, other contaminants such as sulfur and heavy metals further hinder pyrolysis efficiency and reduce the quality of recovered hydrocarbons (Miskolczi et al., 2004). As a result, advanced feedstock purification techniques, catalytic additives, and integrated gas treatment systems are essential for enhancing the economic viability and sustainability of pyrolysis-based recycling.

Future research should prioritize the development of cost-effective decontamination strategies, reactor materials resistant to corrosion, and process optimizations to enhance yield and energy efficiency. Addressing these challenges will be crucial for establishing pyrolysis as a competitive and scalable solution in the transition to a circular plastic economy.

The efficiency of pyrolysis largely depends on the design and operational parameters of the reactor system. Various reactor types have been developed to optimize heat transfer, residence time, and reaction kinetics, each offering distinct advantages. The most commonly utilized pyrolysis reactor configurations include (see Fig. 3):

- Bubbling fluidized bed reactors: These reactors ensure efficient heat and mass transfer by maintaining the feedstock in a fluidized state, leading to uniform heating and enhanced reaction rates (Butler et al., 2011).
- Stirred tank reactors: Commonly used in laboratory-scale pyrolysis studies, these reactors provide precise control over reaction conditions, making them ideal for experimental research.
- Screw/auger reactors: Suitable for continuous feed processing, these reactors facilitate mechanical mixing of plastic waste, enhancing thermal degradation efficiency.

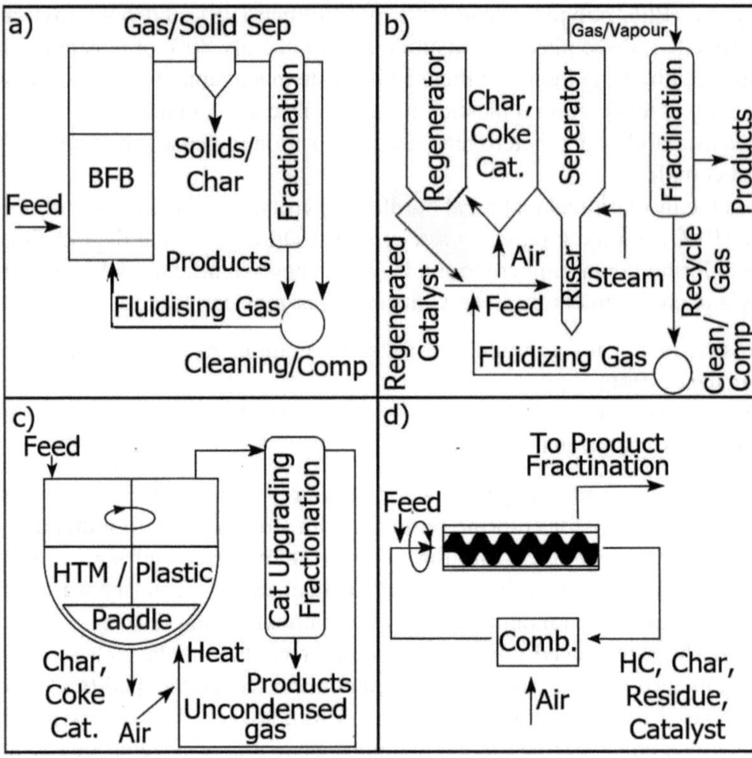


**Fig. 3** Various reactor designs for plastic solid waste pyrolysis (Butler et al., 2011): **a** bubbling fluidized bed, **b** fluid catalytic cracker, **c** stirred tank reactor, and **d** screw/auger reactor

- Fluid catalytic cracking (FCC) reactors: Widely employed in petroleum refining, FCC reactors enable the selective conversion of plastic waste into high-value hydrocarbons, improving product quality.

Recent advancements have introduced a novel reactor design known as the gas–solid vortex reactor in static geometry (GSVR-SG), also referred to as the vortex reactor (Kulkarni et al., 2018). This innovative system enhances convective heat and mass transfer by leveraging rotational motion, allowing for improved efficiency in feedstock conversion. Additionally, the vortex reactor offers greater flexibility in fluidization agents, enabling the use of steam or hydrogen to tailor the pyrolysis process for specific applications.

As pyrolysis continues to evolve, future research should focus on scaling up vortex reactor technology, optimizing energy efficiency, and integrating catalytic enhancements to further improve product selectivity. These advancements will be crucial for overcoming current limitations and establishing pyrolysis as a key technology for sustainable plastic waste management. Figure 3 describes the various reactor designs for plastic solid waste pyrolysis.

A consortium in the Netherlands, including AkzoNobel and Enerkem, is developing a waste-to-chemicals plant in Rotterdam. This facility will convert waste into methanol, a critical chemical for various industrial applications, thereby reducing reliance on fossil fuels. The methanol produced will be used to manufacture valuable chemicals such as acetic acid, thickening agents, and dimethyl ether, contributing to more sustainable waste management in the region.

Syngas, generated from various carbon sources, is essential for producing petroleum-like products, including methanol, through processes such as Fischer–Tropsch synthesis and methanol-to-olefins (MTO). The efficiency and quality of syngas are determined by the $H_2/CO$ ratio, which is crucial for different downstream processes (Wender, 1996).

Gasification is the process by which solid feedstocks are converted into syngas through partial oxidation, typically using steam and oxygen or air. While air is a more economical option, it results in higher emissions and less efficient separation of syngas components (Puig-Arnavat et al., 2010). The syngas produced can then be used for the production of methanol and hydrocarbons. However, the presence of contaminants like ammonia and sulfur presents challenges for catalytic processes, and the purification step can be a significant cost driver.

Plastic waste can also be gasified to generate syngas, but further research is needed to optimize the process. The Texaco gasification process, which involves the liquefaction of plastic waste followed by gasification, is an example of this technology in practice (Brems et al., 2013).

Methanol is synthesized from syngas using a catalytic process. The reaction, which occurs under high pressure and temperature, is exothermic but equilibrium-limited, influencing the reactor design. The methanol produced can then be utilized to create olefins or fuels through the MTO and methanol-to-gasoline (MTG) processes (Spath & Dayton, 2003).

The MTG process converts methanol into hydrocarbons using a zeolite catalyst, producing high-sulfur gasoline that requires additional treatment to prevent engine blockages. The MTO process, on the other hand, focuses on producing light olefins rather than gasoline. While MTO provides more flexibility compared to traditional cracking, it still does not match the efficiency of steam crackers (Keil, 1999).

## *2.4 Upcycling of Thermoplastics*

Upcycling is an advanced and innovative approach to thermoplastic waste management that focuses on enhancing the value and functionality of discarded plastics without chemically breaking down their polymer structures. Unlike traditional recycling methods, which may degrade the material quality over multiple cycles, upcycling preserves or even improves the mechanical, thermal, and aesthetic properties of plastics, allowing for their repurposing into high-value applications. This approach contributes to the principles of the circular economy by extending the lifespan of plastic materials and reducing reliance on virgin resources.

One of the main advantages of upcycling is its ability to generate materials with superior properties compared to their original form. For instance, thermoplastics can be reinforced with additives or natural fibers to enhance strength, durability, and heat resistance, making them suitable for structural components, automotive parts, and high-performance composites. Additionally, upcycled plastics are increasingly used in cutting-edge fields such as 3D printing, construction, and fashion, where material innovation plays a crucial role in sustainability-driven product design.

Despite its benefits, upcycling also presents technical and economic challenges. The variability in waste plastic feedstocks can result in inconsistent material properties, necessitating precise sorting and pretreatment processes. Furthermore, integrating upcycled plastics into industrial production requires compatibility assessments, regulatory compliance, and life cycle evaluations to ensure their safety, performance, and environmental benefits.

Upcycling encompasses various techniques that modify or enhance thermoplastics for high-value applications. These techniques often involve physical and chemical modifications to improve mechanical strength, thermal stability, and aesthetic appeal. Some of the most widely employed methods include:

- Material blending: This technique involves combining waste thermoplastics with performance-enhancing additives, fillers, or reinforcing agents to improve their mechanical, thermal, and chemical properties. For example, blending recycled polyethylene terephthalate (PET) with glass fibers enhances its stiffness and impact resistance, making it suitable for automotive components, aerospace applications, and industrial equipment. Other additives, such as compatibilizers and plasticizers, can be incorporated to improve processability and interfacial adhesion between mixed polymer systems.
- Reinforcement with natural fibers: A growing trend in sustainable materials development involves combining thermoplastic waste with natural fibers such as hemp, flax, jute, or bamboo to produce biocomposites. These materials offer lightweight yet strong alternatives to traditional plastic-based composites and are increasingly used in automotive interiors, furniture, and construction panels. The incorporation of plant-based fibers not only enhances the mechanical properties of the plastic matrix but also contributes to biodegradability and reduced carbon footprint. However, challenges such as fiber-polymer adhesion and moisture resistance require careful material engineering solutions.
- 3D printing applications: Additive manufacturing (3D printing) has emerged as a promising method for upcycling plastic waste into functional products with customized designs. Waste plastics, particularly PET, polylactic acid (PLA), and acrylonitrile–butadiene–styrene (ABS), can be processed into 3D printing filaments and used to fabricate a wide range of consumer goods, prototypes, and industrial parts. Upcycling through 3D printing not only reduces plastic waste accumulation but also enables on-demand manufacturing and material reusability, making it an attractive solution for sustainable production. However, achieving consistent print quality and mechanical performance requires precise control over material extrusion, polymer degradation, and thermal properties.

These upcycling techniques are continuously evolving, with advancements in material science, processing technologies, and design methodologies driving their adoption across diverse industries.

## 3 Recycling of Rubbers and Elastomers

Elastomers are polymeric materials that can be stretched to very high elongation values under applied force (or stress) and return to their original shape once the stress is released. They can be classified as thermoset and thermoplastic elastomers according to their structural properties and processing features. Thermoset elastomers are also divided into two classes: rubbers and thermoplastic vulcanizates (TPVs).

Conventional vulcanized rubbers are characterized by permanent covalent cross-links formed during the curing (vulcanization) process. This covalently cross-linked structure not only provides rubbers with stability and durability under mechanical stress, but also limits their recyclability and reprocessability. TPVs are produced by dynamic vulcanization of immiscible blends of thermoplastics and elastomers, where the elastomer phase is selectively cross-linked during melt processing.

Thermoplastic elastomers (TPEs) can be defined as polymeric materials that combine the elastic behavior of rubbers with the reprocessability of thermoplastics. Due to their recyclability, TPVs are also defined as a subclass of TPEs in some reports (Sahu et al., 2024).

Rubbers are widely used in various applications, especially in the automotive industry. This widespread utilization leads to a significant amount of rubber waste, particularly end-of-life tires. Global estimates suggest that the number of waste tires disposed of each year could exceed 1.2 billion by 2030. This huge amount of waste obviously represents a very serious environmental problem (Leong et al., 2023). Conventional recycling or reprocessing methods that have been effectively used in thermoplastics are generally not suitable for elastomers due to their irreversible three-dimensional cross-linked networks formed during vulcanization reactions. In addition, the complexity of rubber formulations—containing various types of reactive additives such as stabilizers, antioxidants, vulcanizing agents, activators, processing aids, plasticizers, etc.—and relatively high amounts of fillers, such as carbon black and inorganics or minerals, make recycling of waste rubber particularly challenging. Various recycling approaches have been developed to overcome the challenges posed by cross-linked structure of vulcanized rubbers. These approaches can also be categorized into mechanical, thermal, and chemical recycling methods given in Fig. 4, somewhat similar to the aforementioned techniques for thermoplastics.

One of the most common methods is the mechanical grinding process, which turns waste tires into crumb rubber or powder. Grinding is generally classified as a recycling method, when the resulting crumb rubber is directly reused in some applications such as asphalt modification. In some cases, grinding is considered a pretreatment step for preparing waste rubber for more advanced recycling processes

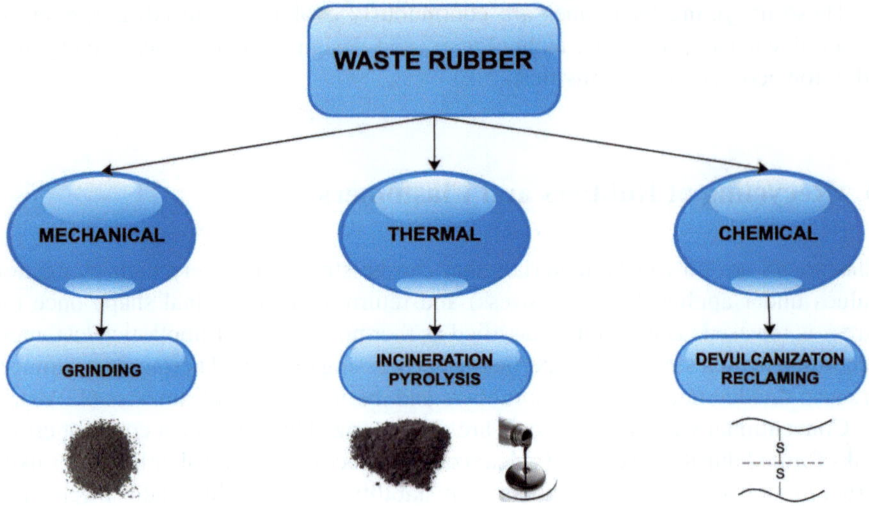


**Fig. 4** Recycling methods of vulcanized rubbers

such as devulcanization and pyrolysis. Classification largely depends on whether the material is directly reused or further processed to change its structure or composition.

Before the grinding process, the waste tire undergoes a pretreatment in which the steel and fabric components are removed from the waste tires. There are different grinding methods such as ambient, cryogenic, and wet grinding.

Crumb rubber, or ground tire rubber (GTR) derived from end-of-life tires, is widely used in construction applications, particularly for the modification of asphalt and concrete (Liang et al., 2020 Vahidi et al., 2014; Sienkiewicz et al., 2017; Boudnani et al., 2024; Aboelkheir et al., 2021; Luo et al., 2023). In addition to its widespread use in construction applications, GTR is also used in the production of recycled products by mixing with other polymers as a filler or reinforcement material, especially in thermoplastic matrices (Archibong et al., 2021; He et al., 2021; Majewska-Laks et al., 2021; Sulcis et al., 2014). Figure 5 schematizes a simple flow chart of the physical recycling process of waste tires.

Mixing GTR with thermoplastic matrices is considered a promising and cost-effective approach for rubber recycling, offering environmental and economic benefits. However, these compounds often exhibit poor mechanical properties due to the inadequate interfacial adhesion between the GTR phase and the thermoplastic matrix. Several formulation and compounding strategies have been employed to enhance the interfacial interactions between thermoplastic matrix and ground rubber fillers, thereby improving dispersion and resulting physical properties. These strategies include the use of compatibilizers or coupling agents, introduction of nanofillers, surface modification of GTR powders as a pretreatment by chemical or physical techniques, high-energy treatments such as gamma radiation, microwaves, plasma, and electron beam radiation, and chemical approaches such as devulcanization of GTR.

**Fig. 5** Schematic representation of crumb rubber production and utilization in recycled products

Conventional approaches used in rubber waste management include landfilling and incineration. It has been accepted that these methods cause significant environmental problems. Incineration is a waste management method that provides energy recovery, but the process is associated with the emission of the greenhouse gases, and its environmental concerns have been increasingly criticized for many years. On the other hand, pyrolysis offers a more advanced alternative, involving the thermal decomposition of rubber in an oxygen-free environment to produce fuel oil, gas, and carbon black (Leong et al., 2023). However, there are also controversial aspects of pyrolysis in waste management. The process requires high-energy input and precise temperature control, making it costly and less sustainable for large-scale operations. Furthermore, the quality and consistency of pyrolytic products, particularly oil and recovered carbon black, can vary significantly depending on the feedstock composition and process parameters. Further purification processes are often required before the resulting materials can be effectively utilized.

The current economic and technical limitations of pyrolysis and different chemical approaches, which target more precise solution from an industrial chemistry point of view, have been investigated to improve the recyclability of vulcanized rubber. One of these approaches is devulcanization which refers to selective breaking of cross-linked units or bonds (C–S and S–S) without scission of polymer backbone. This approach has been considered a particularly promising strategy in recent years, and technical details of devulcanization processes are still being studied. Devulcanization can be achieved through various methods, including thermal, chemical, mechanical, ultrasonic, and microwave-assisted processes. Regardless of the method, devulcanization can result in partial breaking of the main chains, resulting in changes in physical and mechanical properties of waste rubber. Despite such disadvantages, devulcanized rubber can be directly reused or blended with other polymers, since it significantly reduces overall production costs (Asaro et al., 2018).

In recent years, numerous studies have explored the incorporation of devulcanized rubber into elastomeric, thermoplastic, and thermoplastic elastomer formulations to reduce production costs and improve certain mechanical properties of compounds

(Ali et al., 2020; Garcia et al., 2021; Hassan et al., 2013; Hittini et al., 2021; Mnyango et al., 2024; Simon et al., 2018).

In addition to devulcanization, the term "reclaiming" is also used as recycling method of waste rubber. Reclaiming is a less controlled process than devulcanization and affects both cross-linked bonds and C–C bonds in the polymer backbone.

Although recycling of vulcanized rubber involves many challenging issues today and industrially applicable large capacity processes are not yet available, thermoplastic elastomers offer more sustainable and processable alternatives due to their thermoplastic structure. In recent years, TPEs have gradually started to replace thermoset rubbers in many applications, particularly in automotive and medical sectors, by combining the elasticity and flexibility of conventional elastomers with the reprocessability and recyclability of thermoplastics.

TPEs are composed of a hard thermoplastic phase and a soft elastomeric phase, which synergistically combine mechanical performance with processability. They are classified into several types based on their chemical composition and structure, including styrenic block copolymers (SBC), thermoplastic polyolefin elastomers (TPO), thermoplastic polyurethane elastomers (TPU), copolyester elastomers (COPE), and polyamide elastomers (PAE).

Although plastic recycling studies have predominantly focused on commodity thermoplastics such as polypropylene (PP) and polyethylene (PE), polyamides, and thermoplastic polyesters thermoplastic elastomers can also be mechanically recycled using conventional processing techniques including extrusion, injection molding, and compression molding. However, mechanically recycled TPEs (r-TPEs) are typically unsuitable for reuse in their original form due to a significant decrease in mechanical performance caused by thermal aging and constraints during reprocessing. Another difficulty is adjusting the compound formulations, as r-TPE is already a multicomponent material. Instead, r-TPEs can be used as impact modifiers in rigid thermoplastics and/or composites or compatibilizers in thermoplastic compounds.

While mechanical or physical recycling provides several advantages regarding environmental perspective, TPEs can also undergo chemical recycling, particularly in the case of thermoplastic polyurethanes. In mechanical recycling, TPUs are physically reduced into small particles or fine powders without chemical decomposition. Unlike mechanical recycling, chemical recycling involves degradation mechanisms that allow the recovery of monomeric building blocks that can be reused as raw materials for the synthesis of new polyurethane systems (Sahu et al., 2024). Figure 6 shows a schematic representation of the chemical recycling of TPU wastes via glycolysis for polyol recovery, followed by the synthesis of new TPU.

Upcycling of waste materials is also an important technique in waste management. In some advanced strategies, recycled TPEs can be chemically modified by introducing noncovalent or dynamically covalent bonds into polymer chains to improve their mechanical performance. These modified TPEs can also exhibit several functional and superior properties such as self-healing and shape memory. Thus, higher value-added materials could be obtained from recycled TPEs by this kind of functionalization and used in smart elastomer applications. Wang et al. successfully upcycled waste TPU to performance-enhanced PU with covalent adaptable network via a

**Fig. 6** Chemical recycling of TPU via glycolysis for polyol recovery and reuse

chain-breaking cross-linking process on a twin-screw extruder (Wang et al., 2024). This modification significantly enhanced the mechanical properties, thermal stability, creep resistance, and solvent resistance of waste TPU. They concluded that their work pointed out an effective chemical upcycling approach for converting waste TPU into higher-performance, reprocessable, and dynamically cross-linked polymer.

Despite the advantages of TPEs in terms of recyclability and ease of processing, their lack of covalent cross-links makes these elastomers significantly weaker in physical and mechanical properties compared to conventional vulcanized rubbers. Recent studies have focused on development of recyclable elastomers with dynamic covalent or noncovalent interactions (Alanalp et al., 2025; Zhou et al., 2025; Gong et al., 2023; Liu et al., 2022; Cui et al., 2019; Cao et al., 2019). These interactions enhance both mechanical performance and functionality of elastomers while providing reprocessability. Zhang et al. prepared a recyclable polybutadiene (PB) elastomer based on dynamic imine bonds (Zhang et al., 2017). They observed that the resulting elastomer exhibited good thermal stability, adjustable mechanical properties, and solvent resistance. They also concluded that the dynamic nature of imine bonds enabled the material to be reprocessable multiple times and offered a feasible route toward sustainable elastomer recycling.

# 4 Conclusion

Plastics production has consistently increased since the 1970s and now amounts to 400 million tonnes per year. This trend is expected to continue, with projections estimating it could reach over 1 billion metric tons by 2030. The most terrifying aspect of such massive production and consumption is the fact that less than 10% of plastic waste has been recycled.

We believe that it should begin by accepting that the management of plastic waste is not only a technological problem, but also a multidimensional social, political, cultural, and technical equation. Regulations should be made in both education and state policies to ensure the correct and appropriate use of plastic materials in industrial applications and human life. Incentives and, if necessary, penal sanctions should be implemented. First, everyone should be educated about plastic waste separation, starting at the first stage—in homes, workplaces, public and private schools, hospitals, government institutions, etc.—just after these materials are used by consumers. Although this first-step and correct separation strategy is applied in European, North American, and some Asian countries, it has not yet become a globally established culture. Moreover, mismanagement of plastic waste is truly a major problem in low- and middle-income countries that is associated with global environmental threats and affects many natural processes. This situation also reduces the success of plastic waste management globally and makes the subsequent technical steps difficult.

Second, humanity should realize that plastic waste is not just a material feedstock for the recycling industry and that crude oil and other fossil resources, which are the main raw materials for a large portion of synthetic plastics, are limited and unsustainable.

Third, technical initiatives should be taken to eliminate confusions in this area regarding technical definitions such as bioplastics, biodegradable plastics, microplastics, sustainability, service life, life cycle assessment, and carbon footprint, as well as material properties, to ensure that plastic waste management and recycling and/or disposing processes are implemented correctly, and their dissemination should be ensured in relevant industrial areas.

After all these mentioned stages are carried out correctly and in accordance with scientific criteria, developing, implementing, and managing appropriate recycling and disposal processes for plastic waste will become an issue that can be easily solved with the material, process, economy, and engineering approaches of modern technology, as chemists and engineers are studying to develop diverse plastic transformation techniques.

## References

Aboelkheir, M. G., Thomas, S., De Souza JR, F. G., Toledo Filho, R. D., Celestino, R., Thode Filho, S., & de Lima, N. R. B. (2021). Influence of UV-modified GTR on the properties of interlocking concrete paving units. *Journal of Molecular Structure, 1234*, 130110.

Abukasim, S., Zuhria, F. & Saing, Z. (2020). Alternative management of plastic waste. *Journal of Physics: Conference Series* (pp. 012041). IOP Publishing.

Alanalp, M. B., & Durmus, A. (2025). Easy synthesis of self-healing thermoplastic elastomer (TPE) via functionalization of styrene block copolymer (SEBS) with a cyclic amine compound in melt state and rheological assessment of non-covalent dynamic interactions. *Polymer*, 128083.

Ali, M. A., Raslan, H. A., El-Nemr, K. F., & Hassan, M. M. (2020). Thermal and mechanical behavior of SBR/devulcanized waste tire rubber blends using mechano–chemical and microwave methods. *Journal of Polymer Engineering, 40*(10), 815–822.

Al-Salem, S., Lettieri, P., & Baeyens, J. (2009). Recycling and recovery routes of plastic solid waste (PSW): A review. *Waste Management, 29*, 2625–2643.

Al-Salem, S., Lettieri, P., & Baeyens, J. (2010). The valorization of plastic solid waste (PSW) by primary to quaternary routes: From re-use to energy and chemicals. *Progress in Energy Combustion Science, 36*, 103–129.

Angyal, A., Miskolczi, N., & Bartha, L. (2007). Petrochemical feedstock by thermal cracking of plastic waste. *Journal of Analytical Applied Pyrolysis, 79*, 409–414.

Archibong, F. N., Sanusi, O. M., Médéric, P., & Hocine, N. A. (2021). An overview on the recycling of waste ground tyre rubbers in thermoplastic matrices: Effect of added fillers. *Resources, Conservation and Recycling, 175*, Article 105894.

Asaro, L., Gratton, M., Seghar, S., & Hocine, N. A. (2018). Recycling of rubber wastes by devulcanization. *Resources, Conservation and Recycling, 133*, 250–262.

Awan, M. O., Shakoor, A., Rehan, M. S., & Gill, Y. Q. (2021). Development of HDPE composites with improved mechanical properties using calcium carbonate and NanoClay. *Physica B: Condensed Matter, 606*, Article 412568.

Boudnani, Z. E., Bachir, G., & M'hamed, M. (2024). Assessment of ground tyre rubber impact on stiffness modulus and cracking resistance of bituminous mixtures. *Romanian Journal of Transport Infrastructure, 13*(1).

Brems, A., Dewil, R., Baeyens, J., & Zhang, R. (2013). Gasification of plastic waste as waste-to-energy or waste-to-syngas recovery route. *Natural Science, 5*, 695–704.

Browne, M. A., Crump, P., Niven, S. J., Teuten, E., Tonkin, A., Galloway, T., & Thompson, R. (2011). Accumulation of microplastic on shorelines worldwide: Sources and sinks. *Environmental Science Technology, 45*, 9175–9179.

Buekens, A. G., & Huang, H. (1998). Catalytic plastics cracking for recovery of gasoline-range hydrocarbons from municipal plastic wastes. *Resources, Conservation Recycling, 23*, 163–181.

Butler, E., Devlin, G., & McDonnell, K. (2011). Waste polyolefins to liquid fuels via pyrolysis: Review of commercial state-of-the-art and recent laboratory research. *Waste Management Biomass Valorization, 2*, 227–255.

Cao, L., Fan, J., Huang, J., & Chen, Y. (2019). A robust and stretchable cross-linked rubber network with recyclable and self-healable capabilities based on dynamic covalent bonds. *Journal of Materials Chemistry A, 7*(9), 4922–4933.

Chen, X., Wang, Y., & Zhang, L. (2021). Recent progress in the chemical upcycling of plastic wastes. *Chemsuschem, 14*, 4137–4151.

Cui, X., Song, Y., Wang, J. P., Wang, J. K., Zhou, Q., Qi, T., & Li, G. L. (2019). Self-healing polymers with tunable mechanical strengths via combined hydrogen bonding and zinc-imidazole interactions. *Polymer, 174*, 143–149.

Das, P., & Bal, M. (2025). Modeling and optimization study for microplastic removal from aquatic medium using Chroococcidiopsis species. *Chemical Engineering Science, 304*, Article 121085. https://doi.org/10.1016/J.CES.2024.121085

Das, P., Halder, G., & Bal, M. (2023). A critical review on remediation of microplastics using microalgae from aqueous system. *Science of the Total Environment, 898*, Article 166425. https://doi.org/10.1016/J.SCITOTENV.2023.166425

Das, A., Dey, S., Gacem, A., Yadav, K. K., Nath, H., Ram, D. K., & Das, A. P. (2025). In *Revolutionizing solid hydrocarbon sourcing: Plastic's journey from waste to upcycled treasure* (pp. 331–351). https://doi.org/10.1007/978-3-031-81615-4_14

Ding, W., Liang, J., & Anderson, L. L. (1997). Hydrocracking and hydroisomerization of high-density polyethylene and waste plastic over zeolite and silica—Alumina-supported Ni and Ni–Mo sulfides. *Energy and Fuels, 11*, 1219–1224.

Eze, W. U., Umunakwe, R., Obasi, H. C., Ugbaja, M. I., Uche, C. C., & Madufor, I. C. (2021). Plastics waste management: A review of pyrolysis technology. *Clean Technol. Recycl, 1*, 50–69.

Fakhrai, R. & Saadatfar, B. (2015). Feasibility study and quality assurance of the end-product of Alphakat KDV technology for conversion of biomass. KTH Royal Institute of Technology.

García, J. M., & Robertson, M. L. (2017). The future of plastics recycling. *Science, 358*(6365), 870–872. https://doi.org/10.1126/science.aaq0324

Garcia, P. S., de Lima, J. A., Scuracchio, C. H., & Cruz, S. A. (2021). The effect of adding devulcanized rubber on the thermomechanical properties of recycled polypropylene. *Journal of Applied Polymer Science, 138*(29), 50703.

Garforth, A. A., Ali, S., Hernández-Martínez, J., & Akah, A. (2004). Feedstock recycling of polymer wastes. *Current Opinion in Solid State Materials Science, 8*, 419–425.

Garraín, D., Martínez, P., Vidal, R. & Bellés, M. J. (2007). LCA of thermoplastics recycling. In Proceedings 3rd International Conference of Life Cycle Assessment. Zurich, Austria. https://www.lcm2007.org/paper/168. pdf accessed July, 2007. Citeseer, 2008.

George, N., & Kurian, T. (2014). Recent developments in the chemical recycling of postconsumer poly (ethylene terephthalate) waste. *Industrial Engineering Chemistry Research, 53*, 14185–14198.

Ghosh, S., Sinha, J. K., Ghosh, S., Vashisth, K., Han, S., & Bhaskar, R. (2023). Microplastics as an emerging threat to the global environment and human health. *Sustainability, 15*(14), 10821. https://doi.org/10.3390/SU151410821

Gong, C., Guo, J., Xu, P., Lv, J., Li, R., & Li, C. (2023). Recyclable, self-healing, and highly thermal conductive natural rubber nanocomposites enabled by a dynamic covalent network with carboxylated boron nitride nanosheets. *ACS Sustainable Chemistry and Engineering, 11*(24), 9087–9102.

El Haggar, S. (2010). In *Sustainable industrial design and waste management: Cradle-to-cradle for sustainable development*, Academic Press.

Hassan, M. M., Badway, N. A., Elnaggar, M. Y., & Hegazy, E. S. A. (2013). Thermo-mechanical properties of devulcanized rubber/high crystalline polypropylene blends modified by ionizing radiation. *Journal of Industrial and Engineering Chemistry, 19*(4), 1241–1250.

He, M., Gu, K., Wang, Y., Li, Z., Shen, Z., Liu, S., & Wei, J. (2021). Development of high-performance thermoplastic composites based on polyurethane and ground tire rubber by in-situ synthesis. *Resources, Conservation and Recycling, 173*, Article 105713.

Hittini, W., Mourad, A. H. I., & Abu-Jdayil, B. (2021). Utilization of devulcanized waste rubber tire in development of heat insulation composite. *Journal of Cleaner Production, 280*, Article 124492.

Hong, M., & Chen, E.Y.-X. (2017). Chemically recyclable polymers: A circular economy approach to sustainability. *Green Chemistry, 19*, 3692–3706.

Hopewell, J., Dvorak, R., & Kosior, E. (2009). Plastics recycling: Challenges and opportunities. *Philosophical Transactions of the Royal Society b: Biological Sciences, 364*, 2115–2126.

Ignatyev, I. A., Thielemans, W. & Vander Beke, B. (2014). Recycling of polymers: A review. *ChemSusChem, 7*, 1579–1593.

KEIL, F. J. (1999). Methanol-to-hydrocarbons: process technology. *Microporous Mesoporous Materials, 29*, 49–66.

Khan, M. U., Ahring, B., Garcia-Perez, T., & Garcia-Perez, M. (2020). Valorization of municipal solid waste in biorefineries for the creation of a circular economy: Role of emerging technologies. In *Current developments in biotechnology and bioengineering* (pp. 32–347). Elsevier.

Kulkarni, S. R., Vandewalle, L. A., Gonzalez-Quiroga, A., Perreault, P., Heynderickx, G. J., van Geem, K. M., & Marin, G. B. (2018). Computational fluid dynamics-assisted process intensification study for biomass fast pyrolysis in a gas–solid vortex reactor. *Energy and Fuels, 32*, 10169–10183.

Labeckas, G., & Slavinskas, S. (2013). Performance and emission characteristics of a direct injection diesel engine operating on KDV synthetic diesel fuel. *Energy Conversion.*

Lamtai, A., Elkoun, S., Robert, M., Mighri, F. & Diez, C. (2023). Mechanical recycling of thermoplastics: a review of key issues. *Waste* (pp. 860–883) MDPI.

Leong, S. Y., Lee, S. Y., Koh, T. Y., & Ang, D. T. C. (2023). 4R of rubber waste management: Current and outlook. *Journal of Material Cycles and Waste Management, 25*(1), 37–51.

Levin, B. C. & Materials. (1987). A summary of the NBS literature reviews on the chemical nature and toxicity of the pyrolysis and combustion products from seven plastics: Acrylonitrile–butadiene–styrenes (ABS), nylons, polyesters, polyethylenes, polystyrenes, poly (vinyl chlorides) and rigid polyurethane foams. *Fire, 11*, 143–157.

Li, X., Li, J., Zhou, G., Feng, Y., Wang, Y., Yu, G., Deng, S., Huang, J., & Wang, B. (2014). Enhancing the production of renewable petrochemicals by co-feeding of biomass with plastics in catalytic fast pyrolysis with ZSM-5 zeolites. *Applied Catalysis a: General, 481*, 173–182.

Liang, M., Sun, C., Yao, Z., Jiang, H., Zhang, J., & Ren, S. (2020). Utilization of wax residue as compatibilizer for asphalt with ground tire rubber/recycled polyethylene blends. *Construction and Building Materials, 230*, Article 116966.

Linck, M., Felix, L., Marker, T. & Roberts, M. (2016). Integrated biomass hydropyrolysis and hydrotreating: a brief review. In *Advances in bioenergy: The sustainability challenge* (pp. 57–63).

Liu, W., Huang, J., Gong, Z., Fan, J., & Chen, Y. (2022). Healable, recyclable and mechanically robust elastomers with multiple dynamic cross-linking bonds. *Polymer, 252*, Article 124900.

Liu, M., Guo, J., Gu, Y., Gao, J., Liu, F., Yu, S. & Engineering. (2018). Pushing the limits in alcoholysis of waste polycarbonate with DBU-based ionic liquids under metal-and solvent-free conditions. *ACS Sustainable Chemistry, 6*, 13114–13121.

Luo, G., Zhang, J., Zhao, Z., & Sun, M. (2023). Fatigue property evaluation of sustainable porous concrete modified by recycled ground tire rubber/silica fume under freeze-thaw cycles. *Sustainability, 15*(10), 7965.*management,* 66, 173–188.

Majewska-Laks, K., Sykutera, D., & Osciak, A. (2021). Reuse of ground tire rubber (GTR) as a filler of TPE matrix. In*MATEC Web of Conferences* (Vol. 332, pp. 01003). EDP Sciences.

Mallick, K., Sahu, A., Dubey, N. K., & Das, A. P. (2023). Harvesting marine plastic pollutants-derived renewable energy: A comprehensive review on applied energy and sustainable approach. *Journal of Environmental Management, 348*, Article 119371. https://doi.org/10.1016/J.JENVMAN.2023.119371

Marker, T. L., Felix, L. G., Linck, M. B., & Roberts, M. J. (2012). Integrated hydropyrolysis and hydroconversion (IH2) for the direct production of gasoline and diesel fuels or blending components from biomass, part 1: Proof of principle testing. *Environmental Progress Sustainable Energy, 31*, 191–199.

Marker, T. L., Felix, L. G., Linck, M. B., Roberts, M. J., Ortiz-Toral, P., & Wangerow, J. (2014). Integrated hydropyrolysis and hydroconversion (IH2®) for the direct production of gasoline and diesel fuels or blending components from biomass, Part 2: Continuous testing. *Environmental Progress Sustainable Energy, 33*, 762–768.

Mishra, S., Swain, S., Sahoo, M., Mishra, S., & Das, A. P. (2021). Microbial colonization and degradation of microplastics in aquatic ecosystem: A review. *Geomicrobiology Journal*. https://doi.org/10.1080/01490451.2021.1983670

Mishra, S., Singh, R. P., Rout, P. K., & Das, A. P. (2022). Membrane bioreactor (MBR) as an advanced wastewater treatment technology for removal of synthetic microplastics. In *Development in wastewater treatment research and processes* (pp. 45–60). https://doi.org/10.1016/B978-0-323-85583-9.00022-3

Miskolczi, N., Bartha, L., Deak, G., & Jover, B. (2004). Thermal degradation of municipal plastic waste for production of fuel-like hydrocarbons. *Polymer Degradation Stability, 86*, 357–366.

Miskolczi, N., Angyal, A., Bartha, L., & Valkai, I. (2009). Fuels by pyrolysis of waste plastics from agricultural and packaging sectors in a pilot scale reactor. *Fuel Processing Technology, 90*, 1032–1040.

Mnyango, J. I., Hlangothi, B. G., Mabuto, B., Woolard, C. D., & Hlangothi, S. P. (2024). Effect of tulbaghia violacea-derived devulcanized rubber on the properties of natural rubber/styrene-butadiene rubber blends. *Macromolecular Materials and Engineering, 309*(8), 2300410.

Mohanty, J., Dey, S., Bala, K., & Das, A. P. (2024a). Recent advances in microplastic pollution for its sustainable management. In *Sustainable management of urban plastic waste through circular economic approaches* (pp. 27–48). https://doi.org/10.1201/9781003465447-3

Mohanty, J., Dey, S., Bala, K., & Das, A. P. (2024b). Recent advances in microplastic pollution for its sustainable management. In *Sustainable management of urban plastic waste through circular economic approaches* (pp. 27–48). https://doi.org/10.1201/9781003465447-3/recent-advances-microplastic-pollution-sustainable-management-jyotirmayee-mohanty-sudeshna-dey-kiran-bala-alok-prasad-das

Mohanty, M., Tripathy, B., Koteswara Reddy, G., & Das, A. P. (2024a). Greener technologies for establishing restraint over microplastic pollution. In *Sustainable microbial technology for synthetic and cellulosic microfiber bioremediation* (pp. 239–258). https://doi.org/10.1007/978-3-031-62600-5_13

Mohanty, M., Tripathy, B., Koteswara Reddy, G., & Das, A. P. (2024b). Greener technologies for establishing restraint over microplastic pollution. In *Sustainable microbial technology for synthetic and cellulosic microfiber bioremediation* (pp. 239–258). https://doi.org/10.1007/978-3-031-62600-5_13

Okuwaki, A. (2004). Feedstock recycling of plastics in Japan. *Polymer Degradation Stability, 85*, 981–988.

Picuno, C., Alassali, A., Sundermann, M., Godosi, Z., Picuno, P., & Kuchta, K. (2020). Decontamination and recycling of agrochemical plastic packaging waste. *Journal of Hazardous Materials, 381*, Article 120965.

Pinto, F., Costa, P., Gulyurtlu, I., & Cabrita, I. (1999). Pyrolysis of plastic wastes: 2. Effect of catalyst on product yield. *Journal of Analytical Applied Pyrolysis, 51*, 57–71.

Priyadarshini, B., & Das, A. P. (2024). The adverse impact of synthetic microplastics released from disposable face masks pollutants on aquatic microbiota, aquatic species and humans: A review. *Environmental Science and Engineering, Part, F2827*, 67–81. https://doi.org/10.1007/978-3-031-50840-0_5

Priyadarshini, B., Santhosh Kumar, M., & Das, A. P. (2024). Microbial degradation of synthetic plastics released from COVID-19 generated single use disposable face masks. *Environmental Science and Engineering, Part, F2827*, 1–16. https://doi.org/10.1007/978-3-031-50840-0_1

Puig-Arnavat, M., Bruno, J. C., & Coronas, A. (2010). Review and analysis of biomass gasification models. *Renewable Sustainable Energy Reviews, 14*, 2841–2851.

Ragaert, K., Delva, L., & van Geem, K. (2017). Mechanical and chemical recycling of solid plastic waste. *Waste Management, 69*, 24–58.

Ragaert, K., Delva, L. & Van Geem, K. (2022). Important note (September 17).

Ray, S. K., Dey, S., Das, A. P., & John, A. (2024). Microplastic waste management through greener technologies : a review. In *Sustainable management of urban plastic waste through circular economic approaches* (pp. 237–261). https://doi.org/10.1201/9781003465447-14

Ruj, B., Pandey, V., Jash, P., & Srivastava, V. (2015). Sorting of plastic waste for effective recycling. *International Journal Applications Science Engineering Reserach, 4*, 564–571.

Sahu, S., & Das, A. P. (2024a). In *Microplastics degradation mechanism by aquatic plastiomes: A review* (pp. 213–226). https://doi.org/10.1007/978-3-031-50840-0_12

Sahu, S., & Das, A. P. (2024). Microplastics degradation mechanism by aquatic plastiomes: A review. *Environmental Science and Engineering, Part, F2827*, 213–226. https://doi.org/10.1007/978-3-031-50840-0_12

Sahu, P., Oh, J. S., & Isayev, A. I. (2024). Recycling and circular economy of thermoplastic elastomers. In*Advances in thermoplastic elastomers* (pp. 529–561). Elsevier.

SAYED, A. B. L. C. O. H. E. (2023). *Method for the identification of plastics.*

Scheirs, J., & Kaminsky, W. (2006). In *Feedstock recycling and pyrolysis of waste plastics* (Wiley & Sons).

Schyns, Z. O., & Shaver, M. P. (2021). Mechanical recycling of packaging plastics: A review. *Macromolecular Rapid Communications, 42*, 2000415.

Sienkiewicz, M., Borzędowska-Labuda, K., Zalewski, S., & Janik, H. (2017). The effect of tyre rubber grinding method on the rubber-asphalt binder properties. *Construction and Building Materials, 154*, 144–154.

Simon, D. A., Halasz, I. Z., Karger-Kocsis, J., & Barany, T. (2018). Microwave devulcanized crumb rubbers in polypropylene based thermoplastic dynamic vulcanizates. *Polymers, 10*(7), 767.

Singh, M., & Singh, R. (2022). Twin screw extrusion for recycling of thermoplastics.

Spath, P. L. & Dayton, D. C. (2003). Preliminary screening--Technical and economic assessment of synthesis gas to fuels and chemicals with emphasis on the potential for biomass-derived syngas. National Renewable Energy Lab.(NREL), Golden, CO (United States).

Sulcis, R., Lotti, L., Coiai, S., Ciardelli, F., & Passaglia, E. (2014). Novel HDPE/ground tyre rubber composite materials obtained through in-situ polymerization and polymerization filling technique. *Journal of Applied Polymer Science, 131*(10).

Swain, S., Purnima Sudha, M., Dey, S., Badu, Y., Bhoi, S. R., Alreshidi, M. A., Qasim, M. T., Nath, H., Ram, D. K., & Das, A. P. (2025). In *Sources of micro and nano plastic hydrocarbon pollutants in marine bodies and their treatment* (pp. 45–67). https://doi.org/10.1007/978-3-031-81615-4_3

Tripathy, B., Sahoo, P. P., Sundaray, H., & Das, A. P. (2024). In-vitro biodegradation of discarded marine microplastics across the eastern coast of the Bay of Bengal, India using Exiguobacterium sp. *Environmental Chemistry and Ecotoxicology, 6*, 236–247. https://doi.org/10.1016/j.enceco.2024.07.003

Ugduler, S., Van Geem, K. M., Denolf, R., Roosen, M., Mys, N., Ragaert, K. & De Meester, S. J. G. C. (2020). Towards closed-loop recycling of multilayer and coloured PET plastic waste by alkaline hydrolysis. *22*, 5376–5394.

Uzosike, C. C., Yee, L. H., & Padilla, R. V. (2023). Small-scale mechanical recycling of solid thermoplastic wastes: A review of PET, PEs, and pp. *Energies, 16*, 1406.

Vahidi, S., Mogawer, W. S., & Booshehrian, A. (2014). Effects of GTR and treated GTR on asphalt binder and high-RAP mixtures. *Journal of Materials in Civil Engineering, 26*(4), 721–727.

Vermeulen, I., van Caneghem, J., Block, C., Baeyens, J., & Vandecasteele, C. (2011). Automotive shredder residue (ASR): Reviewing its production from end-of-life vehicles (ELVs) and its recycling, energy or chemicals' valorisation. *Journal of Hazardous Materials, 190*, 8–27.

Wang, B., Wang, Y., Li, F., Du, S., Zhu, J., & Ma, S. (2024). Upcycling of thermoplastic polyurethane into covalent adaptable networks via chain breaking-crosslinking strategy. *Journal of Polymer Science, 62*(5), 849–858.

Wender, I. (1996). Reactions of synthesis gas. *Fuel Processing Technology, 48*, 189–297.

Zhang, H., Wang, D., Liu, W., Li, P., Liu, J., Liu, C., & Xu, J. (2017). Recyclable polybutadiene elastomer based on dynamic imine bond. *Journal of Polymer Science Part A: Polymer Chemistry, 55*(12), 2011–2018.

Zhou, W. C., Gao, X. Q., Li, J. H., Ye, C., Wang, Y. Z., & Deng, C. (2025). Robust, tough, ultra-low-creep and reprocessable rubber enabled by weak supramolecular-interaction-dominated yet strong covalent-bond-assisted reverse design paradigm. *Journal of Materials Chemistry A, 13*(3), 1746–1754.

# Innovative Technologies for Heavy Metal Management

**Murat Torun and Nursel Pekel Bayramgil**

## 1 Introduction

Rapid progress in industrialization, mining activities, agricultural fertilizers, and waste management processes causes heavy metal pollution in the environment to increase (Das & Mishra, 2010; Mishra et al., 2022). Heavy metals such as lead (Pb), mercury (Hg), cadmium (Cd), arsenic (As), and chromium (Cr) cause permanent damage to human health and ecosystems through bioaccumulation. Therefore, the control of heavy metals in today's industrial and environmental management processes is of great importance in terms of sustainable development and ecosystem health. The need for new generation technologies for effective monitoring, purification, and recovery of heavy metals is increasing day by day (Das et al., 2012a, 2012b).

Traditional heavy metal removal methods include techniques such as chemical precipitation, coagulation–flocculation, ion exchange, and membrane filtration. However, these methods generally have disadvantages such as high energy requirement, high cost, limited efficiency, and secondary waste production. Today, scientists and engineers are making intense efforts to develop more effective, more efficient, low-cost, and environmentally friendly solutions in heavy metal management, thanks to developing technology (Das et al., 2011; Ghosh & Das, 2018). Heavy metal detection, separation, and disposal are experiencing a technological revolution, thanks to solutions based on nanotechnology, bioremediation methods, monitoring systems supported by artificial intelligence (AI), innovative filtration techniques, combined technologies, and sustainable adsorbent materials. Here, we will go over

M. Torun · N. P. Bayramgil (✉)
Faculty of Science, Department of Chemistry, Hacettepe University, Beytepe, Ankara, Türkiye
e-mail: nursel@hacettepe.edu.tr

M. Torun
e-mail: mtorun@hacettepe.edu.tr

 
A. P. Das and S. Karakuş (eds.), *Synergizing Sustainability for Integrated Waste Management*, Sustainable Environmental Waste Management Strategies,
https://doi.org/10.1007/978-3-032-03718-3_6

the most recent heavy metal management technologies and how they have affected eco-friendliness (Ghosh et al., 2016; Mohanty et al., 2016).

## 2 Traditional Methods Used in Heavy Metal Management

Traditional methods used in heavy metal management are examined in three main categories: physical, chemical, and biological methods.

**Physical methods** aim to separate heavy metals from water or soil through physical processes. These methods are generally fast and effective but can be costly and may involve secondary waste production.

### 2.1 Membrane Filtration

In membrane filtration, selectively permeable membranes are used to separate heavy metals from solution. There are different types of membranes such as reverse osmosis (RO), nanofiltration (NF), and ultrafiltration (UF). In *reverse osmosis* (RO) filtration, heavy metals are separated from water by passing water through a semipermeable membrane under high pressure (Van der Bruggen & Vandecasteele, 2003). In *nanofiltration* (NF), lower pressure is required compared to reverse osmosis and high efficiency is provided in separating certain ions (Mohammad et al., 2015). *Ultrafiltration* (UF), on the other hand, allows larger particles and microorganisms to be retained, but auxiliary agents are usually needed in the separation of heavy metals (Madaeni, 1999).

High separation efficiencies are achieved in membrane filtration, and it is an environmentally friendly technology since chemical use is kept to a minimum. However, membrane clogging is frequently observed, and it is a high-cost technology since its maintenance is difficult.

### 2.2 Ion Exchange

Ion exchange is a widely used physicochemical technique in water and wastewater treatment. This method allows the selective removal of heavy metal ions from solution even at very low concentrations. Solid-phase materials such as ion exchange resins or natural zeolites are generally used. It is highly efficient and effective in removing heavy metals and plays an important role in drinking water treatment, industrial wastewater management, and mining waste treatment (Matis et al., 2004). However, this highly selective method cannot be said to be economical for large amounts of removal, and efficiency may decrease since separation difficulties will be encountered in the presence of other ions.

Research on the regeneration and increase in selectivity of ion exchange resins will contribute to making this technology more sustainable and economical.

## 2.3 Electrokinetic Remediation

Electrokinetic remediation is an environmental technology that uses an electric field to remove heavy metals from contaminated media such as soil, water, and sludge. This method offers an effective alternative, especially for soils with low permeability and heavily contaminated areas. The system works by applying low-voltage direct current to both ends of the contaminated media by placing anode and cathode electrodes. The electric field removes heavy metal ions from the targeted areas by mechanisms such as electromigration, electroosmosis, and electrophoresis. The ions are then collected at the anode or cathode region and separated from water or soil by various chemical or physical processes (Acar & Alshawabkeh, 1993). This method is particularly successful in removing toxic heavy metals such as lead (Pb), cadmium (Cd), mercury (Hg), and arsenic (As). The advantages of electrokinetic remediation include on-site applicability, low chemical consumption, and high removal efficiency, but there are also some technical limitations such as energy consumption and pH changes. In recent years, the effectiveness of the method has been increased by developing hybrid systems supported by nanoelectrodes and biological agents (Reddy & Cameselle, 2009). These developments, despite being an old method, have enabled electrokinetic remediation to become widespread as a sustainable, environmentally friendly, and low-cost heavy metal management method.

**Chemical methods** include methods such as adsorption of heavy metals, precipitation by chemical reactions, oxidation–reduction, and coagulation.

## 2.4 Adsorption

Adsorption is a technique that removes heavy metal ions by binding to a solid surface. Natural and synthetic adsorbents can be used. Activated carbon, due to its high surface area, allows the retention of heavy metals (Babel & Kurniawan, 2003). Zeolites, in their natural or modified form, can retain heavy metals by ion exchange mechanism (Wang & Peng, 2010). Low-cost adsorbents can be used in adsorption, and high efficiencies can be achieved. However, it is selective only for certain ions and the adsorbent must be renewed when saturation is reached (Das et al., 2015b; Ghosh et al., 2023).

## 2.5 Chemical Precipitation

Chemical precipitation separates heavy metal ions from water by converting them into insoluble compounds. Common precipitating agents include hydroxides, carbonates, and sulfates (Fu & Wang, 2011). In *hydroxide precipitation*, metal hydroxides are precipitated by adjusting the pH (Chen et al., 2009). In *sulfide precipitation*, metal sulfides are effective precipitating agents due to their low solubility (Martell et al., 1996).

Chemical precipitation is easy to apply and has low cost. pH control is very important. At the end of precipitation, a large amount of sludge is formed.

## 2.6 Oxidation–Reduction

Oxidation–reduction is the process of reducing heavy metals to less toxic or precipitated forms by converting them to different oxidation levels (Zhao et al., 2012). In chemical oxidation, oxidants such as $O_3$, $Cl_2$, or $H_2O_2$ are used. In chemical reduction, heavy metals are converted to insoluble forms with iron sulfate or liquefied sulfur compounds.

The oxidation–reduction method is an effective and rapid method for the separation of various heavy metals, but in addition to high chemical consumption, the by-products that may be formed also have undesirable environmental effects.

## 2.7 Coagulation

Coagulation is a method that allows the sedimentation or filtration of dissolved heavy metal ions or colloidal particles in water or wastewater by adding chemical substances called coagulants (Duan & Gregory, 2003). Coagulating agents (e.g., metal salts or polyelectrolytes) are added to the solution. These substances neutralize and destabilize heavy metal ions. Thanks to coagulants, the precipitated metal ions come together to form large particles (flocs), and these flocs settle and separate from water.

Coagulation is a widely used, effective, and low-cost method for removing heavy metal ions from water. However, due to its undesirable properties such as the formation of sludge and chemical consumption, studies are ongoing on the development of hybrid systems by combining it with other methods such as ion exchange or membrane filtration. In the future, it is aimed to develop more environmentally friendly and efficient coagulation methods with innovative materials such as nanocoagulants and bio-based polyelectrolytes (Zhao et al.).

**Biological methods** enable the purification of heavy metals using microorganisms, plants, or biomolecules.

### 2.8 Biosorption

Biosorption is the process of adsorption of heavy metals on the surface of microorganisms or biopolymers (Volesky, 2001). Microorganisms such as bacteria, fungi, or algae remove heavy metal ions by binding them with functional groups on their cell surfaces. Heavy metal ions can be adsorbed on the surface of natural polymers such as chitin and chitosan (Gupta & Nayak, 2012).

The biosorption method is an environmentally friendly method and requires low energy. However, its effectiveness is limited to certain types of metals.

### 2.9 Phytoremediation

Phytoremediation is the process of some plants absorbing heavy metals through their roots or tissues (Salt et al., 1998). It is a low-cost method that is environmentally friendly. However, the application period is long and the growth of plants is very dependent on environmental conditions.

After briefly mentioning the traditional heavy metal removal/recovery methods used from the past to the present, the main subject of this section, innovative and advanced technologies in heavy metal removal, is given in detail in the following paragraphs, supported by studies found in the literature.

## 3 Innovative and Advanced Technologies Used in Heavy Metal Management

Although traditional heavy metal removal methods play an important role in the treatment of industrial wastewater and environmental pollutants, these methods generally have disadvantages such as high chemical consumption, sludge formation, and low selectivity (Das & Ghosh, 2018; Das et al., 2015a). Today, innovative and advanced technologies are coming to the forefront in order to overcome these problems and develop more sustainable solutions. Advanced techniques such as nanotechnology-based adsorbents, electrochemical treatment systems, biomimetic membranes, and smart sensor-based monitoring systems offer high efficiency and environmentally friendly solutions in heavy metal management (Das et al., 2012a, 2012b; Das & Ghosh, 2023b). In addition, it is aimed to make heavy metal removal processes faster, less costly, and more energy efficient with artificial intelligence-supported optimization systems. These innovative technologies have great potential in terms of both ensuring environmental sustainability and making industrial water treatment processes more effective.

## 3.1 Bioremediation and Phytoremediation Methods

**Bioremediation**

Bioremediation is an environmentally friendly treatment method that uses biological agents such as microorganisms, plants, and enzymes to convert pollutants in the environment, especially heavy metals, into harmless or less toxic components (Paul et al., 2023; Pradhan et al., 2023). It offers a low-cost, sustainable, and less harmful alternative to traditional physical and chemical methods. Various approaches such as microbial bioremediation, phytoremediation, enzymatic treatment, and mycorrhiza-assisted heavy metal removal are effectively used in the treatment of water, soil, and industrial wastes (Das et al., 2024; Kanungo et al., 2024). In recent years, the effectiveness of bioremediation techniques has increased and spread to a wider range of applications, thanks to the integration of disciplines such as genetic engineering and nanobiotechnology. These methods are among the innovative approaches that both protect the ecological balance and provide economically sustainable solutions in combating heavy metal pollution (Dey et al., 2024; Nag et al., 2023).

*Use of Microorganisms*

The mechanisms that occur in bioremediation involving microorganisms are biosorption, bioaccumulation, and biotransformation. In *biosorption*; microorganisms consisting of bacteria and fungi have cell walls rich in functional groups that retain heavy metals (Das & Ghosh, 2023a; Das et al., 2024). In *bioaccumulation*, some types of microbes absorb and trap heavy metals in the cell, thus reducing the bioavailability effects of heavy metals in the environment. In *biotransformation*, microorganisms enzymatically convert heavy metals into less toxic or more easily precipitated forms. In a study, the treatment of heavy metals such as iron(II) sulfate, aluminum sulfate, and copper(II) sulfate through bioremediation of various fungal species was investigated through adsorption, microbial extraction, and mineralization stages. The results showed that fungi are effective in heavy metal removal (Vaksmaa et al., 2023). The symbiotic relationship between mycorrhizal fungi and plant roots increases phytoremediation capacity. Arbuscular mycorrhizal fungi (AMF) produce proteins that bind to heavy metals, reducing their bioavailability and promoting plant growth in contaminated soils (Chen et al., 2022).

Ayangbenro and Babalola (2017) studied microbial sorbents with high potential in heavy metal treatment. Cai et al. (2019) characterized bacterial strains obtained from electroplating wastewater treatment plants and determined their resistance and potential for bioremediation applications. Yang et al. (2015) investigated the role of extracellular polymer substances obtained from *Klebsiella* sp. J1 in the competitive adsorption of heavy metals and emphasized their importance in bioremediation processes. In another study, *Lysinibacillus sphaericus* JG-A12 strain isolated from uranium mining waste in Germany showed the ability to bind metals such as aluminum, cadmium, copper, lead, and uranium. It was stated that this ability was due to a proteinaceous surface layer covering the cells, which facilitated the achievement of high metal-binding efficiency in saline solutions (Merroun et al.,

2005). *Bacillus* species isolated from chromium-contaminated industrial waste sites have also demonstrated their potential in bioremediation by showing the ability to reduce Cr(VI) ion to its less toxic form (Desai et al., 2008). Another study has shown that *Lemna minor* effectively removes heavy metals such as lead, copper, zinc, and arsenic from water even at low doses (Sasmaz et al., 2015).

Use of Worms

There are studies examining the usability of worms in the removal of heavy metal pollution in soil. The results show that worms are effective in bioremediation techniques and are promising in heavy metal removal (Dada et al., 2016).

**Phytoremediation**

It would be appropriate to say that phytoremediation, one of the traditional methods, has become a popular method again with recent studies (Praveen & Pandey, 2020). This green remediation technology is one of the most attractive research focuses of modern times regarding wastewater treatment. Constructed wetlands are the most effective phytoremediation technique in wastewater treatment. The removal of pollutants from water systems by their accumulation in plant biomass is called rhizofiltration (Rezania et al., 2016). There are different mechanisms in phytoremediation. In *phytoextraction*, plants take heavy metals from the soil and transport them to the aboveground tissues, after which they can be safely disposed of by harvesting the plant. In *explant phytoremediation*, pollutants are transformed/removed only in the rhizosphere by phytochemicals such as carboxylates or microbial enzymes. In *phytostabilization*, plants immobilize heavy metals in the soil, reducing their mobility and bioavailability. In an interesting method, *phytovolatilization*, some plants take heavy metals, convert them into volatile compounds, and release them into the atmosphere. In summary, heavy metal phytoremediation occurs by stabilizing metals in the roots, by removing heavy metals from the environment through the roots and transporting and storing them in the leaves and stems, or by converting them into volatile compounds (Haldar & Ghosh, 2020).

In algae-based phytoremediation, algae are used to remove heavy metals, pesticides, and other waste products (Baghour, 2017; Singh & Tiwari, 2023; Tiwari & Marella, 2019). There are many advantages to using diatom algae in phytoremediation: They are common, nontoxic, cost-effective, environmentally friendly, reusable and recyclable, and highly adaptable to the environment; they can also grow fast, create sustainable systems, and remove heavy metals with low energy input, thanks to their autotrophic properties. These properties have attracted the attention of scientists on a global scale. In a study on phytoremediation in the literature, it was stated that hyperaccumulator plants were used in heavy metal removal. Plants such as *Brassica juncea* (Indian mustard) and *Thlaspi caerulescens* can accumulate heavy metals such as nickel, zinc, and cadmium in the soil (Comis, 1996; Naser et al., 2012).

A study published in the journal *Discover Applied Sciences* (Nedjimi, 2021) presented a detailed analysis of phytoremediation mechanisms, advantages, limitations, and future perspectives. Bouhadi et al. (2021) published a review on plant selection and effectiveness, discussing the bioavailability of heavy metals and the role of bioaccumulating plants in soil cleanup. A review article published in the journal *Chemosphere* discussed traditional phytoremediation strategies, mechanisms, limitations, and modifications required to increase efficiency (Sarwar et al., 2017).

In the literature, it has been determined that some aquatic plants are effective in the bioremediation of heavy metal pollution. For example, plants such as duckweed (*Lemna minor*) and water hyacinth (*Eichhornia crassipes*) have the capacity to absorb heavy metals such as lead and arsenic in aquatic environments (Nassouhi et al., 2018). In these studies, it has been reported that live *L. minor* biomass removes 85–90% of lead nitrate at an initial concentration of 5 mg/L. Due to its fast growth and easy harvest, *L. minor* is used to remove heavy metals and other pollutants in places such as Devils Lake wastewater treatment plant in North Dakota, USA (Goswami et al., 2022). Live *Azolla pinnata* plants have shown the ability to accumulate high levels of heavy metals such as zinc, lead, chromium, mercury, cadmium, copper, and arsenic, and it has been stated that they may be useful in cleaning water bodies with metal pollution (Dwivedi et al., 2010).

Bioremediation and phytoremediation methods are attracting attention as effective, sustainable, and environmentally friendly alternatives for the removal of heavy metal pollution. However, the effectiveness of both methods may vary depending on the type of pollutant, its concentration, environmental conditions, and the characteristics of the organism used. Therefore, detailed research and pilot studies for specific applications and hybrid technologies for high efficiency are needed.

## 3.2 Nanotechnology-Based Solutions

Nanomaterials have been used to remediate environments contaminated with metal(loid) ions due to their high reactivity and higher adsorption capacity compared to the same materials in normal sizes (Fajardo et al., 2020; Peng et al., 2018a, 2018b). These studies have shown that nanomaterials are effective in stabilizing and adsorbing heavy metal(loid)s, thus reducing the mobility and bioavailability of metal(loid) ions. Remediation of environments contaminated with heavy metal(loid)s with nanomaterials is a newly explored research area. Heavy metal(loid)s cannot be degraded into nontoxic substances in sediments (Hu et al., 2013), but can be immobilized or passivated by adding different substances. These additives (such as nanomaterials) usually have high cation exchange capacity and adsorption capacity (Wu et al., 2021).

**Metal Oxide Nanoparticles**

Metal oxide nanoparticles can adsorb heavy metals with high efficiency due to their unique electronic structures that are insulator, semiconductor, or metallic in character. Commonly studied metal oxide nanoparticles include magnesium oxide (MgO), iron oxide ($Fe_3O_4$), silver oxide ($Ag_2O$), zinc oxide (ZnO), and titanium dioxide ($TiO_2$) (Mounier et al., 2023).

MgO nanoparticles exhibit excellent performance in removing heavy metals such as lead and cadmium, as well as organic matter, due to their high adsorption properties. However, they can be difficult to remove from purified water and are generally more effective at high pH levels (Cai et al., 2017).

Magnetic nanoparticles such as magnetite ($Fe_3O_4$) and maghemite ($\gamma$-$Fe_2O_3$) can be easily recovered from the medium using external magnetic fields. They are effective in removing heavy metals and organic dyes from wastewater (Xu et al., 2012).

Recent studies on the use of silver oxide ($Ag_2O$) nanoparticles for heavy metal removal have shown that this material can be effective in removing heavy metal ions from aqueous solutions. In one study, the capacity of nano titanium oxide ($nTiO_2$), nano silver oxide ($nAg_2O$), and $nTiO_2$-coated $nAg_2O$ composite to remove cobalt ($Co^{2+}$) and zinc ($Zn^{2+}$) ions (including Co-60 and Zn-65 isotopes) from water and radioactive wastewater was investigated. The best performance for adsorption was observed at pH 7.0 and 60-min contact time. This study reveals that $Ag_2O$ nanoparticles can be effective in removing heavy metal ions, especially when combined with materials such as $TiO_2$ (Mahmoud et al., 2018).

ZnO nanoparticles have strong photocatalytic activity and antibacterial properties. They are effective in mercury adsorption and can be integrated into nanocomposites to improve their performance (Naseem et al., 2020).

$TiO_2$ nanoparticles are valuable due to their thermal and photostability, low cost, and high photocatalytic activity. They are effective against both organic and inorganic pollutants and can be integrated into filtration membranes (Pascariu et al., 2022).

Mechanisms including electrostatic interactions, effects of coating agents, and weak interfacial bonds such as van der Waals forces play a role in the removal of metals by metal oxide nanoparticles. A significant advantage of metal oxide nanoparticles is their high efficiency at low temperatures due to their unhindered adsorption processes. In contrast to conventional techniques, these properties make them more effective. Recent developments include the development of mixed oxide nanostructures to achieve higher degradation rates of certain pollutants and improve their properties.

## Mercury Nanotrap Water Filtration

Mercury nanotrap water filtration is a nanotechnology-based method for removing mercury, one of the most dangerous metal contaminants in water. Traditional mercury removal methods include chemical precipitation, amalgamation, reverse osmosis, membrane filtration, and photochemical techniques. However, these methods generally face problems such as high cost, time requirement, and low efficiency. Nanofiltration technology overcomes these problems by using nanoparticles with a high surface area/volume ratio and that can be easily chemically functionalized. The ability of nanomaterials to scan large solution volumes in a short time, thanks to Brownian motion, enables this technology to show high efficiency in removing mercury species. Different types of nanoparticles are used to remove mercury species through redox reactions. These include selenium, manganese dioxide nanofibers, carbon nanotube–silver composites, and silver- and gold-based materials. Among these materials, gold nanoparticles coated with citric acid have been one of the most effective materials widely used to capture mercury species from natural water sources (Ganguly et al., 2018; Li et al., 2014).

**New Nanomaterials**

Recently, new nanomaterials with excellent adsorption capacities, high stability, and environmentally friendly performance have enabled great improvements in the capture of heavy metal ions (Yu et al., 2021).

*Metal–Organic Frameworks (MOFs)* stand out as effective adsorbents in heavy metal removal due to their porous structures, high surface areas, and chemically modifiable functional groups. In a study, the potential of MOFs to remove heavy metal ions from water was investigated. It was emphasized that MOFs establish strong interactions with heavy metal ions due to their high surface areas and regular porous structures. It was shown that MOFs enriched with functional groups such as carboxylate, amine, and hydroxyl groups provide stronger binding with heavy metal ions. In the study, it was found that some MOF types can be used repeatedly and can maintain their adsorption capacity significantly. The study is especially important for the removal of toxic heavy metals such as Pb(II), Cd(II), and Hg(II) (Lin et al., 2025). In another study, the use of MOF-5 ($Zn_4O(BDC)_3$) in solid-phase extraction method for heavy metal removal was investigated. It was found that MOF-5 showed high efficiency especially in the removal of Cr(VI) and Pb(II) ions. The reuse potential of MOF-5 was tested, and over 90% recovery rate was achieved. The interaction mechanism of MOF-5 with metal ions was investigated in detail, and it was shown that factors such as electrostatic attraction and complex formation contribute to the adsorption process (Madumo et al., 2023).

*Nanosized zerovalent iron* (nZVI) particles are used as an effective method for heavy metal removal due to their high reactivity and large surface areas. There are various studies on this subject in the literature. In the last 15 years, nZVI has been investigated as a new tool for the treatment of heavy metal-contaminated waters (Tarekegn et al., 2021; Zhu et al., 2022). The removal mechanisms of heavy metals depend on the type of metal and its thermodynamic properties. Metals with redox potentials more negative or close to the reduction potential of Fe(0) are removed by the reduction process, while others are removed by precipitation, complexation, or other adsorption processes. In one study, it was stated that nZVI has a good adsorption capacity for various heavy metal ions. In the study, iron nanoparticles (nZVI) in powder or granular form were used as a permeable reactive barrier, especially for the removal of different pollutants (Galdames et al., 2020). However, undesirable situations also occurred when nZVI was used. The magnetic effect of nZVI in the medium leads to floc formation and loss of reactivity. In another study, the removal of Pb and naphthalene contamination in soil was investigated by integrating nZVI application with bioremediation approaches (biodegradation and biostimulation) (Zafar et al., 2024).

*MXenes* are a family of two-dimensional (2D) materials consisting of transition metal carbides, nitrides, and carbonitrides. Due to their high surface area, conductivity, and functional surface groups, they have attracted attention as potential adsorbents for the removal of heavy metal ions in water treatment. In one study, titanium carbide MXenes were shown to be able to effectively remove mercury ions from water. This process is based on the principle that a solid adsorbent, when placed in contaminated water, attracts and binds heavy metal particles to its surface. Traditional adsorbents, such as activated carbon, clays, and zeolites, generally show limitations such as poor binding tendency with metal ions, low surface area, and pH instability. In contrast, MXenes show the potential to overcome these challenges due to their unique structural properties and abundant surface functional groups (Xia et al., 2019). Again, some studies have shown that pure MXenes can remove approximately 87.6% of lead ions in water within the first 10 min of the treatment process. This rapid adsorption process highlights the potential of MXenes for the effective removal of lead from contaminated water sources (Lv et al., 2021; Yu et al., 2022). A comprehensive review study has revealed that MXene-based nanoarchitectured structures have emerged as promising adsorbents for various toxic metals such as lead (Pb), copper (Cu), zinc (Zn), and chromium (Cr) (Lingamdinne et al., 2024). Their hydrophilic structure, high surface area, and electron-rich surfaces contribute significantly to the high adsorption capacity of heavy metals.

*Graphitic carbon nitride* (g-$C_3N_4$) is a metal-free polymeric semiconductor in which carbon and nitrogen atoms are bonded by covalent bonds. Thanks to these properties, it has attracted attention as a promising photocatalyst for the removal of heavy metal ions in water treatment. The photocatalytic activity of g-$C_3N_4$ can be improved by increasing its surface area and the active sites. For example, petal-like g-$C_3N_4$ structures were synthesized using citric acid, and it was shown that these structures were used in the effective removal of hexavalent chromium [Cr(VI)] ions from aqueous solutions. Citric acid improves the photocatalytic activity of g-$C_3N_4$ by increasing the specific surface area and the surface-active sites (Yu et al., 2023). Modification of g-$C_3N_4$ with metal sulfides can improve its photocatalytic performance. For example, g-$C_3N_4$–$Bi_2S_3$ and g-$C_3N_4$–ZnS composites were found to be effective in photocatalytic degradation of organic pollutants (Mukhtar et al., 2025). Such modifications increase the effectiveness of g-$C_3N_4$ by reducing electron–hole recombination and increasing visible light absorption.

## 3.3 *Artificial Intelligence-Supported Monitoring Systems*

Artificial intelligence (AI) is increasingly being integrated into systems designed for the removal of heavy metals from water and wastewater, increasing the efficiency and enabling optimization of treatment processes. In particular, the application of machine learning (ML) algorithms offers predictive modeling, process optimization, and real-time monitoring of heavy metal removal. In a comprehensive review article, the potential of ML in modeling biochar-based heavy metal removal systems was highlighted, and data sharing methodologies were proposed to be developed with centralized and decentralized approaches (Yuan et al., 2023). In this way, machine learning-supported strategies for heavy metal treatment can be accelerated.

Artificial intelligence techniques, especially artificial neural networks (ANNs) and support vector machines, have been used to analyze large datasets related to adsorption parameters, increasing the prediction accuracy and operational efficiency. In addition, metaheuristic algorithms such as genetic algorithms and simulated annealing have been used to determine the most suitable parameter configurations for adsorption processes. With the integration of AI, real-time monitoring, predictive maintenance, and adaptive process adjustments can be made, thus continuously improving adsorption efficiency. AI, combined with traditional Langmuir and Freundlich isotherm models, offers innovative approaches to improve adsorption kinetics and thermodynamics (Gheibi et al., 2024).

The development of AI models for heavy metal removal from wastewater is being investigated to optimize various treatment techniques. Studies have shown that AI can effectively model complex interactions in wastewater treatment processes, thereby increasing removal efficiency and enabling process optimization. These studies aim to increase the overall efficiency of wastewater treatment systems by focusing on determining the most suitable AI models for specific removal techniques (Bhagat et al., 2020). One study addressed the optimization role of AI in water removal of heavy metals using iron oxide nanoparticles and demonstrated that AI can enhance the adsorption capacity of nano iron oxide materials (Zheng et al., 2023). Beyond water treatment, AI has also been applied to remediate soils contaminated with heavy metals and polycyclic aromatic hydrocarbons (PAHs). AI-assisted systematic review tools have provided in-depth analyses on rehabilitation techniques and evaluated the advantages and limitations of combined remediation methods compared to single-treatment applications.

One study investigated the use of artificial neural networks (ANNs) to model and optimize the biosorption process of heavy metals from aqueous solutions (Elsayed et al., 2022). The study demonstrated that ANNs can effectively predict the removal efficiency of heavy metals under various operational conditions and provide a valuable tool for designing and optimizing biosorption systems. Another paper investigated the application of machine learning techniques including support vector machines and decision trees to predict the removal efficiency of heavy metals in constructed wetlands. It has been highlighted that AI models have the potential to accurately predict treatment performance by aiding in the design and management of constructed wetlands for wastewater treatment (Singh et al., 2024). Again, artificial intelligence techniques such as genetic algorithms and particle swarm optimization have been used to optimize the removal of heavy metals from industrial wastewater (Gheibi et al., 2024). The findings have shown that AI-based optimization can significantly increase treatment efficiency and reduce operating costs. Fuzzy logic models have been applied to predict the removal efficiency of heavy metals in wastewater treatment plants (Chiranjivi et al., 2024). The research has shown that fuzzy logic can address the uncertainties and complexities inherent in wastewater treatment processes and provide reliable predictions of treatment performance. Again, the integration of

AI models with the electrocoagulation process was studied to enhance the removal of heavy metals from wastewater (Shirkoohi et al., 2022). The study showed that AI can effectively predict and optimize the operational parameters of the electrocoagulation process, thereby achieving improved treatment efficiency.

Such studies on the integration of AI into heavy metal removal systems offer a transformative approach to heavy metal remediation processes. By utilizing the predictive and optimization capabilities of AI, these systems can be used to achieve higher efficiency, adaptability, and sustainability.

## 3.4 *Innovative Filtration Systems*

Innovative filtration systems are also being developed for heavy metal removal. Some current examples are presented below:

**DMI-65®Filtration System**

DMI-65® is an advanced catalytic filtration system developed for the removal of heavy metals, especially iron and manganese, in industrial and municipal water treatment applications. The system exhibits stable and satisfactory performance over a wide pH range (5.8–8.6) without the use of potassium permanganate. It also has a long service life of 5–10 years and can operate at high flow rates. DMI-65® is NSF/ANSI 61 and UK Reg. 31 certified for drinking water applications (Musa et al., 2023).

**Kuriflock™Heavy Metal Precipitation Technology**

The Kuriflock™ series, developed by Kurita, offers innovative sedimentation technologies to minimize heavy metal concentrations in wastewater. These products are effective in cases where traditional treatment methods are insufficient due to the partial solubility of metal hydroxides. Kuriflock™ products improve sedimentation and sludge dewatering, requiring less sedimentation time and reducing wastewater sludge production. They also comply with European Union (EU) residue limits and can be used in various industries, such as metallurgy, fossil fuel power plants, and refineries (Carlund, 2023).

**Enrodisc Rotary Disk Micro Filter**

The Enrodisc Rotary Disk Micro Filter, developed by Endemic Process Technologies, is an innovative filtration system designed to provide superior performance in wastewater treatment. With its compact design, high efficiency filtration capacity, and self-cleaning mechanism, it is an ideal solution for effluent filtration, suspended solids removal, and solid–liquid separation in both municipal and industrial applications (endemicprocess.com).

These innovative filtration systems increase environmental compatibility and improve the efficiency of industrial water treatment processes by providing effective solutions for heavy metal removal.

## 3.5 Combined Technologies

The combined remediation refers to the combination of two or more repair methods for soil or wastewater remediation, which can not only improve the efficiency and rate of single contaminated soils, but also overcome the limitations of individual remediation technologies and achieve simultaneous restoration of composite pollutants. According to different repair technologies, the combined repair technology is currently divided into: microbial–plant remediation, microorganism–animal remediation, microorganism–plant–animal remediation, physical/chemical–microbial/plant/animal combined remediation, etc. At present, the application of microbial–phytoremediation is quite common (Xu et al., 2019).

**Chemical–Biological Remediation**

It was stated that some findings that emerged with the use of chemical and biological methods separately were eliminated with the use of a combined method. Higher efficiency was achieved by applying chemical treatment after biological treatment. (Goswami & Mazumdar, 2014). However, an important issue here is the necessity of selecting nontoxic chemicals so that chemical–biological methods do not cause harm. Ahmet et al. (2016) achieved 99.3% Cr(VI) removal from leather industry wastewater by using chemical additives and biological treatment together. Sharma and Malaviya (2014) achieved 62.33% success in total chromium removal with a process where chemical and biological systems were combined using microorganisms called *Fusarium chlamydosporum*.

**Electrochemical–Microbial Remediation**

In this method, heavy metals are removed by using electric current and microorganisms together. This method is an industrial treatment that combines both electrical stimulation and microorganism biodegradation processes for the removal of heavy metals and organic contaminants. Electrochemical breakdown increases the movement of heavy metals, helping microorganisms to break them down more easily. Electrochemical processes can be combined (Zhao et al.,) with sulfate-reducing bacteria to enhance heavy metal removal. *Desulfovibrio desulfuricans* converts sulfate ions to sulfides, forming insoluble metal sulfides (such as PbS, ZnS, CuS) and precipitating heavy metals and removing them from water.

**Electrokinetic–Phytoremediation**

In some circuits, electrokinetic treatment alone is insufficient to increase the heavy metal market in the soil to acceptable levels. The synergistic combination of electrokinetic processes with other remediation technologies has been proven to remove heavy metals from contaminated soils more effectively. An effective (Du et al., 2020) electrokinetic method mobilized Cd(II) and Zn(II) ions in the soil and transported them to the plant roots. *Brassica juncea* exhibited high heavy metal tolerance and accumulated these heavy metals. Cd and Zn concentrations in the soil were reduced by more than 50%.

**Phytobial Remediation**

Phytobial remediation is an advanced environmental cleaning method that is a combination of phytoremediation and bioremediation. This technique uses the symbiotic association between plants and microorganisms in the root zone (rhizosphere) to clean heavy metals and organic decay. In the literature (Zhang et al., 2018), Cd(II) removal can be recovered with *Helianthus annuus* and mycorrhizal fungi. *Helianthus annuus* absorbs Cd(II) through its roots and accumulates it in its biomass. Mycorrhizal fungi establish a symbiotic relationship with the plant roots and promote the uptake of metal ions. It has been reported that Cd(II) removal has increased up to 60% due to the synergistic effect of *Helianthus annuus* and mycorrhizal fungi, even with low annual Cd(II) uptake.

**Other Combined Remediation Methods**

In the *electrokinetic–geosynthetic approach*, geosynthetic features accelerate the electrokinetic treatment process by increasing the mobility of the equipment (Jones et al., 1996). *Microorganism Supported Permeable Reactive Barriers*, solutions provided from polluted waters are created, heavy metals are retained with the help of microorganisms and/or plants (Kober et al., 1999). In *electrokinetic–membrane filtration*, electrokinetic treatment and membrane filtration are combined and over 98% success has been achieved in iron (Fe), zinc (Zn), and calcium (Ca) removal (Peng, W et al., 2018). In the *electrokinetic–soil washing method*, heavy metals were transformed in the soil and removed more effectively. The method was successful in removing metals such as antimony, arsenic, cadmium, chromium, and mercury. However, this method needs to be validated for major changes (Ferrucci et al., 2017).

## 3.6 *Sustainable Adsorbent Materials*

Geopolymers are innovative and economical inorganic polymers that are synthesized from aluminosilicates in soluble silicates, as well as soluble silicates and alkali hydroxide mixtures (Davidovits, 1991; Xu & van Deventer, 2000). Geopolymers have a three-dimensional polymeric structure and are synthesized via polycondensation method at low temperatures (Lin et al., 2006). Geopolymers are good candidates for the removal of heavy metals from wastewaters. The calcium oxide-rich Lidz Donawitz converter slag (LD) was used to produce LD geopolymer to remove $Ni^{2+}$ ions from aqueous environment. The geopolymerization of raw LD increased the surface area from 4.859 $m^2/g$ to 30.84 $m^2/g$ with a decrease in pore volume from 155 A° to 118.6 A° as well as the total pore volume increased from 0.018 $cm^3/g$ to 0.091 $cm^3/g$. The researchers concluded an increase in mechanical strength with a compact mesoporous structure. The geopolymerization of LD increased the maximum adsorption capacity of $Ni^{2+}$ (aq) from 14.345 mg/g to 84.875 mg/g with an increase upto pH 9 and then constant. The adsorption isotherm examinations indicated the Langmuir type adsorption model, meaning a monolayer adsorption. The adsorption kinetic model fit well to the pseudo-second-order model, with a deviation of the theoretical maximum adsorption capacity of 2.8% from the experimental one. The thermodynamic data was examined as well and the standard enthalpy and entropy changes were calculated as 21.735 kJ/mol and 0.093 kJ/(mol.K), respectively. The standard Gibbs energy changes at 298 K, 308 K, and 318 K were determined as − 5.971 kJ/mol, − 7.126 kJ/mol, and − 7.825 kJ/mol, respectively, indicating spontaneous adsorption (Sarkar et al., 2017). Kara et al. examined the removal of $Ni^{2+}$ ions from aqueous solutions using metakaolin-based geopolymer (MKG). They examined the effect of pH on removal efficiency of $Ni^{2+}$ ions from aqueous media, and reversely to Sarkar et al., they observed at about complete adsorption efficiency at acidic media, that is pH 4. The effect of temperature on removal percent was not efficiently changed between 10 °C and 50 °C, that changed from 24.92 mg/g to 29.85 mg/g. The adsorption kinetic model was closer to pseudo-second-order when the plotted data was linearized and processed. The Langmuir isotherm model was concluded for the adsorption of $Ni^{2+}$ on MKG, a monolayer adsorption that had a maximum adsorption capacity of $7.26 \times 10^{-4}$ mol/g. The mean free energy for $Ni^{2+}$ ions was determined as 17.23 kJ/mol, indicating a physisorption interaction between $Ni^{2+}$ ions and MKG. The effect of ionic strength on adsorption capacity was examined in $NaNO_3$ solutions, where a decrease in maximum adsorption capacity was observed with the increase in salt concentration, indicating outer-sphere surface reactions in the adsorption mechanism (Kara et al., 2017).

Another common heavy metal that was examined for adsorption by geopolymers was copper. Al-Harahsheh and co-workers synthesized geopolymer from fly ash in a basic medium. The synthesized geopolymer was investigated for the adsorption of $Cu^{2+}$ ions from aqueous solution. The percent removal efficiency of copper with geopolymer was more than 80%, that is, highest at pH 6. The researchers finished the data at pH 6 since $Cu^{2+}$ will precipitate as $Cu(OH)_2$ at basic media. The maximum adsorption capacity of $Cu^{2+}$ was reached in 60 min after contact with geopolymer in aqueous media. The kinetic data derived from contact time was modeled using pseudo-first-order and pseudo-second-order adsorption kinetics. By comparing correlation coefficients, the pseudo-second-order adsorption kinetics was found to fit well, and the rate constant for this model was calculated as 0.018 g/(mg.min.), suggesting chemisorption as the rate-determining step. The authors also calculated the rate constant for the pseudo-first-order adsorption kinetics as 0.042/min. The maximum adsorption capacity was determined as 44.31 mg/g experimentally, and the theoretical maximum adsorption capacity calculated from the kinetic data was 44.28 mg/g and 44.01 mg/g, respectively. The adsorption capacity was studied at 298 K, 308 K, and 318 K at pH 4, 5, and 6. The adsorption process was more spontaneous when the temperature and pH increased, with the standard Gibbs energy calculated as − 23.09 kJ/mol, the lowest one when compared with other conditions. The standard enthalpy and entropy both stripped at pH 6, and for all pH values, the adsorption process was endothermic and the entropy increased.

Bioadsorbents are promising materials for the adsorption of heavy metals from aquatic environment. Algae are one of the most effective bioadsorbents for heavy metal removal from aquatic environments. The functional groups containing oxygen, nitrogen, phosphorus, and sulfur are the main actors for the adsorption of heavy metals (Ivanova et al., 2012). Algae have the highest maximum adsorption capacity for lead ions in water when compared with plants (Redha, 2020). Biochar is another environmentally friendly material for the removal of heavy metals from aquatic environments. The biomass is carbonized into biochar via biochemical or thermochemical methods. Toxic heavy metal lead can be adsorbed by biochar synthesized from pyrolysis of biomass. The mechanism of the pyrolyzed biochar–lead interaction mainly consists of adsorption, complexation, ion exchange, and precipitation (Bayar et al., 2024). Copper ions were separated from aqueous solutions by comparing commercial charcoal and biochar synthesized from Maiden & Betche (*Macadamia integrifolia*) nuts and guava seeds (*Psidium guajava* L.). The commercial one, however, exhibited a higher adsorption capacity when compared with the biochars derived from biomasses (Rocha et al., 2006). Toxic Cr(VI) was efficiently removed from water by a biochar/iron oxide composite (Zou et al., 2021). Corn straw biochar had a large specific area, that is 2183 $m^2/g$, synthesized via two-step KOH-activated pyrolysis, with an adsorption capacity of 117 mg/g (Qu et al., 2021). Thiol magnetic biochar from beer barley husk was used to remove Hg(II) ions from aquatic environment, and the efficiency of the biochar was concluded by Gheitasi and co-workers. The optimal conditions were reached at pH 6, with a removal efficiency of more than 99% for Hg(II). The maximum adsorption capacity was determined as 109.37 mg/g, obeying Freundlich isotherm modeling and pseudo-second-order kinetics. The adsorption process was concluded to be endothermic and spontaneous (Gheitasi et al., 2022).

Clays are conventional adsorbents for heavy metals because of their economy and widespread use. Clays, their composites, or modified ones were efficiently used to remove $Pb^{2+}$, $Ni^{2+}$, $Cd^{2+}$, $Cu^{2+}$, $Zn^{2+}$, $Cr^{6+}$, $Hg^{2+}$, and $U^{6+}$ from aquatic environments. The mechanism of adsorption of heavy metals on clay minerals involves exchange elements and physical interactions (Xie et al., 2024).

Activated carbon is also one of the most used materials for the adsorption of heavy metals. The wooden granular activated carbon was regenerated to remove $Pb^{2+}$, $Cd^{2+}$, $Cu^{2+}$, $Zn^{2+}$, and $Cr^{6+}$ from aquatic environments (Qiu et al., 2022).

Zeolites are natural minerals found in comparatively large abundance. Zeolites are generally ineffective, and they need modification to increase the adsorption capacity of heavy metals. The modified zeolites have been applied on adsorption studies of As(III), As(V), Cd(II), Pb(II), Hg(II), and Ni(II) from aquatic environments. Zeolites adsorb heavy metals via ion exchange, electrostatic interactions, intrapore adsorption, surface complexation, and surface precipitation (Velarde et al., 2023).

## Our Hopes for the Future

Technological innovations and increasing environmental concerns have made it imperative to take rapid action in heavy metal management. Future research should focus on developing cost-effective, sustainable, and scalable solutions that can be applied across industries and ecosystems. One of the most promising developments here is the combination of nanotechnology with bioremediation to increase the efficiency of heavy metal removal while minimizing environmental impacts. The use of high-tech nanomaterials such as metal–organic frameworks (MOFs) and MXenes is likely to play a significant role in improving adsorption capacity and selectivity.

Another important development is the application of artificial intelligence (AI) and machine learning to monitor and optimize heavy metal removal. AI-driven progressive models can improve real-time detection, automate treatment processes, and provide adaptive responses to changing contamination levels. Integrating AI with smart filtration systems and sensor-based monitoring networks is an important step in removing heavy metals from the environment. Further refinement of recycling and recovery technologies for heavy metals from industrial waste, electronic waste, and wastewater will reduce dependency on raw materials and minimize pollution.

Developments in bioadsorbents and sustainable adsorbent materials will also contribute to environmentally friendly remediation solutions. The future of heavy metal management is likely to be characterized by interdisciplinary collaborations combining chemistry, environmental engineering, biotechnology and data science. Government policies, regulatory frameworks and international collaboration will also be critical in shaping the global deployment of these new technologies.

## 4 Conclusion

Heavy metal pollution remains a significant challenge due to its toxic effects on human health and ecosystems. Conventional remediation techniques such as chemical precipitation, membrane filtration, and ion exchange are widely used but often suffer from limitations such as high costs, secondary waste generation, and limited efficiency. This has led to the development of innovative and advanced technologies aimed at improving the sustainability and efficiency of heavy metal management.

Recent advances in nanotechnology-based adsorbents, bioremediation strategies, and AI-assisted monitoring systems have opened new possibilities to address heavy metal pollution more efficiently. Bioremediation and phytoremediation techniques using microorganisms and hyperaccumulator plants offer environmentally friendly solutions, while emerging filtration systems such as DMI-65® and Kuriflock™ provide improved performance in industrial applications. Furthermore, combined remediation approaches that integrate multiple treatment techniques have shown significant potential to increase heavy metal removal rates.

Looking ahead, continued research and technological innovation will be essential to overcome current challenges and ensure the widespread adoption of sustainable heavy metal removal methods. With interdisciplinary collaborations and supportive policy frameworks, the future of heavy metal management looks bright to create cleaner and healthier environments.

## References


Acar, Y. B., & Alshawabkeh, A. N. (1993). Principles of electrokinetic remediation. *Environmental Science and Technology, 27*(13), 2638–2647. https://doi.org/10.1021/es00049a002

Ahmed, E., Abdulla, H. M., Mohamed, A. H., & El-Bassuony, A. D. (2016). Remediation and recycling of chromium from tannery wastewater using combined chemical–biological treatment system. *Process Safety and Environmental Protection, 104*, 1–10. https://doi.org/10.1016/j.psep.2016.08.004

Ayangbenro, A. S., & Babalola, O. O. (2017). A new strategy for heavy metal polluted environments: A review of microbial biosorbents. *International Journal of Environmental Research and Public Health, 14*(1), 94. https://doi.org/10.3390/ijerph14010094

Babel, S., & Kurniawan, T. A. (2003). Low-cost adsorbents for heavy metals uptake from contaminated water: A review. *Journal of Hazardous Materials, 97*(1–3), 219–243. https://doi.org/10.1016/S0304-3894(02)00263-7

Baghour, M. (2017). Effect of seaweeds in phyto-remediation. In *Biotechnological Applications of Seaweeds*; *Nova Publishers Sciences* (pp. 47–83).

Bayar, J., Ali, N., Dong, Y., Ahmad, U., Anjum, M. M., Khan, G. R., Zaib, M., Jalal, A., Ali, R., & Ali, L. (2024). Biochar-based adsorption for heavy metal removal in water: A sustainable and cost-effective approach. *Environmental Geochemistry and Health, 46*, 428. https://doi.org/10.1007/s10653-024-02214-w

Bhagat, S. K., Tung, T. M., & Yaseen, Z. M. (2020). Development of artificial intelligence for modeling wastewater heavy metal removal: State of the art, application assessment and possible future research. *Journal of Cleaner Production, 250*, Article 119473. https://doi.org/10.1016/j.jclepro.2019.119473

Bouhadi, M., Elkouali, M. H., Talbi, M., & Ainane, T. (2021). Phytoremediation review: Bioavailability of heavy metals and the role of bioaccumulative plants in the remediation of contaminated soils. *Journal of Analytical Sciences and Applied Biotechnology, 3*(1), 40–47. https://doi.org/10.48402/IMIST.PRSM/jasab-v3i1.28238

Cai, Y., Li, C., Wu, D., Wang, W., Tan, F., Wang, X., Wong, P. K., & Qiao, X. (2017). Highly active MgO nanoparticles for simultaneous bacterial inactivation and heavy metal removal from aqueous solution. *Chemical Engineering Journal, 312*, 158–166. https://doi.org/10.1016/J.cej.2016.11.134

Cai, X., Zheng, X., Zhang, D., Iqbal, W., Liu, C., Yang, B., Zhao, X., Lu, X., & Mao, Y. (2019). Microbial characterization of heavy metal resistant bacterial strains isolated from an electroplating wastewater treatment plant. *Ecotoxicology and Environmental Safety, 181*, 472–480. https://doi.org/10.1016/j.ecoenv.2019.06.036

Carlund, A. (2023). Selected PFAS in the water treatment at the waste-to-energy plant in Umeå: Investigating concentration and composition of selected PFAS in different water treatment steps.

Chen, Q., Luo, Z., Hills, C., Xue, G., & Tyrer, M. (2009). Precipitation of heavy metals from wastewater using simulated flue gas: Sequent additions of fly ash, lime and carbon dioxide. *Water Research, 43*(10), 2605–2614. https://doi.org/10.1016/j.watres.2009.03.007

Chen, L., Wang, F., Zhang, Z., Chao, H., He, H., Hu, W., Zeng, Y., Duan, C., Liu, J., & Fang, L. (2023). Influences of arbuscular mycorrhizal fungi on crop growth and potentially toxic element accumulation in contaminated soils: A meta-analysis. *Critical Reviews in Environmental Science and Technology, 53*(20), 1795–1816. https://doi.org/10.1080/10643389.2023.2183700

Chen J, Guo J, Li Z, Liang X, You Y, Li M, He Y, Zhan F. (2022 Jun 30). Effects of an arbuscular mycorrhizal fungus on the growth of and cadmium uptake in maize grown on polluted wasteland, farmland and slopeland soils in a lead-zinc mining area. Toxics. *Environmental Science: Advances. 10*(7):359.

Chiranjivi, M., Suresh, K., Kumar, A. A., Siddartha, M., Teja, M. R., & Rani, M. (2024). Forecasting of effluent from wastewater treatment industries using a novel fuzzy logic system. In *E3S Web of Conferences* (Vol. 547, pp. 02008). EDP Sciences.

Comis, D. (1996). Green remediation: Using plants to clean the soil. *Journal of Soil and Water Conservation, 51*(3), 184–187. https://doi.org/10.1080/00224561.1996.12457067

Dada, E. O., Njoku, K. L., Osuntoki, A. A., & Akinola, M. O. (2016). Heavy metal remediation potential of a tropical wetland earthworm, Libyodrilus violaceus (Beddard).

Das, A. P., & Ghosh, S. (2018). Bioleaching of manganese from mining waste materials. *Materials Today: Proceedings, 5*(1), 2381–2390. https://doi.org/10.1016/J.MATPR.2017.11.459

Das, A. P., Sukla, L. B., Pradhan, N., & Nayak, S. (2011). Manganese biomining: A review. *Bioresource Technology, 102*(16), 7381–7387. https://doi.org/10.1016/J.BIORTECH.2011.05.018

Das, A. P., Sukla, L. B., & Pradhan, N. (2012a). Microbial recovery of manganese using *Staphylococcus Epidermidis*. *International Journal of Nonferrous Metallurgy, 2012*(02), 9–12. https://doi.org/10.4236/IJNM.2012.12002

Das, A. P., Swain, S., Panda, S., Pradhan, N., & Sukla, L. B. (2012b). Reductive acid leaching of low-grade manganese ores. *Geomaterials, 2012*(04), 70–72. https://doi.org/10.4236/GM.2012.24011

Das, A. P., Ghosh, S., Mohanty, S., & Sukla, L. B. (2015). Consequences of manganese compounds: A review. *Toxicological and Environmental Chemistry, 96*(7), 981–997. https://doi.org/10.1080/02772248.2015.1005428

Das, A. P., Nayak, B., & Behera, I. D. (2024a). Applications of microorganisms in advanced geomicrobiological and biotechnological exploration. *Geomicrobiology Journal, 41*(4), 309–311. https://doi.org/10.1080/01490451.2024.2327059/ASSET//CMS/ASSET/DE2A1F67-08C1-475A-ABC7-2B3F4C9E8BCF/01490451.2024.2327059.FP.PNG

Das, A. P., & Ghosh, S. (2023a). Microorganisms in manganese biomining processes. In *Manganese Mining Microorganisms* (pp. 55–61).https://doi.org/10.1016/B978-0-12-822147-1.00013-3

Das, A. P., & Ghosh, S. (2023b). Role of metabolites and metabolomics for understanding of manganese bioleaching. In *Manganese Mining Microorganisms* (pp. 117–127).https://doi.org/10.1016/B978-0-12-822147-1.00011-X

Das, A., & Mishra, S. (2010). Biodegradation of the metallic carcinogen hexavalent chromium Cr (VI) by an indigenously isolated bacterial strain. *Journal of Carcinogenesis, 9*. https://doi.org/10.4103/1477-3163.63584

Das, A. P., Ghosh, S., Mohanty, S., & Sukla, L. B. (2015a). Advances in manganese pollution and its bioremediation. In L. B. Sukla, N. Pradhan, S. Panda, & B. K. Mishra (Eds.), *Environmental microbe biotechnology* (1st ed., pp. 313–328). Springer. https://doi.org/10.1007/978-3-319-19018-1_16

Das, A., Dey, S., Bala, K., Gnanasekaran, R., & Das, A. P. (2024). Microbial bioremediation technology for sustainable treatment and management of synthetic microfiber waste. In *Sustainable microbial technology for synthetic and cellulosic microfiber bioremediation* (pp. 221–238). https://doi.org/10.1007/978-3-031-62600-5_12

Davidovits, J. (1991). Geopolymers. *Journal of Thermal Analysis, 37*, 1633–1656. https://doi.org/10.1007/BF01912193

Desai, C., Jain, K., & Madamwar, D. (2008). Hexavalent chromate reductase activity in cytosolic fractions of Pseudomonas sp. G1DM21 isolated from Cr (VI) contaminated industrial landfill. *Process Biochemistry, 43*(7), 713–721. https://doi.org/10.1016/j.procbio.2008.02.015

Dey, S., Dash, A., & Das, A. P. (2024). Bioremediation as a sustainable and circular solution for mining waste management and bioenergy generation. In *Sustainable management of mining waste and tailings: A circular economy approach* (pp. 107–119). https://doi.org/10.1201/9781003442455-6/bioremediation-sustainable-circular-solution-mining-waste-management-bioenergy-generation-sudeshna-dey-akanshya-dash-alok-prasad-das

Du, J., Guo, Z., Li, R., Ali, A., Guo, D., Lahori, A. H., Wang, P., Liu, X., Wang, X., & Zhang, Z. (2020). Screening of Chinese mustard (Brassica juncea L.) cultivars for the phytoremediation of Cd and Zn based on the plant physiological mechanisms. *Environmental Pollution, 261*, 114213. https://doi.org/10.1016/j.envpol.2020.114213

Duan, J., & Gregory, J. (2003). Coagulation by hydrolysing metal salts. *Advances in Colloid and Interface Science, 100*, 475–502. https://doi.org/10.1016/S0001-8686(02)00067-2

Dwivedi, S., Tripathi, R. D., Srivastava, S., Singh, R., Kumar, A., Tripathi, P., Dave, R., Rai, U. N., Chakrabarty, D., Trivedi, P. K., & Tuli, R. (2010). Arsenic affects mineral nutrients in grains of various Indian rice (Oryza sativa L.) genotypes grown on arsenic-contaminated soils of West Bengal. *Protoplasma, 245*, 113–124. https://doi.org/10.1007/s00709-010-0151-7

Elsayed, A., Moussa, Z., Alrdahe, S. S., Alharbi, M. M., Ghoniem, A. A., El-Khateeb, A. Y., & Saber, W. I. (2022). Optimization of heavy metals biosorption via artificial neural network: A case study of cobalt (II) sorption by pseudomonas alcaliphila NEWG-2. *Frontiers in Microbiology, 13*, Article 893603. https://doi.org/10.3389/fmicb.2022.893603

Fajardo, C., Sánchez-Fortún, S., Costa, G., Nande, M., Botías, P., García-Cantalejo, J., Mengs, G., & Martín, M. (2020). Evaluation of nanoremediation strategy in a Pb, Zn and Cd contaminated soil. *Science of the Total Environment, 706*, Article 136041. https://doi.org/10.1016/j.scitotenv.2019.136041

Ferrucci, A., Vocciante, M., Bagatin, R., & Ferro, S. (2017). Electrokinetic remediation of soils contaminated by potentially toxic metals: Dedicated analytical tools for assessing the contamination baseline in a complex scenario. *Journal of Environmental Management, 203*, 1163–1168. https://doi.org/10.1016/j.envman.2017.02.037

Fu, F., & Wang, Q. (2011). Removal of heavy metal ions from wastewaters: A review. *Journal of Environmental Management, 92*(3), 407–418. https://doi.org/10.1016/j.envman.2010.11.011

Galdames, A., Ruiz-Rubio, L., Orueta, M., Sánchez-Arzalluz, M., & Vilas-Vilela, J. L. (2020). Zero-valent iron nanoparticles for soil and groundwater remediation. *International Journal of Environmental Research and Public Health, 17*(16), 5817. https://doi.org/10.3390/ijerph17165817

Ganguly, M., Dib, S., & Ariya, P. A. (2018). Fast, cost-effective and energy efficient mercury removal-recycling technology. *Scientific Reports, 8*(1), 16255. https://doi.org/10.1038/s41598-018-34172-6

Gheibi, M., Masoomi, S. R., Magala, M. U., Fathollahi-Fard, A. M., Ghazikhani, A., & Behzadian, K. (2024). The application of artificial intelligence (AI) in adsorption process of heavy metals: A systematic review. *Environmental Industry Letters, 2*(2), 57–78. https://doi.org/10.15157/eil.2024.2.2.57-78

Gheitasi, F., Ghammamy, S., Zendehdel, M., & Semiromi, F. B. (2022). Removal of mercury (II) from aqueous solution by powdered activated carbon nanoparticles prepared from beer barley husk modified with Thiol/$Fe_3O_4$. *Journal of Molecular Structure, 1267*, Article 133555. https://doi.org/10.1016/j.molstruc.2021.133555

Ghosh, S., Mohanty, S., Nayak, S., Sukla, L. B., & Das, A. P. (2016). Molecular identification of indigenous manganese solubilising bacterial biodiversity from manganese mining deposits. *Journal of Basic Microbiology, 56*(3), 254–262. https://doi.org/10.1002/JOBM.201500477

Ghosh, S., & Das, A. P. (2018). Metagenomic insights into the microbial diversity in manganese-contaminated mine tailings and their role in biogeochemical cycling of manganese. *Scientific Reports, 8*(1), 1–12. https://doi.org/10.1038/s41598-018-26311-w

Ghosh, S., Tripathy, B., Dey, S., & Das, A. P. (2023). Comparative investigation of fungal and bacterial manganese biomining mechanisms. In *Manganese mining microorganisms* (pp. 99–115). https://doi.org/10.1016/b978-0-12-822147-1.00016-9

Goswami, R., Neog, N., Bhagat, C., Hdeib, R., Mahlknecht, J., & Kumar, M. (2022). Arsenic in the groundwater of the upper Brahmaputra floodplain: Variability, health risks and potential impacts. *Chemosphere, 306*, Article 135621. https://doi.org/10.1016/j.chemosphere.2022.135621

Goswami, K., & Mazumdar, I. (2014). Lead poisoning and some commonly used spices: An Indian scenario.

Gupta, V. K., & Nayak, A. (2012). Cadmium removal and recovery from aqueous solutions by novel adsorbents prepared from orange peel and $Fe_2O_3$ nanoparticles. *Chemical Engineering Journal, 180*, 81–90. https://doi.org/10.1016/j.cej.2011.11.006

Haldar, S., & Ghosh, A. (2020). Microbial and plant-assisted heavy metal remediation in aquatic ecosystems: A comprehensive review. *3 Biotech, 10*(5), 205. https://doi.org/10.1007/s13205-020-02195-4

Hu, Y., Liu, X., Bai, J., Shih, K., Zeng, E. Y., & Cheng, H. (2013). Assessing heavy metal pollution in the surface soils of a region that had undergone three decades of intense industrialization and urbanization. *Environmental Science and Pollution Research, 20*, 6150–6159. https://doi.org/10.1007/s11356-013-1668-z

Ivanova, D., Kadukova, I., Kavulıconva, J., & Horvathova, H. (2012). Determination of the functional groups in Algae Parachlorella Kessleri by potentiometric titrations. *Nova Biotechnologica Et Chimica, 11*(2), 93–99. https://doi.org/10.2478/v10296-012-0010-3

Jones, C. J. F. P., Fakher, A., Hamir, R., & Nettleton I. M. (1996). Geosynthetic materials with improved reinforcement capabilities, In *Proceedings of the International Symposium on Earth Reinforcement*, Fukuoka Kyushu, Japan (Vol. 2, pp. 865–883).

Kanungo, J., Sahoo, T., Bal, M., & Behera, I. D. (2024). Performance of bioremediation strategy in waste lubricating oil pollutants: A review. *Geomicrobiology Journal, 41*(4), 360–373. https://doi.org/10.1080/01490451.2023.2245395

Kara, İ, Yilmazer, D., & Akar, S. T. (2017). Metakaolin based geopolymer as an effective adsorbent for adsorption of zinc(II) and nickel(II) ions from aqueous solutions. *Applied Clay Science, 139*, 54–63. https://doi.org/10.1016/j.clay.2017.01.008

Kober, B., Wessels, M., Bollhofer, A., & Mangini, A. (1999). Pb isotopes in sediments of Lake Constance, Central Europe constrain the heavy metal pathways and the pollution history of the catchment, the lake and regional atmosphere—Constrains on the origin(s) and flux(es) of metals. *Geochim Et Cosmochim Ac, 63*, 1293–1303. https://doi.org/10.1016/S0016-7037(99)00064-2

Li, B., Zhang, Y., Ma, D., Shi, Z., & Ma, S. (2014). Mercury nano-trap for effective and efficient removal of mercury (II) from aqueous solution. *Nature Communications, 5*(1), 5537. https://doi.org/10.1038/ncomms6537

Lin, L., Shaobin, W., & Zhonghua, Z. (2006). Geopolymeric adsorbents from fly ash for dye removal from aqueous solution. *Journal of Colloid and Interface Science, 300*(1), 52–59. https://doi.org/10.1016/j.jcis.2006.03.062

Lin, H., Chen, J., Yao, Y., Lu, G., Huan, W., Ma, N., & Dai, W. (2025). Enhancement capture capability of cadmium ions using a MOF-in-MOF composite. *Journal of Solid State Chemistry, 343*, Article 125149. https://doi.org/10.1016/j.jssc.2024.125149

Lingamdinne, L. P., Kulkarni, R., Koduru, J. R., Karri, R. R., Somala, A. R., Solangi, N. H., Mubarak, N. M., Choi, J. S., Chang, Y. Y., & Dehghani, M. H. (2024). MXenes for advanced energy storage and environmental remediation applications: Synthesis, properties, and challenges. *Journal of Energy Storage, 101*, Article 113806. https://doi.org/10.1016/j.est.2024.113806

Lv, J., Huang, Q., Liu, T., & Pan, Q. (2021). Energy storage device application based on MXenes composites: A mini review. *International Journal of Electrochemical Science, 16*(4), Article 210439. https://doi.org/10.20964/2021/04.13

Madaeni, S. S., Fane, A. G., & Wiley, D. E. (1999). Factors influencing critical flux in membrane filtration of activated sludge. *Journal of Chemical Technology and Biotechnology: International Research in Process, Environmental and Clean Technology, 74*(6), 539–543.

Madumo, T. M., Zikalala, S. A., Gumbi, N. N., Mishra, S. B., Ntsendwana, B., & Nxumalo, E. N. (2023). Development of nitrogen-doped graphene/MOF nanocomposites towards adsorptive removal of Cr (VI) from the wastewater of the Herbert Bickley treatment works. *Environmental Nanotechnology, Monitoring and Management, 20*, Article 100794. https://doi.org/10.1016/j.enmm.2023.100794

Mahmoud, M. E., Saad, E. A., El-Khatib, A. M., Soliman, M. A., Allam, E. A., & Fekry, N. A. (2018). Green solid synthesis of polyaniline-silver oxide nanocomposite for the adsorptive removal of ionic divalent species of Zn/Co and their radioactive isotopes 65 Zn/60 Co. *Environmental Science and Pollution Research, 25*, 22120–22135. https://doi.org/10.1007/s11356-018-2284-8

Martell, A. E., Motekaitis, R. J., Chen, D., Hancock, R. D., & McManus, D. (1996). Selection of new Fe (III)/Fe (II) chelating agents as catalysts for the oxidation of hydrogen sulfide to sulfur by air. *Canadian Journal of Chemistry, 74*(10), 1872–1879.

Matis, K. A., Zouboulis, A. I., Gallios, G. P., Erwe, T., & Blöcher, C. (2004). Application of flotation for the separation of metal-loaded zeolites. *Chemosphere, 55*(1), 65–72. https://doi.org/10.1016/j.chemosphere.2003.11.030

Merroun, M. L., Raff, J., Rossberg, A., Hennig, C., Reich, T., & Selenska-Pobell, S. (2005). Complexation of uranium by cells and S-layer sheets of Bacillus sphaericus JG-A12. *Applied and Environmental Microbiology, 71*(9), 5532–5543. https://doi.org/10.1128/AEM.71.9.5532-5543.2005

Mishra, S., Dash, D., Al-Tawaha, A. R. M. S., & Das, A. P. (2022). A review on heavy metal ion adsorption on synthetic microfiber surface in aquatic environments. *Applied Biochemistry and Biotechnology, 194*(10), 4639–4654. https://doi.org/10.1007/S12010-022-04029-W/METRICS

Mohammad, A. W., Teow, Y. H., Ang, W. L., Chung, Y. T., Oatley-Radcliffe, D. L., & Hilal, N. (2015). Nanofiltration membranes review: Recent advances and future prospects. *Desalination, 356*, 226–254. https://doi.org/10.1016/j.desal.2014.10.043

Mohanty, S., Ghosh, S., Nayak, S., & Das, A. P. (2016). Isolation, identification and screening of manganese solubilizing fungi from low-grade manganese ore deposits. *Geomicrobiological Journal, 34*(4), 309–316. https://doi.org/10.1080/01490451.2016.1189016

Mounier L, Pédrot M, Bouhnik-Le-Coz M, Cabello-Hurtado F. (2023). Impact of iron oxide nanoparticles on a lead-polluted water–soil–plant system under alternating periods of water stress.*Environmental Science: Advances. 2*(5):767–79.

Mukhtar, S., Szabó-Bárdos, E., Őze, C., Juzsakova, T., Rácz, K., Németh, M., & Horváth, O. (2025). G-C3N4 modified with metal sulfides for visible-light-driven photocatalytic degradation of organic pollutants. *Molecules, 30*(2), 253. https://doi.org/10.3390/molecules30020253

Musa, M. A., Mah, S. K., Ng, Y. S., Lim, S., & Chew, C. M. (2023). Iron removal from groundwater through oxidation with DMI-65. *International Journal of Engineering Advanced Research, 5*(1), 1–11.

Nag, M., Lahiri, D., Ghosh, S., Sarkar, T., Pati, S., Das, A. P., Ram, D. K., Bhattacharya, D., & Ray, R. R. (2023). Application of microorganisms in biotransformation and bioremediation of environmental contaminant: A review. *Geomicrobiology Journal.* https://doi.org/10.1080/01490451.2023.2261443

Naseem, T., Waseem, M., Hafeez, M., Din, S. U., & Haq, S. (2020). Reduced graphene oxide/zinc oxide nanocomposite: From synthesis to its application for wastewater purification and antibacterial activity. *Journal of Inorganic and Organometallic Polymers and Materials, 30*, 3907–3919. https://doi.org/10.1007/s10904-020-01529-2

Naser, H. M., Sultana, S., Gomes, R., & Noor, S. (2012). Heavy metal pollution of soil and vegetable grown near roadside at Gazipur. *Bangladesh Journal of Agricultural Research, 37*(1), 9–17. https://doi.org/10.3329/bjar.v37i1.11170

Nassouhi, D., Ergönül, M. B., Fikirdeşici, Ş, Karacakaya, P., & Atasağun, S. (2018). Ağır metal kirliliğinin biyoremediasyonunda sucul makrofitlerin kullanımı. *Süleyman Demirel Üniversitesi Eğirdir Su Ürünleri Fakültesi Dergisi, 14*(2), 148–165.

Nedjimi, B. (2021). Phytoremediation: A sustainable environmental technology for heavy metals decontamination. *SN Applied Sciences, 3*(3), 286. https://doi.org/10.1007/s42452-021-04301-4

Pascariu, P., Cojocaru, C., Homocianu, M., Samoila, P., Dascalu, A., & Suchea, M. (2022). New $La^{3+}$ doped $TiO_2$ nanofibers for photocatalytic degradation of organic pollutants: Effects of thermal treatment and doping loadings. *Ceramics International, 48*(4), 4953–4964. https://doi.org/10.1016/j.ceramint.2021.11.033

Paul, A., Dey, S., Ram, D. K., & Das, A. P. (2023). Hexavalent chromium pollution and its sustainable management through bioremediation. *Geomicrobiology Journal*, 1–11. https://doi.org/10.1080/01490451.2023.2218377

Peng, B., Tang, J., Luo, J., Wang, P., Ding, B., & Tam, K. C. (2018a). Applications of nanotechnology in oil and gas industry: Progress and perspective. *The Canadian Journal of Chemical Engineering, 96*(1), 91–100. https://doi.org/10.1002/cjce.23042

Peng, W., Li, X., Xiao, S., & Fan, W. (2018b). Review of remediation technologies for sediments contaminated by heavy metals. *Journal of Soils and Sediments, 18*, 1701–1719. https://doi.org/10.1007/s11368-018-1921-7

Pradhan, G., Tripathy, B., Ram, D. K., Digal, A. K., & Das, A. P. (2023). Bauxite mining waste pollution and its sustainable management through bioremediation. *Geomicrobiology Journal*, 1–10. https://doi.org/10.1080/01490451.2023.2235353

Praveen, A., & Pandey, V. C. (2020). Pteridophytes in phytoremediation. *Environmental Geochemistry and Health, 42*(8), 2399–2411. https://doi.org/10.1007/s10653-019-00425-0

Qiu, L., Suo, C., Zhang, N., Yuan, R., Chen, H., & Zhou, B. (2022). Adsorption of heavy metals by activated carbon: Effect of natural organic matter and regeneration methods of the adsorbent. *Desalination and Water Treatment, 252*, 148–166. https://doi.org/10.5004/dwt.2022.28160.doi

Qu, J., Wang, Y., Tian, X., Jiang, Z., Deng, F., Tao, Y., Jiang, Q., Wang, L., & Zhang, Y. (2021). KOH-activated porous biochar with high specific surface area for adsorptive removal of chromium (VI) and naphthalene from water: Affecting factors, mechanisms, and reusability exploration. *Journal Hazard Material*, 401, 123292. https://doi.org/10.1016/j.hazmat.2020.123292

Reddy, K. R., Maturi, K., & Cameselle, C. (2009). Sequential electrokinetic remediation of mixed contaminants in low permeability soils. *Journal of Environmental Engineering, 135*(10), 989–998. https://doi.org/10.1061/(ASCE)EE.1943-7870.0000077

Redha, A. A. (2020). Removal of heavy metals from aqueous media by biosorption. *Arab Journal Basic Applications Science, 27*(1), 183–193. https://doi.org/10.1080/25765299.2020.1756177

Rezania, S., Taib, S. M., Din, M. F. M., Dahalan, F. A., & Kamyab, H. (2016). Comprehensive review on phytotechnology: Heavy metals removal by diverse aquatic plants species from wastewater. *Journal of Hazardous Materials, 318*, 587–599. https://doi.org/10.1016/j.jhazmat.2016.07.053

Rocha, W. D., Luz, J. A. M., Lena, J. C., & Bruna-Romero, O. (2006). Adsorçao de Cobre por Carvoes Ativados de Endocarpo de Noz de Macadamia e Semente de Goiaba. *Revista Escola De Minas, 59*(4), 409–414. https://doi.org/10.1590/S0370-44672006000400010

Salt, D. E., Smith, R. D., & Raskin, I. (1998). Phytoremediation. *Annual Review of Plant Biology, 49*(1), 643–668. https://doi.org/10.1146/annurev.arplant.49.1.643

Sarkar, C., Basu, J. K., & Samanta, A. N. (2017). Removal of $Ni^{2+}$ ion from waste water by geopolymeric adsorbent derived from LD Slag. *Journal Water Process Engineering, 17*, 237–244. https://doi.org/10.1016/j.jwpe.2017.04.012

Sarwar, N., Imran, M., Shaheen, M. R., Ishaque, W., Kamran, M. A., Matloob, A., Rehim, A., & Hussain, S. (2017). Phytoremediation strategies for soils contaminated with heavy metals: Modifications and future perspectives. *Chemosphere, 171*, 710–721. https://doi.org/10.3390/plants 11050597

Sasmaz, M., Topal, E. I. A., Obek, E., & Sasmaz, A. (2015). The potential of Lemna gibba L. and Lemna minor L. to remove Cu, Pb, Zn, and As in gallery water in a mining area in Keban, Turkey. *Journal of Environmental Management, 163*, 246–253. https://doi.org/10.1016/j.jemvman.2015.08.029

Sharma, S., & Malaviya, P. (2014). Bioremediation of tannery wastewater by chromium resistant fungal isolate Fusarium chlamydosporium SPFS2-g. *Current World Environment, 9*(3), 721. https://doi.org/10.12944/CWE.9.3.21

Shirkoohi, M. G., Tyagi, R. D., Vanrolleghem, P. A., & Drogui, P. (2022). A comparison of artificial intelligence models for predicting phosphate removal efficiency from wastewater using the electrocoagulation process. *Digital Chemical Engineering, 4*, Article 100043. https://doi.org/10.1016/j.dche.2022.1000043

Singh, P., Pani, A., Mujumdar, A. S., & Shirkole, S. S. (2024). The role of artificial intelligence in drying and biomass valorization in the field of phytoremediation of contaminated soils. *Drying Technology, 42*(6), 955–966. https://doi.org/10.1080/07373937.2024.2345123

Singh, P. K., & Tiwari, A. (2023). New paradigm in algae-based wastewater remediation. In *Bio and nanoremediation of hazardous environmental pollutants* (pp. 44–68). CRC Press.

Tarekegn, M. M., Hiruy, A. M., & Dekebo, A. H. (2021). Nano zero valent iron (nZVI) particles for the removal of heavy metals (Cd 2+, Cu 2+ and Pb 2+) from aqueous solutions. *RSC Advances, 11*(30), 18539–18551. https://doi.org/10.1039/D1RA01427G

Tiwari, A., & Marella, T. K. (2019). Potential and application of diatoms for industry-specific wastewater treatment. In *Application of microalgae in wastewater treatment: Volume 1: Domestic and industrial wastewater treatment* (pp. 321–339).

Vaksmaa, A., Guerrero-Cruz, S., Ghosh, P., Zeghal, E., Hernando-Morales, V., & Niemann, H. (2023). Role of fungi in bioremediation of emerging pollutants. *Frontiers in Marine Science, 10*, 1070905. https://doi.org/10.3389/fmars.2023.1070905

Van der Bruggen, B., & Vandecasteele, C. (2003). Removal of pollutants from surface water and groundwater by nanofiltration: Overview of possible applications in the drinking water industry. *Environmental Pollution, 122*(3), 435–445. https://doi.org/10.1016/S0269-7491(02)00308-1

Velarde, L., Nabavi, M. S., Escalera, E., Antti, M. L., & Akhtar, F. (2023). Adsorption of heavy metals on natural zeolites: A review. *Chemosphere, 328*, Article 138508. https://doi.org/10.1016/j.chemosphere.2023.138508

Volesky, B. (2001). Detoxification of metal-bearing effluents: Biosorption for the next century. *Hydrometallurgy, 59*(2–3), 203–216. https://doi.org/10.1016/S0304-386X(00)00160-2

Wang, S., & Peng, Y. (2010). Natural zeolites as effective adsorbents in water and wastewater treatment. *Chemical Engineering Journal, 156*(1), 11–24. https://doi.org/10.1016/j.cej.2009.10.029

Wu, J., Wang, T., Wang, J., Zhang, Y., & Pan, W. P. (2021). A novel modified method for the efficient removal of Pb and Cd from wastewater by biochar: Enhanced the ion exchange and precipitation capacity. *Science of the Total Environment, 754*, Article 142150. https://doi.org/10.1016/j.scitotenv.2020.142150

Xia, F., Lao, J., Yu, R., Sang, X., Luo, J., Li, Y., & Wu, J. (2019). Ambient oxidation of Ti3C2 MXene initialized by atomic defects. *Nanoscale, 11*(48), 23330. https://doi.org/10.1039/C9NR07236E

Xie, S., Huang, L., Su, C., Yan, J., Chen, Z., Li, M., Du, M., & Zang, H. (2024). Application of clay minerals as adsorbents for removing heavy metals from the environment. *Green and Smart Mining Engineering, 1*(3), 249–261. https://doi.org/10.1016/j.gsme.2024.07.002

Xu, H., & Van Deventer, J. S. J. (2000). The geopolymerisation of alumino-silicate minerals. *International Journal of Mineral Processing, 59*, 247–266. https://doi.org/10.1016/S0301-7516(99)00074-5

Xu, P., Zeng, G. M., Huang, D. L., Feng, C. L., Hu, S., Zhao, M. H., Lai, C., Wei, Z., Huang, C., Xie, G. X., & Liu, Z. F. (2012). Use of iron oxide nanomaterials in wastewater treatment: A review. *Science of the Total Environment, 424*, 1–10. https://doi.org/10.1016/j.scitotenv.2012.02.023

Xu, Y., Lu, Y., Tian, C., & Yang, J. (2019). Application of combined remediation in heavy metal polluted soil. In *IOP conference series: Earth and environmental science* (Vol. 300, No. 3, pp. 032006). IOP Publishing.

Yang, J., Wei, W., Pi, S., Ma, F., Li, A., Wu, D., & Xing, J. (2015). Competitive adsorption of heavy metals by extracellular polymeric substances extracted from Klebsiella sp. J1. *Bioresource Technology, 196*, 533–539. https://doi.org/10.1016/j.biortech.2015.08.011

Yu, G., Wang, X., Liu, J., Jiang, P., You, S., Ding, N., Guo, Q., & Lin, F. (2021). Applications of nanomaterials for heavy metal removal from water and soil: A review. *Sustainability, 13*(2), 713. https://doi.org/10.3390/su13020713

Yu, S., Tang, H., Zhang, D., Wang, S., Qiu, M., Song, G., Fu, D., Hu, B., & Wang, X. (2022). MXenes as emerging nanomaterials in water purification and environmental remediation. *Science of the Total Environment, 811*, Article 152280. https://doi.org/10.1016/j.scitotenv.2021.152280

Yu, K., Li, Y., Cao, X., Wang, R., Zhou, L., Wu, L., He, N., Lei, J., Fu, D., Chen, T., & He, R. (2023). In-situ constructing amidoxime groups on metal-free g-$C_3N_4$ to enhance chemisorption, light absorption, and carrier separation for efficient photo-assisted uranium (VI) extraction. *Journal of Hazardous Materials, 460*, Article 132356. https://doi.org/10.1016/j.jhazmat.2023.132356

Yuan, X., Li, J., Lim, J. Y., Zolfaghari, A., Alessi, D. S., Wang, Y., Wang, X., & Ok, Y. S. (2023). Machine learning for heavy metal removal from water: recent advances and challenges. *ACS ES&T Water*, *4*(3), 820–836. https://doi.org/10.1021/acsestwater.3c00215

Zafar, S., Khan, M. K., Iqbal, N., & Perveen, S. (2024). Nanobioremediation of metal and salt contaminated soils. In *Nano-bioremediation for water and soil treatment* (pp. 259–290). Apple Academic Press.

Zhang, Y., Hu, J., Bai, J., Wang, J., Yin, R., Wang, J., & Lin, X. (2018). Arbuscular mycorrhizal fungi alleviate the heavy metal toxicity on sunflower (Helianthus annuus L.) plants cultivated on a heavily contaminated field soil at a WEEE-recycling site. *Science of the Total Environment, 628*, 282–290. https://doi.org/10.1016/j.scitoenv.2018.01.331

Zhao, H., Xia, B., Qin, J., & Zhang, J. (2012). Hydrogeochemical and mineralogical characteristics related to heavy metal attenuation in a stream polluted by acid mine drainage: A case study in Dabaoshan Mine, China. *Journal of Environmental Sciences, 24*(6), 979–989. https://doi.org/10.1016/S1001-0742(11)60868-1

Zhao, C., Zhou, J., Yan, Y., Yang, L., Xing, G., Li, H., Wu, P., Wang, M., & Zheng, H. (2021). Application of coagulation/flocculation in oily wastewater treatment: A review. *Science of the Total Environment, 765*, Article 142795. https://doi.org/10.1016/j.scitotenv.2020.142795

Zhao, C., Zhou, J., Yan, Y., Yang, L., Xing, G., Li, H., Wu, P., Wang, M., & Zheng, H. (2021a). Application of coagulation/flocculation in oily wastewater treatment: A review. *Science of the Total Environment, 765*, Article 142795. https://doi.org/10.1016/j.scitotenv.2020.142795

Zhao, Q., Li, X., Xiao, S., Peng, W., & Fan, W. (2021). Integrated remediation of sulfate reducing bacteria and nano zero valent iron on cadmium contaminated sediments, *Journal*

Zheng, J., Wang, P., Shi, H., Zhuang, C., Deng, Y., Yang, X., Huang, F., & Xiao, R. (2023). Quantitative source apportionment and driver identification of soil heavy metals using advanced machine learning techniques. *Science of the Total Environment, 873*, Article 162371. https://doi.org/10.1016/j.scitotenv.2023.162371

Zhu, F., Tan, X., Zhao, W., Feng, L., He, S., Wei, L., Yang, L., Wang, K., & Zhao, Q. (2022). Efficiency assessment of ZVI-based media as fillers in permeable reactive barrier for multiple heavy metal-contaminated groundwater remediation. *Journal of Hazardous Materials, 424*, Article 127605. https://doi.org/10.1016/jijhazmat.2021.127605

Zou, H., Zhao, J., He, F., Zhong, Z., Huang, J., Zheng, Y., Zhang, Y., Yang, Y., Yu, F., Bashir, M. A., & Gao, B. (2021). Ball milling biochar iron oxide composites for the removal of chromium (Cr(VI)) from water: Performance and mechanisms. *Journal of Hazardous Materials, 413*, Article 125252. https://doi.org/10.1016/j.hazmat.2021.125252

# Cutting-Edge Technologies for Heavy Metal Waste Management

**S. Charulatha and K. S. Thangamani**

## 1 Introduction

### *1.1 Overview of Heavy Metal Waste*

Heavy metal waste is a major hazard to both human health and the environment, owing to its cytotoxicity and persistence in ecological systems (Das et al., 2011; Ghosh & Das, 2018). Heavy metal contamination is caused by industrial wastewater, mine waste, landfill effluents, and e-waste, the latter of which is especially troublesome due to the rapid development of electronic waste disposal (Ayaz et al., 2023; Houessionon et al., 2021; Ghosh et al., 2016, 2023; Mohanty et al., 2016). Recent studies emphasize the negative impacts of heavy metals in aquatic ecosystems, emphasizing the necessity for efficient removal strategies such as physicochemical methods, utilizing adsorption, advanced oxidation processes (AOPs), and biochar (Aziz et al., 2023).

Landfill leachates are also a significant problem because they contain high quantities of hazardous heavy metals such as cadmium, zinc, and iron, which pose substantial environment and human health risks (Essien et al., 2022; Mohanty et al., 2018). Efforts to create environmentally acceptable remediation solutions are critical, and there is a rising interest in employing heavy metals salvaged from effluent as catalysts in chemical-based synthesis (Das & Poater, 2021). Overall, discussing heavy metal pollution needs a multifaceted strategy that incorporates cost-effective remediation

S. Charulatha (✉)
Department of Chemistry, PSGR Krishnammal College for Women, Coimbatore, Tamil Nadu, India
e-mail: charulatha@psgrkcw.ac.in

K. S. Thangamani
Department of Chemistry, Dr. N.G.P. Institute of Technology, Coimbatore, Tamil Nadu, India

 
A. P. Das and S. Karakuş (eds.), *Synergizing Sustainability for Integrated Waste Management*, Sustainable Environmental Waste Management Strategies,
https://doi.org/10.1007/978-3-032-03718-3_7

procedures and long-term reuse initiatives to offset its detrimental effects (Zhang et al., 2023).

## *1.2 Sources and Environmental Impact*

### 1.2.1 Industrial Processes

Heavy metal contamination from industrial operations is a major cause of environmental deterioration and health hazards. These large-scale manufacturing processes, which frequently release heavy metals into the surrounding environment, comprise a number of physical, chemical, and mechanical phases (Dey et al., 2023; Mohanty et al., 2017). As in gold mining, when mercury is used to extract gold, mining and metallurgical operations produce significant volumes of waste that are heavy-metal-laden, resulting in major contamination and mercury pollution in the atmosphere (Hoàng Ly et al., 2023; Priyadarsini & Das, 2025). Furthermore, steel manufacturing entails smelting iron ore, which can emit heavy metals such as cadmium and lead into the atmosphere and water (Nyiramigisha & Komariah Sajidan, 2021). The chemical production sector is also responsible for heavy metal pollution through operations such as manure and acid manufacture, both of which can produce heavy metals as unwanted by-products (Das et al., 2023). Additionally, plating and metal finishing processes produce hazardous waste including heavy metals, posing threats to both employees and the atmosphere (Balali-Mood et al., 2021). The adverse environmental effect of these industrial operations is serious, since heavy metals may accumulate in ecological systems, threatening the welfare of humans and biodiversity (Piwowarska et al., 2024). Recent research emphasizes the importance of green technology and bioremediation techniques in mitigating these consequences (Adnan et al., 2022) (Fig. 1).

### 1.2.2 Agricultural Practices

Agricultural activities contribute significantly to heavy metal contamination, owing to the use of fertilizers, irrigation, and pesticides with untreated municipal sewage or industrial wastewater (Beauchemin et al., 2020; Nanda, 2023). These techniques can cause heavy metals such as lead, cadmium and chromium to accumulate in agricultural soils, reducing plant development and affecting the food chain, thereby causing problems for humans and wildlife. For example, phosphorus fertilizers may bring cadmium into the soil, and insecticides may include heavy metals as contaminants (Priya et al., 2023; Wan et al., 2024). Irrigation with tainted water can also worsen soil pollution. Recent research emphasizes the need for sustainable farming methods, such as applying organic amendments and precision agriculture techniques, to reduce heavy metal pollution and increase environmental sustainability (Alengebawy et al., 2021; Rashid et al., 2023). However, the listed references do not include

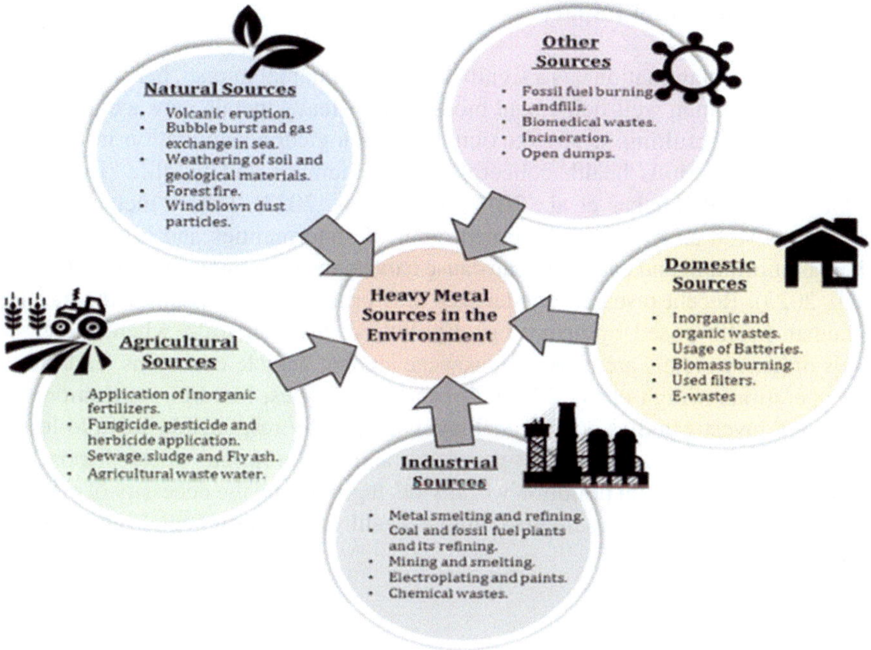


**Fig. 1** Sources of heavy metals. (Jayaumar et al., 2021)

thorough studies on agricultural techniques and their environmental implications. The majority of environmental science literature contains general information about environmental health and long-term sustainability (Bawa, 2023) (Fig. 2).

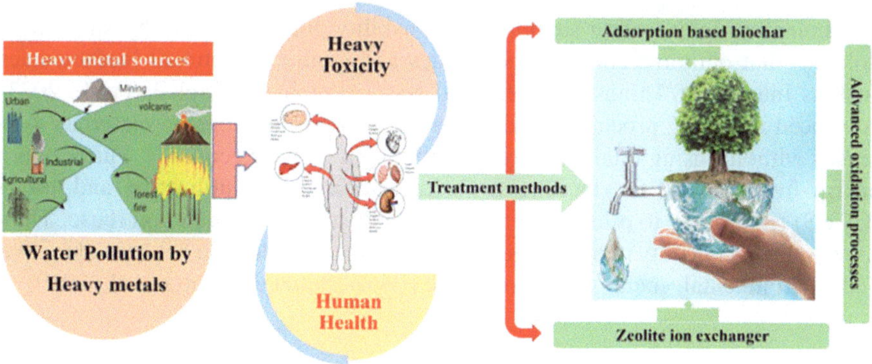


**Fig. 2** An overview of the sources of heavy metal pollution (Aziz et al., 2023)

### 1.2.3 Environmental Impact

Heavy metal contamination has a significant environmental impact, harming ecological systems, human well-being, and biodiversity. Heavy metals may accumulate in soil and water, resulting in phytotoxicity and biological accumulation in the food chain, causing serious health concerns to both humans and wildlife (Liu et al., 2023a, 2023b; Prabhakar et al., 2019; Raj & Das, 2023). Heavy metals such as mercury as well as lead can cause neurological abnormalities and organ damage, whereas chromium and cadmium can cause cancer and respiratory problems (Verma et al., 2023). Recent research emphasizes how critical it is to evaluate the hazards to human health posed by heavy metal pollution in urban soils, where increased levels of heavy metals are caused by sewage sludge, vehicle emissions, and industrial operations (Singh et al., 2024). Green technologies, like bioremediation, are also being investigated as economical and ecologically responsible ways to lessen these effects. Mining, industrial operations, and inappropriate waste disposal all contribute to heavy metal pollution worldwide, highlighting the necessity of efficient management techniques to safeguard human health and environmental sustainability (Ahamad et al., 2024).

## *1.3 Need for Advanced Waste Management Technologies*

The rising amount of waste in the world, which is expected to reach 3.4 billion tonnes per year by 2050, has made the need for sophisticated waste management systems imperative. Innovative solutions are needed to solve environmental and public health issues since traditional approaches are no longer enough (Tanveer et al., 2022). Automation and artificial intelligence (AI) technologies are revolutionizing waste sorting and recycling procedures by increasing accuracy, efficiency, and lowering contamination in recycling streams. Sustainability goals are aided by smart technologies that optimize collection schedules and reduce operational inefficiencies, such as Internet of Things (IoT)-enabled waste containers (Czekała et al., 2023). Advanced recycling processes, such as chemical recycling, enable materials to be reused without losing quality, therefore creating a circular economy. Furthermore, energy-from-waste (EfW) facilities are using carbon capture, utilization, and storage (CCUS) technology to minimize emissions while allowing for carbon reuse. These improvements not only reduce pollution but also meet worldwide net-zero objectives and environmental, social, and governance (ESG) standards (Pintor et al., 2019).

# 2 Biotechnological Approaches

## 2.1 *Bioremediation Using Microorganisms*

Bioremediation is an environmentally benign and cost-effective method that uses microorganisms including fungus, bacteria, and archaea to breakdown, detoxify, or regulate environmental contaminants (Das et al., 2015; Paul et al., 2023). This technique uses microorganisms' metabolic processes to convert dangerous compounds into less toxic or non-toxic forms, therefore repairing the natural environment (Thirumalaivasan et al., 2024). Recent advances emphasize the utilization of bacterial and fungal consortia to improve pollutant breakdown efficiency by combining their complementary metabolic processes. For example, white-rot fungi use their enzyme systems to break down complex organic pollutants such as polycyclic aromatic hydrocarbons (PAHs), whereas bacteria degrade hydrocarbons and heavy metals under a variety of environmental conditions (Sanjana et al., 2024; Ayilara et al., 2023).

Techniques such as ex situ bioremediation (off-site treatment) and in situ bioremediation (on-site treatment) are customized to individual pollutants and site circumstances, with bioventing, bioaugmentation, and biostimulation yielding promising outcomes (Suresh et al., 2024) (Das et al., 2024a; Pradhan et al., 2023). However, considerable obstacles remain, including the existence of refractory contaminants, environmental unpredictability, and the possible formation of hazardous intermediates (Das et al.,). Recent research has highlighted developments such as genetic engineering of microbial strains and immobilization strategies for improving microbial activity and stability in polluted settings (Verma et al., 2024). These developments highlight the crucial significance of bioremediation in solving global pollution concerns in a sustainable manner (Table 1).

## 2.2 *Phytoremediation Techniques*

Phytoremediation is an ecologically friendly and cost-effective technology that uses plants to remove toxins from air, water, and soil through processes including phytotransformation, phytoextraction, phytovolatilization, rhizofiltration, and phytostabilization (Park & Oh, 2023). Plants absorb, accumulate, or degrade contaminants such as heavy metals (cadmium, lead) and organic substances (pesticides, hydrocarbons) through their roots and related microbial populations. For example, hyperaccumulator plants: *Brassica juncea* (Indian mustard) take heavy metals from soil, but aquatic plants such as water hyacinth specialize in rhizofiltration of aquatic pollutants (Delgado-González et al., 2021). Recent improvements emphasize the incorporation of biodiversity protection and carbon sequestration into phytoremediation systems, which address global issues such as pollution management and climate change mitigation (Islam et al., 2024). Despite its benefits, phytoremediation is restricted due to problems such as sluggish plant development and weak root systems. Innovations

**Table 1** Microorganisms and their substrate utilization in different environments

| Microbial species | Occurrence | Organic substrate | References |
|---|---|---|---|
| *Pseudomonas aeruginosa* | Soil and water | Hydrocarbons (e.g. crude oil) | (Hu et al., 2023; Saha et al., 2021; Taffi et al., 2014) |
| *Bacillus subtilis* | Soil and aquatic environments | Hydrocarbons and heavy metals | (Akinsemolu et al., 2024; Karmakar et al., 2024; Wróbel et al., 2023) |
| *Burkholderia* sp. | Polluted soils | Nitrophenolic compounds | (Sharma et al., 2022; Dell' Anno et al., 2021; Liu et al., 2023a, 2023b) |
| *Aspergillus niger* (fungus) | Soil and decaying organic matter | Heavy metals (e.g. cadmium, lead) | (Dinakarkumar et al., 2024; Qader & Shekha, 2023; Vaksmaa et al., 2023) |
| *Alcanivorax borkumensis* | Marine environments | Petroleum hydrocarbons | (Ilhami & Holifah, 2022; Prasad et al., 2023) |
| *Chlorella vulgaris* (algae) | Freshwater | Heavy metals and organic pollutants | (Salgueiro et al., 2016; Sharma et al., 2021 |
| *Pseudomonas putida* | Soil | Toluene, naphthalene | Visvanathan et al., 2024; Maqsood et al., 2024) |
| *Deinococcus radiodurans* | Radioactive waste environments | Ionic mercury, toluene, uranium | (Li et al., 2021; Prakash et al., 2013) |
| *Phanerochaete chrysosporium* (fungus) | Soil and decaying organic matter | Pesticides, PAHs, dioxins | (Kalia et al., 2022; Konan et al., 2024) |
| *Methylibium petroleiphilum* | Contaminated groundwater | Methyl tert-butyl ether (MTBE) | (Hasan et al., 2024; Kane et al., 2007) |
| *Paracoccus denitrificans* | Aquatic and terrestrial environments | Ammonium, nitrate | (Azubuike et al., 2016; Zhao et al., 2020) |

such as plant genetic engineering and increased microbial interactions are being researched to improve efficiency.

# 3 Nanotechnology in Heavy Metal Waste Treatment

## *3.1 Nanomaterials for Heavy Metal Adsorption*

Recent improvements in nanomaterials for sorption of heavy metals highlight their effectiveness owing to their vast surface area, tunable chemistry, and functionalization abilities. Carbon-based materials, like carbon nanotubes (CNTs) and graphene oxide (GO), have remarkable adsorption capacities, particularly when functionalized with groups like carboxyl (−COOH) or amino ($-NH_2$), improving the selectivity for metals such as Hg(II), Cd(II), and Pb(II) through powerful metal–ligand complexes (Olawade et al., 2024). Metal oxide nanoparticles, such as $Fe_3O_4$ and $ZnFe_2O_4$, have good removal efficiency for Pb(II), Cd(II), and Cu(II), whereas polymer-based nanomaterials such as hydrogels and nanofibers provide customizable surface chemistry for selective adsorption. Emerging designs such as nanocages and nanosponges have porous designs that facilitate interaction with contaminants, allowing for faster adsorption kinetics (Silva et al., 2025). Recent research emphasizes multifunctional hybrid solutions, such as GO–magnetic nanoparticle composites, which combine high adsorption capacity with higher regeneration potential, maintaining more than 80% efficiency after repeated cycles (Karnwel & Malik, 2024). These advances highlight the significance of functionalization techniques and scalable synthesis methodologies for practical uses in water purification (Mishra et al., 2022).

## 4 Electrochemical and Advanced Oxidation Processes

### *4.1 Electrocoagulation and Electrochemical Reduction*

Electrocoagulation (EC) is a potential electrochemical technology for the treatment of wastewater that may successfully remove contaminants such as heavy metals, microplastics, and organic pollutants. Recent studies have proved its efficiency in removing over 90% of microbeads from synthetic wastewater under optimized settings, such as pH 7.5 and low current density, which makes it energy-efficient and cost-effective. Furthermore, EC has demonstrated substantial effectiveness in treating industrial wastewater, such as paint production wastewater, with high removal efficiency for chemical oxygen demand (COD) and total organic carbon (TOC) while producing minimal sludge (Cho, 2024). This ecologically friendly approach removes the need for chemical coagulants by producing in situ coagulants, making it ideal for a wide range of applications (Perren et al., 2018).

### *4.2 Photocatalysis for Heavy Metal Removal*

Photocatalysis uses light-activated catalysts to break down contaminants and remove heavy metals from effluent. Titanium dioxide ($TiO_2$) is the most researched photocatalyst. However, its effectiveness under visible light can be increased by doping with metals like silver or non-metals like nitrogen (Song et al., 2023). Recent research has expanded to new materials, such as graphitic carbon nitride (g-$C_3N_4$) and bismuth oxychloride (BiOCl), with greater photocatalytic activity and charge separation under visible light. These developments aid in the removal of heavy metals and organic contaminants from agricultural and industrial wastewater. Photocatalysis is a sustainable and environmentally benign process that uses solar energy and poses no secondary pollution risks (Sanni et al., 2024).

### *4.3 Fenton and Electro-Fenton Techniques*

The Fenton process oxidizes organic pollutants by generating hydroxyl radicals (·OH) from salts of iron and hydrogen peroxide. Electro-Fenton improves the process by continually creating $H_2O_2$ at the cathode, resulting in greater removal effectiveness for COD, biochemical oxygen demand (BOD), and other pollutants compared with traditional Fenton processes. Recent research on industrial wastewater treatment indicated that electro-Fenton accomplished up to 85% BOD removal while being far less expensive than the Fenton method, with operating costs of \$2.06/$m^3$ against \$3.12/$m^3$ (Ahmed et al., 2022). These procedures are extremely efficient, controlled, and environmentally benign, making them perfect for treating complicated industrial effluents.

## 5 Adsorption and Sorption Technologies

### *5.1 Activated Carbon and Biochar Adsorbents*

Activated carbon is widely used in wastewater treatment because of its huge pore size and microporous structure, which allow for effective adsorption of organic pollutants and heavy metals. However, its manufacture is energy-intensive and expensive (Moheb et al., 2025). Biochar, a novel adsorbent formed from biomass pyrolysis, has several environmental benefits, including carbon sequestration and cheaper manufacturing expenses. Recent investigations have shown that modified biochar treated with acids or alkalis improves adsorption capacity by enhancing surface area and functional group density. For example, N-doped biochar synthesized from spruce bark waste has a 22.6% higher adsorption capacity than non-modified biochar for dye removal from effluent (Kosar et al., 2024). Furthermore, sludge biochar utilized in engineered wetlands has accomplished up to 95.85% removal of nitrogen while producing sustainable energy.

### *5.2 Zeolites and Clay Minerals*

Zeolites are crystalline aluminosilicate minerals with strong cation-exchange capacities and interconnected pore architectures, making them good for adsorbing heavy metals, ammonia, and organic pollutants (Kyziol-Komosinska et al., 2025). Synthetic zeolites, such as Na-X, have higher adsorption capacities than natural zeolites like clinoptilolite and bentonite. Modified zeolites treated with surfactants such as hexadecyltrimethylammonium bromide (HDTMA-Br) improve their capacity to adsorb anionic and non-polar organics. These developments position zeolites as adaptable materials for long-term water purification (Manoharmayum & Ningombam, 2024).

### *5.3 Polymer-Based Adsorption Systems*

The capacity of polymer-based adsorbents to extract contaminants selectively and their versatile chemical properties have made them appealing. Polymers such as ion-exchange resins and functionalized hydrogels are designed to target certain contaminants via electrostatic interactions or hydrogen bonding. Recent research has focused on adding nanoparticles or functional groups into matrices of polymers to boost adsorption effectiveness (Alkhaldi et al., 2024). For example, composite polymers containing activated carbon or biochar improve pollutant binding while retaining structural integrity. These systems have features including reusability and scalability, making them ideal for wastewater treatment in manufacturing applications. While investigation into polymer-based systems continues, they are increasingly seen as effective replacements to classic adsorbents such as activated carbon and zeolites (Sheraz et al., 2025).

## 6 Membrane-Based Separation Technologies

### *6.1 Reverse Osmosis (RO) and Ultrafiltration (UF)*

Reverse osmosis (RO) and Ultrafiltration (UF) are two widely used membrane techniques for wastewater treatment. RO is very successful at eliminating dissolved salts and contaminants, resulting in substantial water recovery rates and high-quality water (Li et al., 2024). Recent advances in RO include the creation of more effective membranes and energy retrieval devices, which lower operational costs. In contrast, ultrafiltration is employed as a pre-treatment step to remove bigger particles, bacteria, and viruses, with efficiency often exceeding 90% for suspended solids and organic materials. Both technologies are essential in hybrid systems that mix several membrane types to improve treatment efficiency and flexibility to diverse wastewater types (Duan et al., 2021).

### *6.2 Electrodialysis for Heavy Metal Removal*

Electrodialysis is a membrane-based method that uses an electric field to extract ions, particularly heavy metals, from wastewater. This procedure is effective for specific ion removal and has the potential to be more energy efficient than other methods. Recent research has focused on optimizing electrodialysis systems for heavy metal elimination, particularly the use of ion-selective membrane to improve selectivity while reducing energy usage (Hassan, 2024). Electrodialysis is especially beneficial for treating effluent with high levels of certain ions, providing a cost-effective alternative for industries looking to recover precious metals or comply with severe discharge rules and regulations.

### *6.3 Hybrid Membrane Systems*

Hybrid membrane systems integrate multiple membrane technologies, including ultrafiltration (UF), microfiltration (MF), and reverse osmosis (RO), to provide complete pollutant removal. These systems are very versatile and may be adjusted to treat a range of effluent. Recent innovations have focused on combining membrane bioreactors (MBRs) with different technologies in order to improve biological treatment efficiency and create high-quality wastewater suitable for reuse (Duan et al., 2021; Qasem et al., 2021). Hybrid systems have greater removal efficiency for a variety of contaminants, including organic matter, nutrients, and heavy metals, thereby making them excellent for complicated wastewater treatment applications.

## 7 Chemical Precipitation and Coagulation Techniques

Chemical methods for heavy metal removal from wastewater have been widely used since the early industrial practices. This section explores various chemical-based techniques, including precipitation, flocculation, and coagulation.

Chemical precipitation, also known as coagulation–precipitation, is a widely adopted and well-established method in industrial wastewater treatment. It effectively removes dissolved heavy metal ions by converting them into insoluble solid particles, allowing for easier separation through sedimentation. This process involves the addition of a coagulant, which facilitates metal ion precipitation by altering pH levels, electrochemical potential, or through co-precipitation mechanisms. Once the metals form solid precipitates, they are typically removed through sedimentation or filtration. A schematic representation of the chemical precipitation process is illustrated in the accompanying Fig. 3.

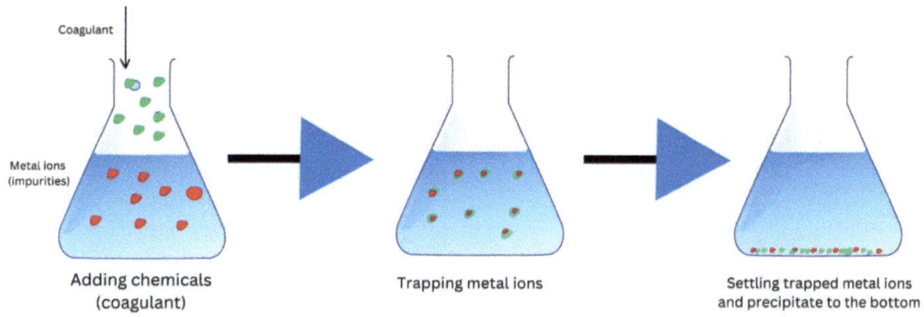


**Fig. 3** Schematic representation of chemical precipitation process

## *7.1 Lime and Sulphide Precipitation*

### 7.1.1 Lime Precipitation

Lime precipitation is a commonly used, cost-effective, and highly efficient technique for removing heavy metals from wastewater. Lime treatment in wastewater results in the formation of calcium carbonate, which functions as a coagulant for removing hardness and suspended particles. However, effective management is essential to regulate pH levels, minimize sludge production, and prevent scaling issues. It is commonly used alongside other coagulants because, when applied alone, it requires significantly larger quantities to achieve optimal effectiveness. Additionally, lime tends to produce a higher volume of sludge compared to other coagulants, making sludge management more challenging. Ensuring a well-designed treatment system and consistent monitoring helps achieve optimal performance across various industrial applications.

### 7.1.2 Sulphide Precipitation

Sulphide precipitation effectively reduces residual metal concentrations to $< 0.01$ mg/L, though it can be affected by the presence of chelating agents. The final metal concentration in this process is closely linked to pH (Fig. 3), with sulphide precipitation proving significantly more efficient than hydroxide precipitation. As illustrated in Fig. 1, sulphides exhibit much lower solubility compared to hydroxides and are also soluble under acidic conditions. A key advantage of sulphide precipitation is its ability to recover metals, as certain metals can be selectively precipitated by stepwise pH adjustments (Lewis, 2010). Table 2 demonstrates the low solubility of most transition metal sulphides, such as CdS, FeS, and ZnS, while lead sulphide is only sparingly soluble. In contrast, most metal hydroxides exhibit only slight solubility (Prokkola et al., 2020).

Sulphide precipitation can be carried out using solid sources (FeS, CaS, and $Na_2S$), aqueous solutions ($Na_2S$, NaHS, and $NH_4S$), or most commonly, gaseous

**Table 2** Theoretical values for metal sulphide and hydroxide precipitation

| Metal | Solubility product of metal sulphide (log $K_{sp}$) | Solubility product of metal hydroxide (log $K_{sp}$) |
|---|---|---|
| Ag | – 49.2 | – 7.9 |
| Cd | – 22.9 | – 14.5 |
| Cu | – 35.9 | – 19.8 |
| Hg | – 52.2 | – 25.4 |
| Pb | – 28.1 | – 19.9 |
| Sn | – 27.5 | – 26.3 |
| Mn | – 13.3 | – 12.7 |
| Ni | – 21.0 | – 15.3 |
| Fe | – | – 38.6 |
| Co | – 22.1 | – 14.5 |
| Fe(II) | – 18.8 | – 16.3 |
| Zn | – 24.5 | – 16.1 |

sulphide sources such as hydrogen sulphide ($H_2S$) (Li et al., 2020; Marchioretto et al., 2005). To enhance the techno-economic feasibility of metal purification, metals can be more effectively precipitated using a hydrogen sulphide solution (HS); it has the potential to emit foul-smelling and poisonous hydrogen sulphide gas. For these reasons, sulphide precipitation is suggested to be carried out at a neutral pH (Wimalaweera et al., 2024) The highly toxic $H_2S$ gas traditionally used for metal sulphide precipitation can be replaced by its conversion into soluble $H_2S$ or solid $Na_2S$, offering a safer alternative. A key aspect of sustainable development in this process is metal recovery and recycling, which helps minimize environmental impact. The use of $HS^-$ ions generated through anaerobic processes and recycled for metal precipitation from acidic wastewater before anaerobic treatment is currently under investigation in several studies. This approach has the potential to improve material efficiency in wastewater treatment.

## 7.2 *Flocculation and Coagulation Methods*

Coagulation involves destabilizing colloidal particles by neutralizing the repulsive forces that prevent them from clustering together, while flocculation refers to the process of aggregating these destabilized particles into larger clusters (Ibarra-Rodríguez et al., 2017). This process involves the addition of coagulants, such as aluminium sulphate, ferrous sulphate, and ferric chloride, which are commonly used to neutralize the charges of dissolved ions, promoting particle destabilization. Flocculation further enhances this process by binding the destabilized particles into larger

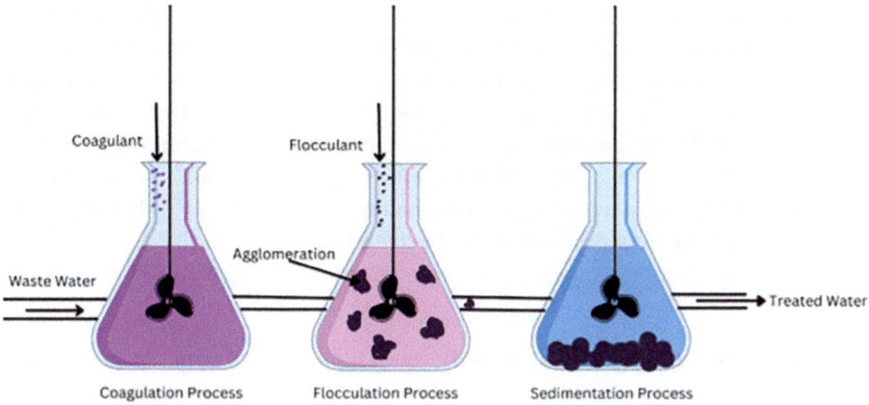


**Fig. 4** Schematic representation of coagulation and flocculation process

agglomerates with the aid of flocculants like polyferric sulphate (PFS), polyaluminium chloride (PAC), polyacrylamide (PAM), and other high-molecular-weight compounds (Ayach et al., 2024). Among various flocculants, polyethylene (PE) has been recognized as highly effective; however, the resulting sludge may contain toxic components. Additionally, most synthetic flocculants are non-biodegradable, raising environmental concerns regarding their long-term impact (Nourani et al., 2016) (Fig. 4).

Some drawbacks of using inorganic coagulants include their potential toxicity and health risks, as well as the generation of large volumes of sludge that require proper disposal. Additionally, these coagulants are selective in removing specific metals and may not be effective against emerging contaminants. They can also contribute to increased effluent colouration, reducing overall water clarity. While natural coagulants offer an eco-friendly alternative, they often exhibit lower efficiency in certain applications. Furthermore, scaling up the coagulation process can be challenging due to its complexity and operational limitations.

## 8 Solidification and Stabilization Methods

Solidification/stabilization (S/S) is a frequently used method to combat heavy metal pollution. Currently, it is the most efficient method to treat heavy metal pollution in solid wastes (Luo et al., 2020). Solidification refers to the physical process of converting waste into a solid form by incorporating it into a binder matrix. This process reduces the surface area exposed to the environment and minimizes the potential for contaminant dispersion. Historically, S/S treatment has been the last step in pollutant treatment before their disposal. Such methods reduce heavy metal efficacy by trapping or binding them to a physical mass (solidification) and structurally weakening them (stabilization). Stabilization involves chemical reactions between the contaminants and added reagents. These reactions often result in the formation of insoluble compounds, effectively reducing the solubility and mobility of the contaminants. Various bioavailable materials such as fly ash and rice husk ash have been employed for stabilization (Li et al., 2022). Synthetic compounds like ordinary Portland cement (OPC), along with natural and synthetic zeolites in different proportions, are also used (Ahmad et al., 2024). Non-metallic minerals often find use as stabilizers because of their low cost and non-pollutive nature. Using a variety of S/S processes, a high degree of leaching prevention can be achieved.

## *8.1 Cement-Based Stabilization*

Cement-based stabilization is a subset of S/S methods that exploits the well-known hydration reactions of cement. When water is added to cement, a series of exothermic hydration reactions occur, leading to the formation of calcium silicate hydrate (C–S–H) and other crystalline compounds. These alkaline compounds excel in adsorbing heavy metals through hydrolysis. These compounds not only provide mechanical strength but also serve to chemically bind contaminants within a rigid matrix. OPC has been widely used as a S/S agent due to its wide availability and strength. Hardened cement forms a physical barrier around heavy metals and contaminants that prevent leaching and diffusion into the environment. Soils treated using OPC have shown promising levels in leaching for Cu, Cd, Pb, and Ni, even 17 years after stabilization. However, production of cement is a major stakeholder in greenhouse gas emissions, which contributes to almost 5–7% of the global $CO_2$ emissions. Sustainable attempts have been made to reduce the use of OPC by substituting it with additives like red mud and gypsum (Geng et al., 2019; Zeng et al., 2022). Other minerals with equivalent stabilizing properties, viz. cement kiln dusts and Portland limestone cements (PLC), have shown promising results in usage (Jihad et al., 2023; Bazne et al., 2017; Arulrajah et al., 2017). Such supplementary cementitious materials (SCMs) can help optimize the S/S process by decreasing porosity and enhancing density, which directly makes the matrix less vulnerable to leaching. Cement-based stabilization has also been used in situ for neutralizing nuclear contamination in soil (Sukmak et al., 2017). Further studies into nanoparticle additives have revealed highly positive results (Ye et al., 2022). Nanoparticles can fill nanovoids within the matrix, further reinforcing the mechanical strength.

## *8.2 Encapsulation of Heavy Metals*

Encapsulation is a crucial process in the S/S method, which aims to bind heavy metal ions and isolate them in a matrix. Encapsulation can take place either physically or chemically within a medium depending on how the contaminants are entrapped. Physical entrapment methods involve cement- or polymer-based binders that surround the ions. This process hinders pore connectivity and diffusion paths within the matrix. Chemical encapsulation takes this a step further by introducing chemical reactions between the binding medium and the contaminant. For example, high pH environment from the hydration of cement acts on soluble ions and convert them to less soluble forms (Chen et al., 2023). Recent developments propose a hybrid approach by incorporating additives like fly ash and slag into the binder. These additives strengthen the physical structure while also promoting chemical stabilization via pozzolanic reactions. Geopolymers, formed by the alkali-activation of aluminosilicate materials, have also gained attention as sustainable alternatives to conventional cement. Their three-dimensional network structure efficiently traps impurities while offering higher chemical resistance and lower permeability than typical cementitious systems.

### *8.3 Long-Term Stability and Environmental Concerns*

The effectiveness of S/S methods has been under intense scrutiny. Comprehensive research of laboratory experiments combined with in situ field studies is required to map the feasibility. Compressive stress tests on the medium are crucial to determine its structural integrity after solidification. Integrity can be improved by introducing SCMs as discussed previously. Toxicity characteristic leaching procedure (TCLP) is a characteristic leaching test used to determine the level of immobilization of metal ions within the matrix (Chen et al., 2023). Long-term field studies suggest that fluctuations in groundwater chemistry, physical stress, and chemical exposure contribute to the detriment of stabilization mediums.

## 9 Waste-To-Resource Technologies

The recovery and recycling of heavy metals from industrial waste are crucial for environmental protection and resource sustainability. Industrial processes often generate waste containing valuable metals such as silver (Ag), molybdenum (Mo), nickel (Ni), cobalt (Co), gold (Au), copper (Cu), chromium (Cr), and zinc (Zn). Improper disposal of these wastes can lead to environmental contamination and health hazards; therefore, implementing effective recovery methods is essential.

### *9.1 Heavy Metal Recovery and Recycling*

The primary methods for recovering metals from industrial waste are:

Pyrometallurgical methods: These involve high-temperature processes such as calcining, roasting, smelting, and refining to extract metals from waste materials.

Hydrometallurgical methods: These utilize aqueous solutions containing lixiviants to leach metals from waste, followed by concentration and purification techniques like precipitation, cementation, solvent extraction, and ion exchange.

Biohydrometallurgical methods: This novel methodology uses microorganisms to extract metals from waste materials, providing an environmentally benign alternative to standard procedures (Priyadarsini & Das, 2024).

### 9.1.1 Benefits of Metal Recovery

Recovering metals from industrial waste offers several advantages:

- Environmental protection: It inhibits the discharge of hazardous metals into the environment, lowering soil and water pollution.
- Resource conservation: Recycling metals decreases the demand for mining new ores, preserving natural resources.
- Economic gains: Extracted metals can be reused in industrial processes, reducing raw material costs and contributing to a circular economy.

## *9.2 Circular Economy Approaches*

The circular economy paradigm is critical for decreasing waste and optimizing resources, particularly when ore grades fall. Advances in recycling strategies for critical materials like rare earth elements (REEs), cobalt, lithium, and nickel emphasize sustainable resource management to meet growing clean energy demands.

Traditional extraction and recycling rely on costly, environmentally harmful chemicals. By reintegrating recovered minerals into sectors such as electronics and renewable energy, a closed-loop system may improve scalability and efficiency while mitigating challenges like acid mine drainage. Recovering minerals from mine tailings and post-consumer recycling reduces reliance on virgin materials, promoting sustainability.

Despite progress, challenges remain in scaling up recycling to meet demand. The International Energy Agency highlights the need for increased supplies of critical materials for clean energy. Efficient closed-loop supply chains, technological advancements, and industry standards for eco-design are essential to overcoming barriers in extracting metals from complex components.

### *9.3 Industrial Applications of Recovered Metals*

Recovered metals are essential across various industries, promoting sustainability and reducing dependence on virgin materials. Recycled rare earth elements (REEs), gold, silver, and copper are crucial for circuit boards, semiconductors, and batteries, supporting the demand for smartphones, laptops, and other devices. Lithium, cobalt, and nickel from recovered sources power lithium-ion batteries for energy storage and electric vehicles (EVs), while REEs are vital for wind turbine magnets and solar panels. Recycled aluminium, steel, and platinum-group metals (PGMs) contribute to vehicle production and emission control. Recycled titanium, aluminium, and high-strength alloys lower costs and environmental impact in aircraft manufacturing, while REEs support defence technologies like radar and precision weapons.

By integrating recovered metals into industrial supply chains, industries can reduce environmental impact, lower production costs, and enhance resource efficiency, contributing to a more sustainable and circular economy.

## 10 Challenges and Future Directions

Despite the benefits, problems remain, such as developing cost-effective and ecologically friendly metal recovery technology. Ongoing research aims to enhance recovery efficiency and develop sustainable methods to address these challenges. In conclusion, the recovery and recycling of heavy metals from industrial waste are vital for sustainable development. By adopting advanced recovery methods and addressing existing challenges, industries can mitigate environmental impacts and promote resource efficiency.

## 11 Conclusion

The rapid emergence of cutting-edge technology has substantially increased the effectiveness as well as the sustainability of heavy metal waste handling. Nanotechnology, bioremediation, electrochemical techniques, and improved membrane filtration are among the emerging technologies for heavy metal pollution reduction. These methods improve metal removal efficiency while simultaneously promoting recovery of resources, cost-effectiveness, and protection of the environment. However, obstacles remain, including scalability, operating expenses, and regulatory compliance, necessitating further research and development. The combination of several methodologies, together with artificial intelligence and automated processes, shows considerable potential for optimizing waste treatment operations. Future efforts should concentrate on creating eco-friendly, economically feasible, and massive deployable alternatives for ensuring long-term environmental sustainability. Finally, improved hazardous waste management methods must be implemented in order to protect ecosystem and public health. Continued innovation and multidisciplinary collaboration will propel us towards a cleaner, more sustainable future.

## References

Adnan, M., Xiao, B., Xiao, P., Zhao, P., & Bibi, S. (2022). Heavy metal, waste, COVID-19, and rapid industrialization in this modern era—Fit for sustainable future. *Sustainability, 14*(8), 4746. https://doi.org/10.3390/su14084746

Ahamad, M. I., Yao, Z., Ren, L., Zhang, C., Li, T., & Lu, H. et al. (2024). Impact of heavy metals on aquatic life and human health: A case study of River Ravi Pakistan. *Frontiers in Marine Science, 11*. https://doi.org/10.3389/fmars.2024.1374835

Ahmad, S., Mohd, Syed, M., & Al-Osta, M. A. (2024) Utilization of fly ash with and without secondary additives for stabilizing expansive soils: A review. *Results in Engineering*, 102079–102079. https://doi.org/10.1016/j.rineng.2024.102079

Ahmed, M., Mavukkandy, M. O., Giwa, A., Elektorowicz, M., Katsou, E., & Khelifi, O. et al. (2022). Recent developments in hazardous pollutants removal from wastewater and water reuse within a circular economy. *npj Clean Water, 5*(1). https://doi.org/10.1038/s41545-022-00154-5

Akinsemolu, A. A., Onyeaka, H., Odion, S., & Adebanjo, I. (2024). Exploring bacillus subtilis: Ecology, biotechnological applications, and future prospects. *Journal of Basic Microbiology, 64*(6). https://doi.org/10.1002/jobm.202300614

Alengebawy, A., Abdelkhalek, S. T., Qureshi, S. R., & Wang, M.-Q. (2021). Heavy metals and pesticides toxicity in agricultural soil and plants: Ecological risks and human health implications. *Toxics, 9*(3), 42. https://doi.org/10.3390/toxics9030042

Alkhaldi, H., Alharthi, S., Alharthi, S., AlGhamdi, H. A., AlZahrani, Y. M., Mahmoud, S. A., et al. (2024). Sustainable polymeric adsorbents for adsorption-based water remediation and pathogen deactivation: A review. *RSC Advances, 14*(45), 33143–33190. https://doi.org/10.1039/d4ra05269b

Arulrajah, A., Mohammadinia, A., D'Amico, A., & Horpibulsuk, S. (2017). Effect of lime kiln dust as an alternative binder in the stabilization of construction and demolition materials. *Construction and Building Materials, 152*, 999–1007. https://doi.org/10.1016/j.conbuildmat.2017.07.070

Ayach, J., El Malti, W., Duma, L., Lalevée, J., Al Ajami, M., Hamad, H., et al. (2024). Comparing conventional and advanced approaches for heavy metal removal in wastewater treatment: An in-depth review emphasizing filter-based strategies. *Polymers, 16*(14), 1959. https://doi.org/10.3390/polym16141959

Ayaz, H., Nawaz, R., Nasim, I., Irshad, M. A., Irfan, A., & Khurshid, I., et al. (2023) Comprehensive human health risk assessment of heavy metal contamination in urban soils: Insights from selected metropolitan zones. *Frontiers in Environmental Science, 11*. https://doi.org/10.3389/fenvs.2023.1260317

Ayilara, M. S., & Babalola, O. O. (2023). Bioremediation of environmental wastes: The role of microorganisms. *Frontiers in Agronomy, 5*. https://doi.org/10.3389/fagro.2023.1183691

Aziz, K., Mustafa, F., Omer, K. M., Hama, S., Hamarawf, R. F., & Rahman, K. O. (2023). Heavy metal pollution in the aquatic environment: Efficient and low-cost removal approaches to eliminate their toxicity: A review. *Heavy Metal Pollution in the Aquatic Environment: Efficient and Low-Cost Removal Approaches to Eliminate Their Toxicity: A Review, 13*(26), 17595–17610. https://doi.org/10.1039/d3ra00723e

Azubuike, C. C., Chikere, C. B., & Okpokwasili, G. C. (2016). Bioremediation techniques–classification based on site of application: Principles, advantages, limitations and prospects.

*World Journal of Microbiology and Biotechnology, 32*(11). https://doi.org/10.1007/s11274-016-2137-x
Balali-Mood, M., Naseri, K., Tahergorabi, Z., Khazdair, M. R., & Sadeghi, M. (2021). Toxic mechanisms of five heavy metals: Mercury, lead, chromium, cadmium, and arsenic. *Frontiers in Pharmacology, 12*(643972). https://doi.org/10.3389/fphar.2021.643972.
Bawa, U. (2023). Heavy metals concentration in food crops irrigated with pesticides and their associated human health risks in Paki, Kaduna State, Nigeria. *Cogent Food and Agriculture, 9*(1). https://doi.org/10.1080/23311932.2023.2191889
Bazne, M. O. A., Howard, I. L., & Vahedifard, F. (2017). Stabilized very high–moisture dredged soil: Relative behavior of Portland-limestone cement and ordinary Portland cement. *Journal of Materials in Civil Engineering, 29*(9). https://doi.org/10.1061/(asce)mt.1943-5533.0001970
Beauchemin, M., Cohn, E., & Shelton, R. C. (2020). Implementation of clinical practice guidelines in the health care setting. *Advances in Nursing Science, 42*(4), 307–324. https://doi.org/10.1097/ans.0000000000000263
Chen, L., Nakamura, K., & Hama, T. (2023). Review on stabilization/solidification methods and mechanism of heavy metals based on OPC-based binders. *332*, 117362–117362. https://doi.org/10.1016/j.jenvman.2023.117362
Cho, K. T. (2024). Mini-review on electro-coagulation for wastewater treatment. *Civil Engineering Research Journal, 14*(3). https://doi.org/10.19080/cerj.2024.14.555889
Czekała, W., Drozdowski, J., & Łabiak, P. (2023). Modern technologies for waste management: A review. *Applied Sciences, 13*(15), 8847. https://doi.org/10.3390/app13158847
Das, T. K., & Poater, A. (2021). Review on the use of heavy metal deposits from water treatment waste towards catalytic chemical syntheses. *International Journal of Molecular Sciences, 22*(24), 13383. https://doi.org/10.3390/ijms222413383
Das, A. P., Sukla, L. B., Pradhan, N., & Nayak, S. (2011). Manganese biomining: A review. *Bioresource Technology, 102*(16), 7381–7387. https://doi.org/10.1016/J.BIORTECH.2011.05.018
Das, A. P., Ghosh, S., Mohanty, S., & Sukla, L. B. (2015). Advances in manganese pollution and its bioremediation. In L. B. Sukla, N. Pradhan, S. Panda, & B. K. Mishra (Eds.), *Environmental microbe biotechnology* (1st ed., pp. 313–328). Springer. https://doi.org/10.1007/978-3-319-19018-1_16
Das, S., Sultana, K. W., Ndhlala, A. R., Mondal, M., & Chandra, I. (2023). Heavy metal pollution in the environment and its impact on health: Exploring green technology for remediation. *Environmental Health Insights, 17*. https://doi.org/10.1177/11786302231201259.
Das, A., Dey, S., Bala, K., Gnanasekaran, R., & Das, A. P. (2024b). Microbial bioremediation technology for sustainable treatment and management of synthetic microfiber waste. *Sustainable microbial technology for synthetic and cellulosic microfiber bioremediation* (pp. 221–238). https://doi.org/10.1007/978-3-031-62600-5_12
Das, A., Dey, S., Bala, K., Gnanasekaran, R., & Das, A. P. (2024c). Microbial bioremediation technology for sustainable treatment and management of synthetic microfiber waste. *Sustainable microbial technology for synthetic and cellulosic microfiber bioremediation* (pp. 221–238). https://doi.org/10.1007/978-3-031-62600-5_12
Das, A., Dey, S., Bala, K., Gnanasekaran, R., & Das, A. P. (2024a). Microbial bioremediation technology for sustainable treatment and management of synthetic microfiber waste. *Sustainable microbial technology for synthetic and cellulosic microfiber bioremediation* (pp. 221–238). https://doi.org/10.1007/978-3-031-62600-5_12
Delgado-González, C. R., Madariaga-Navarrete, A., Fernández-Cortés, J. M., Islas-Pelcastre, M., Oza, G., Iqbal, H. M. N., et al. (2021). Advances and applications of water phytoremediation: A potential biotechnological approach for the treatment of heavy metals from contaminated water. *International Journal of Environmental Research and Public Health, 18*(10), 5215. https://doi.org/10.3390/ijerph18105215
Dell' Anno, F., Rastelli, E., Sansone, C., Brunet, C., Ianora, A., & Dell' Anno, A. (2021) Bacteria, fungi and microalgae for the bioremediation of marine sediments contaminated by petroleum

hydrocarbons in the omics era. *Microorganisms, 9*(8), 1695. https://doi.org/10.3390/microorganisms9081695
Dey, S., Tripathy, B., Kumar, M. S., & Das, A. P. (2023). Ecotoxicological consequences of manganese mining pollutants and their biological remediation. *Environmental Chemistry and Ecotoxicology*. https://doi.org/10.1016/J.ENCECO.2023.01.001
Dinakarkumar, Y., Gnanasekaran, R., Koteswara Reddy, G., Vasu, V., Balamurugan, P., & Murali, G. (2024). Fungal bioremediation: an overview of the mechanisms, applications and future perspectives. *Environmental Chemistry and Ecotoxicology, 6*. https://doi.org/10.1016/j.enceco.2024.07.002
Duan, X., Wang, C.-W., Wang, T., Xie, X., Zhou, X., & Ye, Y. (2021). Removal of metal ions in phosphoric acid by electro-electrodialysis with cross-linked anion-exchange membranes. *ACS Omega, 6*(48), 32417–32430. https://doi.org/10.1021/acsomega.1c03720
Essien, J. P., Ikpe, D. I., Inam, E. D., Okon, A. O., Ebong, G. A., & Benson, N. U. (2022). Occurrence and spatial distribution of heavy metals in landfill leachates and impacted freshwater ecosystem: An environmental and human health threat. *PLoS ONE, 17*(2), Article e0263279. https://doi.org/10.1371/journal.pone.0263279
Geng, Y., Wang, Z., Shen, L., & Zhao, J. (2019). Calculating of $CO_2$ emission factors for Chinese cement production based on inorganic carbon and organic carbon. *Journal of Cleaner Production, 217*, 503–509. https://doi.org/10.1016/j.jclepro.2019.01.224
Ghosh, S., Mohanty, S., Nayak, S., Sukla, L. B., & Das, A. P. (2016). Molecular identification of indigenous manganese solubilising bacterial biodiversity from manganese mining deposits. *Journal of Basic Microbiology, 56*(3), 254–262. https://doi.org/10.1002/JOBM.201500477
Ghosh, S., & Das, A. P. (2018). Metagenomic insights into the microbial diversity in manganese-contaminated mine tailings and their role in biogeochemical cycling of manganese. *Scientific Reports, 8*(1), 1–12. https://doi.org/10.1038/s41598-018-26311-w
Ghosh, S., Tripathy, B., Dey, S., & Das, A. P. (2023). Comparative investigation of fungal and bacterial manganese biomining mechanisms. In *Manganese mining microorganisms* (pp. 99–115). https://doi.org/10.1016/b978-0-12-822147-1.00016-9
Hasan, A., Elumalai, A., & Kumar, A. (2024). Advancements in bioremediation: Harnessing nanomaterials and microorganisms for environmental restoration by degradation of pollutants. *Environmental Quality Management, 34*(3). https://doi.org/10.1002/tqem.70018
Hassan, H. (2024). Machine learning application in municipal wastewater treatment to enhance the performance of a sequencing batch reactor wastewater treatment plant. *Environmental Science: Advances*. https://doi.org/10.1039/d4va00285g
Hoàng Ly, N., Joo, S.-W., Choo, J., Vasseghian, Y., Cho, J., & Rezania, S. (2023). Sustainable cutting-edge techniques for gold valorization from electronic wastes. *Chemical Engineering Journal, 471*, Article 144324. https://doi.org/10.1016/j.cej.2023.144324
Houessionon, M. G. K., Ouendo, E.-M.D., Bouland, C., Takyi, S. A., Kedote, N. M., Fayomi, B., et al. (2021). Environmental heavy metal contamination from electronic waste (E-Waste) recycling activities worldwide: A systematic review from 2005 to 2017. *International Journal of Environmental Research and Public Health, 18*(7), 3517. https://doi.org/10.3390/ijerph18073517
Hu, F., Wang, P., Li, Y., Ling, J., Ruan, Y., Yu, J., et al. (2023). Bioremediation of environmental organic pollutants by Pseudomonas aeruginosa: Mechanisms, methods and challenges. *Environmental Research, 239*, Article 117211. https://doi.org/10.1016/j.envres.2023.117211
Ibarra-Rodríguez, D., Lizardi-Mendoza, J., López-Maldonado, E. A., & Oropeza-Guzmán, M. T. (2017). Capacity of "nopal" pectin as a dual coagulant-flocculant agent for heavy metals removal. *Chemical Engineering Journal, 323*, 19–28. https://doi.org/10.1016/j.cej.2017.04.087
Ilhami, M. A., & Holifah, H. (2022). Marine bioremediation using Alcanivorax Borkumensis SK2 as a waste prevention oil industry on the Tunda Island to the impact of flood in Banten Region. *GMPI Conference Series* (Vol. 1, pp. 92–103). https://doi.org/10.53889/gmpics.v1.89

Islam, M. M., Saxena, N., & Sharma, D. (2024). Phytoremediation as a green and sustainable prospective method for heavy metal contamination: A review. *RSC Sustainability*. https://doi.org/10.1039/d3su00440f

Jayakumar, M., Surendran, U., Raja, P., Kumar, A., & Senapathi, V. (2021). A review of heavy metals accumulation pathways, sources and management in soils. *Arabian Journal of Geosciences, 14*(20). https://doi.org/10.1007/s12517-021-08543-9

Jihad, S. N., & Saeed, K. A. (2023). A review of solidification/stabilization of heavy metal contaminated soil. *Journal of Engineering and Sustainable Development, 27*(2), 227–244. https://doi.org/10.31272/jeasd.27.2.7

Kalia, A., Sharma, S., Semor, N., Babele, P. K., Sagar, S., & Bhatia, R. K., et al. (2022) Recent advancements in hydrocarbon bioremediation and future challenges: A review. *3 Biotech, 12*(6). https://doi.org/10.1007/s13205-022-03199-y

Kane, S. R., Chakicherla, A. Y., Chain, P. S. G., Schmidt, R., Shin, M. W., Legler, T. C., et al. (2007). Whole-genome analysis of the methyltert-butyl ether-degrading beta-proteobacterium methylibium petroleiphilum PM1. *Journal of Bacteriology, 189*(5), 1931–1945. https://doi.org/10.1128/jb.01259-06

Karmakar, D., Magotra, S., Negi, R., Kumar, S., Rustagi, S., & Singh, S. et al. (2024) Bacillus species for sustainable management of heavy metals in soil: Current research and future challenges. *Journal of Applied Biology and Biotechnology*. https://doi.org/10.7324/jabb.2024.157765

Karnwal, A., & Malik, T. (2024). Nano-revolution in heavy metal removal: Engineered nanomaterials for cleaner water. *Frontiers in Environmental Science, 12*. https://doi.org/10.3389/fenvs.2024.1393694

Konan, D., Ndao, A., Koffi, E., Elkoun, S., Robert, M., & Rodrigue, D. et al. (2024). Biodecomposition with Phanerochaete chrysosporium: A review. *AIMS Microbiology, 10*(4), 1068–1101. https://doi.org/10.3934/microbiol.2024046

Kosar, Fatah, N. M., & Muhammad, K. T. (2024) Advancements in application of modified biochar as a green and low-cost adsorbent for wastewater remediation from organic dyes. *Royal Society Open Science, 11*(5). https://doi.org/10.1098/rsos.232033

Kyziol-Komosinska, J., Pająk, M., Dzieniszewska, A., & Panek, R. (2025) Comparative study on cadmium ion removal from acidic solutions using zeolites and bentonite. *Scientific Reports, 15*(1). https://doi.org/10.1038/s41598-025-94903-4

Lewis, A. E. (2010). Review of metal sulphide precipitation. *Hydrometallurgy, 104*(2), 222–234. https://doi.org/10.1016/j.hydromet.2010.06.010

Li, X., Zhang, Q., & Yang, B. (2020). Co-precipitation with $CaCO_3$ to remove heavy metals and significantly reduce the moisture content of filter residue. *Chemosphere, 239*, Article 124660. https://doi.org/10.1016/j.chemosphere.2019.124660

Li, X., Zhang, X., Zhao, M., Zheng, X., Wang, Z., & Fan, C. (2024). Application of decentralized wastewater treatment technology in rural domestic wastewater treatment. *Sustainability, 16*(19), 8635–8635. https://doi.org/10.3390/su16198635

Li, S., Zhu, Q., Luo, J., Shu, Y., Guo, K., & Xie, J. et al. (2021). Application progress of Deinococcus radiodurans in biological treatment of radioactive uranium-containing wastewater. *Indian Journal of Microbiology, 61*(4). https://doi.org/10.1007/s12088-021-00969-9

Li, C., Wang, H., Liao, X., Xiao, R., Liu, K., & Bai, J. et al. (2022). Heavy metal pollution in coastal wetlands: A systematic review of studies globally over the past three decades. *Journal of Hazardous Materials, 424*(Part A), 127312. https://doi.org/10.1016/j.jhazmat.2021.127312

Liu, C.-J., Deng, S.-G., Hu, C., Gao, P., Khan, E., Yu, C.-P., et al. (2023a). Applications of bioremediation and phytoremediation in contaminated soils and waters: CREST publications during 2018–2022. *Critical Reviews in Environmental Science and Technology, 53*(6), 723–732. https://doi.org/10.1080/10643389.2023.2168365

Liu, D., Liu, Z., Wang, Y., & Zhou, L. (2023). Editorial: Understanding heavy metal pollution and control in the environment around metal tailings. *Frontiers in Environmental Science, 11*. https://doi.org/10.3389/fenvs.2023.1168949

Luo, Z., Tang, C., Hao, Y., Wang, Z., Yang, G., Wang, Y., et al. (2020). Solidification/stabilization of heavy metals and its efficiency in lead–zinc tailings using different chemical agents. *Environmental Technology, 43*(11), 1613–1623. https://doi.org/10.1080/09593330.2020.1845817

Manoharmayum, V. S., & Ningombam, D. (2024). Zeolites as versatile material for sustainable water purification: A review. *ResearchGate, 65*, 25–34. https://doi.org/10.6092/issn.2281-4485/20297

Maqsood, Q., Waseem, R., Sumrin, A., Wajid, A., Tariq, M. R., & Ali, S. W. et al. (2024). Recent trends in bioremediation and bioaugmentation strategies for mitigation of marine based pollutants: Current perspectives and future outlook. *Discover Sustainability, 5*(1). https://doi.org/10.1007/s43621-024-00607-6

Marchioretto, M. M., Bruning, H., & Rulkens, W. (2005). Heavy metals precipitation in sewage sludge. *Separation Science and Technology, 40*(16), 3393–3405. https://doi.org/10.1080/01496390500423748

Mishra, S., Dash, D., Al-Tawaha, A. R. M. S., & Das, A. P. (2022). A review on heavy metal ion adsorption on synthetic microfiber surface in aquatic environments. *Applied Biochemistry and Biotechnology, 194*(10), 4639–4654. https://doi.org/10.1007/S12010-022-04029-W/METRICS

Mohanty, S., Ghosh, S., Nayak, S., & Das, A. P. (2016). Isolation, identification and screening of manganese solubilizing fungi from low-grade manganese ore deposits. *Geomicrobiological Journal, 34*(4), 309–316. https://doi.org/10.1080/01490451.2016.1189016

Mohanty, S., Ghosh, S., Nayak, S., & Das, A. P. (2017). Bioleaching of manganese by Aspergillus sp. isolated from mining deposits. *Chemosphere, 172*, 302–309. https://doi.org/10.1016/J.CHEMOSPHERE.2016.12.136

Mohanty, S., Ghosh, S., Bal, B., & Das, A. P. (2018). A review of biotechnology processes applied for manganese recovery from wastes. *Reviews in Environmental Science and Bio/Technology, 17*(4), 791–811. https://doi.org/10.1007/S11157-018-9482-1

Moheb, M., El-Wakil, A. M., & Awad, F. S. (2025). Highly porous activated carbon derived from the papaya plant (stems and leaves) for superior adsorption of alizarin red s and methylene blue dyes from wastewater. *RSC Advances, 15*(1), 674–687. https://doi.org/10.1039/d4ra07957d

Nanda, S. (2023). Integrating traditional and contemporary systems for health and well-being. *Annals of Neurosciences, 30*(2), 77–78. https://doi.org/10.1177/09727531231185648

Nourani, M. R., Baghdadi, M., Javan, M., & Bidhendi, G. N. (2016). Production of a biodegradable flocculant from cotton and evaluation of its performance in coagulation-flocculation of kaolin clay suspension: Optimization through response surface methodology (RSM). *4*(2), 1996–2003. https://doi.org/10.1016/j.jece.2016.03.028

Nyiramigisha, P., Komariah, & Sajidan. (2021). Harmful impacts of heavy metal contamination in the soil and crops grown around dumpsites. *Reviews in Agricultural Science, 9*(0), 271–282. https://doi.org/10.7831/ras.9.0_271

Olawade, D. B., Wada, O. Z., Egbewole, B. I., Fapohunda, O., Ige, A. O., & Usman, S. O. et al. (2024). Metal and metal oxide nanomaterials for heavy metal remediation: Novel approaches for selective, regenerative, and scalable water treatment. *Frontiers in Nanotechnology, 6*. https://doi.org/10.3389/fnano.2024.1466721

Park, J. K., & Oh, K. (2023). Advancements in phytoremediation research for soil and water resources: Harnessing plant power for environmental cleanup. *Sustainability, 15*(18), 13901. https://doi.org/10.3390/su151813901

Paul, A., Dey, S., Ram, D. K., & Das, A. P. (2023). Hexavalent chromium pollution and its sustainable management through bioremediation. *Geomicrobiology Journal*, 1–11. https://doi.org/10.1080/01490451.2023.2218377

Perren, W., Wojtasik, A., & Cai, Q. (2018). Removal of microbeads from wastewater using electrocoagulation. *ACS Omega, 3*(3), 3357–3364. https://doi.org/10.1021/acsomega.7b02037

Pintor, L. L., Godezano, J. B., & Abris, M. I. V. (2019) A review of waste management technologies towards a circular economy in the Philippines. *Philippine Social Development and Resource Management Journal*, 2(1), 15. https://doi.org/10.36292/psdrmj.v2i1.52

Piwowarska, D., Kiedrzyńska, E., & Jaszczyszyn, K. (2024). A global perspective on the nature and fate of heavy metals polluting water ecosystems, and their impact and remediation. *Critical Reviews in Environmental Science and Technology*, 1–23. https://doi.org/10.1080/10643389.2024.2317112.

Prabhakar, A., Mishra, S., & Das, A. P. (2019). Isolation and identification of lead (Pb) solubilizing bacteria from automobile waste and its potential for recovery of lead from end of life waste batteries. *Geomicrobiological Journal, 36*(10), 894–903. https://doi.org/10.1080/01490451.2019.1654044

Pradhan, G., Tripathy, B., Ram, D. K., Digal, A. K., & Das, A. P. (2023). Bauxite mining waste pollution and its sustainable management through bioremediation. *Geomicrobiology Journal*, 1–10. https://doi.org/10.1080/01490451.2023.2235353

Prakash, D., Gabani, P., Chandel, A. K., Ronen, Z., & Singh, O. V. (2013). Bioremediation: A genuine technology to remediate radionuclides from the environment. *Microbial Biotechnology, 6*(4), 349–360. https://doi.org/10.1111/1751-7915.12059

Prasad, M., Obana, N., Lin, S.-Z., Zhao, S., Sakai, K., & Carles Blanch-Mercader et al. (2023). Alcanivorax borkumensis biofilms enhance oil degradation by interfacial tubulation. *Science, 381*(6659), 748–753. https://doi.org/10.1126/science.adf3345

Priya, A.K., Muruganandam, M., Ali, S. S., & Kornaros, M. (2023). Clean-up of heavy metals from contaminated soil by phytoremediation: A multidisciplinary and eco-friendly approach. *Toxics, 11*(5), 422–422. https://doi.org/10.3390/toxics11050422

Priyadarsini, S., & Das, A. P. (2024). Analytical and structural characterization of waste lithium-ion batteries for their effective recycling strategy. *Environmental Chemistry and Ecotoxicology*. https://doi.org/10.1016/j.enceco.2024.12.004

Priyadarsini, S., & Das, A. P. (2025). Lithium bioleaching: A review on microbial-assisted sustainable technology for lithium bio-circularity. *Journal of Water Process Engineering*, *69*, 106744. https://doi.org/10.1016/j.jwpe.2024.106744

Prokkola, H., Nurmesniemi, E.-T., & Lassi, U. (2020) Removal of metals by sulphide precipitation using Na2S and HS−-solution. *ChemEngineering*, *4*(3), 51–51. https://doi.org/10.3390/chemengineering4030051

Qader, M. Q., & Shekha, Y. A. (2023). Bioremediation of heavy metals by using Aspergillus Niger and Candida Albicans. *ZANCO Journal of Pure and Applied Sciences, 35*(3), 180–186. https://doi.org/10.21271/ZJPAS.35.3.16

Qasem, N. A. A., Mohammed, R. H., & Lawal, D. U. (2021). Removal of heavy metal ions from wastewater: A comprehensive and critical review. *npj Clean Water, 4*(1), 1–15. https://doi.org/10.1038/s41545-021-00127-0

Raj, K., & Das, A. P. (2023). Lead pollution: Impact on environment and human health and approach for a sustainable solution. *Environmental Chemistry and Ecotoxicology*, *5*, 79–85. https://doi.org/10.1016/J.ENCECO.2023.02.001

Rashid, A., Schutte, B. J., Ulery, A., Deyholos, M. K., Sanogo, S., Lehnhoff, E. A., et al. (2023). Heavy metal contamination in agricultural soil: Environmental pollutants affecting crop health. *Agronomy, 13*(6), 1521. https://doi.org/10.3390/agronomy13061521

Saha, L., Tiwari, J., Bauddh, K., & Ma, Y. (2021). Recent developments in microbe–Plant-based bioremediation for tackling heavy metal-polluted soils. *Frontiers in Microbiology, 12*. https://doi.org/10.3389/fmicb.2021.731723

Salgueiro, J. L., Pérez, L., Maceiras, R., Sánchez, A., & Cancela, A. (2016) Bioremediation of wastewater using chlorella vulgaris microalgae: Phosphorus and organic matter. *International Journal of Environmental Research, 10*(3), 465–470.

Sanjana, M., Prajna, R., Urvi, S. K., & Kavitha, R. V. (2024). Bioremediation—The recent drift towards sustainable environment. *Environmental Science Advances, 3*(8), 1097–1110. https://doi.org/10.1039/d3va00358b

Sanni, S. E., Oni, B. A., Okoro, E. E., & Pandya, S. (2024). Recent advances in the use of biogenic nanomaterials and photocatalysts for wastewater treatment: challenges and future prospects. *Frontiers in Nanotechnology, 6*. https://doi.org/10.3389/fnano.2024.1469309

Sharma, P., Bano, A., Singh, S. P., Dubey, N. K., Chandra, R., & Iqbal, H. M. N. (2022). Recent advancements in microbial-assisted remediation strategies for toxic contaminants. *Cleaner Chemical Engineering, 2*, Article 100020. https://doi.org/10.1016/j.clce.2022.100020

Sharma, R., Mishra, A., Pant, D., & Malaviya, P. (2021). Recent advances in microalgae-based remediation of industrial and non-industrial wastewaters with simultaneous recovery of value-added products. *Bioresource Technology, 126129*. https://doi.org/10.1016/j.biortech.2021.126129

Sheraz, M., Sun, X.-F., Siddiqui, A., Hu, S., & Song, Z. (2025). Research advances in natural polymers for environmental remediation. *Polymers, 17*(5), 559. https://doi.org/10.3390/polym17050559

Silva, M. D., Cao, G., & Michael. (2025). Nanomaterials for the removal and detection of heavy metals: A review. *Environmental Science Nano*. https://doi.org/10.1039/d4en01041h

Singh, V., Ahmed, G., Vedika, S., Kumar, P., Chaturvedi, S. K., Rai, S. N., et al. (2024). Toxic heavy metal ions contamination in water and their sustainable reduction by eco-friendly methods: Isotherms, thermodynamics and kinetics study. *Scientific Reports, 14*(1), 7595. https://doi.org/10.1038/s41598-024-58061-3

Song, J., Ashtar, M., Yang, Y., Liu, Y., Chen, M., & Cao, D. (2023). Photocatalytic removal of heavy metal ions and antibiotics in agricultural wastewater: A review. *Journal of Semiconductors, 44*(11), Article 111701. https://doi.org/10.1088/1674-4926/44/11/111701

Sukmak, P., Sukmak, G., Horpibulsuk, S., Setkit, M., Kassawat, S., & Arulrajah, A. (2017). Palm oil fuel ash-soft soil geopolymer for subgrade applications: Strength and microstructural evaluation. *Road Materials and Pavement Design, 20*(1), 110–131. https://doi.org/10.1080/14680629.2017.1375967

Suresh, R., Rajendran, S., Chen, W.-H., Soto-Moscoso, M., Sundaram, T., & Jalil, A. A. et al. (2024). A review on algae-mediated adsorption and catalytic processes for organic water pollution remediation. *Frontiers in Materials, 11*. https://doi.org/10.3389/fmats.2024.1432704

Taffi, M., Paoletti, N., Angione, C., Pucciarelli, S., Marini, M., & Liò, P. (2014). Bioremediation in marine ecosystems: a computational study combining ecological modeling and flux balance analysis. *Frontiers in Genetics, 5*. https://doi.org/10.3389/fgene.2014.00319

Tanveer, M., Khan, S. A. R., Umar, M., Yu, Z., Sajid, M. J., & Haq, I. U. (2022). Waste management and green technology: Future trends in circular economy leading towards environmental sustainability. *Environmental Science and Pollution Research, 29*(53), 80161–80178. https://doi.org/10.1007/s11356-022-23238-8

Thirumalaivasan, N., Gnanasekaran, L., Kumar, S., Durvasulu, R., Sundaram, T., & Rajendran, S. et al. (2024) Utilization of fungal and bacterial bioremediation techniques for the treatment of toxic waste and biowaste. *Frontiers in Materials, 11*. https://doi.org/10.3389/fmats.2024.1416445

Vaksmaa, A., Guerrero-Cruz, S., Ghosh, P., Zeghal, E., Hernando-Morales, V., & Niemann, H. (2023). Role of fungi in bioremediation of emerging pollutants. *Frontiers in Marine Science, 10*. https://doi.org/10.3389/fmars.2023.1070905

Verma, D. K., Verma, C., & Mahish, P. K.(2023). Heavy metals in the environment: Management strategies for global pollution. *American Chemical Society*. https://doi.org/10.1021/bk-2023-1456

Verma, S., Sahu, D., & Almutairi, B. O. (2024). Production and characterization of biodiesel fuel produced from third-generation feedstock. *Frontiers in Materials, 11*. https://doi.org/10.3389/fmats.2024.1454120

Visvanathan, P., Kirubakaran, D., Selvam, K., Manimegalai, P., Rajkumar, M., & Vasantharaj, K. (2024). Recent strategies for natural bioremediation of emerging pollutants: Development of a green and sustainable environment. ACS Symposium Series (pp. 1–19). https://doi.org/10.1021/bk-2024-1476.ch001

Wan, Y., Liu, J., Zhuang, Z., Wang, Q., & Li, H. (2024). Heavy metals in agricultural soils: Sources, influencing factors, and remediation strategies. *Toxics, 12*(1), 63–63. https://doi.org/10.3390/toxics12010063

Wang, L., Zeraatpisheh, M., Wei, Z., Xu, M. (2022). Heavy metal pollution and risk assessment of farmland soil around abandoned domestic waste dump in Kaifeng City. *Frontiers in Environmental Science, 10*. https://doi.org/10.3389/fenvs.2022.946298

Wimalaweera, I. P., Wei, Y., Zuo, F., Tang, Q., Ritigala, T., & Wang, Y. et al. (2024). Enhancing rubber industry wastewater treatment through an integrated AnMBR and A/O MBR system: Performance, membrane fouling analysis, and microbial community evolution. *Membranes, 14*(6), 130–130. https://doi.org/10.3390/membranes14060130

Wróbel, M., Śliwakowski, W., Kowalczyk, P., Kramkowski, K., & Dobrzyński, J. (2023). Bioremediation of heavy metals by the genus bacillus. *International Journal of Environmental Research and Public Health, 20*(6), 4964. https://doi.org/10.3390/ijerph20064964

Ye, S., Wang, L., & Liu, T. (2022). Study of solidification and stabilization of heavy metals by passivators in heavy metal-contaminated soil. *Open Chemistry, 20*(1), 1–9. https://doi.org/10.1515/chem-2021-0101

Zeng, J., Li, C., Wang, J., Tang, L., Wu, C., & Xue, S. (2022). Pollution simulation and remediation strategy of a zinc smelting site based on multi-source information. *Journal of Hazardous Materials, 433*, 128774–128774. https://doi.org/10.1016/j.jhazmat.2022.128774

Zhang, P., Yang, M., Lan, J., Huang, Y., Zhang, J., Huang, S., et al. (2023). Water quality degradation due to heavy metal contamination: Health impacts and eco-friendly approaches for heavy metal remediation. *Toxics, 11*(10), 828. https://doi.org/10.3390/toxics11100828

Zhao, Y., Lu, W., Liu, Y., Wang, J., Zhou, S., Mao, Y., et al. (2020). Efficient total nitrogen removal from wastewater by Paracoccus denitrificans DYTN-1. *Letters in Applied Microbiology, 70*(4), 263–273. https://doi.org/10.1111/lam.13268

# Innovative Strategies for Sustainable Waste Disposal

**Kiran Bala, Renuka Harit, Arushi Dogra, Raj Kumar, and Alok Prasad Das**

## 1 Introduction

For the conservation of nature and natural resources, proper management of waste disposal is an alarming need of the hour (Prabina et al., 2024; Satapathy et al., 2024). Population expansion threatens the natural environment and accelerates urbanization and industrialization (Kadakia et al., 2023). According to the World Bank's 2018 report on waste management, the global annual production of solid waste by humans is 2.01 billion metric tons. Waste generation per capita, that is, waste generated per individual per day, averages 0.74 kg but ranges widely, from 0.11 to 4.54 kg, pandemically, of which waste generation per capita from India ranges from 0.50 to 0.99 kg. By 2050, the amount of waste is expected to surge to 3.40 billion tonnes, that is a threat to the mother Earth and its life forms (Fig. 1).

Three broad categories of waste based on their physical states are liquid, solid, and gaseous waste (Amasuomo et al., 2016) (Fig. 2). Further classifications are stated in different settings of the world, such as based on source of waste (household/

K. Bala
Department of Zoology, Deshbandhu College, University of Delhi, New Delhi, India

R. Harit (✉)
ICMR-National Institute of Malaria Research, Dwarka, New Delhi, India
e-mail: renuka06witty@gmail.com

Academy of Scientific and Innovative Research (AcSIR), Ghaziabad, Uttar Pradesh, India

A. Dogra
Faculty of Life Sciences and Biotechnology, South Asian University, Delhi, India

R. Kumar
Department of Zoology, S. N. Sinha College, Tekari, Gaya, Bihar, India

A. P. Das
Department of Life Sciences, Rama Devi Women's University, Bhubaneswar, Odisha, India



A. P. Das and S. Karakuş (eds.), *Synergizing Sustainability for Integrated Waste Management*, Sustainable Environmental Waste Management Strategies,
https://doi.org/10.1007/978-3-032-03718-3_8

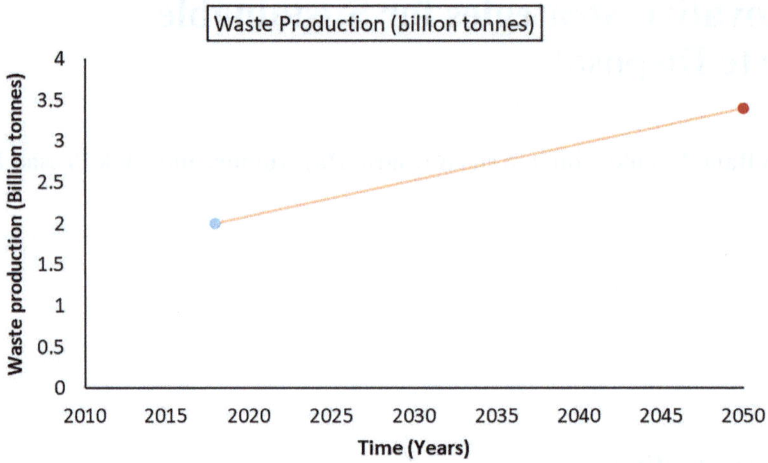


**Fig. 1** Graphical representation of waste production in billion tonnes from 2018 to 2050

domestic waste, industrial waste, agricultural waste, commercial waste, demolition and construction waste, mining waste) and environmental influence of the waste (hazardous and nonhazardous waste) (Amasuomo et al., 2016; Bal et al., 2018; Priyadarsini & Das, 2024).

Waste disposal is a critical component for environmental management. In developing countries, the most common practice of waste processing and management includes (i) landfilling, (ii) incineration, and (iii) vermicomposting. Sustainable waste disposal strategies involve the waste minimization and reuse or recycling of the resources (Mohammed et al., 2020; Visvanathan et al., n.d.). Compared to the

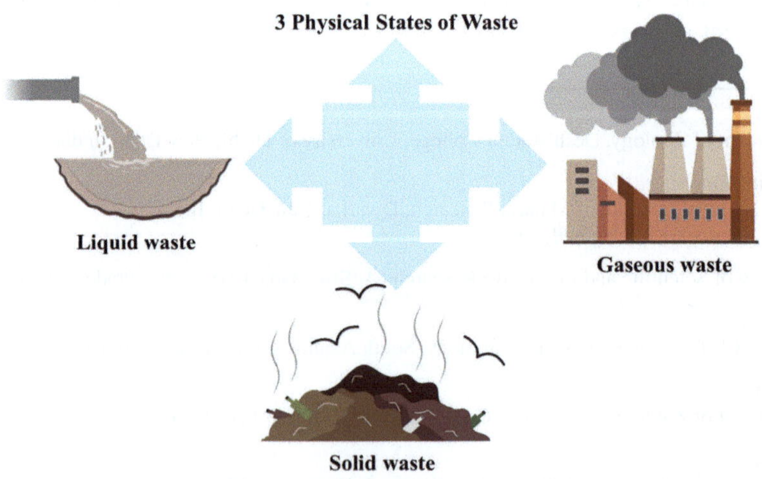


**Fig. 2** Pictorial representation of the three broad physical categories of waste

traditional linear economic model that dictates the concept of take-make-consume-throw away pattern, the circular economy emphasizes the sharing, leasing, reusing, repairing, refurbishing, and recycling existing materials and products to a greater extent (Voukkali et al., 2023; Das et al., 2024; Dey et al., 2024).

## 2 Waste Disposal and Sustainable Waste Disposal

Waste disposal refers to removing unwanted or useless items, materials, and substances that are no longer needed (Das et al., 2025; Kanungo et al., 2023). The different steps in the waste disposal ladder are collection, transportation, treatment, and safe disposal. Following Merriam-Webster's online dictionary definition, the noun "waste" refers to loss resulting from excessive consumption of something valuable or from its usage in an inefficient or unnecessary manner; something that is used or done that causes something important to be lost needlessly; and something that is useful is either not being used at all or is being used in an inappropriate or ineffective way (Thürer et al., 2017). Sustainable waste disposal refers to waste management that lowers and suppresses its environmental impact, restores resources, and aids long-term ecological balance (John & Das, 2024; Sahu et al., 2024).

## 3 Waste Disposal Methods

From conventional to contemporary methods of waste disposal, reducing, reusing, recycling, composting, incineration, and landfilling have remained constant. However, new innovative approaches that focus on the segregation of waste in the color-coded plastic bags and bins and usage of artificial intelligence (AI)-encoded smart bins are grabbing attention worldwide (Prabhakar et al., 2019; Pradhan et al., 2023).

### *3.1 Conventional Waste Disposal Methods*

#### 3.1.1 3R (Reduce, Reuse, and Recycle)

The Kyoto workshop on 3R (reduce, reuse, and recycle) and waste management was conducted in Kyoto, Japan, in October 2009 and aimed to gather comparative data on various institutional approaches to 3R and waste management. It featured the participation of policymakers and researchers from diverse regions, including the European Commission, Germany, England, Italy, Denmark, Sweden, the United States, Korea, China, Vietnam, and Japan (Sakai et al., 2011). Waste collection costs in many developing countries have faced significant challenges due to escalating overheads; the scarcity of land for landfills near urban areas is a pressing issue, compounded by high transportation costs and environmental concerns that restrict landfill development in remote locations (Memon et al., 2012). To address these challenges effectively, the most impactful solution is to reduce waste generation. Reuse is a nondestructive approach that aims at using of materials multiple times without the change of state. The materials can be used among acquaintances and be put to others in market for resale. The reuse of existing materials may facilitate the circumvention of hazardous processes within both societal and industrial contexts (Cooper et al., 2017). Recycle is the third pillar of 3R and is "the act of procuring and transforming materials that would otherwise be discarded as junk and transfigure them into new articles" (Ma et al., 2019). Recycling is a substantive means to have a sustainable planet (Ma et al., 2019). Jacobs states "reversal of waste to resources is possible only when external energy source is subjected." "Recycling" is not a self- driven process; it has to be powered by an energy source. This is because the embodied energy (known as energy) used in production is lost during the recycling process (King et al., 2006). One of the significant challenges associated with recycling is related to the first two laws of thermodynamics. According to these laws, no new creation and destruction of energy are possible; only transformation from one form to another is feasible. Furthermore, the second law indicates that energy transformations often require additional energy input, which can lead to increased energy consumption in the recycling process (King et al., 2006). Figure 3 represents the 3R followed in order to save ecosystem.

### 3.1.2 Composting

Composting is a mineralization process wherein a diverse range of aerobic and facultative microorganisms decompose organic materials into inorganic compounds. A significant end product of this process is stabilized humic acid or humus. Furthermore, during composting, various compounds such as carbon dioxide ($CO_2$), nitrates ($NO_3$), sulfates ($SO_4$), and phosphates ($PO_4$) are released, primarily in gaseous forms, under aerobic conditions (Ayilara et al., 2020). Different ways of composting exist, namely, Indian Bangalore composting, in-vessel composting, windrow composting, vermicomposting, static composting, sheet composting, Indian Indore composting, and Berkley rapid composting, and each has its pros and cons (Ayilara et al., 2020). Composting is a long-used technology for waste management, but it has challenges such as pathogen detection, low nutrient status, and long composting duration. Another crucial element to take into account for future development is compost process management. More information on managing enzymes, microbial communities, substrates, and processing conditions is necessary to efficiently limit methane and nitrous oxide emissions, since the potential for greenhouse gas formation during composting drives the need for appropriate management procedures.

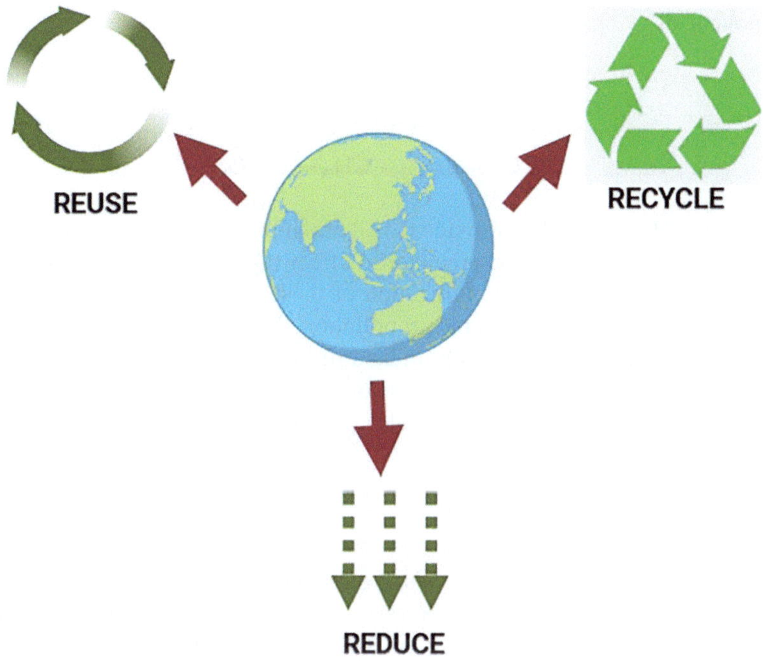


**Fig. 3** 3R (reduce, reuse, and recycle)

### 3.1.3 Landfills

Landfills are of the earliest methods of waste management. The origin of landfills dates to 1900s, with the advent of industrial revolution and urbanization. Landfills refer to the dumping and containment of the waste materials into the pits. Landfills can be broadly divided into the two categories: dumps and regulated landfills (Themelis et al., 2007). In the developing countries, the most common type is dump; the waste is deposited into the pits until it reaches the desirable maximum height and is covered with the soil. The regulated landfills, on the other hand, involve the covering of pits using some impermeable liners and caps. As the landfills reach to a maximum height, it is covered with an impervious layer that restricts the entry of rainwater into the pits and allows undisturbed bioreaction within the pits (Themelis et al., 2007). Soon after the waste deposition in the landfills, the biochemical reactions commence. Acetogenesis and methanogenesis are the two major anaerobic chemical reactions that take place inside the landfill (Themelis et al., 2007).

Acetogenesis

$$C_6H_{12}O_6 \rightarrow 2C_2H_5OH + 2CO_2$$

Methanogenesis

$$CH_3COOH \rightarrow CH_4 + CO$$

$$CO_2 + 4H_2 \rightarrow CH_4 + 2H_2O$$

Methane ($CH_4$) and carbon dioxide ($CO_2$) are the two major inflammable landfill gases (LFGs) and contribute to greenhouse effect (Qin et al., 2001; Danthurebandara et al., 2012). However, the utilization of these LFGs in a productive manner conserves the nature and reduces the air pollution. The LFGs are used in generating electricity. There are many drawbacks of landfills, namely requirement of large abandoned area for its construction, the release of volatile and nonvolatile gases, and the contamination of groundwater by leachates (Danthurebandara et al., 2012).

### 3.1.4 Incineration

Incineration involves four primary levels: pretreatment, combustion, energy recovery, and emission control. Incineration is a process involving the thermal transformation of materials at temperatures ranging from 800 to 1000 °C. This process oxidizes the organic components of waste into inorganic substances, primarily carbon dioxide and water. It has been extensively utilized in waste management, particularly in regions like Europe and Japan. However, on a global level, a relatively small fraction of municipal solid waste is treated through incineration, with landfilling still being the predominant method for over 80% of the world's population. The primary objectives of incineration are to eliminate organic materials through combustion for sanitation purposes and to alleviate the pressure on landfill space in densely populated urban areas (Brunner, 2012).

## *3.2 Contemporary Waste Disposal Methods*

### 3.2.1 Circular Economy

The circular economy (CE) is first appeared as an "umbrella" idea in 2010 to develop a more resource-efficient economic structure (Blomsma et al., 2017; Avilés-Palacios et al., 2021). It has been suggested as a remedy for the traditional, unsustainable "take-make-dispose" economic paradigm (Tisserant et al., 2017). Some definitions of CE misinterpret it as being restricted to recycling alone, whereas the larger goal is to concurrently create social fairness, economic success, and environmental quality. The main goal of CE is to cut down to disentangle the consumption of natural resources from economic growth. Reducing material use and waste production is essential to achieving this (Priyadarsini & Das, 2025). The 3R principles—reduce, reuse, and recycle—have a close relationship with the CE notion (Tisserant et al., 2017; Ghisellini et al., 2016). A country's boundaries are not the only place where a circular economy may exist. From a consumption-based viewpoint, the picture is frequently different even if a nation's national economy may exhibit high rates of recycling and recovery since many imported goods represent substantial volumes of non-recycled garbage (Tisserant et al., 2017).

## 4 Innovative Strategies of Waste Disposal

Government officials now face significant hurdles because of the massive volume of garbage generated. The enormous quantity of rubbish produced by cities cannot be disposed of by the present waste disposal facilities. To make the current trash disposal schemes extremely appropriate for smart cities and effective enough to handle the formation of enormous amounts of waste, it is necessary to use various technologies, such as sensors, embedded systems, and Internet of Things (IoT) (Dey et al., 2023; Mishra et al., 2023). This chapter narrates different approaches to garbage disposal by effectively leveraging a number of more recent technologies to create a more intelligent waste disposal plan appropriate for smart cities.

### *4.1 Smart Bins and Internet of Things (IoT) Integration*

The smart bins are the need of the present-day waste management strategy. It is equipped with Global Positioning System (GPS), load, MQ-136 gas, and ultrasonic sensors (Ravi et al., 2021). An ultrasonic sensor is used to measure the fullness of the garbage. A load sensor is used to determine the weight of the trash in the smart bin. The GPS sensor is used to determine the location of the smart bin. Hydrogen sulfide, which is created when bacteria break down organic materials in the smart bin waste, is detected by the MQ-136 gas sensor.

The Internet of Things (IoT) refers to a growing network of Internet-connected devices used worldwide (Vishnu et al., 2021). Through real-time process management and monitoring, IoT is essential to improving smart city applications (Vishnu et al., 2021). Some of the monitoring systems included short-range wireless networking technology, such as Bluetooth, infrared, ZigBee, and Wi-Fi, to track the bins (Vishnu et al., 2021). The two sorts of end sensing nodes that comprise Home Bin Level Monitoring Unit (HBLMU) and Public Bin Level Monitoring Unit (PBLMU), are responsible for detecting bins in home and public locations that make up an IoT system (Vishnu et al., 2021). The trash bin's unfilled level and position data are measured by the PBLMUs and HBLMUs, which then process and send the data to a central monitoring station for analysis and storage. The garbage collection authority may observe and assess each trash can's unfilled status attributable to an intuitive graphical user interface (GUI).

## *4.2 Recycling Automation*

Automated recycling is an innovative approach toward waste management. It involves the employment of AI, robotics, and magnetic separators for the sorting of waste materials that are recyclable without much utilization of human labor. It is quite possible and time-efficient to apply intelligent automation techniques for trash management. In order to increase system flexibility, operating systems have experience implementing machine algorithms. Clear and implicit objects can be distinguished from one another using early sensor data from light transmission and sound testing. In addition, they can use sound waves to deftly separate plastic, metal, and glass. The actual metal must be filtered using electromagnetic sensors. Numerous businesses have created robots that reuse machine learning. ZenRobotics, Biotech, and Intel are a few of the possible businesses involved in the official investment and technological waste disposal programs (Ahmed et al., 2020).

## *4.3 Sustainable Packaging Solutions*

In our daily lives, packaging is crucial, particularly when purchasing goods from a retail or wholesale establishment. When goods are delivered all over the world, which has been a daily occurrence in recent years, packaging becomes crucial. Even though packaging guarantees the products' longevity and safety and represents the seller's brand, it contributes to waste accumulation because the majority of these traditional packaging materials cannot be recycled. A significant amount of plastic trash, including Styrofoam, cardboard, and paper, is used in the process of moving items from factories to warehouses and then to retailers, as well as the tremendous number of online purchases that are placed every day (Rajesh et al., 2021). Recycling materials reduces waste during production or raw material processing, which is how sustainable packaging is created. Packaging should prolong the healthy life of products and provide high standards of product cleanliness, particularly for food and beverages (Emblem et al., 2012; Varžinskas et al., 2020).

# 5 Future Perspectives

Sustainable waste management depends on cutting-edge technologies, sensible regulations, and cross-sector cooperation. Waste production can be greatly decreased by switching to circular economy models that place an emphasis on upcycling, recycling, and reusing. Waste management is about to undergo a revolution thanks to artificial intelligence (AI)- and Internet of Things (IoT)-enabled solutions that optimize sorting, collecting, and recycling while increasing efficiency. Sharing best practices, technology, and expertise requires international cooperation, particularly to solve the gaps between developed and developing countries. Finally, for comprehensive solutions, interdisciplinary research combining engineering, environmental science, economics, and social sciences is essential. When combined, these tactics can propel a long-term shift toward a resource-efficient, waste-free society.

## 6 Conclusion

In conclusion, in order to handle the increasing problems brought on by urbanization, population increase, and resource depletion, creative approaches to sustainable trash disposal are essential. We can turn garbage from a burden into a resource by combining the concepts of the circular economy with cutting-edge technologies and biological discoveries. Both public involvement and policy support are essential for promoting behavior change and cultivating a sustainable culture. Multidisciplinary research and international cooperation will guarantee that the answers are significant, scalable, and equitable. When combined, these strategies open the door to a future in which resources are conserved for future generations, waste is efficiently handled, and environmental effects are reduced.

## References


Ahmed, A. A. A.., & Asadullah, A. (2020). Artificial intelligence and machine learning in waste management and recycling. *Engineering International, 8*(1), 43–52. https://doi.org/10.18034/ei.v8i1.498

Amasuomo, E., & Baird, J. (2016). The concept of waste and waste management. *Journal of Management and Sustainability, 6*(4), 88. https://doi.org/10.5539/jms.v6n4p88

Avilés-Palacios, C., & Rodríguez-Olalla, A. (2021). The sustainability of waste management models in circular economies. *Sustainability (Switzerland), 13*(13). https://doi.org/10.3390/su13137105

Ayilara, M. S., Olanrewaju, O. S., Babalola, O. O., & Odeyemi, O. (2020). Waste management through composting: Challenges and potentials. In *Sustainability (Switzerland)* (Vol. 12, Issue 11). MDPI. https://doi.org/10.3390/su12114456

Bal, B., Ghosh, S., & Das, A. P. (2018). Microbial recovery and recycling of manganese waste and their future application: A review. *Geomicrobiological Journal, 36*(1), 85–96. https://doi.org/10.1080/01490451.2018.1497731

Blomsma, F., & Brennan, G. (2017). The emergence of circular economy: A new framing around prolonging resource productivity. *Journal of Industrial Ecology, 21*(3), 603–614. https://doi.org/10.1111/jiec.12603

Cooper, D. R., & Gutowski, T. G. (2017). The environmental impacts of reuse: A review. *Journal of Industrial Ecology, 21*(1), 38–56. https://doi.org/10.1111/jiec.12388

Danthurebandara, M., Passel, S. V., Nelen, D., Tielemans, Y., & Acker, K. V. (2012). *Environmental and socio-economic impacts of landfills.* http://hdl.handle.net/1942/14494

Das, A., Dey, S., Bala, K., Gnanasekaran, R., & Das, A. P. (2024). Microbial bioremediation technology for sustainable treatment and management of synthetic microfiber waste. *Sustainable Microbial Technology for Synthetic and Cellulosic Microfiber Bioremediation, 221–238.* https://doi.org/10.1007/978-3-031-62600-5_12

Das, A., Dey, S., Gacem, A., Yadav, K. K., Nath, H., Ram, D. K., & Das, A. P. (2025). Revolutionizing solid hydrocarbon sourcing: Plastic's journey from waste to upcycled treasure. *Environmental Science and Engineering, Part F218*, 331–351. https://doi.org/10.1007/978-3-031-81615-4_14

Dey, S., Das, A., Mallick, K., Sahu, A., & Das, A. P. (2023). Environmental petroleum waste: Pollution. *Toxicity, Sustainable Remediation, 159–175*. https://doi.org/10.1007/978-3-031-48220-5_7

Dey, S., Dash, A., & Das, A. P. (2024). Bioremediation as a sustainable and circular solution for mining waste management and bioenergy generation. *Sustainable Management of Mining Waste and Tailings: A Circular Economy Approach, 107–119*. https://doi.org/10.1201/9781003442455-6

Emblem, A., & Emblem, H. (2012). *Packaging technology*. Woodhead Publishing Limited. https://doi.org/10.1533/9780857095701

Ghisellini, P., Cialani, C., & Ulgiati, S. (2016). A review on circular economy: The expected transition to a balanced interplay of environmental and economic systems. *Journal of Cleaner Production, 114*, 11–32. https://doi.org/10.1016/j.jclepro.2015.09.007

John, A., & Das, A. P. (2024). Government strategies for achieving circular economy for plastic waste management in Southeast Asia. *Sustainable Management of Urban Plastic Waste through Circular Economic Approaches, 72–82*. https://doi.org/10.1201/9781003465447-6/

Kadakia, K. T., & Galea, S. (2023). Urbanization and the Future of Population Health. *Milbank Quarterly, 101*, 153–175. https://doi.org/10.1111/1468-0009.12624

Kanungo, J., Sahoo, T., Bal, M., & Behera, I. D. (2023). Performance of bioremediation strategy in waste lubricating oil pollutants: A review. *Geomicrobiology Journal*. https://doi.org/10.1080/01490451.2023.2245395

King, A. M., Burgess, S. C., Ijomah, W., & McMahon, C. A. (2006). Reducing waste: Repair, recondition, remanufacture or recycle? *Sustainable Development, 14*(4), 257–267. https://doi.org/10.1002/sd.271

Ma, B., Li, X., Jiang, Z., & Jiang, J. (2019). Recycle more, waste more? When recycling efforts increase resource consumption. *Journal of Cleaner Production, 206*, 870–877. https://doi.org/10.1016/j.jclepro.2018.09.063

Memon, M. A. (2012). Integrated solid waste management based on the 3R approach. In *Sustainable solid waste management* (pp. 133–149). Apple Academic Press. https://doi.org/10.1007/s10163-009-0274-0

Mishra, S., Ghosh, S., van Hullebusch, E. D., Singh, S., & Das, A. P. (2023). A critical review on the recovery of base and critical elements from electronic waste-contaminated streams using microbial biotechnology. *Applied Biochemistry and Biotechnology, 2023*, 1–30. https://doi.org/10.1007/S12010-023-04440-X

Mohammed, M., Shafiq, N., Abdallah, N. A. W., Ayoub, M., & Haruna, A. (2020). A review on achieving sustainable construction waste management through application of 3R (reduction, reuse, recycling): A lifecycle approach. *IOP Conference Series: Earth and Environmental Science, 476*(1). https://doi.org/10.1088/1755-1315/476/1/012010

Brunner, P. H. (2012). Incineration, an essential component of municipal solid waste management. Incenerimento, una componente essenziale della gestione dei rifiuti urbani. Ambiente Rischio Comunicazione 2 – febbraio

Prabhakar, A., Mishra, S., & Das, A. P. (2019). Isolation and identification of lead (Pb) solubilizing bacteria from automobile waste and its potential for recovery of lead from end of life waste batteries. *Geomicrobiological Journal, 36*(10), 894–903. https://doi.org/10.1080/01490451.2019.1654044

Prabina, K. E., Benya, A., Mohanty, S., Van Hullebusch, E. D., Das, A. P., & Singh, S. (2024). Recovery of multi-functional carbon fiber from COVID-19 generated waste for an effective management through circular economy. *Environmental Science and Engineering, Part F2827*, 83–97. https://doi.org/10.1007/978-3-031-50840-0_6

Pradhan, G., Tripathy, B., Ram, D. K., Digal, A. K., & Das, A. P. (2023). Bauxite mining waste pollution and its sustainable management through bioremediation. *Geomicrobiology Journal, 1–10.* https://doi.org/10.1080/01490451.2023.2235353

Priyadarsini, S., & Das, A. P. (2024). Analytical and structural characterization of waste lithium-ion batteries for their effective recycling strategy. *Environmental Chemistry and Ecotoxicology.* https://doi.org/10.1016/j.enceco.2024.12.004

Priyadarsini, S., & Das, A. P. (2025). Lithium bioleaching: A review on microbial-assisted sustainable technology for lithium bio-circularity. *Journal of Water Process Engineering, 69*, 106744. https://doi.org/10.1016/j.jwpe.2024.106744

Qin, W., Egolfopoulos, F. N., & Tsotsis, T. T. (2001). Fundamental and environmental aspects of landfill gas utilization for power generation. *Chemical Engineering Journal, 82.* https://doi.org/10.1016/S1385-8947(00)00366-1

Rajesh, P., & Subhashini, V. (2021). *Sustainable packaging from waste material: A review on innovative solutions for cleaner environment* (pp. 259–270). https://doi.org/10.1007/978-3-030-64122-1_18

Ravi, V. R., Hema, M., SreePrashanthini, S., & Sruthi, V. (2021). Smart bins for garbage monitoring in smart cities using IoT system. *IOP Conference Series: Materials Science and Engineering, 1055*(1), 012078. https://doi.org/10.1088/1757-899x/1055/1/012078

Sahu, A., Dey, S., Mallick, K., Priyadarsini, S., Behera, I. D., & Das, A. P. (2024). A review on advanced technology for sustainable management of synthetic microplastic waste. *Sustainable Management of Urban Plastic Waste through Circular Economic Approaches, 262–279.* https://doi.org/10.1201/9781003465447-15

Sakai, S., Yoshida, H., Hirai, Y., Asari, M., Takigami, H., Takahashi, S., Tomoda, K., Peeler, M. V., Wejchert, J., Schmid-Unterseh, T., Douvan, A. R., Hathaway, R., Hylander, L. D., Fischer, C., Oh, G. J., Jinhui, L., & Chi, N. K. (2011). International comparative study of 3R and waste management policy developments. *Journal of Material Cycles and Waste Management, 13*(2), 86–102. https://doi.org/10.1007/s10163-011-0009-x

Satapathy, S., Kar, D., Priyadarsini, S., Das, A. P., & Kuanar, A. (2024). Application of microorganisms in emerging waste management strategies and remediation of synthetic and cellulosic microfibers. *Sustainable Microbial Technology for Synthetic and Cellulosic Microfiber Bioremediation, 183–206*. https://doi.org/10.1007/978-3-031-62600-5_10

Themelis, N. J., & Ulloa, P. A. (2007). Methane generation in landfills. *Renewable Energy, 32*(7), 1243–1257. https://doi.org/10.1016/j.renene.2006.04.020

Thürer, M., Tomašević, I., & Stevenson, M. (2017). On the meaning of 'Waste': review and definition. In *Production planning and control* (Vol. 28, Issue 3, pp. 244–255). Taylor and Francis Ltd. https://doi.org/10.1080/09537287.2016.1264640

Tisserant, A., Pauliuk, S., Merciai, S., Schmidt, J., Fry, J., Wood, R., & Tukker, A. (2017). Solid waste and the circular economy: A global analysis of waste treatment and waste footprints. *Journal of Industrial Ecology, 21*(3), 628–640. https://doi.org/10.1111/jiec.12562

Varžinskas, V., & Markevičiūtė, Z. (2020). Sustainable food packaging: Materials and waste management solutions. *Environmental Research, Engineering and Management, 76*(3), 154–164. https://doi.org/10.5755/j01.erem.76.3.27511

Vishnu, S., Jino Ramson, S. R., Senith, S., Anagnostopoulos, T., Abu-Mahfouz, A. M., Fan, X., Srinivasan, S., & Kirubaraj, A. A. (2021). IoT-enabled solid waste management in smart cities. *Smart Cities, 4*(3), 1004–1017. https://doi.org/10.3390/smartcities4030053

Visvanathan, C., & Norbu, T. (n.d.). *Reduce, reuse, and recycle: The 3Rs in South Asia.*

Voukkali, I., Papamichael, I., Loizia, P., Lekkas, D. F., Rodríguez-Espinosa, T., Navarro-Pedreño, J., & Zorpas, A. A. (2023). Waste metrics in the framework of circular economy. *Waste Management and Research, 41*(12), 1741–1753. https://doi.org/10.1177/0734242X231190794

# Waste Management's Effects on the Environment and Health

**Yasemin Yilmazoğlu and Emre Yilmazoğlu**

## 1 Introduction

The relationship between the world and the human population contains interesting contradictions. The human population has been increasing rapidly over the last few centuries, consuming the world's resources at an even faster rate. Just like bacteria multiplying with an exponential function, the rate of consumption of resources by human beings is at a dramatic rate compared to population growth. On the other hand, the world is not given time to renew itself, and human beings exploit the world like greedy heirs. In addition, the waste that human beings create both personally and industrially is narrowing down the livable areas in the world. As a result, most of the population, which continues to grow despite being squeezed into a smaller area and raw materials being depleted, is more tense, more ambitious, and more eager to get a bigger share of the pie. It seems that if scientific developments were sufficient for human beings to move all together to another planet, the world would have been exhausted and abandoned long ago. On the other hand, there are also people who think that the solution to this important bottleneck is to get rid of ambitions, return to friendship with nature, produce more efficiently, consume consciously, and ensure the proper recycling of waste. There are concepts that emerge in this context.

Y. Yilmazoğlu (✉)
Department of Electrical and Electronics Engineering, Marmara University, Faculty of Technology, Kadikoy, Istanbul 34722, Türkiye
e-mail: yasemin.yilmazoglu@marmara.edu.tr

E. Yilmazoğlu
Department of Chemical Engineering, Istanbul University-Cerrahpasa, Faculty of Engineering, Avcilar, Istanbul 34320, Türkiye
e-mail: emre.yilmazoglu@iuc.edu.tr

 
A. P. Das and S. Karakuş (eds.), *Synergizing Sustainability for Integrated Waste Management*, Sustainable Environmental Waste Management Strategies,
https://doi.org/10.1007/978-3-032-03718-3_9

## 1.1 Sustainability

Sustainability is a concept developed on the efficient use of resources on a global scale and to prevent future generations from encountering potential risks (Bala et al., 2024; Dey et al., 2023; Singh Jadaun et al., 2022). Using renewable energy sources as an alternative to depleting energy sources is one of the most common sustainability practices in the industrial sense (Dey et al., 2024a, 2024b; Mohanty et al., 2024; Paul et al., 2023). Thus, a more ecological environment is created by reducing greenhouse gas emissions and the carbon footprint of processes. For the same purpose, developing innovative approaches instead of using harmful chemicals to reduce waste, choosing components of products from materials that are as environmentally friendly as possible, designing production processes in a way that will produce maximum ecological and social benefits, using material resources in a way that will prevent not only current but also future problems, and disposing of waste generated in the process in a way that will create the least risk to nature provide sustainability (Dey et al., 2023; Mishra et al., 2019, 2023). However, sustainability is a subject that concerns not only production processes but also individuals and governments. When everyone considers sustainability in production–consumption processes, the resources left to us from previous generations will be consumed less and more will be inherited by the generations after us.

## 1.2 3R (Reduce, Reuse, and Recycle)/5R (Reduce, Reuse, Recycle, Refuse, and Repurpose)

Waste management includes not only the process after waste is created but also the behaviors and preferences during consumption. Accordingly, the 3R understanding includes **reduce**, **reuse**, and **recycle** behaviors, while the 5R understanding includes **refuse** and **repurpose** behaviors (Fig. 1).

Refuse is the rejection of the use of products that may directly or indirectly harm nature. Especially in terms of household goods and consumables, instead of products made of materials that are very difficult to decompose in nature, such as plastics (Dey et al., 2024a, 2024b; Priyadarshini et al., 2024), products made of recyclable materials such as glass and metal should be preferred. Using cloth bags or paper bags instead of plastic bags is a necessity of thinking about nature. White goods with low energy efficiency can be purchased cheaper, but they consume more energy, both increasing the electricity bill and indirectly causing energy resources to deplete quickly. Reuse is not throwing away a broken item, repairing it if possible, or using it as a source of spare parts if not possible (Priyadarsini & Das, 2024, 2025). Every material has a service life. The energy efficiency and operating quality of a device that has been used for a long time decrease significantly. This creates dissatisfaction in the user. Repairing a broken device is often cheaper than buying a new device. Therefore, when an item breaks down, it should be considered whether it can be repaired before buying

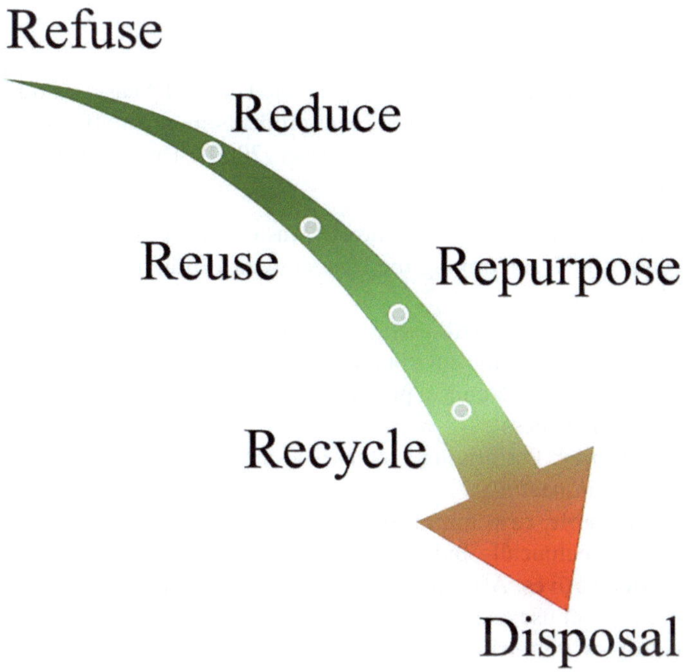


**Fig. 1** 5R practices that can be evaluated before disposal of an item

a new one. Today, thanks to resources such as the Internet, many unqualified people can do small-scale repairs themselves. Devices and items that cannot be repaired even by a master should not be thrown away, and their usable parts should be evaluated. Reduce also applies practices such as reducing the waste created, carbon footprint, and energy consumption. For example, you may have rejected a fuel-powered car and bought an electric car. In this case, the concept of refuse is applied. However, if you use public transportation while you have a car, the concept of reduce is applied. In fact, using a bicycle or walking for short distances reduces the damage to nature to zero.

Recycle is the most common concept in waste management and refers to the management of "waste" in the classical sense. Accordingly, waste should be separated as recyclable and nonrecyclable, especially by on-site separation, and recyclable waste should be recovered without being released into nature (Noor et al., 2020). In summary, instead of wild storage, controlled and planned storage should be provided, and the highest possible level of recyclable material should be included in production. The concept of repurpose is the evaluation of the use of a material that cannot be recycled or repaired for a different purpose. For example, instead of a plastic bag, refuse is applied by using a cloth bag, reuse is applied by sewing a torn cloth bag, and repurpose is applied by making a cloth pencil case from the solid parts of this bag.

## 1.3 Green Chemistry

The concept of green chemistry is a concept created by Paul Anastas and John Warner in 1998 to conduct chemical processes and products more safely and to minimize their environmental impact (Anastas & Warner, 2000). The 12 principles included in this framework are listed below:

- **Prevention**: Preventing the generation of waste instead of cleaning it.
- **Atom economy**: Transforming the least amount of raw material into a product with high efficiency.
- **Less harmful chemical synthesis**: Use and production of chemicals that are least harmful to humans and the environment.
- **Safe product design**: Chemical products that are least harmful and maintain their functional effectiveness.
- **Energy efficiency**: Performing syntheses at the lowest energy consumption and environmental impact.
- **Use of renewable raw materials**: Raw materials should be renewable, not exhaustible, if technically and economically possible.
- **Avoiding derivatives**: Avoiding chemical modifications unless necessary due to additional reagent use and waste generation.
- **Use of catalysts**: Preferring highly selective catalytic reagents over stoichiometric reagents.
- **Degradation**: Degradation of chemical products in a way that does not harm the environment when they are no longer used.
- **Real-time waste prevention**: Developing analysis and control that will enable monitoring during the process before hazardous substances are formed.
- **Safer chemistry for accident prevention**: Selecting the type and form of chemicals in a way that minimizes the risk of accidents during the process.

## 1.4 Industry 4.0

Industry 4.0 is an expression of an effort to increase digitalization in industrial production. Industry 4.0, which is seen as the biggest innovation in production processes after the use of steam power, the transition to mass production and automation, includes features such as the cyber-connectivity of equipment and processes used in production, the ability to manage over the Internet, the storage of large data in cloud environments, the efficient use of production for the optimization of production, and the ability of production tools to provide production on their own (autonomously). The importance of all this in terms of waste management is the minimization of the amount of waste generated as a result of more efficient and more controlled production processes. Production processes are optimized thanks to the more accurate interpretation of people's consumption behaviors. As the decision-making actions of automated systems provide high accuracy, their use in the separation of waste becomes more advantageous. The budget surplus that occurs thanks to the lower cost of producing products can be used in the management of waste or in other areas (Dahunsi, 2025b).

## 2 Waste Sources and Classification

Today, people living in rural areas can adapt more easily to technological developments thanks to globalization and access to the Internet. However, when we look at the distribution of the world's population, we see that the vast majority live in cities. While urbanization brings easier access to services and products, it also causes an increase in waste production per capita. Apart from personal waste, which changes in content and quantity due to changes in daily routines, the construction of new homes and workplaces, ensuring the adequacy of sewage systems, and the increase in energy and water requirements cause planning difficulties (Das & Mishra, 2009; Mishra et al., 2024; Mishra & Das, 2023).

The socioeconomic development of countries where both domestic waste and industrial waste are produced is an important indicator. According to studies, waste production per capita in developed and developing countries is considerably higher than in underdeveloped countries. In fact, some studies state that waste production does not increase in direct proportion to economic growth and that the increase in waste production in these countries is greater than their growth rates. The increase in food waste is striking, especially due to the low purchasing costs of products, especially electronic devices, and habits such as consuming ready-made meals. Until the recent past, developed countries mainly exported their waste to developing and underdeveloped countries. These countries generally produced energy by burning waste in thermal power plants (Das & Ghosh, 2022; Das et al., 2024). However, due to rapid industrialization efforts in developing countries, waste production has increased rapidly in recent years. This leads to more uncontrolled processes in waste disposal. This situation can occur in different ways. In countries where legal regulations are inadequate or corruption cannot be prevented, there is a possibility that waste disposal is not carried out as it should be, by both the state and the private sectors. If more waste is sent to underdeveloped countries, it is difficult for the economically weak states to take the necessary precautions for the disposal of waste. This situation increases the exposure to hazardous chemicals for both the local people and the natural life in the region. Therefore, the export of waste from developed countries to other countries is a process that must be kept under control from beginning to end (Mohanty et al., 2018; Tripathy et al., 2024). Nevertheless, the most appropriate solution is to only apply to waste exports that are too large to be recycled on site and to provide global support for underdeveloped countries for issues such as energy production and waste management (United Nations Environment Programme, 2024).

When waste is classified according to its source, it is seen that although they are formed in different processes, all of them stand out with certain organic or inorganic chemical compounds. In general, waste released because of life in homes is considered domestic waste (Mishra et al., 2024; Mishra & Das, 2021). The total waste mass formed in areas where people live together is defined as municipal waste. Its content is very diverse and is thrown out in solid, liquid, and gaseous forms, such as food residues, fabric, plastic (Mallick et al., 2023; Ray et al., 2024; Sahu et al., 2024), metal wastes (Das et al., 2011, 2015; Ghosh & Das, 2018; Ghosh et al., 2016), sewage wastes (Kumar et al., 2023a, 2023b; Mishra & Das, 2022), and chimney smoke formed after heating. While domestic waste contains various contents in small amounts, industrial waste contains a small number of components in very high amounts. The waste generated in businesses is larger in mass than domestic waste and often contains more harmful components. Also, the physical state of the waste released from the business varies greatly depending on the industrial area (Fig. 2). For example, while hazardous solid waste stands out in a health institution, the content of liquid waste is generally important in a pharmaceutical production facility. In rolling mills of metallurgical facilities, it is more possible to find harmful metals in the vapor phase due to high heat, and in slaughterhouses and agricultural businesses that use pesticides, it is more likely to find harmful chemicals in water (Dahunsi, 2025a; Singh et al., 2022; Tripathi et al., 2021).

When medical waste from hospitals is divided into two according to their hazard class, 80–85% of them are considered nonhazardous, and the rest is considered hazardous (Khatun et al., 2023a, 2023b; Sadasivuni et al., 2019). Nonhazardous medical wastes are products containing paper, fabric, plastic, food, etc., for general use. These can be processed in the same way as domestic waste. Hazardous medical wastes include metabolic wastes (feces, urine, blood, etc.), body parts such as tissues and organs from surgeries, disposable tools and equipment, and medicines. Sharp-cutting tools used and discarded in surgeries also pose a physical risk. The risk they will create is considered in the special classification of medical waste. Accordingly, in addition to chemical, pathological, sharp, radioactive, and drug contents, liquid and pressurized wastes are also evaluated in separate boundaries. These should be stored separately from other wastes in medical waste storage units and disposed of appropriately (Fig. 3). If this distinction is not made, nonhazardous waste will also be biologically/chemically contaminated, and the risk of disease will increase. Storage in relevant facilities must be done safely, unauthorized personnel must not be allowed to enter the waste area, and waste must be delivered to the disposal facility at certain times without allowing pollutants to leak into the environment in various ways. Since not taking adequate safety precautions poses serious risks to environmental health, waste management must be implemented very seriously in these facilities (Krishna & Sachan, 2023; Yuliana et al., 2022).

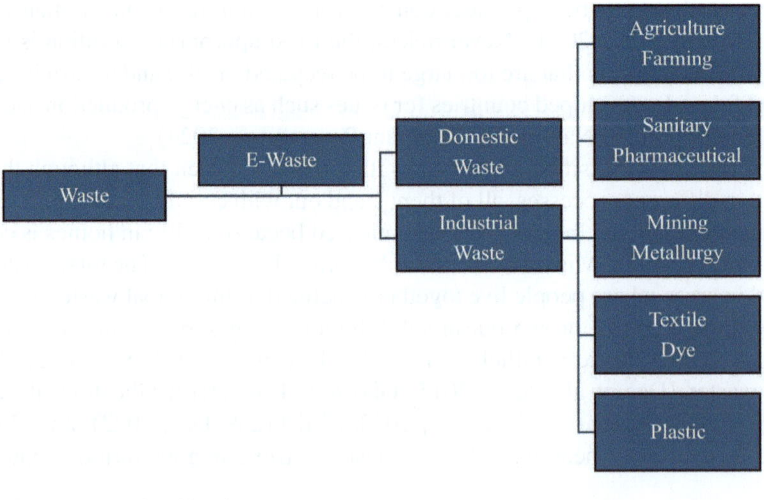


**Fig. 2** Simple waste classification

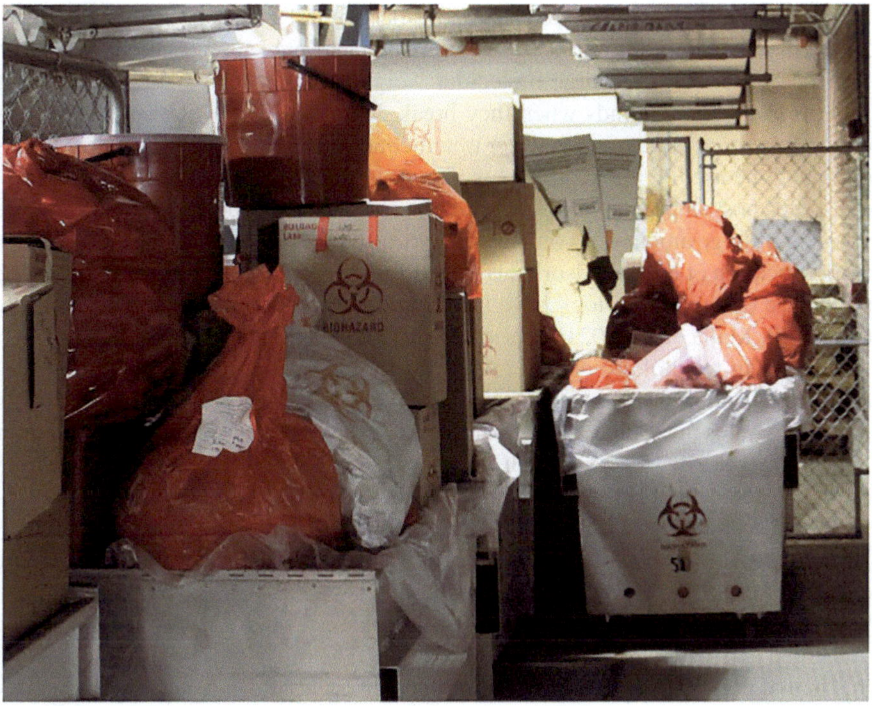

**Fig. 3** Medical waste storage area in a hospital

However, wastes, which are usually disposed of by burning like medical waste, also pose serious risks, especially in terms of the volatile organic compounds (VOCs) they contain. Special attention should be paid to these compounds in the disposal of plastic products and waste chemicals, and the necessary precautions should be taken to prevent them from mixing with nature. These compounds, some of which are carcinogenic or mutagenic, can negatively affect not only the individuals exposed to them but also their children and grandchildren through their genetic structure. These types of hazardous chemicals, which mix into the air through uncontrolled burning processes and into water during irregular storage, are among the most threatening factors in terms of climate change. Aliphatic and aromatic hydrocarbons and their derivatives released as a result of different industrial productions mix into the air and increase the greenhouse gas effect.

In order to feed the increasing population, businesses producing agricultural and animal products engage in various risky initiatives. The most important of these are artificial fertilizers, plant protection chemicals (pesticides, fungicides, insecticides, etc.), and hormones. These additives threaten living health in two ways: directly for living things that feed on them and indirectly for substances mixed into water and soil. Among these, elements such as nitrogen, phosphorus, and sulfur, especially when contained in fertilizers, cannot mix with the natural cycle when excessively present and pollute water. This causes chemical accumulation from microorganisms to aquatic plants, fish, and other living things that feed on them, threatening life.

For this reason, the use of composts and other natural products instead of chemical fertilizers in agriculture increases efficiency and prevents pollution. Similarly, pesticides generally have large molecular weights and can cause the formation of various reactive compounds when they break down. The removal of pesticides has been an important issue in wastewater treatment for years. While traditional methods continue to be used in this regard, new materials and new methods are being developed. The most important problem in carrying out these studies is to provide the treatment of large amounts of wastewater at a sufficiently low cost. For this purpose, the modification of natural adsorbents in various ways, various filtration applications and the use of nanotechnology have increased their importance (Bano, 2024; Little et al., 2008; Malla et al., 2021; Sharma et al., 2021; Srivastav, 2020).

Changes in technology create a need for rapid updating of electronic devices. Consumers' common behavior is to replace their old devices with more modern ones. Sometimes these changes are encouraged with campaigns such as "bring your old device—get a new device at a discount." As part of such campaigns, old devices collected by companies facilitate recycling. Electronic devices found in homes, workplaces, and cars, and, in short, in every area of our lives are largely made up of plastic and metal. Thanks to these recycled components, both production costs are contributed to and environmental pollution is prevented. Electronic devices collected by illegal businesses create environmental problems when they are not properly dismantled and separated into their components. The most important of these problems is that collectors want to get rid of plastics because they do not make money by burning these devices. Burning plastic causes harmful chemicals to be released into the air. These devices may also contain harmful metals such as lead, mercury, and cadmium. During burning, treatment with acid, cyanide, or halide compounds, these metals and the resulting by-products may enter the environment. The smoke created by the burning process mixes chemicals into the air that can cause respiratory system disorders and diseases such as cancer. Biomedical devices contain radioactive components as well as these components. This increases the environmental risk during uncontrolled operations. Therefore, when electronic devices become unusable, they must be taken to the manufacturer or given to authorized collectors (Patel et al., 2023; Radhakrishnan et al., 2023; Sudhakar et al., 2024).

The term "heavy metal" does not have a definite definition and includes elements in different periodic groups. Transition metals, semimetals, lanthanides, and actinides are examples of heavy metals in general. The definition of heavy metal is related to density in metallurgy, atomic weight in physics, and health effects in biology and chemistry. Although metals with a physical density greater than 5 g/ml are defined as heavy metals, this definition is different in terms of their effects on the environment. Some of these directly (carcinogenic, radioactive, mutagenic, etc.) and indirectly (ecosystem degradation, etc.) endanger public health. While some have a poisonous effect in the organism at every concentration, some are necessary for metabolism at certain levels but are harmful at high concentrations. In addition to the areas listed above, heavy metals that are released in different industries in a way that poses a health risk and are likely to be released into the environment are summarized in Table 1 (Chernysh et al., 2024; Frazer-Williams & Sankoh, 2024; Mukherjee et al., 2024).

**Table 1** Industrial branches that are sources of heavy-metal-containing waste

| Industry | Al | As | Be | Cd | Co | Cr | Cu | Fe | Hg | Li | Mn | Mo | Ni | Pb | Zn |
|---|---|---|---|---|---|---|---|---|---|---|---|---|---|---|---|
| Metal | X | X | | X | X | X | X | | | X | X | X | X | X | X |
| Battery and fuel cell | | | X | | X | | | | X | X | X | X | X | X | X |
| Agriculture | | X | | | | | | | X | | X | X | | | |
| Ceramic and glass | X | X | | | X | | | | | X | | X | X | X | |
| Chemistry, pharmaceutical, and dentistry | X | X | | X | X | X | X | | X | X | | X | | X | X |
| Plating | X | | | X | | X | X | | | | | | | X | X |
| Electronic device | X | | X | | X | | | | X | X | | X | | | |
| Fertilizer | | X | | X | | X | | | X | | X | X | X | X | X |
| Fossil fuel | | X | | X | | | | | X | | | | | X | |
| Mining | | X | X | X | | X | X | | X | X | X | | X | X | X |
| Dye and pigment | | X | | X | X | X | X | | | | | | X | X | X |
| Petroleum refinery | | X | | | | X | X | | | | | | X | X | X |
| Machinery | X | | | | X | | X | X | | X | | X | | X | |
| Plastic | X | | | X | X | | | | | X | | X | | X | |
| Paper | X | | | | | X | X | | X | | | | | X | |
| Textile | | X | | | | X | X | | | | | | | | |

The hazard codes for the heavy metals listed above were compiled from the related documents and are given in Table 2 (US Occupational Safety and Health Administration, 2017).

Nuclear waste occurred from particular medical devices and nuclear plants can pose risks ranging from low to high hazards. These wastes are classified by taking into account both the intensity of their radioactivity and their decay time. Equipment and consumables used in environments where radioactive chemicals are used show low radioactivity. These wastes, which are not radioactive but still pose a health risk, must be stored in concrete-covered storage. When the environmental elements and cooling fluids of reactors where nuclear work is carried out are to be taken out of operation, great care must be taken. Since they have the potential to exhibit moderate radioactivity, they cannot be stored in aboveground storages. They are stabilized with materials such as cement or bitumen in specially designed, thick concrete storages and buried underground. These storages are placed in deep wells in places containing solid minerals or clay. Spent nuclear fuel and reactor parts emit intense radiation and generate heat. The danger they create can last hundreds or even thousands of years due to their half-life. For this reason, these wastes are considered one of the most difficult to treat among all wastes. The primary preference is to pass the highly radiative waste through various chemical processes and reduce it to lower radiation levels. It is very important to store the processed nuclear waste in stable environments for long periods of time (Kumar et al., 2023a, 2023b; Laraia, 2015; Rahman, 2024).

**Table 2** Hazard codes of some heavy metals according to the Global Harmonization System and the explanations of these codes

| Metal | Hazard codes and explanation thereof |
|---|---|
| Aluminum | H250: May cause flash fire on contact with air<br>H261: Emits flammable gases on contact with water |
| Arsenic | H301: Toxic if swallowed<br>H331: Toxic if inhaled<br>H400: Very toxic to aquatic life<br>H410: Very toxic to aquatic life with long lasting effects |
| Copper | H302: Harmful if swallowed<br>H319: Causes serious eye irritation<br>H331: Toxic if inhaled<br>H400: Very toxic to aquatic life<br>H410: Very toxic to aquatic life with long-lasting effects<br>H411: Toxic to aquatic life with long-lasting effects<br>H412: Harmful to aquatic life with long-lasting effects |
| Beryllium | H301: Toxic if swallowed<br>H315: Causes skin irritation<br>H317: May cause an allergic skin reaction<br>H319: Causes serious eye irritation<br>H330: Fatal if inhaled<br>H335: May cause respiratory irritation<br>H350i: May cause cancer if inhaled<br>H372: Causes damage to organs through prolonged or repeated exposure |
| Mercury | H330: Fatal if inhaled<br>H360D: May damage the unborn child<br>H372: Causes damage to organs through prolonged or repeated exposure<br>H400: Very toxic to aquatic life<br>H410: Very toxic to aquatic life with long-lasting effects |
| Zinc | H250: May cause flash fire when in contact with air<br>H260: Releases flammable gases which may ignite spontaneously in contact with water<br>H400: Very toxic to aquatic life<br>H410: Very toxic to aquatic life with long-lasting effects |
| Iron | H228: Flammable solid<br>H251: Self-heating; flammable<br>H319: Causes serious eye irritation |
| Cadmium | H250: May cause flash fire on contact with air<br>H330: Fatal if inhaled<br>H341: Suspected of causing genetic defects<br>H350: May cause cancer<br>H361fd: Suspected of damaging fertility. Suspected of damaging the unborn child<br>H372: Causes damage to organs through prolonged or repeated exposure<br>H400: Very toxic to aquatic life<br>H410: Very toxic to aquatic life with long lasting effects |
| Cobalt | H317: May cause an allergic skin reaction<br>H334: May cause breathing difficulties, asthma attacks, or allergy if inhaled<br>H413: May cause long-lasting harmful effects to aquatic life |

(continued)

**Table 2** (continued)

| Metal | Hazard codes and explanation thereof |
|---|---|
| Chromium | H317: May cause an allergic skin reaction<br>H319: Causes serious eye irritation<br>H334: May cause breathing difficulties, asthma attacks, or allergy if inhaled<br>H400: Very toxic to aquatic life<br>H410: Very toxic to aquatic life with long-lasting effects |
| Lead | H360FD: May damage fertility. May damage the unborn child<br>H362: May harm breast-fed children |
| Lithium | H260: Emits flammable gases which may spontaneously ignite in contact with water |
| Manganese | H228: Flammable solid<br>H319: Causes serious eye irritation<br>H411: Toxic to aquatic life with long-lasting effects<br>H412: Harmful to aquatic life with long-lasting effects |
| Molybdenum | H225: Highly flammable liquid and vapor<br>H228: Flammable solid<br>H361: Suspected of damaging the unborn child or fertility |
| Nickel | H317: May cause an allergic skin reaction<br>H351: Suspected of causing cancer<br>H372: Causes damage to organs through prolonged or repeated exposure<br>H412: Harmful to aquatic life with long-lasting effects |

# 3 Waste Disposal Methods

## *3.1 Landfill*

The collection of waste in an area without special pools or tanks is landfilling, unsanitary/wild storage, or uncontrolled dumping. In this method, the leakage of wastewater is absorbed by the soil and the wastewater that finds its way combines with other waters and pollutes the basins. This creates a threat that accumulates in the metabolism from the plants used to feed the living beings in the environment to aquatic life and living beings higher on the food pyramid in the ecosystem. Since the activities of bacteria continue in the organic contents of the solid waste stored in the wild, the formation of gases such as methane can be seen. In some cases, since the gases formed are trapped in closed volumes among the waste, explosions occur due to heat or chemical initiators. Although explosions are not always seen, the gases caused by the decay of the waste contents and the vapors of the leakage water spread into the environment as the garbage smell created by the waste storage areas. Wild storage areas are areas where sufficient security measures are not taken in many countries (Fig. 4). This causes the local people, especially the poor, to enter these areas and search for valuables. The risk of people spending time among the wastes getting sick and transmitting various diseases to their social circles is very high. In addition, in these areas, animal populations such as insects and mice, again due

**Fig. 4** A landfill area that is close to accommodation units and water source (Khattak, 2023)

to organic content, increase uncontrollably. All of these threaten life in the region, restrict social and economic development, and disrupt recycling processes (Oliveri Conti et al., 2024; Robin et al., 2023).

## *3.2 Sanitary Landfill*

Approximately two-thirds (62%) of household waste are disposed of in a controlled manner. Half of this is landfilled, while the rest is recycled or burned for energy production. In Western and Northern European countries, space constraints have ceased to be a problem and have become a reason for the recycling of waste and energy production alternatives. When the use of regular storage areas, recycling of waste, and energy production are considered together, the most efficient waste management process is designed. In this context, storage units are created in large areas for the waste that is desired to be stored regularly. The floors and walls of these units are made of inert material that will not let wastewater through and are stable and physically strong. There is a system for collecting and purifying the leakage water of garbage. The water that is sufficiently cleaned after purification can be used in processes such as agricultural irrigation. The pollution removed from the leakage water according to the purification method also has the opportunity to be used. For example, by processing this water in a system using supercritical fluid, the pollutants are broken down and combustible gas is obtained. In addition, methane gas formed due to the conditions in the sanitary landfill unit is collected with a system and used in energy production. Another advantage of sanitary landfill is that a greater amount of waste can be stored in unit area and unit volume compared to uncontrolled landfill thanks to the fact that the volumes are filled more densely. By covering the storage units, pollution of the environment by air is prevented (Prasad, 2024).

## 3.3 *Thermal Methods*

Uncontrolled storage of waste from facilities such as hospitals poses a risk of infectious diseases. Since these types of medical waste often have organic content, they have high energy potential. Thermal processes are applied in the disposal of such nonrecyclable waste. In addition, waste from research laboratories and chemical manufacturing businesses are also burned so that they do not harm the environment. Thermal processes applied to wastes vary depending on the content of the waste and the product desired to be obtained (Escámez et al., 2025; Handa & Rajamani, 2023).

### 3.3.1 Hydrothermal Carbonization

If the waste to be heat treated has a high moisture content, instead of a pretreatment for moisture removal, it is preferred to apply a hydrothermal transformation to the waste. In the hydrothermal carbonization method, the aqueous waste is exposed to an average temperature of 200 °C under 2–6 MPa pressure without feeding air into the environment. As a result of carbonization under the effect of the applied conditions, a solid fuel with a lower energy content compared to fossil fuels and a liquid that can be used as a food additive or biofuel are obtained. Despite the advantages such as being formed at low temperatures and being used in the disposal of moist waste, the disadvantages are that it requires high pressure, tar formation can be seen, and the wastes that can be processed are limited.

### 3.3.2 Torrefaction

The method called torrefaction is also mostly applied to agricultural and other organic wastes. It is carried out at 200–300 °C and in an oxygen-free environment. However, no pressure is applied in this process and the main purpose is to obtain a product with a higher energy content because of partial combustion. The energy content of the product is equivalent to that of fossil fuels such as coal on average. The reason why this method has come to the fore in recent years is the possibility of using waste as a source for carbon storage. On the other hand, there are limitations such as the waste to be subjected to the torrefaction process being biomass with high carbon content.

### 3.3.3 Pyrolysis

One of the methods frequently used in the thermal disposal of waste is pyrolysis. Wastes heated in an oxygen-free environment, usually at temperatures above 500 °C, turn into solid, liquid, and gas products at the end of the process. This process is basically the partial breakdown of hydrocarbons and other structures under the effect of heat. Apart from biomass, it is used especially in the disposal of waste that may pose a risk with the gases released when subjected to heat treatment, such as rubber/plastic. Therefore, the use of special systems brings high costs. In addition, the waste should not contain moisture. The biochar, biofuel, and synthesis gas produced at the end of the process can be used in heat and electricity production.

### 3.3.4 Gasification

The gasification process takes place at higher temperatures (700–1500 °C) and under partial oxygen than pyrolysis. Under the effect of the conditions created, biochar and biofuel are also broken down and converted into synthesis gas. Many businesses that dispose of waste with thermal processing provide high amounts of heat and electricity production with the gasification process. It requires slightly higher technology than other methods to carefully control the thermal degradation reactions and the content of the gas formed. When the operating cost is compared to the value of the products formed, it is not preferred for wastes with low carbon content.

### 3.3.5 Incineration

The execution of the thermal process in a way that ensures complete combustion with sufficient oxygen is called incineration. Although it is carried out at approximately the same temperatures as gasification, the main purpose of the incineration process is not to produce energy but to eliminate waste that cannot be used in processes such as gasification. Incineration is preferred especially in the disposal of domestic and medical wastes because it cannot be used in other applications. At the end of the process, ash formation and carbon dioxide ($CO_2$) emission occur together with the heat release. If the incineration process is carried out in the open air, the ash and smoke formed may threaten environmental health. This situation should be prevented by carrying out the incineration process in closed environments and by applying filtration processes suitable for flue gas. The high heat generated in the process is generally used in converting water into steam. The steam generated is used in heating, electricity generation with turbines, or other processes needed in the business (drying, endothermic reactions, etc.).

### 3.3.6 Plasma-Supported Gasification

The last method used in the disposal of waste containing heavy metals or hazardous chemicals is the plasma arc gasification process. In this method, inert gases such as argon and nitrogen are transformed into the fourth state of matter, plasma, with the help of an electric arc. The very high energy of the ions is directed to the waste in the plasma reactor with a torch. Due to the effect of the temperature of thousands of °C, even the difficult components contained in the waste undergo thermal transformation. At the end of the process, a syngas containing hydrogen ($H_2$) and carbon monoxide (CO) and a solid containing inorganic components are formed. The formed gas is used in electricity production. After the metal contained in the solid product is recovered, the remaining glassy part is used to contribute to building materials. The formation of toxic chemicals is prevented by high temperatures. However, the reformation of dioxins is observed during the cooling phase of the product. Since the installation cost is very high and the energy obtained is not high compared to the energy consumed, it is preferred to be applied only for waste that needs to be eliminated under special conditions (Barman et al., 2024; Krishna & Sachan, 2023).

## *3.4 Composting*

Food and garden wastes constitute more than half of household waste worldwide. The wastes of enterprises such as agriculture and animal husbandry, which contain high organic content, are a very suitable breeding and living environment for bacteria and some types of worms. The biomass called compost is formed when the wastes are digested by microorganisms and worms. Thanks to the composting process, while the waste is disposed of, a highly nutritious fertilizer that does not contain artificial chemicals for plants is obtained. The nutritional value and water retention capacity of the soil added to the compost increase, and the plants growing in this soil are healthier. The composting process is one of the waste disposal methods that support sustainability the most. For the produced compost to have a suitable content, the wastes that are carbon sources (inkless newspaper/cardboard wastes, wood pieces, etc.) should be three times as much as the plant and animal wastes that are nitrogen sources (vegetable/fruit wastes, leaves, animal manure, tea/coffee wastes, eggshells, etc.) (Fig. 5). The mixture should have a moisture content of around 60%, waste that is broken into small pieces should be used whenever possible, the compost pile should be mixed and aerated well, and the temperature should be kept at 45–55 °C to eliminate pathogens. Depending on these conditions and the size of the piles, composting can take from a few months to a year. This process, which can also be done in a home or garden environment, does not cause problems such as insect infestation or odor if it is continued under appropriate conditions. If the heating caused by the digestion of bacteria in the pile has stopped, compost with an earthy odor, a dark brown-black color, and a homogeneous structure is obtained. Since the

**Fig. 5** An efficient compost includes brown and green ingredients in a proper ratio

composting method, which requires a long time, cannot be applied to all waste types and requires labor, it is only functional for certain situations (Liang et al., 2024).

### 3.5 *Anaerobic Digestion*

Another disposal method that uses bacteria, similar to composting, is anaerobic digestion. This method is carried out in a closed environment, and the main purpose is to produce biogas from organic waste. As mentioned in the section on wild storage, anaerobic microorganisms survive in the absence of oxygen in closed volumes or in the airless inner areas of the waste pile. The methane and carbon dioxide produced by these bacteria are withdrawn from the pile and stored to be used in energy production. Although the remaining part is not as nutritious as compost, it can be used as fertilizer. It takes much less time than composting (from a few days to a month, depending on conditions such as temperature and humidity). Since oxygen is not required in the process, the pile is not mixed. This allows much larger piles to be used. In addition, wastewater containing organic pollution (from textiles, pharmaceuticals, slaughterhouses, etc.) can also be used in biogas production with this method. At the beginning of the process, large organic molecules are hydrolyzed into smaller molecules. Then, small molecules are converted into organic acids, alcohols, $CO_2$, and $H_2$ by acidogenic bacteria. After the conversion of molecules such as organic acids and alcohols into acetate, $CO_2$, and $H_2$ by acetogenic bacteria, the acetate ($C_2H_3O_2^-$) and $H_2$ in the environment are converted into methane ($CH_4$) and $CO_2$ in the methanogenesis phase and the biogas production process is completed. The wastewater with organic content remaining in the system can be used as liquid fertilizer, or it can be used as solid fertilizer after the sludge moisture separated from this liquid by filtration is removed. Although the anaerobic digestion method is suitable for a wider variety of waste than the composting method, it is costly because it requires specially designed reactors and the adaptation of the microorganisms used to the process may take time (Choudhary et al., 2024).

## *3.6 Chemical Methods*

### 3.6.1 Oxidation

Chemical compounds such as hydrogen peroxide ($H_2O_2$), ozone ($O_3$), chlorine ($Cl_2$), hypochlorite (NaOCl), potassium permanganate ($KMnO_4$), and persulfate ($S_2O_8{}^{-2}$) are strong oxidizers, and when they react with various compounds, they break down the pollution into small molecules such as $CO_2$ and water. Particularly, pollution that cannot be broken down in biological treatment processes can be removed with such processes. Oxidizing compounds can be used in the treatment of soil, wastewater, and flue gases. The oxidation process is useful in petroleum-based pollution, phenolic contents, and stubborn chemicals such as pesticides and disinfection to kill microorganisms. Some of these oxidizers may be disadvantageous due to pH adjustment, activation time, by-product formation, or low stability. The chemical oxidation process can be developed with other applications that support the process by using catalysts, ultraviolet (UV) light, or free radical formation. It is a method that is frequently used, especially in wastewater, and can be used for more than one pollution.

### 3.6.2 Reduction

Molecules such as sodium sulfite ($Na_2SO_3$), sodium borohydride ($NaBH_4$), iron ($Fe^{+2}$, $Fe^0$), sodium dithionite ($Na_2S_2O_4$), sulfur compounds ($Na_2S$, $H_2S$), and hydrazine ($N_2H_4$) provide a reducing effect and enable the transformation of various pollutants into less harmful compounds. The difference between the reduction process and the oxidation process is that in the oxidation process, the molecules that create pollution are oxidized by giving electrons, while in the reduction process, these molecules receive electrons. The reduction process is frequently used in the separation of heavy metals from wastewater. In general, reducers that enable the precipitation of pollution in wastewater sometimes enable the transition of pollutant molecules to the gas phase. For example, cyanide ions ($CN^-$) react with sulfur dioxide ($SO_2$) and air to form $CO_2$ and ammonia ($NH_3$). The by-products formed after reduction must be processed appropriately. High chemical consumption will increase costs and may also cause undesirable reactions to occur.

### 3.6.3 Neutralization

Neutralization is the process of neutralizing the pH level of the environment by treating acidic impurities with basic compounds and basic impurities with acidic compounds. At the end of this process, salt and water are formed depending on the type of acid and base. Commonly used bases for this purpose include caustic soda and lime and sulfuric or hydrochloric acid among acids.

### 3.6.4 Precipitation

Pollutants such as metal ions in wastewater can be separated from the dissolved state by turning them into precipitates. The process is based on the precipitation of the chemical that causes pollution by reacting with a suitable compound. It can usually be carried out with fast and simple reactions. However, in order for this method to be applied, the dissolved chemical must have a compound that forms a precipitate. Apart from heavy metals, this method can also be used for phosphate and sulfate compounds. The precipitate formed after the process must be removed from the environment (Nandakumar et al., 2024).

### 3.6.5 Ion Exchange

In the ion exchange method, ionic functional groups found on the surfaces of materials such as ion exchange resins react with molecules that create wastewater pollution and bind these molecules to themselves. They are used to remove ions that create hardness in water, heavy metals, and ionic compounds such as nitrate, sulfate, or ammonium from the environment. The most important advantage of the ion exchange method is that it can be reused by removing the dirt on the resin surface with suitable solutions. It is easier to control since it does not require chemical additives. Although it is more effective in terms of selectivity, it is difficult to use in the treatment of industrial-sized amounts of wastewater (Gupta et al., 2024).

## *3.7 Solidification*

Liquid and solid wastes that are likely to mix with water can change to a slurry or solid form by chemically or physically using various binding agents. In this way, leaks from cracks and flows into water resources are tried to be prevented. For this purpose, lime, cement, and clay derivatives are frequently used as binders. Especially in waste pools where lime is poured, microorganism growth is prevented, and decay and rot are prevented. The material used as binder must be harmless to nature, have high adsorption capacity, and be able to solidify when mixed with water (Xiong et al., 2021).

**Fig. 6** Waste bins in different colors for on-site sorting of waste (colors may vary depending on country or facility)

# 4 Alternative Solutions

## *4.1 On-Site Recycling*

On-site recycling is one of the most commonly implemented waste management strategies on the global scale. This strategy facilitates recycling by ensuring that waste is classified on dumpster area (Fig. 6). Since public participation in this process is low when it is based on a voluntary basis, participation is usually encouraged through a reward mechanism. Waste from industries such as textile, glass, plastic, and metal left at collection units allows the participant to obtain money, coupons, or discounts according to their number or weight. This collection of recyclable waste lightens the burden on disposal facilities. In addition, although informal waste workers, who are seemingly an unsecured and unofficial business, may suffer financial losses, this problem can be overcome if these people can be employed in different jobs and these people, whose number reaches 20 million worldwide, will be kept away from unhealthy environments (Chen, 2023).

## *4.2 Natural Alternatives for Synthetic Plastics*

Synthetic plastics are petroleum-based structures. Their unsustainability, limited resources, costly recycling processes, and the fact that they remain in nature for a very long time without breaking down create major problems. In addition, the high mortality rates seen in marine organisms due to plastics dumped into water resources today, the decrease in biodiversity, and the disorders caused by microplastics accumulating in the metabolism of humans and animals that feed on marine organisms are among the problems that are taken seriously (Fig. 7). Using natural polymers instead of these is an approach that combines the concepts of green chemistry and sustainability by reducing both plant and plastic-based waste (Kumar et al., 2024; Mangla et al., 2024).

Natural polymers can be suitable alternatives to synthetic materials with their biodegradable and biocompatible properties. The use of natural materials, especially in films used in food packaging, facilitates the disposal of waste and reduces the

**Fig. 7** Floating waste island consists of mostly plastics

pollution accumulated in nature. Among these alternative films, edible food protective films are also an important field of study. Various natural structures stand out as strong alternatives due to their appropriate levels of moisture, $O_2$ and $CO_2$ permeability, antimicrobial and antioxidant activities against food spoilage, flexibility/durability, and their ability to be consumed with food or used as compost. In addition to protective films, the amount of medical waste can be reduced with natural polymers used in the biomedical field (wound dressing, suture thread, etc.).

## *4.3 Microorganisms*

Another solution applied to the problems mentioned above, created by synthetic plastics, is to use microorganisms at the dispose of these wastes (Prabhakar et al., 2019; Pradhan et al., 2023). In addition to microorganisms that provide the breakdown of organic wastes through composting or oxidative decay, plastic materials with large molecular weights, some of which contain chlorine, bromine, heavy metals, phthalates, phenol derivatives, and organic components, can also be broken down by some special microorganisms (Das et al., 2025; John & Das, 2024; Sahu et al., 2024). Microorganisms that provide partial or complete breakdown of plastics such as polyethylene terephthalate (PET), polyethylene (PE), polypropylene (PP), and polyvinyl chloride (PVC), which are proportionally at the top of the list among wastes, are compiled in the publication written in 2022 (Ekanayaka et al., 2022).

Apart from plastics, there are also bacteria and fungi that break down molecules contained in some heavy metals, organic solvents, and oil. In particular, applications that provide practical benefits in the use of microorganisms for cleaning oils mixed into the sea are reflected in the news.

The biggest advantage of using microorganisms on a large scale for the disposal of organic or inorganic waste is that energy is not required to be spent during the transformation of by-products into less harmful by-products and that the wastes formed when these organisms die after the wastes they can feed on run out do not pose a biological risk.

## *4.4 Nanotechnology*

The advanced activities of nano-sized materials in terms of physical, chemical, biological, optical, or electronic properties increase the usability of these materials in innovative studies in different fields. According to the generally accepted view, nanomaterials are one-, two-, or three-dimensional structures with each dimension below 100 nm. These structures can be formed from a single component or by using organic and/or inorganic components together. Developments in nanotechnology provide great advantages in waste management as in other fields. These advantages can be divided into two classes. The first of these is the reduction of waste production by using nanomaterials with various properties such as biodegradability, long life, and corrosion prevention instead of traditional materials. The second advantage is the use of nanomaterials as catalysts, adsorbents, membranes, etc. structures by utilizing the extremely high surface area per unit volume in the removal of waste.

### 4.4.1 Nanotechnology in Reducing Waste

Sonochemical synthesis is a technique used to obtain nano- and micro-sized materials. In this method, sound waves are used at frequencies beyond the limits of sound that can be heard by the human ear (> 20 kHz). It provides the disintegration of solid particles in this medium by the vibration of liquid molecules, and the particles reduced to small sizes create a more homogeneous distribution in the medium. In addition, this method is very suitable for the green chemistry approach. It is beneficial in terms of both the use of biological materials as reductants and the elimination of the need for the use of alternative chemical solvents. Plant materials used for this purpose show activity with the reductant components they contain. In addition to reduction, they are also used for capping, emulsification, and stabilization functions. In these ways, agglomeration of nanoparticles is prevented, inorganic molecules are kept in elemental form, and nano-scale dimensions are preserved. Some plant products can also be used for purposes such as homogeneous distribution of lipophilic structures in the aqueous phase.

Thanks to their superior properties, nanoparticles help reduce waste by ensuring that products last longer and work more efficiently. For example, the use of high-performance batteries produced using nanotechnology instead of traditional batteries reduces the amount of electronic waste. When thermal properties are improved with nanostructures added to interior and exterior paints, the energy consumed for heating purposes and the carbon footprint are reduced. Materials with biodegradable properties are often in nano- or microscale (Gupta & Dutta, 2024; Kumar, 2024; Sharma & Singh, 2024; Tiwari et al., 2024; Villagrán et al., 2024).

### 4.4.2 Nanotechnology in Waste Disposal

Nanoparticles are suitable materials for applications such as adsorption and filtration because they have very large surface areas in very small volumes. Their selectivity can be increased by customizing their surface morphology and functional groups. With the same properties, they show high performance as catalysts and can perform photocatalytic processes with high efficiency. They can be used to separate contaminants such as metals from the environment by precipitation, oxidation, and other chemical methods. They can be customized to ensure the breakdown of biological wastes, and they can provide easier separation and disposal by biological (enzymatic, etc.) or chemical (encapsulation, etc.) methods. Nanoparticles, which are frequently used in sensor applications, can detect the presence of certain components in various environments with sensitivity (Gola et al., 2021; Mathur & Sanyal, 2021; Rani et al., 2023).

## 5 Conclusion

Due to the increase in human population, the migration of large crowds from villages to cities has required an increase in production and this has increased consumption. The rise in production increases the amount of industrial waste, and changes in human consumption behavior increase the amount of domestic waste. Petrochemical, mining, textile, metallurgical, pharmaceutical, and food industries are the leading sources of pollution. Pollution can be released into the environment as gaseous, liquid, or solid waste. For example, if gaseous wastes contain hydrocarbon compounds, they can cause the greenhouse gas effect. Gases containing elements such as nitrogen or phosphorus can trigger acid rain. When contaminated water from the plants enters water systems, it can damage the aquatic ecosystem and then the terrestrial ecosystem. Substances such as heavy metals, dyes, or pharmaceuticals contained in wastewater can be a threat to the life of all organisms if they get into the soil. Domestic waste is mostly solid and contains a large proportion of plastic and organic matter. Since it takes centuries for plastics to dissolve in nature, they must be disposed of properly. In addition, if these substances are mixed into water resources in micro-sized pieces, the accumulation of plastics is observed in the metabolism of all living organisms feeding in those water resources. Another important waste type among domestic wastes is oil. The oil used in the cooking area should not be discharged directly into the sewerage system but should be disposed of in a distinct way. To minimize the negative effects of both factories and consumers on the environment, the first thing to do is to reduce the amount of waste. For this purpose, more durable and long-lasting materials should be used instead of disposable materials. The second stage in waste management is to ensure the reuse of waste material. If the service life of the material has expired and the possibility of reuse is over, it is preferred to recycle the material and use it as raw material in production. For achieving this, wastes must be separated at the source or, if this is not possible, at gathering centers. Waste materials that cannot be utilized in the production process are used for other purposes. For example, many metallic wastes such as iron, aluminum, chrome, and bronze can be melted down and reused. Wood waste can be used as fuel or in the production of items of varied sizes. Considering all these, the concept of sustainability should be considered for waste management and minimizing negative impacts on the environment. Sustainability is an understanding that covers the entire process from the planning of production to the recycling of waste in order to protect the ecosystem on earth. Accordingly, since it is aimed at realizing production with the lowest energy requirement and to utilize waste as much as possible, it is ensured that the total cost is low and economic efficiency is high.

## References


Anastas, P. T., & Warner, J. C. (2000). *Green chemistry*. Oxford University Press. https://doi.org/10.1093/oso/9780198506980.001.0001

Bala, K., Anshul, Verma, S., Dogra, A., Das, A. P., Kumar, R., & Nandini. (2024). Sustainable management for urban plastic waste generation. *Sustainable Management of Urban Plastic Waste Through Circular Economic Approaches*, 211–236. https://doi.org/10.1201/9781003465447-13

Bano, A. (2024). Direct environmental pollution from solid waste. In *Waste-to-energy: Sustainable approaches for emerging economies* (pp. 249–269). Elsevier. https://doi.org/10.1016/B978-0-443-22356-3.00011-7

Barman, A., Roy, S., Saha, P., Dey, S., Patel, S., Meena, D. K., Bera, A., Singh, S. V., Mandal, S., Nath, S., & Bagrecha, S. (2024). Sustainable management and diversification of problematic wastes: Prospects and challenges. In *Biochar production for green economy: Agricultural and environmental perspectives* (pp. 57–79). Elsevier. https://doi.org/10.1016/B978-0-443-15506-2.00020-1

Chen, S. (2023, December 28). *Unsung heroes: Four things policymakers can do to empower informal waste workers*. https://www.undp.org/Blog/Unsung-Heroes-Four-Things-Policymakers-Can-Do-Empower-Informal-Waste-Workers

Chernysh, Y., Yakhnenko, O., Bataltsev, Y., Plyatsuk, L., & Balintova, M. (2024). Heavy metals contamination of world soils and their perspectives for remediation through ecotechnologies. In *Bio-organic amendments for heavy metal remediation: Water, soil and plant approaches and technologies* (pp. 299–315). Elsevier. https://doi.org/10.1016/B978-0-443-21610-7.00011-2

Choudhary, M., Singh, D., Parihar, M., Choudhary, K. B., Nogia, M., Samal, S. K., & Mishra, R. (2024). Impact of municipal solid waste on the environment, soil, and human health. In *Waste management for sustainable and restored agricultural soil* (pp. 33–58). Elsevier. https://doi.org/10.1016/B978-0-443-18486-4.00011-7

Dahunsi, O. S. (2025a). Environmental impacts of waste generation and improper management. In *Biorefinery* (pp. 15–62). Elsevier. https://doi.org/10.1016/B978-0-443-24129-1.00003-X

Dahunsi, O. S. (2025b). Policies and strategies for wastes management/valorization. In *Biorefinery* (pp. 647–683). Elsevier. https://doi.org/10.1016/B978-0-443-24129-1.00024-7

Das, A., Dey, S., Gacem, A., Yadav, K. K., Nath, H., Ram, D. K., & Das, A. P. (2025). Revolutionizing solid hydrocarbon sourcing: Plastic's journey from waste to upcycled treasure. *Environmental Science and Engineering, Part F218*, 331–351. https://doi.org/10.1007/978-3-031-81615-4_14
Das, A., & Ghosh, S. (2022). Role of microorganisms in extenuation of mining and industrial wastes. *Geomicrobiology Journal, 39*, 173–175. https://doi.org/10.1080/01490451.2022.2038953
Das, A. P., Ghosh, S., Mohanty, S., & Sukla, L. B. (2015). Consequences of manganese compounds: A review. *Toxicological & Environmental Chemistry, 96*(7), 981–997. https://doi.org/10.1080/02772248.2015.1005428
Das, A. P., & Mishra, S. (2009). Hexavalent chromium [Cr (VI)]: yellow water pollution and its remediation. *Sarovar Saurabh ENVIS Newsl Wetl Ecosyst, 5*(2).
Das, A. P., Sukla, L. B., Pradhan, N., & Nayak, S. (2011). Manganese biomining: A review. *Bioresource Technology, 102*(16), 7381–7387. https://doi.org/10.1016/J.BIORTECH.2011.05.018
Das, A., Ray, S. K., Mohanty, M., Mohanty, J., Dey, S., & Das, A. P. (2024). Ecotoxicity of microplastic wastes and their sustainable management: A review. *Environmental Chemistry and Ecotoxicology, 6*, 144–152. https://doi.org/10.1016/J.ENCECO.2024.05.003
Dey, S., Bissoyi, A., Koteswara Reddy, G., & Das, A. P. (2024a). Synthetic microfibers: An urban emerging pollutant demanding for sustainable environmental solutions. *Sustainable Microbial Technology for Synthetic and Cellulosic Microfiber Bioremediation, 107–121.* https://doi.org/10.1007/978-3-031-62600-5_7
Dey, S., Das, A., Mallick, K., Sahu, A., & Das, A. P. (2023). Environmental petroleum waste: Pollution. *Toxicity, Sustainable Remediation, 159–175.* https://doi.org/10.1007/978-3-031-48220-5_7
Dey, S., Dash, N., Ray, S. K., Priyadarsini, S., Chhotaray, C., & Das, A. P. (2024b). Occurrence and quantification of tire wear plastics pollutants from highway city dust of Bhubaneswar City of Odisha. *Sustainable Microbial Technology for Synthetic and Cellulosic Microfiber Bioremediation, 77–89.* https://doi.org/10.1007/978-3-031-62600-5_5
Ekanayaka, A. H., Tibpromma, S., Dai, D., Xu, R., Suwannarach, N., Stephenson, S. L., Dao, C., & Karunarathna, S. C. (2022). A review of the fungi that degrade plastic. *Journal of Fungi*, 8(8). MDPI. https://doi.org/10.3390/jof8080772
Escámez, A., Sánchez-Lozano, D., Aguado, R., Vera, D., & Jurado Melguizo, F. (2025). Energy from waste. In *Towards future smart power systems with high penetration of renewables* (pp. 53–74). Elsevier. https://doi.org/10.1016/B978-0-443-29871-4.00003-8
Frazer-Williams, R., & Sankoh, A. (2024). Soil contamination resulting from inefficient solid waste management. In *Environmental pollution and public health* (pp. 251–264). Elsevier. https://doi.org/10.1016/b978-0-323-95967-4.00010-6
Ghosh, S., & Das, A. P. (2018). Metagenomic insights into the microbial diversity in manganese-contaminated mine tailings and their role in biogeochemical cycling of manganese. *Scientific Reports 2018, 8*(1), 1–12. https://doi.org/10.1038/s41598-018-26311-w
Ghosh, S., Mohanty, S., Nayak, S., Sukla, L. B., & Das, A. P. (2016). Molecular identification of indigenous manganese solubilising bacterial biodiversity from manganese mining deposits. *Journal of Basic Microbiology, 56*(3), 254–262. https://doi.org/10.1002/JOBM.201500477
Gola, D., Adhikary, A., Bharti, R. K., Dey, P., Gola, S., Chauhan, N., Srivastava, S. K., & Mehrotra, R. (2021). Advances in nanotechnologies of waste water treatment: Strategies and emerging opportunities. In *Development in wastewater treatment research and processes: Removal of emerging contaminants from wastewater through bio-nanotechnology* (pp. 123–138). Elsevier. https://doi.org/10.1016/B978-0-323-85583-9.00018-1
Gupta, A. P., & Dutta, J. (2024). Nanotechnological applications in agro-waste management with special reference to Indian agricultural sector. In *Waste management for sustainable and restored agricultural soil* (pp. 295–310). Elsevier. https://doi.org/10.1016/B978-0-443-18486-4.00005-1
Gupta, A. R., Joshi, V. C., & Sharma, S. (2024). Ion-exchange polymers: versatile sorbents for water remediation. In *Advances in drinking water purification: Small systems and emerging issues* (pp. 181–200). Elsevier. https://doi.org/10.1016/B978-0-323-91733-9.00009-X

Handa, A., & Rajamani, P. (2023). Waste management and environment. In *Basic biotechniques for bioprocess and bioentrepreneurship* (pp. 391–413). Elsevier. https://doi.org/10.1016/B978-0-12-816109-8.00027-1

John, A., & Das, A. P. (2024). Government strategies for achieving circular economy for plastic waste management in Southeast Asia. *Sustainable Management of Urban Plastic Waste through Circular Economic Approaches, 72–82*. https://doi.org/10.1201/9781003465447-6

Khattak, R. (2023, March 14). *What can we do to minimise landfill waste?* https://earth.org/Minimising-Landfill-Waste/.

Khatun, R., Singh, S., Dubey, N. K., & Prasad Das, A. (2023a). A review on marine-based phytochemicals and their application in biomedical research. *Recent Frontiers of Phytochemicals: Applications in Food, Pharmacy, Cosmetics, and Biotechnology, 383–395*. https://doi.org/10.1016/B978-0-443-19143-5.00013-X

Khatun, R., Singh, S., Dubey, N. K., & Prasad Das, A. (2023b). A review on marine-based phytochemicals and their application in biomedical research. In *Recent frontiers of phytochemicals: Applications in food, pharmacy, cosmetics, and biotechnology* (pp. 383–395). Elsevier. https://doi.org/10.1016/B978-0-443-19143-5.00013-X

Krishna, D., & Sachan, H. K. (2023). Hazardous biomedical waste management scenario in developing countries. In *Waste management and resource recycling in the developing world* (pp. 505–519). Elsevier. https://doi.org/10.1016/B978-0-323-90463-6.00015-4

Kumar, A., Agrawal, M., Singh, A. K., Singh, H., & Kumar, P. (2024). Microplastics contamination in marine system. In *Microplastics: Origins, risks, and mitigation* (pp. 213–230). Elsevier. https://doi.org/10.1016/B978-0-443-29804-2.00008-1

Kumar, M., Pandit, S., Patel, V., Khanna, N., Nag, M., Lahiri, D., Ray, R. R., Das, A., & Das, D. (2023a). Maximization of energy recovery from starch processing wastewater by thermophilic dark fermentation coupled with microbial fuel cell technology. *Geomicrobiology Journal, 1–12*. https://doi.org/10.1080/01490451.2023.2209555

Kumar, P. (2024). Nanomaterial fabrication from industrial solid waste: A sustainable approach for resource recovery. In *Sustainable management of agro-food waste: Fundamental aspects and practical applications* (pp. 303–321). Elsevier. https://doi.org/10.1016/B978-0-443-23679-2.00021-5

Kumar, P., Yadav, S., & Yadav, A. (2023). Radioactive waste minimization and management. In *Green chemistry approaches to environmental sustainability: Status, challenges and prospective* (pp. 165–184). Elsevier. https://doi.org/10.1016/B978-0-443-18959-3.00003-3

Laraia, M. (2015). Radioactive contamination and other environmental impacts of waste from nuclear and conventional power plants, medical and other industrial sources. In *Environmental remediation and restoration of contaminated nuclear and norm sites* (pp. 35–56). Elsevier Inc. https://doi.org/10.1016/B978-1-78242-231-0.00002-8

Liang, W., Liu, T., Bao, H., Kumar, V., Sun, X., Sindhu, R., Binod, P., Zhang, Z., Pandey, A., & Awasthi, M. K. (2024). Novel animal waste management techniques. In *Engineering applications in livestock production* (pp. 243–264). Elsevier. https://doi.org/10.1016/B978-0-323-98385-3.00006-2

Little, D. C., Milwain, G. K., & Price, C. (2008). Pesticide contamination in farmed fish: assessing risks and reducing contamination. In *Improving farmed fish quality and safety* (pp. 71–96). Elsevier. https://doi.org/10.1533/9781845694920.1.71

Malla, M. A., Gupta, S., Dubey, A., Kumar, A., & Yadav, S. (2021). Contamination of groundwater resources by pesticides. In *Contamination of water: Health risk assessment and treatment strategies* (pp. 99–107). Elsevier. https://doi.org/10.1016/B978-0-12-824058-8.00023-2

Mallick, K., Sahu, A., Dubey, N. K., & Das, A. P. (2023). Harvesting marine plastic pollutants-derived renewable energy: A comprehensive review on applied energy and sustainable approach. *Journal of Environmental Management, 348*, 119371. https://doi.org/10.1016/J.JENVMAN.2023.119371

Mangla, S., Pathak, G., & Bhan, V. (2024). Microplastics contamination in soil, water, air, and environmental health risk assessment. In *Microplastics: Origins, risks, and mitigation* (pp. 3–30). Elsevier. https://doi.org/10.1016/B978-0-443-29804-2.00001-9

Mathur, P., & Sanyal, D. (2021). Removal of emerging contaminants from pharmaceutical waste through application of bio nanotechnology. In *Development in wastewater treatment research and processes: Removal of emerging contaminants from wastewater through bio-nanotechnology* (pp. 485–500). Elsevier. https://doi.org/10.1016/B978-0-323-85583-9.00028-4

Mishra, S., & Das, A. P. (2021). Current treatment technologies for removal of microplastic and microfiber pollutants from wastewater. *Wastewater Treatment: Cutting-Edge Molecular Tools, Techniques and Applied Aspects, 237–251.* https://doi.org/10.1016/B978-0-12-821881-5.00011-8

Mishra, S., & Das, A. P. (2022). Treatment of the wastewater polluted with synthetic microfiber released from washing machine. In B. B. Das, H. Hettiarachchi, P. K. Sahu, & S. Nanda (Eds.), *Lecture notes in civil engineering* (Vol. 207, pp. 109–117). Springer Science and Business Media Deutschland GmbH. https://doi.org/10.1007/978-981-16-7509-6_9/COVER

Mishra, S., & Das, A. P. (2023). Microbial remediation of synthetic microfiber contaminated wastewater. In M. P. Shah (Ed.), *Microbial technologies in industrial wastewater treatment* (pp. 337–350). Springer. https://doi.org/10.1007/978-981-99-2435-6_17

Mishra, S., Dash, D., & Das, A. P. (2024). Isolation and molecular characterization of polyester-degrading bacterial strains from municipal wastewater and their application in microplastic degradation in contaminated wastewater. *Geomicrobiology Journal*. https://doi.org/10.1080/01490451.2024.2376101

Mishra, S., Ghosh, S., van Hullebusch, E. D., Singh, S., & Das, A. P. (2023). A critical review on the recovery of base and critical elements from electronic waste-contaminated streams using microbial biotechnology. *Applied Biochemistry and Biotechnology, 2023*, 1–30. https://doi.org/10.1007/S12010-023-04440-X

Mishra, S., Rout, P. K., & Das, A. P. (2019). Solar photovoltaic panels as next generation waste: A review. *Biointerface Research in Applied Chemistry*, *9*(6).

Mohanty, J., Dey, S., Bala, K., & Das, A. P. (2024). Recent advances in microplastic pollution for its sustainable management. *Sustainable Management of Urban Plastic Waste through Circular Economic Approaches, 27–48.* https://doi.org/10.1201/9781003465447-3

Mohanty, S., Ghosh, S., Bal, B., & Das, A. P. (2018). A review of biotechnology processes applied for manganese recovery from wastes. *Reviews in Environmental Science and Bio/Technology, 17*(4), 791–811. https://doi.org/10.1007/S11157-018-9482-1

Mukherjee, A., Sarkar, S., Chakraborty, P., Bose, S., & Das, P. (2024). Application of different techniques for platinum and gold recovery from electronic waste. In *Metal value recovery from industrial waste using advanced physicochemical treatment technologies* (pp. 97–139). Elsevier. https://doi.org/10.1016/B978-0-443-21884-2.00020-4

Nandakumar, N. P., Naveenkumar, M., & Senthilkumar, K. (2024). Heavy metal harmony: Sustainable solutions for industrial effluent treatment. In *Metal value recovery from industrial waste using advanced physicochemical treatment technologies* (pp. 55–69). Elsevier. https://doi.org/10.1016/B978-0-443-21884-2.00004-6

Noor, T., Javid, A., Hussain, A., Bukhari, S. M., Ali, W., Akmal, M., & Hussain, S. M. (2020). Types, sources and management of urban wastes. In *Urban ecology: Emerging patterns and social-ecological systems* (pp. 239–263). Elsevier. https://doi.org/10.1016/B978-0-12-820730-7.00014-8

Oliveri Conti, G., Pulvirenti, E., Cristaldi, A., & Ferrante, M. (2024). Organic waste management and health. In *Occurrence and behavior of emerging contaminants in organic wastes and their control strategies* (pp. 227–239). Elsevier. https://doi.org/10.1016/B978-0-443-13585-9.00013-6

Patel, A., Kumar, S., & Awasthi, A. (2023). Hydrometallurgy: Urban mining of E-waste and its future implications. In *Global E-waste management strategies and future implications* (pp. 105–123). Elsevier. https://doi.org/10.1016/B978-0-323-99919-9.00007-6

Paul, A., Dey, S., Ram, D. K., & Das, A. P. (2023). Hexavalent chromium pollution and its sustainable management through bioremediation. *Geomicrobiology Journal, 1–11*. https://doi.org/10.1080/01490451.2023.2218377

Prabhakar, A., Mishra, S., & Das, A. P. (2019). Isolation and identification of lead (Pb) solubilizing bacteria from automobile waste and its potential for recovery of lead from end of life waste batteries. *Geomicrobiological Journal, 36*(10), 894–903. https://doi.org/10.1080/01490451.2019.1654044

Pradhan, G., Tripathy, B., Ram, D. K., Digal, A. K., & Das, A. P. (2023). Bauxite mining waste pollution and its sustainable management through bioremediation. *Geomicrobiology Journal, 1–10*. https://doi.org/10.1080/01490451.2023.2235353

Prasad, M. N. V. (2024). Integrated municipal solid waste management for energy recovery and pollution prevention. In *Environmental materials and waste: Circular economy and pollution abatement* (pp. 135–192). Elsevier. https://doi.org/10.1016/B978-0-443-22069-2.00018-8

Priyadarshini, B., Santhosh Kumar, M., & Das, A. P. (2024). Microbial degradation of synthetic plastics released from COVID-19 generated single use disposable face masks. *Environmental Science and Engineering, Part F2827*, 1–16. https://doi.org/10.1007/978-3-031-50840-0_1

Priyadarsini, S., & Das, A. P. (2024). Analytical and structural characterization of waste lithium-ion batteries for their effective recycling strategy. *Environmental Chemistry and Ecotoxicology*. https://doi.org/10.1016/j.enceco.2024.12.004

Priyadarsini, S., & Das, A. P. (2025). Lithium bioleaching: A review on microbial-assisted sustainable technology for lithium bio-circularity. *Journal of Water Process Engineering, 69*, 106744. https://doi.org/10.1016/j.jwpe.2024.106744

Radhakrishnan, L., Mary, J. S., Sweta, K., Jee, A. A., Maurya, N. S., Nema, A., & Sharma, D. (2023). Occupational health hazards associated with E-waste handling, treatment, management, and case studies. In *Global E-waste Management Strategies and Future Implications* (pp. 153–181). Elsevier. https://doi.org/10.1016/B978-0-323-99919-9.00014-3

Rahman, R. O. A. (2024). Prospective on management of nuclear waste containing contaminants of emerging concerns. In *Nuclear waste management facilities* (pp. 507–518). Elsevier. https://doi.org/10.1016/b978-0-323-96008-3.00012-7

Rani, D., Rana, V., Rani, A., Malyan, S. K., Kumar, A., Dhaka, R. K., & Rana, A. (2023). Microbial contamination in municipal water: Potential sources, analytical methods and remediation strategies. In *Algae based bioelectrochemical systems for carbon sequestration, carbon storage, bioremediation and bioproduct generation: Bioelectrochemical systems: The way forward* (Vol. 3, pp. 125–141). Elsevier. https://doi.org/10.1016/B978-0-323-91023-1.00009-0

Ray, S. K., Dey, S., Das, A. P., & John, A. (2024). Microplastic waste management through greener technologies: A review. *Sustainable Management of Urban Plastic Waste through Circular Economic Approaches, 237–261*. https://doi.org/10.1201/9781003465447-14

Robin, Kaur, P., Kaur, J., & Bhat, S. A. (2023). Solid waste management: Challenges and health hazards. In *Recent trends in solid waste management* (pp. 171–195). Elsevier. https://doi.org/10.1016/B978-0-443-15206-1.00002-5

Sadasivuni, K. K., Rattan, S., Waseem, S., Brahme, S. K., Kondawar, S. B., Ghosh, S., Das, A. P., Chakraborty, P. K., Adhikari, J., Saha, P., & Mazumdar, P. (2019). Silver nanoparticles and its polymer nanocomposites—Synthesis, optimization, biomedical usage, and its various applications. *Lecture Notes in Bioengineering, 331–373*. https://doi.org/10.1007/978-3-030-04741-2_11

Sahu, A., Dey, S., Mallick, K., Priyadarsini, S., Behera, I. D., & Das, A. P. (2024). A review on advanced technology for sustainable management of synthetic microplastic waste. *Sustainable Management of Urban Plastic Waste through Circular Economic Approaches, 262–279*. https://doi.org/10.1201/9781003465447-15

Sharma, S., Saxena, S., Mudgil, B., & Vats, S. (2021). Advances in biodegradation and bioremediation of environmental pesticide contamination. In *Biological approaches to controlling pollutants* (pp. 79–106). Elsevier. https://doi.org/10.1016/B978-0-12-824316-9.00009-4

Sharma, S., & Singh, N. (2024). Nanobiotechnology for sustainable food waste management. In *Nanobiotechnology for food processing and packaging* (pp. 265–284). Elsevier. https://doi.org/10.1016/B978-0-323-91749-0.00001-0

Singh Jadaun, J., Bansal, S., Sonthalia, A., Rai, A. K., & Singh, S. P. (2022). Biodegradation of plastics for sustainable environment. *Bioresource Technology, 347*, 126697. https://doi.org/10.1016/J.BIORTECH.2022.126697

Singh, N., Poonia, T., Siwal, S. S., Srivastav, A. L., Sharma, H. K., & Mittal, S. K. (2022). Challenges of water contamination in urban areas. In *Current directions in water scarcity research* (pp. 173–202). https://doi.org/10.1016/b978-0-323-91838-1.00008-7

Srivastav, A. L. (2020). Chemical fertilizers and pesticides: Role in groundwater contamination. In *Agrochemicals detection, treatment and remediation: Pesticides and chemical fertilizers* (pp. 143–159). Elsevier. https://doi.org/10.1016/B978-0-08-103017-2.00006-4

Sudhakar, P. P., Dave, H., & Kulwant, M. (2024). Electronic waste management in developing economies: Challenges and approaches. In *Solid waste management for resource-efficient systems* (pp. 43–63). Elsevier. https://doi.org/10.1016/b978-0-443-23775-1.00013-8

Tiwari, A. K., Jha, S., Awasthi, R. R., Singh, A. K., Ojha, R. P., Dikshit, A., Rai, S., & Bhardwaj, A. K. (2024). Nanomaterial production using agro-waste: a current status and future outline. In *Sustainable management of agro-food waste: Fundamental aspects and practical applications* (pp. 249–261). Elsevier. https://doi.org/10.1016/B978-0-443-23679-2.00017-3

Tripathi, G., Husain, A., Ahmad, S., Hasan, Z., & Farooqui, A. (2021). Contamination of water resources in industrial zones. In *Contamination of water: Health risk assessment and treatment strategies* (pp. 85–98). Elsevier. https://doi.org/10.1016/B978-0-12-824058-8.00017-7

Tripathy, B., Dey, S., Dubey, N. K., Lo, W.-C., & Das, A. P. (2024). *An omics-based approach for managing the risk of COVID-19-generated wastes*. 51–66. https://doi.org/10.1007/978-3-031-50840-0_4

United Nations Environment Programme. (2024). *Global waste management outlook 2024—Beyond an age of waste: Turning rubbish into a resource*. https://doi.org/10.59117/20.500.11822/44939

U.S. Occupational Safety and Health Administration. (2017). *Globally harmonized system*.

Villagrán, Z., Anaya-Esparza, L. M., Velázquez-Carriles, C. A., Silva-Jara, J. M., Ruvalcaba-Gómez, J. M., Aurora-Vigo, E. F., Rodríguez-Lafitte, E., Rodríguez-Barajas, N., Balderas-León, I., & Martínez-Esquivias, F. (2024). Plant-based extracts as reducing, capping, and stabilizing agents for the green synthesis of inorganic nanoparticles. In *Resources* (Vol. 13, Issue 6). Multidisciplinary Digital Publishing Institute (MDPI). https://doi.org/10.3390/resources13060070

Xiong, X., Zhang, Y., Wang, L., & Tsang, D. C. W. (2021). Overview of hazardous waste treatment and stabilization/solidification technology. In *Low carbon stabilization and solidification of hazardous wastes* (pp. 1–14). Elsevier. https://doi.org/10.1016/B978-0-12-824004-5.00031-1

Yuliana, M., Santoso, S. P., Putro, J. N., Wijaya, C. J., & Lunardi, V. B. (2022). Environmental impact, health hazards and waste management of medical textile products. In *Medical textiles from natural resources* (pp. 829–864). Elsevier. https://doi.org/10.1016/B978-0-323-90479-7.00006-3

# Integrated Waste Management: Importance, Legislation, and Research Strategies Towards Sustainability

**V. P. Sharma, Pragya Sharma, and Alok Prasad Das**

## 1 Introduction

Integrated waste management (IWM) is a crucial approach for environmental sustainability, combining various waste management methods such as landfill management, recycling, composting, and waste-to-energy systems (Mishra et al., 2023). Community engagement is crucial for citizen responsibility and improved environmental outcomes. Advanced waste management technology is revolutionising the sector, paving the way for environmental sustainability.

One of the most significant approaches for addressing the growing challenges of waste management and the sustainability of the environment is integrated management of waste, or IWM. To minimise waste and increase resource recovery, it combines several waste administration methodologies, including landfill management, recycling, composting, and waste-to-energy systems (Mohanty et al., 2024).

Community engagement's importance: Because it enables citizens to take responsibility for their waste management habits, community engagement is essential to waste management. Involving the community leads to more relevant waste solutions, which increases participation rates and improves environmental outcomes. Active citizens are more inclined to support initiatives in the neighbourhood, develop environmentally friendly habits, and help create an environmentally conscious world.

V. P. Sharma (✉)
ASSIST Division, CSIR-Indian Institute of Toxicology Research, Lucknow, UP, India
e-mail: vpsitrc1@rediffmail.com; vpsiitr@gmail.com

Chief Scientist & Professor AcSIR (Former), Lucknow, India

P. Sharma
National Artist As Dastan Go and Former Research Scholar, BBAU, Lucknow, India

A. P. Das
Department of Life Sciences, RDW University, Bhubaneswar, Orissa, India

 
A. P. Das and S. Karakuş (eds.), *Synergizing Sustainability for Integrated Waste Management*, Sustainable Environmental Waste Management Strategies,
https://doi.org/10.1007/978-3-032-03718-3_10

Thus, advanced waste management technology is revolutionising the sector and paving the way to environmental sustainability. The approaches may be subject to consideration by the related technologies for microplastics and other plastic additives and waste generated from consumer products (Ageel, 2022; Akyildiz, 2024; Ali, 2023; Browne, 2013; Caterino, 2018; Vianello, 2019; Yu, 2008; Zalasiewiez, 2016; Zhang, 2023; Zhao, 2023; Ziani, 2023).

## 2 Recent Approaches for Waste Management

Current approaches for waste management involve the use of blockchain-based technology, chemical recycling, and artificial intelligence (AI) (Priyadarsini et al., 2024, 2025). AI-powered technologies are revolutionising waste sorting and efficiency. Recovery rates may be maximised through the integration of robotic arms with machine learning algorithms that accurately identify, characterise, and sort items (Alimba, 2019; Brandon, 2019; OECD, 1998; World Risk Report, 2025).

Complex and contaminated waste, including multi-layer polymers, is broken down into its initial monomers using advanced chemical recycling procedures. Thereby, it is feasible to produce recycled materials of superior quality (Mohanty et al., 2024; Prabina et al., 2024). The principles of blockchain are used for material monitoring to guarantee both transparency and accountability. It improves compliance and recycling efficiency by generating unchangeable transaction records.

## 3 How is the Concept of Circular Economy Advancing and Why?

One strategy that has gained momentum lately is the circular economy (Figge & Thorpe, 2023). The circular economy is an economic paradigm that seeks to reduce waste and pollution, rehabilitate natural systems, and protect goods and resources that are in use. A circular economy may be beneficial for the environment, the economy, and general well-being. For instance, it might reduce the release of greenhouse gases, increase the effectiveness of resources (World Economic Forum, 2014), establish new business opportunities, improve human welfare, and promote integration into society (World Economic Forum Report, 2025).

In contrast to linear models, the circular economy concepts promote reduction, reuse, and regenerative methods to address urgent issues, including waste management, environmental pollution, and climate change. Additionally, waste entrepreneurship, which is in line with the ideas of the circular economy, turns garbage into useful resources that support social and economic development as well as environmental sustainability. Reducing waste and pollution, reusing materials and products at their best, and regenerating natural products are the goals of the circular economy idea (Bal et al., 2018; Satapathy et al., 2024). By highlighting the different ways that organic and non-organic waste can be valorised, **Waste to Wealth** provides a path towards a sustainable future. Making up 15–30% of lignocellulose, lignin is a complex aromatic polymer that has substantial potential for valorisation.

Enhancing their metabolic capabilities to encompass a broader range of waste materials will facilitate a more efficient conversion of lignin feedstock. Although certain polyhydroxyalkanoate (PHA) manufacturers may utilise specific aromatic monomers derived from lignin for the growth and synthesis of PHA, ferulic and coumaric acids remain the primary components of lignin (Mishra & Das, 2023; Mishra et al., 2021). The biodegradation of biomass involves fermentation methods that transform raw biomass into polymers, such as polyester, PHA produced by bacteria in nutrient-deficient conditions, and feedstocks like lactic acid for polylactic acid (PLA) production. In comparison to traditional plastic beverage packaging, biodegradable polymers offer a more environmentally friendly alternative.

Degradable polymers provide a greener option compared to conventional plastic beverage packaging (Jagatee et al., 2024; Singh et al., 2020). Plant-based polysaccharides such as cellulose and starches can be used to create biodegradable packaging bottles, films, and/or sheets (Mishra et al., 2020). For packaging applications, cellulose nanomaterials generated from wood, cotton, and other plant sources, such as cellulose nanofibers (CNF) and cellulose nanocrystals (CNC), have favourable characteristics.

PLA, a commercially available bioplastic made by lactic acid polymerisation, is ideal for beverage bottles and films because of its high mechanical strength, low toxicity, and efficient gas, water vapour, and odour barrier qualities (Mishra et al., 2024; Satapathy et al., 2024). PHAs are polyesters produced by bacterial fermentation, including polyhydroxybutyrate (PHB) and poly(3-hydroxybutyrate-co-3-hydroxyvalerate) (PHBV). Traditional polymers like polyethylene (PE) and polyvinyl chloride (PVC) raise serious concerns about chemical migration into food products; PHA's non-toxic and biocompatible qualities avoid this issue.

## 4 Green Banks Concept and Carbon Credit

Green banks are often established as standalone entities, internal windows within larger public development banks, or even hybrid country platforms. By tackling high perceived investment risks and resolving information asymmetries, their design offers the financial and technical assistance required to expand low-carbon, climate-resilient projects across a range of geographical areas. Other bioconversion techniques and combinations of techniques include the following:

Utilising microbial enzymes to convert organic waste into biofuels, bioplastics, and other products with added value is known as microbial bioconversion.

Advanced circular economy frameworks involve increasing the efficiency and transparency of waste management by integrating digital tracking systems and blockchain technology.

Creating environmentally friendly catalysts and solvents to improve waste-to-resource conversion while reducing environmental effects is known as "green chemistry innovation".

Moreover, we may utilise a combination of physical, chemical, and biological processes, and hybrid waste valuation approaches aim to recover as much valuable material as possible from waste streams (De Falco, 2019; Dufton, 1998; Duty, 2003; Gong, 2020; OECD, 1998; Prata, 2018).

## 5 Strategies for Effective Engagement in the Food Packaging Sector

Interest in biodegradable materials has increased due to rising environmental pollution concerns and the need for sustainable food packaging. This study provides a thorough analysis of the procedures for converting food waste into the biodegradable polymer polyhydroxyalkanoates (PHA). Bioplastics based on PHA, particularly those made from affordable carbon sources such as waste oils and volatile fatty acids (VFAs), present a viable way to cut down on plastic waste and improve the sustainability of food packaging. The manufacture of PHA can achieve notable efficiency gains through the adjustment of microbial fermentation processes; under the right circumstances, yields can reach up to 87% PHA content.

**Education and Awareness: Training for Trainers (ToT), Workshops, and Programmes**: Training for Trainers (ToT), workshops, and programmes for education and Awareness. Programmes, viz, community workshops, can teach locals about composting, recycling, and trash reduction. Educational initiatives can change behaviour over time by fostering environmental awareness at a young age. There are numerous ways to implement educational programmes, ranging from community seminars that showcase composting methods to school programmes that teach kids the value of recycling. By educating people with the information and resources they need to make more ecologically friendly decisions in their daily lives, these educational programmes are essential for long-term behavioural modifications.

We need to educate that plastics are vital and indispensable materials in the present era, and many products manufactured from plastics are a boon to public health but also simultaneously pose health risks, and thus, the 2025 UN theme for World Environment Day was *Beat Plastic Pollution*. Of principal concern are endocrine-disrupting properties, as triggered, for example by bisphenol A and di-(2-ethylhexyl) phthalate (DEHP) (Halden, 2010).

**Informative Initiatives**: The important role of appropriate garbage elimination and the effects of waste upon the ecosystem may be brought to the public's consciousness through online platforms, posters, and community celebrations.

To fulfil the industrial sector's increasing demands for sustainability, technological advances, and autonomous operations, waste management organisations are adopting the latest modes. Drones monitor landfills to assess air quality and identify radiation, while robotic reuse facilities that sort waste using algorithms powered by artificial intelligence (AI) are being developed to assist entrepreneurs in automating waste sorting, collection, and disposal processes. Cutting-edge recycling and creative composting are two instances of the waste management trends of the future. In addition, businesses are looking into ways to support a closed-loop supply chain by including circular design of goods, biodegradable packaging, sustainable materials, and beyond. Materials lifecycles are tracked by blockchain-powered technologies to increase recycling effectiveness and keep trash out of landfills. The primary focus is on achieving sustainability with the least amount of pollution, as well as strengthening the general efficiency of waste management initiatives.

## 6 Community-Based Initiatives

Although it creates a sense of transparency, increases awareness, and motivates active participation in environmentally friendly waste practices, community engagement is essential to waste management and recycling.

## 7 Recycling Initiatives

Developing recycling programmes that are easy to use and effective promotes participation from the local population (Ghosh et al., 2016; Sahu et al., 2024). Participation is increased when there is clear knowledge of what can be recycled and convenient locations for collection. Setting measurable targets and scheduling waste audits (ISO 17025: 2017) are essential elements of a systematic approach to zero-waste facility administration. We must have approaches or concepts in life related to the 5 Rs, i.e. reduce, reuse, recycle, rethink, and restrain, for minimising pre- and postnatal exposures to potentially harmful components of plastics and other waste categories.

## 8 Composting

Composting programmes in the community transform organic waste into materials that are beneficial. Community responsibility can be reinforced by providing decomposition information and facilities for compost collection. A useful way to reduce organic waste is by composting, which turns yard trash and kitchen leftovers into soil with abundant nutrients that is beneficial for community gardens and green areas. Residents are better able to view waste as a valuable resource that supports local agriculture and lessens the need for synthetic fertilisers attributable to these activities. The scale of communal composting operations varies, from backyard bins to neighbourhood composting centres for each subject, illustrating the adaptability and diversity of composting as a waste management method.

**Active participation**: Organising neighbourhood clean-up drives encourages community spirit and active participation, making debris control initiatives a shared duty.

**Mechanisms of feedback**: Community involvement is strengthened when pathways are established for locals to express their concerns and ideas. Surveys and open forums can provide insightful information and promote teamwork.

**Partnerships and cooperatives**: Government–private collaborations may work together with nearby companies to improve trash management initiatives. The supplies and technological solutions required for efficient waste management can be acquired through these kinds of relationships.

**Government support**: Municipal governments can assist community initiatives by funding the infrastructure that is required, such as recycling facilities, recycling bins, and educational resources.

**Clean-up drives**: Organising neighbourhood clean-up events encourages active participation and fosters community spirit, making waste management efforts a collective responsibility.

**Feedback mechanisms**: Creating channels for residents to voice their concerns and suggestions strengthens community involvement. Surveys and public forums can yield valuable insights and foster cooperation.

**Sustainability initiatives**: Incorporating upcycling into everyday activities is an excellent means to have fun as well as help create a more sustainable future. Sustainable waste management is a global commitment by communities to protect our world for future generations, not just an environmental project. This strategy includes reducing pollution, utilising landfills less frequently, and implementing more environmentally friendly recycling and garbage disposal methods. Apart from being a legal need, industrial disposal of waste is a strategic necessity for sustainable development (UN Sustainable Development Goals).

**Zero waste techniques**: Communities can minimise waste creation through zero-waste efforts by proactively reducing, reusing, and recycling waste. Effective case studies show that residents can make a substantial contribution to achieving these objectives when they participate in the process.

## 9 Guidelines and Legislation(s) Supporting Integrated Waste Management

Laws and policies have recently been put in place by nations all over the world to support sustainable waste management. In India, recycling, trash segregation, and scientific waste disposal have been given priority by the Solid Waste Management Rules, 2016, and their revisions. Regular updates are made to the standards for managing garbage made of plastic. The Basel Convention, one of the international structures, governs the transboundary movement of hazardous waste. Hazardous and Other Wastes (Management and International Movement) Rules, 2016, govern the treatment and management of dangerous materials. Investigating and implementing policies is the responsibility of the Bureau of Indian Standards, the Central Pollution Control Boards, and the Ministry of Environment, Forests, and Climate Change (MOEFCC) in India. The OECD countries that have signed have been working closely with the World Health Organisation (WHO) and the United States Environmental Protection Agency (EPA) (BIS, FSSAI, Regulation EC as amended from time to time).

## 10 Innovative in Research and Development

The niche areas of biotechnology are focusing on waste-to-energy technologies and converting waste into usable energy. The latest big data and Internet of Things (IoT) help in the enhancement of waste collection and management efficiency. The most recent environmentally friendly methods are focused on creating environmentally friendly procedures and bioremediation methods for the benefit of future generations (Roslan, 2024; Kloas; 2020).

## 11 Renewable Energy from Plastic Waste

Universities have created a patented pyrolysis catalytic reactor that generates high-grade fuel oil using potentially dangerous plastic trash, providing a sustainable solution to solving the challenges that accompany urban plastic waste while simultaneously recovering valuable energy (Dey et al., 2024; Paul et al., 2024). Indore, in the state of Madhya Pradesh, has become a model city for environmentally friendly debris handling. The municipal corporation has put in place cutting-edge waste processing facilities, such as a waste-to-energy plant, door-to-door collection of garbage, and source segregation. In addition to incorporating international strategies, Lucknow, the state seat of Uttar Pradesh, is making an effort to adhere to the same model. The latest eco-friendly approaches are directed towards developing bioremediation techniques and sustainable practices for the next generation's betterment. Efforts are towards understanding aspects of microplastics and other waste entities in different matrices (Hirai, 2011; Hussain, 2023; Jenner, 2021; Kabir, 2023; Prata, 2018; Kaprzak, 2022; Lyche, 2009; Marturano, 2017; Miyagawa, 2016).

Countries like Singapore have taken steps towards integrated waste management into their urban planning, focusing on waste segregation, recycling, energy recovery, etc. The city-state has implemented advanced waste-to-energy plants and strict policies to minimise landfill use. In San Francisco, USA, it has achieved high recycling rates through its "Zero Waste" initiatives, and it takes steps for waste segregation, composting, and recycling, aiming to divert all waste from landfills.

The United Nations Environment Programme (UNEP) and World Wildlife Fund (WWF) have launched Plastic Reboot, another major global effort to mitigate plastic pollution by focusing on upstream solutions, viz redesigning packaging and promoting circular economy practices.

## 12 Conclusion

Diverting debris from landfills and nurturing creative, environmentally conscious habits require promoting the creative reuse of resources through upcycling projects. Cutting throwaway bottles in half and turning them into vertical herb gardens is one efficient way to upcycle them. This method gives your kitchen a revitalising splash of green while also saving space. Upcycled crafts are a potent representation of ecological consciousness and creativity. Through the resurgence of dismissed materials, this transformative art transforms what the general public typically views as trash into priceless gems. Beyond only basic craftsmanship, it makes a powerful statement against the disposable society and encourages everyone to an innovative world where inventiveness as well as environmental responsibility coexist. Good Laboratory Practices (GLP) provide a structured framework to ensure the quality, integrity, and reliability of laboratory studies. This would lead to confidence via data integrity and reliability. We would be able to ensure that laboratory data is accurate, reproducible, and traceable with reduced errors and inconsistencies in experimental results. The regulatory compliance would help laboratories to meet national and international regulatory requirements (FDA, EPA, etc.). It would facilitate the acceptance of study results across different regulatory bodies with improved safety and efficient risk management (OECD, 1998).

It is crucial for local governments, organisations like Nagar Nigam and Jal Sansthan, urban local bodies (ULBs), and community leaders to prioritise community engagement in their waste management strategies. By doing so, we may help to cultivate a strong culture of environmental responsibility. Environmental preservation must also be our top responsibility. Pollution reduction through promoting regeneration and energy recovery, and minimising the proportion of debris discharged in landfills. Resource efficiency might contribute to the transformation of waste into useful resources. In doing so, a circular economy would be promoted by producing energy or compost. Effective waste management can ultimately safeguard public health by reducing health hazards related to inappropriate or insufficient disposal.

The enormous volumes of food waste produced each year and the environmental damage brought on by traditional plastics can both be addressed imaginatively and effectively by turning food waste into polyhydroxyalkanoates (PHA). We must comply with the law of the land and follow the polluter pays principle from our internal consciousness.

**Acknowledgements** The authors are thankful to the Director, CSIR-IITR, and the facilities of the Knowledge Resource Centre (KRC) of the institute for access to rich literature sources and databases.

**Conflict of Interest** There is no conflict of interest with any organisations or contributors.

## References

Ageel, H. K., Harrad, S., & Abdallah, M. A. E. (2022). Occurrence, human exposure, and risk of microplastics in the indoor environment. *Environmental Science: Processes & Impacts, 24*, 17–31.

Akyildiz, S. H., Fiore, S., Bruno, M., et al. (2024). Release of microplastic fibres from synthetic textiles during household washing. *Environmental Pollution, 357*, 124455. https://doi.org/10.1016/j.envpol.2024.124455

Ali, T., Habib, A., Muskan, F., Mumtaz, S., & Shams, R. (2023). Health risks posed by microplastics in tea bags: Microplastic pollution—A truly global problem. *International Journal of Surgery, 109*(3), 515–516. https://doi.org/10.1097/JS9

Alimba, C. G., & Faggio, C. (2019). Microplastics in the marine environment: Current trends in environmental pollution and mechanisms of toxicological profile. *Environmental Toxicology and Pharmacology, 68*, 61–74.

Andrady, A. L., & Rajapakse, N. (2016). Additives and chemicals in plastics. Hazardous chemicals associated with plastics in the marine environment. In H. Takada, H. Karapanagioti (Eds.), *The handbook of environmental chemistry* (p. 78). Springer.

Bal, B., Ghosh, S., & Das, A. P. (2018). Microbial recovery and recycling of manganese waste and their future application: A review. *Geomicrobiological Journal, 36*(1), 85–96. https://doi.org/10.1080/01490451.2018.1497731

Brandon, J. A., Jones, W., Ohman, M. D., Brandon, J. A., Jones, W., & Ohman, M. D. (2019). Multidecadal increase in plastic particles in coastal ocean sediments. *Science Advances, 5*, eaax0587.

Browne, M. A., Niven, S. J., Galloway, T. S., Rowland, S. J., & Thompson, R. C. (2013). Microplastic moves pollutants and additives to worms, reducing functions linked to health and biodiversity. *Current Biology, 23*, 2388–2392.

Catarino, A. I., Macchia, V., Sanderson, W. G., Thompson, R. C., & Henry, T. B. (2018). Low levels of microplastics (MP) in wild mussels indicate that MP ingestion by humans is minimal compared to exposure via household fibres fallout during a meal. *Environmental Pollution, 237*, 675–684.

Chandrakanthan, K., Fraser, M. P., & Herckes, P. (2023). Airborne microplastics in a suburban location in the desert southwest: Occurrence and identification challenges. *Atmospheric Environment, 298*, 119617.

Chen, W., Pan, S., Cheng, H., Sweetman, A., Zhang, H., & Jones, K. (2018). Diffusive gradients in thin-films (DGT) for in situ sampling of selected endocrine disrupting chemicals (EDCs) in waters. *Water Research, 137*, 211–219.

Cingotti, N., & Jensen, G. K. (2019) *Health and Environment Alliance (HEAL).* Food Contact Materials and Chemical Contamination; Health and Environment Alliance.

Cook, T. (2019). *How are microplastics transported to polar regions?* (p. 100). EOS.

De Falco, F., Di Pace, E., Cocca, M., & Avella, M. (2019). The contribution of washing processes of synthetic clothes to microplastic pollution. *Science and Reports, 9*(1), 6633. https://doi.org/10.1038/s41598-019-43023-x

Dey, S., Bissoyi, A., Koteswara Reddy, G., & Das, A. P. (2024). Synthetic microfibers: An urban emerging pollutant demanding for sustainable environmental solutions. *Sustainable Microbial Technology for Synthetic and Cellulosic Microfiber Bioremediation, 107–121.* https://doi.org/10.1007/978-3-031-62600-5_7

Dufton, P. W. (1998). *Functional additives for the plastics industry: A report from Rapra's Industry Analysis Group.* CRC Press.

Duty, S. M., Silva, M. J., Barr, D. B., Brock, J. W., Ryan, L., Chen, Z., Herrick, R. F., Christiani, D. C., & Hauser, R. (2003). Phthalate exposure and human semen parameters. *Epidemiology, 14*, 269–277.

Figge, F., Thorpe, A., Gutberlet, M. (2023). Definitions of the circular economy—Circularity matters. *Ecological Economics, 208.*

Ghosh, S., Mohanty, S., Akcil, A., Sukla, L. B., & Das, A. P. (2016). A greener approach for resource recycling: Manganese bioleaching. *Chemosphere, 154*, 628–639. https://doi.org/10.1016/J.CHEMOSPHERE.2016.04.028

Gong, J., & Xie, P. (2020). Research progress in sources, analytical methods, eco-environmental effects, and control measures of microplastics. *Chemosphere, 254*, 126790.

Halden, R. U. (2010). Plastics and health risks. *Annual Review of Public Health, 31*, 179–194.

Hauser, R., & Calafat, A. M. (2005). Phthalates and human health. *Occupational and Environmental Medicine, 62*, 806–818.

Hirai, H., Takada, H., Ogata, Y., Yamashita, R., Mizukawa, K., Saha, M., Kwan, C., Moore, C., Gray, H., Laursen, D., et al. (2011). Organic micropollutants in marine plastics debris from the open ocean and remote and urban beaches. *Marine Pollution Bulletin, 62*, 1683–1692.

Hussain, K. A., Romanova, S., Okur, I., et al. (2023). Assessing the release of microplastics and nanoplastics from plastic containers and reusable food pouches: Implications for human health. *Environmental Science and Technology, 57*(26), 9782–9792. https://doi.org/10.1021/acs.est.3c01942

International Standardisation Organisation ISO/IEC 17025: 2017. (2017). *Guidelines for testing and calibration of laboratories.*

Jagatee, S., Priyadarshini, S., Rath, C. C., & Das, A. P. (2024). Synthetic microfiber: An enduring environmental problem linked to sustainable development. *Environmental Science and Engineering, Part F2333*, 93–112. https://doi.org/10.1007/978-3-031-51792-1_6

Jenner, L. C., Sadofsky, L. R., Danopoulos, E., & Rotchell, J. M. (2021). Household indoor microplastics within the Humber region (United Kingdom): Quantification and chemical characterisation of particles present. *Atmospheric Environment, 259*, 118512.

Jobling, S., Reynolds, T., White, R., Parker, M. G., & Sumpter, J. P. (1995). A variety of environmentally persistent chemicals, including some phthalate plasticizers, are weakly estrogenic. *Environmental Health Perspectives, 103*, 582–587.

Kabir, M. S., Wang, H., Luster-Teasley, S., Zhang, L., & Zhao, R. (2023). Microplastics in landfill leachate: Sources, detection, occurrence, and removal. *Environ Sci Ecotechnol., 16*, 100256. https://doi.org/10.1016/j.ese.2023.100256

Kacprzak, S., & Tijing, L. D. (2022). Microplastics in indoor environment: Sources, mitigation and fate. *Journal of Environmental Chemical Engineering, 10*, 107359.

Kadac-Czapska, K., Knez, E., & Grembecka, M. (2024). Food and human safety: The impact of microplastics. *Critical Reviews in Food Science and Nutrition, 64*, 3502–3521.

Lu, L., et al. (2018). Polystyrene microplastics induce gut microbiota dysbiosis and hepatic lipid metabolism disorder in mice. *Science of the Total Environment*.

Levy, J. I. (1998). Impact of residential nitrogen dioxide exposure on personal exposure: An international study. *Journal of the Air & Waste Management Association, 48*, 553–560.

Li, J., Shan, E., Zhao, J., Teng, J., & Wang, Q. (2023). The factors influencing the vertical transport of microplastics in marine environment: A review. *Science of the Total Environment, 870*, 161893.

Lian, J., et al. (2020). Impact of polystyrene nano plastics (PSNPs) on seed germination and seedling growth of wheat (*Triticum aestivum* L.). *Journal of Hazardous Materials*.

Liao, Z., Ji, X., Ma, Y., Lv, B., Huang, W., Zhu, X., Fang, M., Wang, Q., Wang, X., Dahlgren, R., et al. (2021). Airborne microplastics in indoor and outdoor environments of a coastal city in Eastern China. *Journal of Hazardous Materials, 417*, 126007.

Lyche, L. J., Gutleb, A. C., Bergman, Å., Eriksen, G. S., Murk, A. T. J., Ropstad, E., Saunders, M., & Skaare, J. U. (2009). Reproductive and developmental toxicity of phthalates. *Journal of Toxicology & Environmental Health Part b: Critical Reviews, 12*, 225–249.

Marturano, V., Cerruti, P., & Ambrogi, V. (2017). Polymer additives. *Physical Review Journals, 2*.

Mishra, S., & Das, A. P. (2023). Microbial remediation of synthetic microfiber contaminated wastewater. In M. P. Shah (Ed.), Microbial technologies in industrial wastewater treatment (pp. 337–350). Springer. https://doi.org/10.1007/978-981-99-2435-6_17

Mishra, S., Dash, D., & Das, A. P. (2024). Microbial degradation of polyester microfibers using indigenously isolated bacterial strain *Exiguobacterium* Sp. *CLEAN—Soil, Air, Water, 52*(12), e202300343. https://doi.org/10.1002/CLEN.202300343

Mishra, S., Rout, P. K., & Das, A. P. (2021). Emerging microfiber pollution and its remediation. In R. Prasad (Ed.), *Environmental and microbial biotechnology* (pp. 247–266). Springer. https://doi.org/10.1007/978-981-15-5499-5_9

Mishra, S., Singh, R. P., Rath, C. C., & Das, A. P. (2020). Synthetic microfibers: Source, transport and their remediation. *Journal of Water Process Engineering, 38*, 101612. https://doi.org/10.1016/J.JWPE.2020.101612

Miyagawa, S., Sato, T., Iguchi, T. (2016). Subchapter 101C—Bisphenol A. In Y. Takei, H. Ando, & K. Tsutsui (Eds.), *Handbook of hormones* (pp. 577–578). Academic Press. ISBN 9780128010280.

Mohanty, J., Dey, S., Bala, K., & Das, A. P. (2024). Recent advances in microplastic pollution for its sustainable management. *Sustainable Management of Urban Plastic Waste through Circular Economic Approaches, 27–48*. https://doi.org/10.1201/9781003465447-3

Nakano, H., Alfonso, M. B., Jandang, S., Imai, K., & Arakawa, H. (2024). Microplastic pollution indexes in the coastal and open ocean areas around Japan. *Regional Studies in Marine Science, 69*, 103287.

Nihart, A. J., Garcia, M. A., El Hayek, E., et al. (2025). Bioaccumulation of microplastics in decedent human brains. *Nature Medicine, 31*(4), 1114–1119. https://doi.org/10.1038/s41591-024-03453-1

OECD. (1998). *OECD series on principles of good laboratory practices and compliance monitoring*. NGCMA, ENV/MC/CHEM(98)17.

Paul, A., Priyadarsini, S., Karakuş, S., Behera, J. K., Gnanasekaran, R., & Das, A. P. (2024). Fate and accumulation of synthetic microfibers in the environment. *Sustainable Microbial Technology for Synthetic and Cellulosic Microfiber Bioremediation, 61–76*. https://doi.org/10.1007/978-3-031-62600-5_4

Prabina, K. E., Benya, A., Mohanty, S., Van Hullebusch, E. D., Das, A. P., & Singh, S. (2024). Recovery of multi-functional carbon fiber from COVID-19 generated waste for an effective management through circular economy. *Environmental Science and Engineering, Part F2827*, 83–97. https://doi.org/10.1007/978-3-031-50840-0_6

Prata, J. C. (2018). Airborne microplastics: Consequences to human health? *Environmental Pollution, 234*, 115–126.

Prata, J. C. (2023). Microplastics and human health: Integrating pharmacokinetics. *Critical Reviews in Environment Science and Technology, 53*, 1489–1511.

Priyadarsini, S., & Das, A. P. (2024). Analytical and structural characterization of waste lithium-ion batteries for their effective recycling strategy. *Environmental Chemistry and Ecotoxicology*. https://doi.org/10.1016/j.enceco.2024.12.004

Priyadarsini, S., & Das, A. P. (2025). Lithium bioleaching: A review on microbial-assisted sustainable technology for lithium bio-circularity. *Journal of Water Process Engineering, 69*, 106744. https://doi.org/10.1016/j.jwpe.2024.106744

Regulation, E.C. No 1272/2008 of the European Parliament and of the Council of 16 December 2008 on classification, labelling and packaging of substances and mixtures, amending and repealing directives 67/548/EEC and 1999/45/EC, and amending regulation (EC) No 1907/2006. Off. J. Eur. Union 2008, 2008b L, 353.

Rehse, S., Kloas, W., & Zarfl, C. (2018). Microplastics reduce short-term effects of environmental contaminants. Part I: Effects of bisphenol A on freshwater zooplankton are lower in presence of polyamide particles. *International Journal of Environmental Research and Public Health, 15*, 280.

Rezania, S., Park, J., Din, M. F. M., Taib, S. M., Talaiekhozani, A., Yadav, K. K., & Kamyab, H. (2018). Microplastics pollution in different aquatic environments and biota: A review of recent studies. *Marine Pollution Bulletin, 133*, 191–208.

Ricciardi, M., Pironti, C., Motta, O., Miele, Y., Proto, A., Montano, L. (2023). Correction: Ricciardi et al. Microplastics in the aquatic environment: Occurrence, persistence, analysis, and human exposure. *Water, 15*, 1718.

Rillig, M. C., Ingraffia, R., & De Souza Machado, A. (1805). Microplastic incorporation into soil in agroecosystems. *Frontiers in Plant Science, 2017*, 8.

Rillig, M. C., & Lehmann, A. (2020, 26 June). Microplastic in terrestrial ecosystems. *Science, 368*(6498), 1430–1431. https://doi.org/10.1126/science.abb5979. PMC 7115994. PMID 32587009.

Roscam, A. M. (2019). *Plastic soup: An atlas of ocean pollution.* Island Press. ISBN-10: 1642830089.

Roslan, N. S., Lee, Y. Y., Ibrahim, Y. S., et al. (2024). Detection of microplastics in human tissues and organs: A scoping review. *Journal of Global Health, 14*, 04179. https://doi.org/10.7189/jogh.14.04179

Sahu, A., Mallick, K., & Das, A. P. (2024). *Environmental impact of single-use synthetic face mask and its recycling: A sustainable approach* (pp. 197–212). https://doi.org/10.1007/978-3-031-50840-0_11

Satapathy, S., Kar, D., Priyadarsini, S., Das, A. P., & Kuanar, A. (2024). Application of microorganisms in emerging waste management strategies and remediation of synthetic and cellulosic microfibers. *Sustainable Microbial Technology for Synthetic and Cellulosic Microfiber Bioremediation, 183–206*. https://doi.org/10.1007/978-3-031-62600-5_10

Shelby, M. D. (2008). NTP-CERHR monograph on the potential human reproductive and developmental effects of bisphenol A. *NTP CERHR MON, 22*, 1–64.

Singh, R. P., Mishra, S., & Das, A. P. (2020). Synthetic microfibers: Pollution toxicity and remediation. *Chemosphere, 257*, 127199. https://doi.org/10.1016/j.chemosphere.2020.127199

The REACH Regulation (Regulation (EC) No 1907/2006) on Registration, Evaluation and Authorisation and Restriction of Chemicals. Available online: https://osha.europa.eu/it/legislation/directives/regulation-ec-no-1907-2006-of-the-european-parliament-and-of-the-council. Accessed on December 2, 2019.

Torres, S., Compa, M., Box, A., Pinya, S., & Sureda, A. (2024). Presence and potential effects of microplastics in the digestive tract of two small species of shark from the Balearic Islands. *Fishes, 9*, 55.

U.S. Environmental Protection Agency. How do I recycle common recyclables? U.S. Food and Drug Administration. Microplastics and nanoplastics in foods.

United Nations UN Sustainability Development Goals SDGs 2030.

Vianello, A., Jensen, R. L., Liu, L., & Vollertsen, J. (2019). Simulating human exposure to indoor airborne microplastics using a Breathing Thermal Manikin. *Science and Reports, 9*, 8670.

Vom Saal, F. S., & Myers, J. P. (2008). Bisphenol A and risk of metabolic disorders. *JAMA, 300*, 1353–1355.
Wagner, M., & Lambert, S. (2018). Freshwater microplastics. Emerging environmental contaminants? In *The handbook of environmental chemistry* (Vol. 58). Springer Open.
Waring, R. H., Harrisa, R. M., & Mitchell, S. C. (2018). Plastic contamination of the food chain: A threat to human health? *Maturitas, 115*, 64–68.
Wei, W., Huang, Q. S., Sun, J., Wang, J. Y., Wu, S. L., & Ni, B. J. (2019). Polyvinyl chloride microplastics affect methane production from the anaerobic digestion of waste activated sludge through leaching toxic Bisphenol-A. *Environmental Science and Technology, 53*, 2509–2517.
World Economic Forum January 15th. (2025). In M. Elsner, G. Atkinson, & S. Zahidi (Eds.), *Global risk report-2025* (20th ed.).
Yu, Z., Peldszus, S., & Huck, P. M. (2008). Adsorption characteristics of selected pharmaceuticals and an endocrine disrupting compound-Naproxen, carbamazepine and nonylphenol on activated carbon. *Water Research, 42*, 2873–2882.
Yuan, W., Liu, X., Wang, W., Di, M., & Wang, J. (2019). Microplastic abundance, distribution and composition in water, sediments, and wild fish from Poyang Lake, China. *Ecotoxicology and Environmental Safety, 170*, 180–187.
Zalasiewicz, J., Waters, C. N., Do Sul, I. J. A., Corcoran, P. L., Barnosky, A. D., Cearreta, A., Edgeworth, M., Gałuszka, A., Jeandel, C., Leinfelder, R., et al. (2016). The geological cycle of plastics and their use as a stratigraphic indicator of the Anthropocene. *Anthropocene, 13*, 4–17.
Zhang, K., & Wu, C. (2023). Formation of airborne microplastics. In *Comprehensive analytical chemistry* (Vol. 100, pp. 1–16). Elsevier.
Zhang, Q., Xu, E. G., Li, J., Chen, Q., Ma, L., Zeng, E. Y., & Shi, H. (2020). A review of microplastics in table salt, drinking water, and air: Direct human exposure. *Environmental Science and Technology, 54*, 3740–3751.
Zhao, X., & You, F. (2024). Microplastic human dietary uptake from 1990 to 2018 grew across 109 major developing and industrialized countries but can be halved by plastic debris removal. *Environmental Science and Technology, 58*, 8709–8723.
Zhao, X., Zhou, Y., Liang, C., Song, J., Yu, S., Liao, G., Zou, P., Tang, K. H. D., & Wu, C. (2023). Airborne microplastics: Occurrence, sources, fate, risks and mitigation. *Science of the Total Environment, 858*, 159943.
Ziani, K., Ioniță-Mîndrican, C.-B., Mititelu, M., Neacșu, S. M., Negrei, C., Moroșan, E., Drăgănescu, D., & Preda, O.-T. (2023). Microplastics: A real global threat for environment and food safety: A state of the art review. *Nutrients, 15*, 617.

# Leveraging AI for Intelligent Waste Management in Smart Cities

**Ritah Nabawagga, Jeanritah Roenah Birungi, Shikta Singh, and Satya Ranjan Dash**

## 1 Introduction

### *1.1 Background*

Cities are growing at an unprecedented rate, and so is the waste they produce. Every day, tons of garbage pile up in urban areas, overwhelming waste management systems that were never designed to handle such massive loads (Akankshya Das et al., 2024; Dey et al., 2023). Studies suggest that by 2050, global waste production could reach a staggering 3.4 billion metric tons, putting immense pressure on already stretched municipal services (World Bank, 2022). The problem isn't just about the sheer volume of waste. Traditional waste collection methods are inefficient; garbage trucks follow fixed schedules instead of responding to real-time waste levels, leading to overflowing bins in some areas and unnecessary pickups in others. Landfills continue to expand, emitting methane and other harmful greenhouse gases (Kaza et al., 2018) due to inadequate infrastructure, insufficient funding, and resources, leading to illegal

R. Nabawagga
School of Technology and Innovations, University of Vaasa, Vaasa, Finland

J. R. Birungi
School of Computer Science and Engineering, University of Bologna, Bologna, Italy

S. Singh (✉)
School of Management, KIIT University, Bhubaneswar, India
e-mail: shiktasingh@ksom.ac.in

S. R. Dash
School of Computer Applications, KIIT University, Bhubaneswar, India
e-mail: sdashfca@kiit.ac.in



A. P. Das and S. Karakuş (eds.), *Synergizing Sustainability for Integrated Waste Management*, Sustainable Environmental Waste Management Strategies,
https://doi.org/10.1007/978-3-032-03718-3_11

dumping and environmental pollution (Bundhoo, 2018). Recycling, despite its benefits, remains poorly optimized in many cities due to poor waste segregation and outdated sorting processes (Bundhoo, 2018).

To address these challenges, artificial intelligence (AI) offers smart, data-driven solutions that can revolutionize urban waste management. Machine learning (ML) algorithms can predict waste patterns, Internet of Things (IoT)-powered smart bins can notify collection teams when they are full, and robotic sorting systems can automate recycling with remarkable accuracy (Zhang et al., 2023). Even though these technologies are currently applied in smart cities, there is a pressing need to evolve beyond fragmented solutions toward a holistic and responsive waste management model. Current systems, while innovative, often lack seamless integration, scalability, and adaptability to the growing demands of urban populations. A more unified framework is required—one that integrates advanced technologies with sustainability practices and community engagement to create resilient, efficient, and eco-friendly waste management systems tailored to the dynamic needs of future cities. This shift underscores the importance of aligning technological applications with broader sustainability goals to address the multifaceted challenges posed by rapid urbanization (Szpilko et al., 2023).

The benefits of the proposed framework include:

1. Improved waste management efficiency: AI-powered waste management systems can optimize waste collection routes, reduce waste disposal costs, and increase recycling rates.
2. Enhanced sustainability: The framework promotes energy recovery, reduces greenhouse gas emissions, and minimizes environmental harm.
3. Increased economic viability: The intelligent system model enables cities to reduce waste management costs, generate revenue through recycling and energy recovery, and create new business opportunities.
4. Better decision-making: The framework provides real-time data and analytics, enabling cities to make informed decisions about waste management strategies and infrastructure investments.

## 1.2 Problem Statement

Despite technological advancements, many cities still struggle with outdated and inefficient waste management practices. Overflowing landfills, inefficient collection routes, and low recycling rates continue to pose environmental, economic, and social challenges (Dey et al., 2023; Mishra et al., 2023). The problem is not just about collecting and disposing of waste; it's about managing it intelligently. Most current waste management systems rely on reactive approaches, meaning they address waste problems only after they occur. Without predictive analytics or real-time monitoring, municipalities end up spending more resources than necessary, while environmental pollution worsens (Al-Sabbagh et al., 2022). Moreover, poor waste segregation at the source makes recycling more complex and less cost-effective, leading to valuable materials ending up in landfills instead of being repurposed (Alok Das et al., 2022; Mishra et al., 2019; Mohanty et al., 2018).

AI offers a way to shift from a reactive to a proactive waste management model. However, while AI technologies exist, their application in urban waste management remains underutilized. Many cities lack the knowledge, infrastructure, or policies needed to implement AI-driven solutions at scale. This paper aims to bridge the gap between current waste management challenges and the potential of AI technologies. By analyzing existing practices and proposing innovative, AI-driven solutions, this research charts a path toward smarter, more sustainable urban environments.

## *1.3 Research Objectives*

This study seeks to:

1. Examine the current challenges in urban waste management and identify inefficiencies.
2. Explore the role of AI technologies—such as machine learning, IoT, and robotics—in optimizing waste collection, sorting, and disposal.
3. Propose a comprehensive AI-powered waste management framework tailored for smart cities.
4. Assess the potential benefits of AI-driven waste management, including cost reductions, improved sustainability, and enhanced operational efficiency.
5. Identify barriers to AI adoption in waste management and suggest policy and implementation strategies to overcome them.

## *1.4 Significance of the Study*

The adoption of AI in waste management isn't just about keeping streets clean—it's about creating a sustainable future. AI-driven waste tracking and sorting systems can significantly reduce landfill waste, lower carbon emissions, and improve recycling efficiency (Gupta & Pathak, 2021). Economically, optimized collection routes and automated sorting can cut operational costs for municipalities while opening up new revenue streams through better resource recovery. Socially, cleaner urban environments contribute to better public health and improved quality of life.

This research is particularly valuable for city planners, policymakers, and technology developers who are looking for feasible ways to integrate AI into smart city initiatives. By providing a structured approach to AI-based waste management, this study contributes to the broader goal of making urban spaces more livable, sustainable, and technologically advanced.

### 1.5 Structure of the Paper

To explore these ideas in detail, this paper is structured as follows:

- Section 2 provides an overview of how waste management currently functions in smart cities, highlighting existing challenges.
- Section 3 dives into AI-driven solutions, exploring how technologies like machine learning, IoT, and automation are transforming waste management.
- Section 4 introduces a proposed AI-powered waste management framework, detailing its architecture and real-world applications.
- Section 5 discusses the key benefits AI brings to waste management, from economic advantages to sustainability improvements.
- Section 6 addresses the barriers and strategic solutions to AI implementation in urban waste systems.
- Section 7 explores future directions and research opportunities in AI-based waste management.
- Section 8 concludes with key takeaways and recommendations for smart cities moving forward.

## 2 Overview of Waste Management in Smart Cities

Efficient waste management plays a crucial role in sustainable urban development, especially in smart cities that utilize technology to enhance urban services (Mishra & Das, 2023; Tripathy et al., 2024). As cities rapidly expand, driven by industrial growth and population increases, waste production has surged. Effective waste management is essential to reducing environmental hazards, lowering costs, and improving the quality of life for urban residents. However, traditional waste management systems still rely on outdated and inefficient methods, making it difficult to handle the growing volume of waste in urban areas (Mishra et al., 2024, 2021). These outdated systems contribute to environmental degradation and financial strain (Mohabuth et al., 2024). A major concern is the overcapacity of landfills, which leads to methane emissions, soil and water contamination, and health risks for nearby communities (Olawade et al., 2024). Additionally, finding new landfill sites in urban areas has become increasingly difficult, intensifying the waste crisis (Szpilko et al., 2023). Operational inefficiencies further exacerbate the situation. Waste collection often follows fixed schedules, leading to excessive fuel use and inefficient operations as trucks travel predetermined routes, ignoring real-time waste levels (Ghahramani et al., 2021). Recycling efforts also fall short due to inadequate waste segregation and insufficient infrastructure, resulting in the loss of valuable recyclable resources (Behera et al., 2020). Although data-driven route optimization has shown promise in reducing costs and improving efficiency, such advancements are not commonly implemented in traditional systems.

Smart cities offer a solution by overcoming the limitations of traditional waste management through the integration of innovative technologies, such as IoT systems, machine learning, and blockchain. These technologies enable real-time waste level monitoring via smart sensors, allowing for dynamic route optimization that reduces fuel consumption and operational inefficiencies (Ghahramani et al., 2021). Furthermore, AI and robotic systems enhance waste sorting and recycling processes, increasing the recovery of valuable materials and reducing reliance on landfills (Behera et al., 2020). Despite these advancements, challenges remain, including high initial implementation costs, limited scalability, and fragmented data systems, which hinder seamless integration across cities (Szpilko et al., 2023). To overcome these obstacles, a comprehensive and collaborative approach is necessary, combining technological innovation with effective policies and community engagement to create sustainable and adaptive waste management solutions for the future.

## 2.1 *Types of Waste*

Urban areas produce various waste types, each requiring specialized management strategies. Solid waste, which includes household, commercial, and construction debris, poses a significant challenge, particularly in cities where improper disposal remains prevalent (Jha et al., 2022; Szpilko et al., 2023) (Mishra et al., 2022a, 2022b; Mishra and Das, 2022). A study by Ayadinuno et al. (2021) on waste management in Awka, Nigeria, highlighted the accumulation of solid waste in streets and markets due to widespread open dumping. Similar practices are observed in other Nigerian cities, where waste is discarded in drains, rivers, abandoned buildings, and public spaces, creating environmental and health hazards. Hazardous waste, comprising medical, industrial, and electronic waste, adds complexity to urban waste management due to its potential risks (Jha et al., 2022) (Ibrahim et al. 2024). Electronic waste, or e-waste, including discarded computers, mobile phones, and appliances, contains hazardous materials like lead and mercury, requiring specialized handling and recycling processes to prevent environmental contamination. Liquid waste (Mekonnen, 2016), including sewage, industrial runoff, and stormwater, can pollute water sources if left untreated. Organic waste (Sharma et al., 2024), such as food scraps and yard debris, presents an opportunity for composting and soil enrichment (Behera et al., 2020). Additionally, recyclable materials like paper, plastics, glass, and metals can be reprocessed into new products, helping to reduce landfill waste and preserve natural resources (Fang et al., 2023).

## 2.2 Environmental, Economic, and Social Impacts of Poor Waste Management

The negative effects of ineffective waste management are extensive:

- **Environmental impact**: Improper waste disposal leads to groundwater contamination, land degradation, and declining air and soil quality, particularly when hazardous waste is mismanaged (Mohabuth et al., 2024; Alahi et al., 2023; Lipianina-Honcharenko et al., 2024). Furthermore, burning solid waste without considering biomass-related carbon emissions can misrepresent greenhouse gas reduction efforts (Fang et al., 2023). Improper waste disposal can also disrupt ecosystems, leading to habitat destruction and loss of biodiversity, particularly in areas near landfills and waste disposal sites (Bala et al., 2024; Kumar et al., 2023).
- **Health risks**: Overflowing waste bins create breeding grounds for disease-carrying pests, while exposure to unmanaged waste has been linked to increased risks of cancer, higher child mortality rates, and congenital disorders (Olawade et al., 2024; Ghahramani et al., 2021; Sosunova et al., 2022). The health risks associated with poor waste management lead to increased healthcare costs, further straining economic resources (Das et al., 2018; Mishra et al., 2022a, 2022b).
- **Economic burden**: Inefficiencies in waste disposal contribute to rising operational costs, particularly in transportation, which can range from USD 20 to 50 per ton. Poor management also leads to resource misallocation (Olawade et al., 2024). Noncompliance with waste management regulations can result in hefty fines and legal penalties, adding to the economic burden.
- **Social challenges**: Public awareness about effective waste management remains low, and ethical concerns surrounding AI-driven waste solutions, such as job displacement and environmental justice, need to be addressed to ensure equitable implementation (Olawade et al., 2024; Sosunova et al., 2022). Engaging communities is crucial to aligning AI-based waste initiatives with societal needs (Olawade et al., 2024). Poor waste management often disproportionately affects marginalized communities, exacerbating social inequalities and environmental injustice (Ray et al., 2024; Sahu et al., 2024).

Addressing these challenges requires innovative approaches beyond conventional waste management. The next section explores how AI-driven technologies—including computer vision, machine learning, IoT, and robotics are transforming waste collection, sorting, recycling, and monitoring, leading to more efficient and sustainable urban waste management systems in smart cities.

# 3 AI Technologies in Waste Management

AI-driven solutions are significantly transforming waste management in smart cities, offering innovative approaches to collection, sorting, recycling, and monitoring (Olawade et al., 2024). These solutions leverage technologies such as machine learning, the Internet of Things (IoT), computer vision, and automation (including robotics) to optimize efficiency, reduce costs, increase recycling rates, and minimize environmental harm (Fang et al., 2023). The integration of AI is crucial for enhancing sustainability and operational effectiveness in urban waste management (Nwokediegwu et al., 2024).

## *3.1 Internet of Things (IoT)*

The foundation of many AI-driven waste management systems lies in the Internet of Things (IoT) (Sosunova et al., 2022). IoT devices, such as sensors embedded in smart bins and waste collection vehicles, collect real-time data on various parameters like waste levels, temperature, humidity, and location (Vishnu et al., 2021). This constant stream of data provides a comprehensive overview of waste generation and management processes (Sosunova et al., 2022). Sensor-equipped smart bins enable real-time waste level monitoring, allowing for optimized collection planning and preventing overflows (Olawade et al., 2024; Sutikno, 2024). AI-driven route optimization analyses historical data, traffic conditions, and waste generation patterns to minimize travel time, fuel consumption, emissions, and operational costs (Olawade et al., 2024; Yao et al., 2024). Additionally, dynamic scheduling informed by real-time data and demand prediction models further enhances collection efficiency and resource allocation (Namoun et al., 2022; Olawade et al., 2024).

## *3.2 Machine Learning*

Machine learning (ML) algorithms play a vital role in analyzing the vast quantities of data generated by IoT sensors (Olawade et al., 2024). ML enables systems to identify patterns, predict trends, and make data-driven decisions without explicit programming (Fang et al., 2023; Lipianina-Honcharenko et al., 2024). They demonstrate the use of machine learning algorithms, such as k-means clustering, to process and analyze data from urban waste streams. When integrated with IoT sensors, these algorithms help identify waste generation patterns across various regions, aiding in the dynamic allocation of collection resources and forecasting future waste production trends. This data-driven approach streamlines operations, reduces inefficiencies, and supports sustainable waste management practices.

## 3.3 Computer Vision

AI significantly advances waste sorting through automated systems that integrate computer vision, high-resolution cameras, hyperspectral imaging, and advanced sensors. Deep learning models, particularly convolutional neural networks (CNNs), accurately classify waste based on visual features (Olawade et al., 2024). These AI-driven systems can distinguish between different types of waste, ensuring proper categorization and reducing contamination in recycling streams. This improves sorting efficiency and enhances material recovery rates (Czekała et al., 2023; Sutikno, 2024).

## 3.4 Robotics and Automation

AI-powered robotics further enhances sorting efficiency, performing tasks with greater speed and precision while adapting to varying waste streams. In recycling, AI enhances material identification and sorting, increasing the purity of recycled materials and overall efficiency (Olawade et al., 2024). It also reduces costs and improves quality control through AI-powered systems that assess material quality (Olawade et al., 2024; Sutikno, 2024). Additionally, AI-driven robotics and automation streamline recycling operations, improving throughput and minimizing manual labor requirements (Fang et al., 2023; Olawade et al., 2024).

## 3.5 AI for Waste Monitoring

AI plays a vital role in waste monitoring through real-time systems that analyze sensor data from waste levels, collection routes, and disposal points, enabling proactive responses to operational challenges (Olawade et al., 2024; Sutikno, 2024). Predictive analytics forecast waste generation rates, facilitating improved planning and resource allocation (Lipianina-Honcharenko et al., 2024). AI-driven data analysis supports informed decision-making by identifying trends and patterns, while the integration of IoT and sensor networks provides comprehensive monitoring for continuous optimization (Sosunova & Porras, 2022).

Building upon these AI advancements, the next section presents a proposed model for AI-driven waste management in smart cities, showcasing how the described AI technologies can be applied in an integrated system. This model outlines the flow of waste from collection to disposal, emphasizing key components such as smart bins, citizen engagement, predictive analytics, and waste-to-energy (WTE) conversion, ensuring a comprehensive and sustainable waste management strategy.

## 4 Proposed AI-Powered Waste Management Framework

The proposed AI-driven waste management model, as shown in Fig. 1, leverages advanced technologies such as artificial intelligence, computer vision, machine learning, Internet of Things (IoT), and robotics to create a seamless and efficient waste management ecosystem. This model encourages public participation while optimizing collection, sorting, and recycling processes to maximize resource recovery and minimize environmental pollution.

The process begins with citizens sorting their waste into intelligent bins equipped with AI-driven recognition systems. These bins use computer vision and machine learning to identify the type of waste deposited. In cases of incorrect disposal, the system rejects the item and provides real-time guidance on proper sorting through an interactive screen. Cloud storage solutions are used to ensure real-time waste data is securely stored and accessible for future analytics and decision-making. To enhance compliance, cameras installed around disposal areas monitor sorting patterns and human behavior, deterring improper waste disposal and providing valuable data for optimizing waste management policies. A dedicated mobile application further facilitates citizen participation. The app incentivizes users by awarding digital tokens, cash incentives, tax discounts, or reduced garbage collection fees based on sorting accuracy. Additional features, such as leaderboards, educational resources, and a community engagement section, foster a gamified approach that enhances civic responsibility and promotes a culture of responsible waste disposal.

Industries contribute to the recycling process by deploying reverse vending machines in strategic locations. These machines allow individuals to deposit recyclable materials, such as plastic bottles and aluminum cans, in exchange for incentives. Such a system encourages recycling at the source and reduces the burden on centralized waste management facilities. Empirical evidence supports the effectiveness of this approach. In Spain, 62.6% of participants identified financial incentives

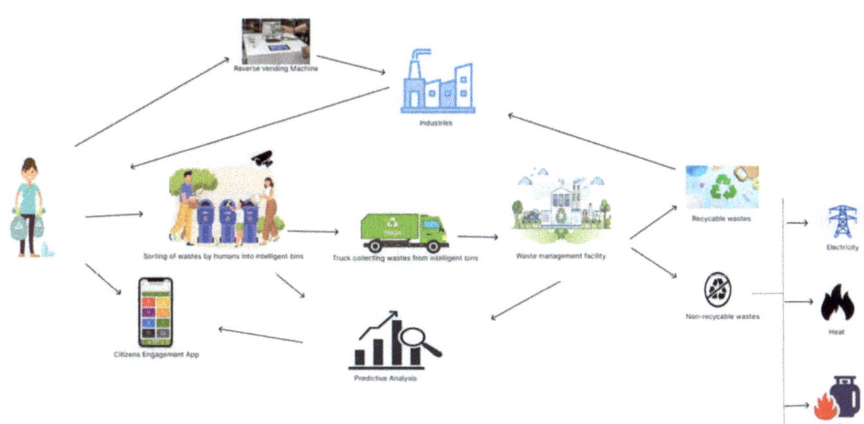


**Fig. 1** A proposed AI-driven waste management model with integrated public participation

as the primary motivator for recycling, while 52.8% recognized these rewards as enhancing their knowledge of waste recycling (Gibovic et al., 2021). Similarly, in Shanghai, a 2021 pilot project implementing an IoT-based smart incentive system led to a 229.3% increase in recyclable waste collection, demonstrating the importance of responsive feedback and efficient resource allocation (Zhou et al., 2021).

Once intelligent bins reach their capacity, IoT-enabled sensors supported by 5G send instant alerts to waste collection services. These alerts are transmitted and processed using cloud computing, allowing for real-time tracking and coordination of waste collection trucks. Predictive analytics is then applied to optimize waste collection routes, ensuring timely collection while reducing fuel consumption and operational costs. 5G connectivity ensures that waste collection trucks receive live updates on route optimization, allowing them to adjust dynamically based on real-time conditions. The AI model also predicts when bins are likely to reach capacity again based on user behavior, seasonal trends, and other factors. A study by Ferrer et al. (2019) introduced BIN-CT, a software system that plans optimal waste collection routes using computational learning algorithms. The system leverages historical and predictive data to minimize the distance travelled by collection trucks, thereby reducing fuel consumption and operational costs. Cloud storage is used to retain this historical data, enabling continuous learning and refinement of route optimization algorithms. The implementation of BIN-CT in a Spanish city demonstrated that accurate fill-level predictions prevent container overflows and unnecessary collection trips, leading to cost reductions and decreased emissions. Moreover, a 2023 study explored the application of particle swarm optimization (PSO) in AI-based predictive waste management. The research highlighted that integrating PSO algorithms with AI technologies can revolutionize sustainability and efficiency in waste collection, sorting, recycling, and overall waste optimization. This integration enables more accurate predictions of waste generation patterns, facilitating proactive decision-making and resource allocation (Gandhimathinathan, 2023).

Upon arrival at waste management facilities, advanced robotics technology is used to enhance the precision of waste sorting processes. While intelligent bins facilitate initial waste categorization, robots equipped with artificial intelligence (AI) and machine learning algorithms further separate recyclables from nonrecyclable materials. These systems are particularly adept at identifying and recovering recyclables that may have been incorrectly disposed of, thereby improving the overall efficiency of recycling operations. Studies have demonstrated that robotic sorting systems can significantly outperform manual methods in accuracy. For instance, certain waste-sorting robots are capable of classifying items into over 100 categories of recyclables with up to 99% accuracy, a level of precision unattainable by human workers (Construction Management Association of America (CMAA), 2024). Additionally, AI-driven platforms like Recycle OS have achieved sorting accuracies exceeding 95%, further highlighting the potential of robotics in waste management (Kesari, 2024). The integration of robotics not only enhances sorting accuracy but also addresses challenges associated with manual sorting, such as labor shortages and safety concerns. Robots provide reliability, predictability, and consistency, eliminating variables like worker absenteeism and operational disruptions

(Recycling Inside, 2024). Moreover, robotic systems equipped with AI and advanced sensors can accurately identify various waste materials, including plastics and metals, thereby streamlining the recycling process and contributing to more sustainable waste management practices (AZoRobotics, 2024).

The sorted recyclable waste is transported to industries where it serves as raw material for manufacturing new products, closing the loop in the circular economy. Nonrecyclable waste, on the other hand, is utilized for energy recovery. Modern waste-to-energy (WTE) technologies convert this waste into electricity, heat, and biogas through processes such as incineration, gasification, and anaerobic digestion. Research supports that WTE facilities can significantly reduce landfill dependency and contribute to sustainable energy generation. For instance, Kumar et al.(2024) analyzed the efficiency of modern WTE technologies and found that these systems offer environmentally sound and economically viable alternatives to conventional fossil fuel-based power plants. Additionally, comprehensive evaluations of various WTE technologies, including pyrolysis and anaerobic digestion, highlight their potential in sustainable waste management and energy production (Sharma et al., 2024). To enforce proper waste management, the model integrates regulatory measures such as fines and penalties for noncompliance. Industrial entities are subjected to a "Pay-As-You-Throw" scheme, where waste generation beyond a certain threshold incurs additional charges. Public awareness campaigns and community service programs for repeat offenders further reinforce responsible waste disposal habits.

The integration of cloud computing, cloud storage, and 5G significantly enhances the efficiency of the waste management model by streamlining operations and improving responsiveness. Cloud computing enables scalable data processing, facilitating waste classification, predictive analytics, and route optimization with greater accuracy and speed. Meanwhile, cloud storage ensures real-time access to essential waste management data, such as bin fill levels, AI model updates, and compliance reports, enabling better monitoring and decision-making. Additionally, 5G connectivity supports ultrafast communication between IoT devices, AI models, and waste collection services, reducing latency and enhancing system responsiveness to dynamic waste management demands. Together, these technologies create a seamless and intelligent system that optimizes performance at every stage of the process.

In conclusion, this AI-driven waste management model offers a comprehensive framework that combines advanced technologies with active community participation and regulatory enforcement to address modern waste management challenges. Beyond its functionalities, this approach unlocks significant benefits, including improved efficiency, enhanced sustainability, economic viability, and informed decision-making. The following section delves into these advantages, highlighting the transformative potential of this intelligent system to revolutionize waste management for urban environments.

# 5 Key Benefits of AI in Waste Management

The proposed AI-powered waste management model offers a comprehensive approach to addressing the challenges of urban waste management. This section explores the key benefits of implementing this model, focusing on improved efficiency, sustainability, economic viability, and decision-making.

## *5.1 Improved Waste Management Efficiency*

Our model leverages AI to significantly enhance waste management efficiency. By integrating predictive analytics and real-time data monitoring, the model optimizes waste collection routes, ensuring timely pickups and reducing unnecessary trips. This dynamic scheduling minimizes fuel consumption and operational costs, while also preventing waste overflow (Ferrer et al., 2019). The use of sensor-equipped smart bins within the model allows for precise tracking of waste levels, further enhancing collection efficiency. Additionally, the model uses advanced AI technologies, such as computer vision and robotics, to automate waste sorting processes. This automation not only increases the accuracy of recyclable material identification but also accelerates the sorting process, leading to higher recycling rates and reduced reliance on manual labor (Kesari, 2024).

## *5.2 Enhanced Sustainability*

The proposed model promotes sustainability by facilitating energy recovery and reducing greenhouse gas emissions. Through optimized waste management practices, the model diverts waste from landfills to recycling and waste-to-energy facilities (Ghosh et al., 2017; Pradhan et al., 2023). This diversion minimizes methane emissions from landfills, contributing to a cleaner environment (Kumar et al., 2024). Moreover, the model supports waste-to-energy conversion through advanced technologies like incineration and gasification. By converting nonrecyclable waste into usable energy, the model provides a sustainable alternative to traditional waste disposal methods, reducing reliance on fossil fuels and promoting cleaner energy production (Sharma et al., 2024).

## *5.3 Increased Economic Viability*

Implementing our AI-powered waste management model offers substantial economic benefits. By optimizing collection routes and reducing disposal costs, cities can achieve significant savings in operational expenses. The enhanced recycling capabilities of the model also create opportunities for generating revenue through the sale of recycled materials to industries (Zhou et al., 2021). Furthermore, the model's waste-to-energy solutions provide an additional avenue for economic viability. By converting waste into energy, cities can reduce their dependence on external energy sources, lowering energy costs and enhancing energy security. This economic benefit is further amplified by the potential for job creation in the waste management and energy sectors, contributing to local economic development.

### 5.4 *Better Decision-Making*

The model's AI-driven analytics provide real-time data and insights that support informed decision-making in waste management strategies and infrastructure investments. By analyzing historical data and current trends, the model can forecast waste generation patterns, enabling proactive planning and resource allocation (Gandhimathinathan, 2023). Additionally, the model facilitates data-driven decision-making by identifying inefficiencies and opportunities for improvement within the waste management system. This insight is invaluable for policymakers and city planners, enabling them to make strategic investments in infrastructure and technology that enhance waste management outcomes.

However, despite these promising advantages, significant barriers still exist that may hinder the widespread adoption and implementation of such advanced systems. Addressing these obstacles is critical to fully realizing the potential of AI-driven waste management. The following section examines these barriers in detail and explores strategies to overcome them effectively.

## 6 Barriers to Adoption and Strategic Solutions

### 6.1 *Barriers*

While the proposed AI-driven waste management model demonstrates significant potential in revolutionizing waste management practices, several barriers might impede its successful implementation. These challenges, rooted in technological, financial, and societal factors, must be carefully addressed to ensure widespread adoption and effectiveness.

The effectiveness of AI in waste management heavily depends on high-quality, standardized, and interoperable data. However, waste management data often suffers from fragmentation, inconsistencies, and a lack of integration among stakeholders, limiting the accuracy and reliability of AI predictions (Olawade et al., 2024). Additionally, real-time monitoring of waste disposal behaviors through IoT-enabled sensors and cameras raises concerns about data privacy and security (Mohabuth et al., 2024). Substantial investment in specialized hardware, software, and computational resources is required (Olawade et al., 2024; Yao et al., 2024). The deployment of intelligent bins with computer vision, predictive analytics for waste collection optimization, and robotic sorting systems all demand significant financial commitments. Many municipalities and waste management agencies may find these costs prohibitive, particularly in regions with budgetary constraints. Despite advancements in AI, challenges persist in adapting existing models to waste management-specific needs. Many current AI applications rely on preexisting models rather than custom-built systems tailored to waste categorization and recycling processes (Fang et al., 2023). Additionally, AI algorithms often function as "black boxes," meaning their decision-making processes are not easily interpretable. This lack of transparency can lead to uncertainty in regulatory adoption and operational trust (Szpilko et al., 2023).

For seamless adoption, the proposed AI solutions must integrate with traditional waste collection infrastructure. However, interoperability challenges may arise due to the lack of standardized data formats and communication protocols across different waste management facilities (Olawade et al., 2024). Without proper alignment, AI-driven sorting systems and predictive analytics may struggle to function efficiently within existing frameworks. Scaling AI-driven waste management systems across diverse urban and rural settings presents another major challenge. Processing large volumes of real-time data from intelligent bins, IoT sensors, and robotic sorting facilities requires significant computational power and optimized algorithms (Olawade et al., 2024). Ensuring these technologies remain efficient and cost-effective at a city-wide or national scale remains a key hurdle. The success of the proposed model relies on active public participation in waste sorting and recycling incentives. However, limited environmental awareness and resistance to adopting new technologies can undermine engagement (Czekała et al., 2023; Szpilko et al., 2023). Even with financial incentives, behavioral shifts in waste disposal habits require sustained public education and regulatory enforcement. The widespread adoption of AI and robotics in waste management could lead to job displacement for manual waste workers, raising ethical concerns about automation's impact on employment (Czekała et al., 2023; Olawade et al., 2024). The lack of universal standards and policy norms for AI-enabled waste management presents a major barrier to implementation. Inconsistent regulatory frameworks across different regions can delay technology deployment and hinder cross-sector collaboration (Szpilko et al., 2023). Additionally, ensuring compliance with available data privacy laws and waste disposal regulations adds another layer of complexity. The proposed model's reliance on real-time monitoring, data collection, and AI-driven processes requires strict adherence to legal and ethical standards to align with regulations and maintain public trust (Sutikno, 2024).

## 6.2 Strategic Solutions

Addressing the challenges of AI-driven waste management requires a multi-faceted approach that enhances technological capabilities, fosters collaboration, and ensures regulatory compliance.

One of the primary barriers—data fragmentation and lack of standardization—can be overcome through improved data accessibility and quality initiatives. Waste management authorities and technology developers must collaborate to establish standardized data formats, develop integrated data platforms, and promote data-sharing practices. Leveraging IoT and sensor advancements will enhance real-time data collection, ensuring AI models receive accurate and comprehensive datasets for optimized decision-making (Olawade et al., 2024). Privacy and security concerns, particularly regarding real-time monitoring and AI-based surveillance, must be addressed through robust security measures. Implementing privacy-enhancing techniques, anonymizing sensitive data, and ensuring compliance with data protection regulations will help build public trust. Additionally, evolving regulatory frameworks should focus on establishing ethical guidelines that prioritize data security and equitable AI adoption in waste management operations (Olawade et al., 2024). The high cost of AI infrastructure remains a significant barrier, but strategic investment from governments, organizations, and the private sector can drive scalable implementation. Funding initiatives should prioritize the deployment of smart sensors, IoT devices, and high-performance computing resources to support the proposed AI-driven waste sorting and predictive analytics. Financial incentives, such as subsidies for recycling facilities and tax benefits for AI adoption, can further ease the financial burden on municipalities and industries (Olawade et al., 2024). Addressing technological limitations, including the "black box" problem and the lack of customized AI models for waste management, requires sustained research and innovation. Collaboration between academia, industry, and policymakers can lead to the development of AI algorithms specifically tailored to the proposed processes. Multidisciplinary discussions involving ethicists, policymakers, and community representatives can facilitate responsible AI adoption, ensuring that automation does not disproportionately affect employment opportunities (Olawade et al., 2024).

Seamless integration with existing waste management infrastructure necessitates the development of universal standards and policy norms. Regulatory bodies must establish clear guidelines for AI deployment, focusing on interoperability, data security, and system compatibility. Policy initiatives should encourage data-sharing agreements among stakeholders while maintaining ethical considerations around AI-driven decision-making (Szpilko et al., 2023). Strengthening security measures within AI-based systems will further ensure reliable and compliant waste management practices. Lastly, since public engagement and education are critical for the success of the proposed model, overcoming resistance to technology adoption requires comprehensive awareness campaigns that highlight the benefits of AI in waste management (Olawade et al., 2024). Before implementation, gamified incentives, financial rewards, and community-driven initiatives can be introduced as pilot programs to encourage active citizen participation and test the effectiveness of engagement strategies. These efforts, combined with collaboration among waste management agencies, technology providers, and policymakers, lay the groundwork for the successful deployment of sustainable and efficient urban waste solutions.

# 7 Future Directions and Research Opportunities in AI Waste Management

As AI-driven waste management continues to evolve, future research must focus on refining existing technologies and addressing emerging challenges. The proposed model, which integrates artificial intelligence, IoT, robotics, and public participation, presents several avenues for further exploration and innovation before and during implementation. Advancements in these areas will enhance the model's efficiency, scalability, and adaptability while ensuring a more sustainable and data-driven waste management ecosystem.

One promising research direction involves improving AI-based waste recognition and sorting. While computer vision and machine learning have demonstrated high accuracy in waste classification (CMAA, 2024; Kesari, 2024), future studies should explore more sophisticated deep-learning models capable of identifying a wider variety of waste materials, including complex and composite waste. Research into AI-powered multimodal sensors that combine visual, chemical, and infrared analysis could further enhance sorting precision. Additionally, developing self-learning AI models that continuously improve their accuracy through real-time feedback from waste streams will contribute to more adaptive waste management systems. The integration of blockchain technology into AI waste management presents another compelling research opportunity. Blockchain could enhance data transparency, traceability, and accountability in waste disposal and recycling processes (Yao et al., 2024). Future studies should investigate how decentralized ledger systems can be used to track waste generation at the source, verify proper disposal, and incentivize recycling efforts through secure and tamper-proof digital reward systems. Combining blockchain with the existing incentive-based approach in the proposed model could further increase public participation and trust in AI-driven waste management.

Enhancing predictive analytics for waste collection and route optimization is another critical area for future research. While the proposed model already incorporates predictive AI for waste collection scheduling, more advanced algorithms—such as reinforcement learning and hybrid AI techniques—could improve route efficiency by dynamically adjusting to real-time traffic, weather conditions, and seasonal waste generation patterns. Research should also explore AI-driven decision-support systems that allow municipalities to simulate various waste management policies and assess their environmental and economic impacts before implementation. Another crucial research avenue is the development of ethical and regulatory frameworks for AI waste management. As AI-based waste monitoring systems involve real-time surveillance and data collection, ensuring compliance with data privacy laws and ethical AI principles is essential. Future studies should focus on designing privacy-preserving AI models that anonymize user data while maintaining system effectiveness. Additionally, research into policy harmonization across different jurisdictions could facilitate standardized AI regulations for waste management, ensuring smooth implementation and international cooperation. Public engagement and behavioral analysis using AI-driven insights represent another key future direction. While the proposed model encourages citizen participation through gamification and incentives, future research should examine how AI can analyze behavioral patterns to tailor waste management strategies to specific communities. AI-driven sentiment analysis, combined with psychological and behavioral studies, could help policymakers design more effective awareness campaigns and interventions to promote responsible waste disposal habits.

Furthermore, expanding AI applications in waste-to-energy (WTE) conversion processes is a vital research opportunity. AI can play a crucial role in optimizing energy recovery from waste by predicting the most efficient incineration or anaerobic digestion methods based on waste composition. Future studies should explore how AI-driven simulations can improve the operational efficiency of WTE plants, reduce emissions, and enhance energy output. Integrating AI with emerging sustainable technologies, such as hydrogen energy production from waste, could also contribute to the development of cleaner energy solutions for smart cities. Lastly, research should focus on scaling AI-driven waste management solutions for diverse environments, including rural and underdeveloped regions. While the current model is optimized for urban settings, AI technologies must be adapted to areas with limited infrastructure and resources. Investigating cost-effective AI solutions, such as low-power IoT devices and decentralized AI systems, could facilitate broader implementation and ensure equitable access to smart waste management technologies.

By exploring these future directions, researchers and policymakers can enhance the capabilities of the proposed AI-driven waste management, making it more intelligent, ethical, and sustainable. These advancements will not only optimize waste collection and recycling but also contribute to the broader goals of smart cities and circular economies.

## 8 Conclusion

The integration of AI technologies in waste management represents a transformative shift toward a more efficient, sustainable, and intelligent urban environment. This paper has explored the current challenges in urban waste management, highlighting the inefficiencies and environmental impacts of traditional practices. By leveraging AI-driven solutions such as machine learning, IoT, computer vision, and robotics, cities can optimize waste collection, sorting, recycling, and monitoring processes, leading to significant improvements in efficiency and sustainability. The proposed AI-powered waste management framework offers a comprehensive approach to addressing the multifaceted challenges of urban waste management. By enabling real-time monitoring, predictive analytics, and automated sorting, this framework enhances resource recovery, reduces environmental harm, and lowers operational costs. The benefits of this model are far-reaching, including improved waste management efficiency, enhanced sustainability through energy recovery and reduced emissions, increased economic viability through cost savings and revenue generation, and better decision-making through data-driven insights.

However, while the proposed framework presents substantial theoretical and technological advantages, its real-world effectiveness depends on overcoming key barriers such as data fragmentation, high implementation costs, technological limitations, and societal resistance. Validation through real-world trials is essential to ensure adaptability across diverse urban environments and waste management infrastructures. Implementing pilot projects will provide empirical evidence on performance, efficiency, and sustainability outcomes. Additionally, active engagement with key stakeholders, including policymakers, municipal authorities, industry leaders, and the public, is crucial. Their involvement will facilitate regulatory frameworks, ethical AI deployment, and community participation. Addressing public concerns regarding data privacy, automation's impact on employment, and behavioral adaptation to AI-powered systems is vital for fostering long-term success.

Future research should focus on refining AI technologies for waste recognition and sorting to improve accuracy, integrating blockchain for transparency and accountability in waste disposal and recycling, enhancing predictive analytics for route optimization, developing ethical and regulatory frameworks, and scaling AI solutions for diverse environments, ensuring accessibility beyond urban smart cities. As technology evolves, continuous refinement of AI models is necessary to overcome emerging challenges, such as integrating new waste classification techniques, enhancing AI-driven route optimization, and improving energy recovery from waste. By fostering a cycle of validation, collaboration, and improvement, AI-driven waste management can be continuously optimized for greater efficiency and sustainability. In conclusion, AI offers a promising pathway to revolutionize waste management in smart cities, contributing to the development of resilient, livable, and environmentally responsible urban environments. However, its full potential can only be realized through real-world validation, stakeholder collaboration, and iterative advancements. By embracing AI technologies and fostering partnerships between researchers, policymakers, and industry leaders, cities can unlock a cleaner, healthier, and more sustainable future.

## References

Alahi, M. E. E., Sukkuea, A., Tina, F. W., Nag, A., Kurdthongmee, W., Suwannarat, K., & Mukhopadhyay, S. C. (2023). Integration of IoT-enabled technologies and artificial intelligence (AI) for smart city scenario: Recent advancements and future trends. *Sensors, 23*(11), 5206. https://doi.org/10.3390/s23115206

Al-Sabbagh, M., Al-Omari, A., & Al-Khatib, I. A. (2022). Artificial intelligence in waste management: Current trends and future directions. *Waste Management & Research, 40*(2), 145–162. https://doi.org/10.1177/0734242X211048201

Ayadinuno, E., Njoku, B., & Akakuru, O. (2021). Sustaining a healthy environment whilst utilizing wastes as energy sources: Physicochemical analysis of solid wastes in Awka. *Nigeria., 36*(1), 75–87.

AZoRobotics. (2024). AI-powered robotics transforming waste management. AZoRobotics News & Articles. *Retrieved from* https://www.azorobotics.com/Article.aspx?ArticleID=695

Bala, K., Anshul, Verma, S., Dogra, A., Das, A. P., Kumar, R., & Nandini. (2024). Sustainable management for urban plastic waste generation. *Sustainable Management of Urban Plastic Waste Through Circular Economic Approaches*, 211–236. https://doi.org/10.1201/9781003465447-13

Behera, S. K., Aouthithiye Barathwaj, Y., Vasundhara, L., Saisudha, G., Haariharan, C. N., & Sathya Priya, J. (2020). AI-based waste classifier with thermo-rapid composting. In *2020 International Conference on Power, Energy, Control and Transmission Systems (ICPECTS)*, pp. 1–4. https://doi.org/10.1109/ICPECTS49113.2020.9337012

Bundhoo, Z. M. A. (2018). Solid waste management in least developed countries: Current status and challenges faced. *Journal of Material Cycles and Waste Management, 20*, 1867–1877. https://doi.org/10.1007/s10163-018-0728-3

Construction Management Association of America (CMAA). (2024). Robotics in construction and demolition waste recycling: Future trends and challenges. *Retrieved from* https://www.cmaanet.org/sites/default/files/resource/Robotics_CD_Waste_CE.pdf

Czekała, W., Drozdowski, J., & Łabiak, P. (2023). Modern technologies for waste management: A review. *Applied Sciences, 13*(1), 8847. https://doi.org/10.3390/app13158847

Das, A. P., & Ghosh, S. (2018). Bioleaching of manganese from mining waste materials. *Materials Today: Proceedings, 5*(1), 2381–2390. https://doi.org/10.1016/J.MATPR.2017.11.459

Das, A., Ray, S. K., Mohanty, M., Mohanty, J., Dey, S., & Das, A. P. (2024). Ecotoxicity of microplastic wastes and their sustainable management: A review. *Environmental Chemistry and Ecotoxicology, 6*, 144–152. https://doi.org/10.1016/J.ENCECO.2024.05.003

Das, A., & Ghosh, S. (2022). Role of microorganisms in extenuation of mining and industrial wastes. *Geomicrobiology Journal, 39*, 173–175. https://doi.org/10.1080/01490451.2022.2038953

Dey, S., Das, A., Mallick, K., Sahu, A., & Das, A. P. (2023a). Environmental petroleum waste: Pollution, toxicity, sustainable remediation, pp. 159–175. https://doi.org/10.1007/978-3-031-48220-5_7

Fang, B., Yu, J., Chen, Z., Osman, A., Farghali, M., Ihara, I., Hamza, E., Rooney, D., & Yap, P. S. (2023). Artificial intelligence for waste management in smart cities: A review. *Environmental Chemistry Letters, 21*, 1–14. https://doi.org/10.1007/s10311-023-01604-3

Ferrer, J., & Alba, E. (2019). BIN-CT: Urban waste collection based on predicting the container fill level. *Bio Systems, 186*, Article 103962. https://doi.org/10.1016/j.biosystems.2019.04.006

Gandhimathinathan, R. (2023). Application of particle swarm optimization (PSO) in AI-based predictive waste management. *Journal of Environmental Science and Technology, 56*(2), 123–135.

Ghahramani, M., Zhou, M., Mölter, A., & Pilla, F. (2021). IoT-based route recommendation for an intelligent waste management system. *IEEE Internet of Things Journal,* PP(1), 1–1. https://doi.org/10.1109/JIOT.2021.3132126

Ghosh, S., & Das, A. P. (2017). Bioleaching of manganese from mining waste residues using Acinetobacter sp. *Geology, Ecology, and Landscapes, 1*(2), 77–83. https://doi.org/10.1080/24749508.2017.1332847

Gibovic, D., & Bikfalvi, A. (2021). Incentives for plastic recycling: how to engage citizens in active collection empirical evidence from Spain. *Recycling, 6*(2), 29. https://doi.org/10.3390/RECYCLING6020029

Gupta, A., & Pathak, D. (2021). Smart waste management using AI: Challenges and opportunities. *Journal of Environmental Science & Technology, 55*(3), 112–129. https://doi.org/10.1007/s11356-021-15042-3

Ibrahim, I. A., & Baballe, M. A. (2024). A comprehensive review of how the smart waste management system works. *Journal* homepage: https://gjrpublication.com/gjrecs *4*(05).

Jha, R., Dwivedi, S., & Modhera, B. (2022). Measurement and practices for hazardous waste management. In D. Yadav, P. Kumar, P. Singh, & D. A. Vallero (Eds.), *Hazardous waste management* (pp. 89–115). Elsevier. https://doi.org/10.1016/B978-0-12-824344-2.00011-2

Kaza, S., Yao, L., Bhada-Tata, P., & Van Woerden, F. (2018). What a waste 2.0: a global snapshot of solid waste management to 2050. *World Bank Publications.*

Kesari, G. (2024, May 31). Turning trash into treasure: How AI is revolutionizing waste sorting.*Forbes*.https://www.forbes.com/sites/ganeskesari/2024/05/31/turning-trash-into-treasure-how-ai-is-revolutionizing-waste-sorting/

Kumar, M., Pandit, S., Patel, V., Khanna, N., Nag, M., Lahiri, D., Ray, R. R., Das, A., & Das, D. (2023). Maximization of energy recovery from starch processing wastewater by thermophilic dark fermentation coupled with microbial fuel cell technology. *Geomicrobiology Journal*, 1–12. https://doi.org/10.1080/01490451.2023.2209555

Kumar, R., Sharma, S., Kumar, A., et al. (2024). Sustainable energy recovery from municipal solid wastes: An in-depth analysis of waste-to-energy technologies and their environmental implications in India. *Energy Exploration and Exploitation, 43*(1), 3–28. https://doi.org/10.1177/01445987231210 32

Lipianina-Honcharenko, K., Komar, M., Osolinskyi, O., Shymanskyi, V., Havryliuk, M., & Semaniuk, V. (2024). Intelligent waste-volume management method in the smart city concept. *Smart Cities, 7*(1), 78–98. https://doi.org/10.3390/smartcities7010004

Mekonnen, F. H. (2016). Liquid waste management: The case of Bahir Dar, Ethiopia. The Ethiopian Journal of Health Development, *26*(1). *Retrieved from* https://ejhd.org/index.php/ejhd/article/view/193

Mishra, S., & Das, A. P. (2021). Current treatment technologies for removal of microplastic and microfiber pollutants from wastewater. *Wastewater Treatment: Cutting-Edge Molecular Tools, Techniques and Applied Aspects*, 237–251. https://doi.org/10.1016/B978-0-12-821881-5.00011-8

Mishra, S., & Das, A. P. (2022). Treatment of the wastewater polluted with synthetic microfiber released from washing machine. In B. B. Das, H. Hettiarachchi, P. K. Sahu, & S. Nanda (Eds.), Lecture Notes in Civil Engineering (Vol. 207, pp. 109–117). Springer Science and Business Media Deutschland GmbH. https://doi.org/10.1007/978-981-16-7509-6_9/COVER

Mishra, S., & Das, A. P. (2023). Microbial remediation of synthetic microfiber contaminated wastewater. In M. P. Shah (Ed.), *Microbial technologies in industrial wastewater treatment* (pp. 337–350). Singapore: Springer, Singapore. https://doi.org/10.1007/978-981-99-2435-6_17

Mishra, S., Dash, D., & Das, A. P. (2022a). Detection, characterization and possible biofragmentation of synthetic microfibers released from domestic laundering wastewater as an emerging source of marine pollution. *Marine Pollution Bulletin, 185*, Article 114254. https://doi.org/10.1016/J.MARPOLBUL.2022.114254

Mishra, S., Dash, D., & Das, A. P. (2024). Isolation and molecular characterization of polyester-degrading bacterial strains from municipal wastewater and their application in microplastic degradation in contaminated wastewater. *Geomicrobiology Journal*. https://doi.org/10.1080/01490451.2024.2376101

Mishra, S., Ghosh, S., van Hullebusch, E. D., Singh, S., & Das, A. P. (2023). A critical review on the recovery of base and critical elements from electronic waste-contaminated streams using microbial biotechnology. *Applied Biochemistry and Biotechnology, 2023*, 1–30. https://doi.org/10.1007/S12010-023-04440-X

Mishra, S., Rout, P. K., & Das, A. P. (2019). Solar photovoltaic panels as next generation waste: A review. *Biointerface Research in Applied Chemistry*, *9*(6).

Mishra, S., Singh, R. P., Rout, P. K., & Das, A. P. (2022). Membrane bioreactor (MBR) as an advanced wastewater treatment technology for removal of synthetic microplastics. *Development in Wastewater Treatment Research and Processes*, 45–60. https://doi.org/10.1016/B978-0-323-85583-9.00022-3

Mohabuth, A. Q., & Sunnassy, N. R. (2024). Prototyping waste collection by the implementation of a smart bin using AI techniques. *International Conference on Smart Systems for Applications in Electrical Sciences (ICSSES), 2024*, 1–6. https://doi.org/10.1109/ICSSES62373.2024.10561251

Mohanty, S., Ghosh, S., Bal, B., & Das, A. P. (2018). A review of biotechnology processes applied for manganese recovery from wastes. *Reviews in Environmental Science and Bio/Technology 2018 17:4*, *17*(4), 791–811. https://doi.org/10.1007/S11157-018-9482-1

Namoun, A., Hussein, B. R., Tufail, A., Alrehaili, A., Syed, T. A., & BenRhouma, O. (2022). An ensemble learning based classification approach for the prediction of household solid waste generation. *Sensors, 22*(9), 3506. https://doi.org/10.3390/s22093506

Nwokediegwu, Z., Ugwuanyi, E., Majemite, M., & Obaigbena, A. (2024). AI-driven waste management systems: A comparative review of innovations in the USA and Africa. *Engineering Science & Technology Journal, 5*(2), 507–516. https://doi.org/10.51594/estj.v5i2.828

Olawade, D. B., Fapohunda, O., Wada, O. Z., Usman, S. O., Ige, A. O., Ajisafe, O., & Oladapo, B. I. (2024). Smart waste management: A paradigm shift enabled by artificial intelligence. *Waste Management Bulletin, 2*(2), 244–263. https://doi.org/10.1016/j.wmb.2024.05.001

Pradhan, G., Tripathy, B., Ram, D. K., Digal, A. K., & Das, A. P. (2023). Bauxite mining waste pollution and its sustainable management through bioremediation. *Geomicrobiology Journal*, 1–10. https://doi.org/10.1080/01490451.2023.2235353

Ray, S. K., Dey, S., Das, A. P., & John, A. (2024). Microplastic waste management through greener technologies : a review. *Sustainable Management of Urban Plastic Waste Through Circular Economic Approaches*, 237–261. https://doi.org/10.1201/9781003465447-14

Recycling Inside. (2024). The customizable power of AI sorting in waste management. Recycling Inside. *Retrieved from* https://recyclinginside.com/recycling-technology/separation-and-sorting-technology/the-customizable-power-of-ai-sorting/

Sahu, A., Dey, S., Mallick, K., Priyadarsini, S., Behera, I. D., & Das, A. P. (2024). A review on advanced technology for sustainable management of synthetic microplastic waste. *Sustainable Management of Urban Plastic Waste Through Circular Economic Approaches*, 262–279. https://doi.org/10.1201/9781003465447-15uns

Sharma, S., Mallikarjuna Reddy, V., Raj, G. R., Nijhawan, G., Yadav, D. K., Raed, R., & Tyagi, L. K. (2024). Leveraging waste-to-energy technologies for sustainable development: A comprehensive review. *E3S Web of Conferences*, 529, 02010. https://doi.org/10.1051/e3sconf/202452902010

Sosunova, I., & Porras, J. (2022). IoT-enabled smart waste management systems for smart cities: A systematic review. *IEEE Access, 10*, 73326–73363. https://doi.org/10.1109/ACCESS.2022.3188308

Sutikno, T. (2024). The future of artificial intelligence-driven robotics: applications and implications. *IAES International Journal of Robotics and Automation (IJRA)*, 13, 361–372. https://doi.org/10.11591/ijra.v13i4.pp361-372

Szpilko, D., de la Torre Gallegos, A., Jimenez Naharro, F., Rzepka, A., & Remiszewska, A. (2023). Waste management in the smart city: Current practices and future directions. *Resources, 12*(1), 115. https://doi.org/10.3390/resources12100115

Tripathy, B., Dey, S., Dubey, N. K., Lo, W.-C., & Das, A. P. (2024). *An omics-based approach for managing the risk of COVID-19-generated wastes*. 51–66. https://doi.org/10.1007/978-3-031-50840-0_4

Vishnu, S., Ramson, S. R. J., Senith, S., Anagnostopoulos, T., Abu-Mahfouz, A. M., Fan, X., Srinivasan, S., & Kirubaraj, A. A. (2021). IoT-enabled solid waste management in smart cities. *Smart Cities, 4*(3), 1004–1017. https://doi.org/10.3390/smartcities4030053

World Bank. (2022). Trends in solid waste management. Retrieved from https://www.worldbank.org/en/topic/urbandevelopment/brief/solid-waste-management

Yao, Y., Weng, J., He, C., Gong, C., & Xiao, P. (2024). AI-powered strategies for optimizing waste management in smart cities in Beijing. *World Journal of Innovation and Modern Technology.*

Zhang, Y., Chen, W., & Zhao, L. (2023). AI-powered waste management for sustainable smart cities: A systematic review. *Sustainable Cities and Society, 92*, Article 104587. https://doi.org/10.1016/j.scs.2023.104587

Zhou, J., Jiang, P., Yang, J., & Liu, X. (2021). Designing a smart incentive-based recycling system for household recyclable waste. *Waste Management, 123*, 142–153. https://doi.org/10.1016/j.wasman.2021.01.030

# Waste-to-Energy Conversion

**Selcan Karakuş and Alok Prasad Das**

## 1 Introduction

An essential component of national development, success, and economic growth is energy. It has become one of the most critical global issues in recent years because of rising consumer demand, limited resources, and sustainability concerns (Ghosh et al., 2023; Priyadarshini et al., 2024). The requirement for environmentally friendly and living environments clashes with the necessity for larger landfills due to worldwide waste production's increasing growth. In addition to lowering the value of land, this expansion jeopardizes habitats by contaminating water supplies, exhausting assets, and posing a health risk to the general public. It is now more important than ever to make the shift to a circular economy in light of these critical concerns. In contrast to the conventional mathematical structure of "take-make-consume-dispose," circular economy encourages an alternative approach that guarantees the continued use of resources while minimizing waste.

Growing waste output and rising global energy use have made environmentally friendly renewable energy options even more essential (Dey et al., 2023). Because it tackles waste management and generation of energy at the same time, the waste-to-energy (WtE) strategy has drawn a lot of interest as an imaginative approach that promotes energy safety and preservation of the environment. Globally, the generation of municipal solid waste (MSWt) has been rising quickly, and this trend will probably persist as a result of evolving lifestyles brought on by industrialization and growth

S. Karakuş (✉)
Department of Chemistry, Istanbul University-Cerrahpaşa, Istanbul, Türkiye
e-mail: selcan@iuc.edu.tr

Health Biotechnology Joint Research and Application Center of Excellence, Esenler, Istanbul, Türkiye

A. Prasad Das
Department of Life Sciences, Rama Devi Women's University, Bhubaneswar, Odisha, India



A. P. Das and S. Karakuş (eds.), *Synergizing Sustainability for Integrated Waste Management*, Sustainable Environmental Waste Management Strategies,
https://doi.org/10.1007/978-3-032-03718-3_12

in population (Mohanty et al., 2024; Prabina et al., 2024). The yearly generation of waste is currently billions of tons, and estimates suggest that this amount will rise significantly over the following decade or so. Environment and health are put at risk by the sheer amount and variety of these waste materials. Hence, in order to maintain the health of the planet, improve the standard of life, and protect natural resources, proper waste oversight has to be given the highest level of priority (Priyadarsini et al., 2024; Satapathy et al., 2024). However, carbon emissions continue to be an important issue, even if cost-effective energy access is crucial for the continued growth of the country. A workable answer to these problems is the conversion of waste into energy, which offers green energy production together with minimizing waste. It is feasible to fulfill the world's energy needs and reduce the ecological footprint by incorporating modern waste-to-energy solutions (Fig. 1).

Waste generation rises quickly as a result of ineffective waste management, posing major ecological and financial issues (Bal et al., 2018; Das et al., 2024). According to estimates, waste output could outpace population growth and reach concerning levels by 2050. Because it puts tension on the public's resources and raises the price of waste disposal, removal, and medical treatment, poorly managed waste places an enormous cost on society. In addition to impeding sustainable growth, this economic pressure also waste materials that may be better used in different fields. Numerous methods, as well as domestic and industrial operations, generate a variety of waste materials. This consist of waste from construction and demolition (C&D), municipal

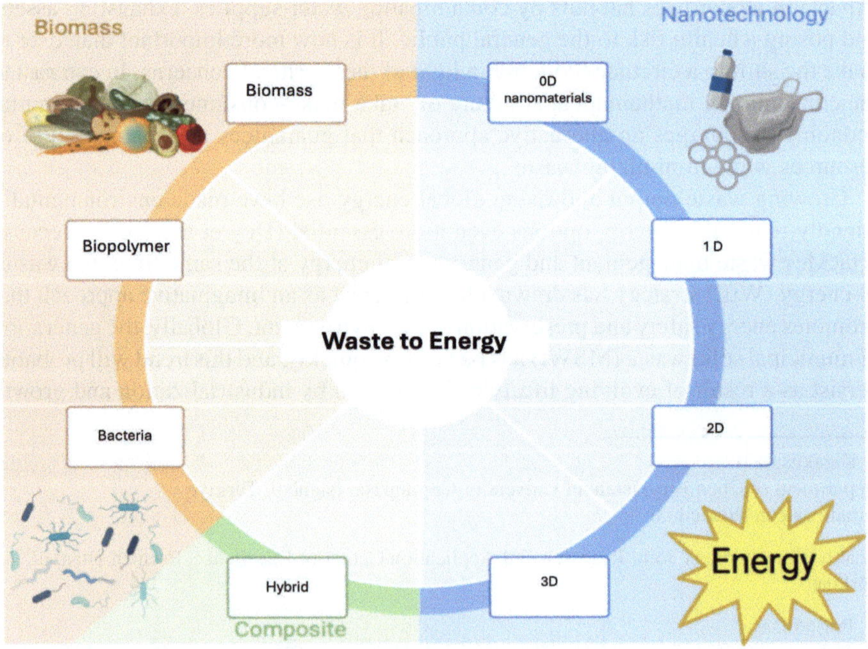


**Fig. 1** Modern waste-to-energy solutions

liquid waste (MLWt), MSWt, toxic waste (Khatun et al., 2023a, 2023b; Sadasivuni et al., 2019), technological waste (Priyadarsini et al., 2025), biowaste, radioactive materials, mining waste (Das et al., 2011, 2015; Ghosh et al., 2016; Mohanty et al., 2016), waste from agriculture, and waste made of plastic (Bala et al., 2024; Das et al., 2025; Singh Jadaun et al., 2022).

Strategies to waste disposal include a variety of waste product procedures, each with specific advantages and disadvantages (Dey et al., 2024). Among these, thermal method is frequently employed to handle waste and produce energy in the form of fuel, heat, or electricity. For waste disposal and waste-to-value manufacturing, techniques like pyrolysis, gasification, plasma gasification, and incineration are frequently used (Hsu et al., 2024). Derived from organic resources like wood, industrial, and agricultural wastes, biomass offers a sustainable energy source. The use of biomass is rendered possible by techniques such as gasification, fermentation, and the combustion process, which lessen reliance on fossil fuels and emissions of greenhouse gases. In addition to biomass, the circular economy requires an innovative waste management strategy. In addition to reducing environmental damage, practices like composting, recycling, and safe disposal raise the worth of waste products financially. As a result, embracing the circular economy and environmentally friendly methods provides a comprehensive approach for maintaining an equilibrium between social development, ecological preservation, and financial growth (Islam et al., 2024).

## 2 Waste-to-Energy Strategy

According to the WtE principle, many different types of waste, including agricultural residues, municipal solid trash, and industrial by-products, can be transformed into forms of useful energy. A variety of WtE technologies have been created to improve energy recovery effectiveness while reducing environmental effects. Among the most popular approaches are the following:

### *2.1 Incineration*

Methods of managing waste include incineration, composting, treatment with biological agents, and landfilling. Nevertheless, a large number of landfills are still antiquated and do not have the required layout to prevent the spread of hazardous wastewater. Because of this, organizations and governments around the world are spending additional funds on incineration methods in an effort to decrease their reliance on waste. One of the most popular methods for WtE is incineration, which involves burning waste at high temperatures to produce heat that may be turned into energy. Gasification and fluidized bed combustion are two examples of modern incineration applications that have decreased greenhouse gas emissions and improved productivity. Future society plans that incorporate emergency energy-sharing approaches may find that small-scale incinerators for energy dispatch functions are advantageous. The various energy requirements associated with centralized community energy systems can be efficiently met by these networks by integrating gas-powered turbines and gas reactors. For future viability and improved energy efficiency, such advanced technologies' dispatch must be enhanced.

Carbon dioxide ($CO_2$) and methane ($CH_4$) emissions from landfill-generated gases are substantially decreased by the incineration process, which is an essential part of modern waste management. In addition to minimizing the amount of solid waste materials, incineration is an efficient waste removal technique that allows the recovery of energy in the form of thermal energy or electricity. Nevertheless, as waste amounts increase and heat values vary, municipal solid waste incinerator (MSWI) plants face performance and capacity issues that lead to burning instability. Advanced incineration systems need to be enhanced in order to solve these problems. Incineration systems must be modernized and updated in order to handle increasing waste loads efficiently, improve energy recovery, and lessen contamination in the natural environment.

AI-based technology has been used extensively in green protection in recent years, including solid waste incinerator management, wastewater treatment, and air quality control. Proportional-integral-derivative (PID) control performs effectively for fuels with constant heat values, but standard PID systems are ineffective when dealing with the complex incineration results caused by the variable composition and insufficient blending of waste. In order to overcome these issues, ML has been utilized more and more to handle real-time outputs and complicated input values, decreasing the need for lengthy tests and simulation-based modeling while enhancing precision and efficacy. For instance, a compelling study using ML to enhance WtE decision-making was presented by Hou et al. (2024). They improve predictive accuracy by combining oversampling and stacked ensemble models. This method offered flexible site suggestions and minimizes dependence on rigid assumptions, especially in the region of Hunan, China. The analytical creativity, which successfully captures intricate interactions, is a significant strength. To balance the performance of models, however, variations in recall and accuracy require for more research. Although the study shows real-world relevance, its influence would be increased by talking about problems such as accessibility of data and model interpretability. All things considered, this study demonstrated ML's promise for environmentally friendly waste disposal. For better decision-making, future research could improve class balancing, include real-time considerations, and investigate hybrid AI techniques. By combining actual operational details with extra measurement instruments, Lim et al. developed a physics-based algorithm to forecast the composition of waste and heating rate in incineration systems (Lim et al., 2024). With this approach, the unique features of every incinerator can be reflected more accurately. In order to optimize processes and evaluate the impact of energy recovery, a simulator model was also introduced. In comparison with traditional operations, the study suggests an optimal approach that increases energy recovery by approximately ten percent by integrating the composition of waste, temperature value, and process analysis. The shortcomings of conventional temperature value calculation are addressed by this approach, which offers a more accurate and flexible way to enhance waste incineration systems' environmental impact and energy consumption. The results provide a helpful structure for improving different incinerator functions.

## 2.2 *Anaerobic Digestion*

Anaerobic digestion (AD) produces digestate, liquid, and biogas, a mixture high in $CO_2$ and $CH_4$, by breaking down biomass in an atmosphere without oxygen. AD depends on a bio-process at relatively cooler temperatures, in contrast to pyrolysis, which thermally breaks down material at high temperatures. Another use for the digestate waste product is as a fertilizer that is high in nutrients. After being purified, digestate is a beneficial soil nutrient, and biogas can be utilized as energy. In order to maximize energy and energy performance and save operating expenses, some research has investigated into turning biogas power plants into trigeneration power, cooling, and heating facilities. Given the advantages of conversion innovations, no single process is suitable for every kind of biowaste. As an example, AD is inadequate for processing lignocellulosic material, which is mainly made up of lignin, hemicellulose, cellulose, and fiber. To overcome the issue, combining various innovations has been proposed as a possible approach. One approach that is receiving interest is the integration of pyrolysis and AD, where biochar is introduced to the A-digester to improve the chemical reaction (Ebrahimi et al., 2025). The critical requirement to transition from carbon-intensive sources of energy to zero-carbon alternatives in order to reduce climate change was emphasized by Ebrahimi et al. in a newly published research. Energy-from-biowaste comes out as an attractive option for tackling a growing issue of waste management while also generating sustainable energy. Among these technologies, AD and pyrolysis have drawn a lot of interest, especially because of their possible combination. In this work, three systems—pyrolysis, AD, and integrated configuration—are subjected to a thorough exergoeconomic examination and computational model to investigate the financial feasibility of merging these methods. The findings demonstrated that although digestate processing is still expensive, the combined approach reduces the expense of producing biogas and bio-oil, as opposed to separate methods. The economic benefits gained from integration are further reinforced by a sensitivity analysis, which takes into account changes in the price of biogas, electricity, and NaOH. These results support the prospective use of pyrolysis and AD in waste disposal and renewable energy production by offering insightful information about their feasibility.

### 2.3 Pyrolysis and Gasification

Under controlled conditions, organic substances, such as waste and biomass, can be broken down into valuable energy-producing materials via the thermochemical transformation processes of gasification and pyrolysis.

Pyrolysis: Biochar, bio-oil, and syngas are the end results of pyrolysis, which takes place in an oxygen-limited or oxygen-free condition. It is frequently employed in the production of carbon materials and biofuel. In order to improve the production of biofuel and biochar, Fardi et al. (2024) investigated the waste-to-energy potential of co-pyrolyzing potato peel with the marine biomass Sargassum angustifolium (Fardi et al., 2024). A common thermochemical technique called pyrolysis was used to maximize the quality of the biochar and bio-oil that were obtained. According to the findings, combining marine and terrestrial biomass enhanced the synthesis of bio-oil while lowering its pH level. The biochar's capacity to serve a variety of uses was enhanced by the presence of alkali and alkaline soil metals as well as key chemical groups. The study emphasizes how co-pyrolysis is a sustainable method of turning biowaste into useful energy resources and chemical compounds with benefits for the environment as well as the economy.

Gasification: Materials derived from carbon are gasified with a regulated amount of oxygen or steam to create syngas, a flammable gas combination primarily composed of carbon monoxide, hydrogen, and methane. Energy recovery from garbage and biomass is a common use for this method.

## 3 Biomass Energy

Sustainable development requires global collaboration, and lowering the release of greenhouse gases requires switching to renewable energy sources. The use of clean energy sources to replace petroleum and coal greatly aids in the decrease of emissions. Given its regenerative origin and wide range of applications, biomass—which is made from alive or recently existing biomaterial—stands out as a possible alternative. It functions as a renewable form of fuel and can produce both electricity and heat through heat-/power-based technologies. Biomass promotes renewable energy production and safety by lowering dependency on fossil fuels. Biomass continues to play a significant role in decarbonizing energy as research progresses, hence promoting sustainable development targets. The adaptability of biomass is further demonstrated by the growing number of nations using it for the generation of energy, from large-scale energy to small-scale heating systems.

## 3.1 *Carbon-Fiber-Reinforced Polymer Waste*

The energy, armed forces, automobile, and electronic sectors all prefer carbon fiber/polymer-based materials because of their high stability, durability, lightweight, and flexibility properties. Comparing these composites to traditional mild steel, the weight of the product can be reduced by fifty to seventy percent. For sustainable and effective resource management, it is essential to comprehend the environmental costs of waste creation and treatment in comparison with the advantages of use. Purpose-grown biomass, biogenic municipal solid waste, and residues from farms and forests constitute significant resources of biomass for generating energy. They also reduce our dependence on traditional fuels while promoting the field's development. But there are obstacles to efficiently using the plant matter, especially when it comes to controlling distribution system expenses. Performance is greatly impacted by variables such as feedstock supply, price, quality, transportation, and logistics. A well-optimized structure is necessary to improve the entire sustainability and efficiency during operation while additionally enhancing the biomass delivery chain's economic, social, and environmental impact (Dai et al., 2025).

In 2024, microwave molten salt pyrolysis–oxidation was presented by Shu et al. as a successful method for recycling waste made of carbon fiber–polymer composites (Yatim et al., 2024). They improved recovery efficiency by optimizing the microwave molten salt pyrolysis–oxidation method using a composite molten solution of sodium carbonate ($NaCO_3$), potassium carbonate ($KCO_3$), and lithium carbonate ($LiCO_3$) that was microwave irradiated. Their findings showed that the regenerated carbon fibers had a high recovery rate, a noticeable rise in tensile modulus, and almost all of the virgin carbon fibers' tensile strength. In addition, the process yielded important by-products like carbon monoxide (CO), hydrogen ($H_2$), and aromatic chemicals such as phenols. For carbon fiber–polymer composites, the microwave molten salt pyrolysis–oxidation technique offers a quick and efficient recycling alternative that promotes environmental responsibility and the development of the circular economy.

## 3.2 *Some Applications in the Literature in the Field of WtE*

### 3.2.1 Conducting Polymers-Based Materials

Polymers, which are of increasing interest especially in the field of energy, are macromolecular structures consisting of repeating subunits and have been considered as insulating materials for many years. However, with the discovery of certain polymeric materials, the existence of conductive polymers has also been revealed and these structures have attracted attention as organic structures that can conduct electricity due to the presence of free and mobile electrons, as in metal materials. They are generally non-thermoplastic organic structures and do not have some important properties such as elasticity and mechanical strength.

The conjugated π-electron systems of conductive polymers, which are a family of organic materials, enable them to conduct electricity. The most commonly used conductive polymers are poly(3,4-ethylenedioxythiophene) (PEDOT), polyaniline (PANI), and polypyrrole (PPy), all of which have been extensively studied for use in flexible electronic devices, detectors, and batteries for energy. The adjustable optical and electrical properties of polythiophene (PTh) and its derivatives, including poly(2,5-thienylenevinylene) (PTV), polythiophene vinylene (PTh-V), poly(3-octylthiophene-3-methylthiophene) (POTMT), and poly(3-alkylthiophene) (PAT), make them especially valuable. Poly-p-phenylene-sulfide (PPS), poly(p-phenylene) (PPP), and poly(p-phenylenevinylene) (PPV) are also appropriate for applications in optical electronics due to their excellent mechanical strength and thermal stability. Several remarkable conductive polymers with distinct molecular and electrical characteristics are polyacetylene (PAc), poly(isothianaphthene) (PITN), poly(p-phenylene-terephthalamide) (PPTA), polyazulene (PAZ), polyfuran (PFu), and poly(a-naphthylamine) (PNA). Furthermore, the elastomeric nature of polybutadiene (PBD) and polyisoprene (PIP) advances the area by improving the ductility and durability of materials based on conductive polymers. These many polymers are still essential to the development of high-tech generation (Shahabuddin et al., 2022). In addition to the odd phenomenon known as quantum hall, graphene itself can be obtained in a number of forms, such as nanosheets, nanoribbons, nanoplates, and three-dimensional (3D) nanoformulation. Every structure has excellent uses. Additionally, because of their distinct chemical reactivity toward foreign molecules, other nanocarbons, such as carbon nanotubes (CNTs) and carbon-based additives, cannot replace it, unlike bulky material such as charcoal.

In 2017, Yao et al. presented a polymeric method that combines rapid pyrolysis and hydrothermal carbonization to turn waste polyvinyl chloride (PVC) into energy (Yao et al., 2017). The study investigated how, in relatively moderate situations, the temperature during hydrothermal carbonization might greatly speed up the dechlorination of polymer. The resultant hydrochar had a content of carbon similar to sub-bituminous coal. Hydrocarbons such as toluene, benzene, diphenyl, phenanthrene, and naphthalene were found in the evolved goods made from hydrochars after experimental investigation. The study demonstrated that aromatic hydrocarbons increased with temperature. The study shows how to efficiently turn polymeric waste into useful energy sources by maximizing the conditions required for pyrolysis and hydrothermal carbonization. Graphene-supported gold-silver oxide polyindole nanocomposites (rGO–Au–$Ag_2O$/PIn NCs) were studied by Shoeb et al. to examine their combined function in supercapacitor and antimony (Sb) adsorption uses (Shoeb et al., 2025). Under ideal circumstances, the researchers showed that the prepared NCs, which they had hydrothermally produced, were efficient at extracting Sb(III) from wastewater. For better sustainability, they addressed additional contamination problems by repurposing the spent adsorbent for electrochemical energy storage. With a significant unique capacitance and sustained durability during long cycles, the resultant supercapacitor device displayed encouraging electrochemical performance. This study revealed a novel approach for turning toxic waste into a useful resource by effectively combining adsorption and energy technology. The results aided in the invention of green responsible for win–win WtE applications. In another study, Zhu et al. developed a nano-tin/graphite@carbon (nano-Sn/G@C) composite anode in order to examine the high-value use of waste carbon residue from wasted lithium-ion batteries (Zhu et al., 2022). The research turned waste carbon into a high-performance electrode, addressing economic and green issues. Utilizing the porous and oxidative character of waste carbon, tin(II) chloride ($SnCl_2$) was employed as a reducing agent to generate a nano-precursor. Via orthogonal and single-factor studies, the researchers improved the molten-salt roasting mediums, successfully eliminating impurities including nickel (Ni), cobalt (Co), manganese (Mn), aluminum (Al), silicon (Si), and iron (Fe) while attaining a tin loading of over forty percent. Using density functional theory (DFT) simulations, the underlying atomic-scale chemical processes were investigated. A sandwich-like shape that improved electrochemical efficiency was the result of the subsequent carbothermal reduction of the generated nano-based anode. The anode maintained a high reversible specific capacity even after prolonged charge–discharge cycles, demonstrating remarkable cycling stability and strong initial Coulombic efficiency. This research showed how to recycle battery waste in a sustainable way, providing an economical approach for converting waste into resources for energy applications.

### 3.2.2 Biowaste and Polymer Composite for Efficient Energy Applications

The advantages of biomass over the advanced energy materials included cost-effectiveness, environmental sustainability, and a large number of pore structures (Das et al., 2025; Sahu et al., 2024). In energy storage applications, these green materials showed promise and were widely used in the production of supercapacitors and electrodes. In order to efficiently harvest energy, Gaur et al. developed a hybrid material using biowaste orange peel and poly(vinylidene fluoride) (PVDF) (Gaur et al., 2020). Orange peel is one type of biowaste that greatly increases landfills and contamination in the natural environment. There are possible financial and environmental advantages to using such biowastes. Biowastes naturally have piezoelectric effects since its components are mainly made up of cellulose and other protein. By combining this biowaste with polymer, the hybrid's piezoelectric qualities are improved. The powdered biowaste serves as a filler, inducing an electroactive phase of approximately seventy percent. An energy harvesting structure that does not require traditional high-voltage poling was developed using this mixture of material. When activated by basic physical actions including tapping the finger, the device produced a high open-circuit voltage and power, demonstrating amazing ability. Furthermore, a sliding door's motions or different human actions, such as bending, twisting, and walking, could power light-emitting diodes using the hybrid nanogenerator. The composition, shape, and thermal properties of the hybrid were shown to be responsible for its improved piezoelectric characteristics, which shed light on the mechanism behind them. This green piezoelectric energy harvester is a promising way to harvest energy and use biowaste in an environmentally way. In another study,

Prabu et al. synthesized manganese carbonate nanocomposites ($MnCO_3$ NCs) and activated carbon derived from mango peel as possible electrodes for outstanding performance solid-state energy storage (Prabu et al., 2025). Asymmetric supercapacitors are showing promise as energy storage technologies because they mix redox-type and electric double-layer surfaces. The need for effective, affordable, and sustainable electrodes was the focus of this study. The researchers developed the NCs by pyrolyzing $MnCO_3$ NCs with activated carbon derived from biowastes. Excellent specific capacitance and exceptional stability during cycles were displayed by the resultant nano-based electrodes, suggesting its potential for long-term application in energy. Activated carbon and nanoformulations were employed to produce the asymmetric device, which improved energy density and power density by utilizing two different electrochemical potential. Following stability testing, the device demonstrated exceptional capacity retention, confirming the dependability and longevity of the electrode materials. These results imply that NCs are potential materials for electrode-active nanostructures in energy systems, providing a high-performing and environmentally friendly alternative for upcoming energy technologies. Khan et al. investigated into the possibility of using leftover longan peel as an economical as well as green material for selenium-based cathodes in lithium-selenium (Li-Se) and sodium-selenium (Na-Se) batteries in 2024 (Khan et al., 2024). A porous carbon matrix (LP2) with a hierarchical structure that improves electrochemical activity was designed using an improved activation method. Melt diffusion was used to add selenium to the LP2 matrix. Together, these results highlighted the Se@LP2 composite's potential for long-term energy storage strategies. The potential of lemon peel oil as an alternative energy source for diesel and gasoline engines was investigated by Abdellatief et al. Lemon peel oil is a completely organic and environmentally beneficial substitute for fossil fuels since it is high in terpenes like limonene (Abdellatief et al., 2024). The method of production, important chemical characteristics, and its function as a bio-based component to improve combustion performance and lower emissions were all highlighted in the study. Tests of engine performance showed that it may increase energy efficiency while still adhering to emission standards. The study highlighted the economic and environmental advantages of using lemon peel oil, which promote minimized waste and a circular economy, in addition to its effects on fuel systems. The revolutionary potential of lemon peel oil as a renewable source of energy for transportation is highlighted by this study. Using a mixture of enzymatic and hydrodistillation processes, Banerjee et al. investigated a sustainable and environmentally friendly approach to the simultaneous extraction of pectin and essential oil from Assam lemon peel (Banerjee et al., 2024). The study demonstrated that citrus species contain a substance called Khusilal. A biologically powered triboelectric nanogenerator (TENG) was also created for effective energy production, exhibiting promise for use in small wearable technology. The usefulness of citrus trash in recovering materials and green energy applications is demonstrated by this study. In 2024, Ahmad et al. examined into using pre-treated melon peel waste to produce high-energy–density semi-solid-state supercapacitors (Ahmad et al., 2024). In order to develop a hierarchical porous material appropriate for energy storage applications, the study concentrated on improving the activation of biowaste through chemical and physical processes. A new gel polymer electrolyte based on ionic liquid was reported; it has a wide stability range. The resultant supercapacitors offered a balance between high specific energy and power density, exhibiting encouraging electrochemical characteristics. This study demonstrates how substances obtained from biowaste can be used to develop sustainable and reasonably priced energy storage solutions. In order to remove chloramphenicol from water-based solutions, Aziz et al. investigated the conversion of Nephelium lappaceum L. peel into activated carbon using energy from microwaves (Aziz et al., 2024). With a sizable surface area and mesopore content, the resultant rambutan-peel-derived activated carbon (RPAC) demonstrated an advanced porous structure. According to adsorption investigations, the chloramphenicol–RPAC system exhibited pseudo-second-order kinetic model and the Langmuir isotherm model. The Boyd plot verified that film diffusion was the rate-limiting stage, and the thermodynamic evaluation showed that the adsorption process was exothermic and spontaneous. These results show how microwave-assisted activation can be used to create sustainable and reasonably priced adsorbents for remediation of the environment. In another study, for solid-state supercapacitor applications, Taer et al. presented a novel way of producing sulfur-oxygen self-dual-doped hierarchical porous carbon nanofibers (CNFs) using mangosteen peel (Taer et al., 2024). Nitrogen ($N_2$) and carbon dioxide ($CO_2$) were employed in this study's dual-gas pyrolysis process to generate S and O self-doped carbon nanofibers that were effectively enriched. With its wormhole-like nanostructure, large number of pores, and equal micropore-to-mesopore proportion, the developed nanomaterial demonstrated exceptional electrochemical characteristics. When included into a dual-cylinder supercapacitor, the improved nanoelectrode showed outstanding electrochemical characteristics, as evidenced by its high capacitance, strong rate ability, and superior Coulomb efficiency with low resistance within it. A symmetric supercapacitor arrangement produced additional energy due to the greater faradaic effect, which was ascribed to pseudocapacitance contributions. Mangosteen peel waste has the potential to be an environmentally friendly carbon source for the formation of S- and O-rich self-doped CNFs, opening the door to more sustainable and efficient energy storage options. In 2025, a novel approach for using jackfruit peel/n-docosane composite phase change materials (PCMs) based on biomass from waste for storage of thermal energy and battery thermal management was developed by Luo et al. (Luo et al., 2025). In this study, scrap jackfruit peel was combined with n-docosane (ND) to produce shape-stable PCMs. The approach offered an efficient and environmentally friendly alternative to store and convert energy. From low levels to a much higher level, the activated peel's specific surface area grew drastically, indicating an important increase in the pores. The produced composite shows significant enhancements in its thermal characteristics. The experimental results demonstrated that the cooling efficiency was considerably increased by the composite in comparison with traditional air cooling solutions. The battery pack's highest temperature and difference of temperature were significantly lowered, increasing safety and facilitating better thermal control. In addition to providing a sustainable method of handling biowaste, this work shows how shape-stable PCMs can be used in thermal systems, especially for battery and energy-storing devices. A promising approach to improving thermal storage and maximizing energy savings in a variety of applications is the combination of biowastes and phase change materials.

In order to maximize the drying process of avocado peels, Santos et al. developed a novel pretreatment approach that uses carbonation–ultrasonication to improve rheological qualities, energy consumption, and antioxidant retention (Santos et al., 2024). The goal of this study was to optimize the method of drying in order to add value to biomass processing waste products, especially peels, which are frequently thrown away as waste material. Biowastes (avocado peels) were treated with dry ice and sonication and then allowed to dry at different temperatures using convective air. Energy consumption, antioxidant capacity, water adsorption behavior, rehydration rate, and rheological characteristics were evaluated in order to compare the results of this combination preparation with traditional methods of drying. The results showed that while the presence of $CO_2$ helped retain bioactive components and antioxidant qualities, the application greatly decreased drying time as well as energy usage. The pretreatment materials also showed altered viscosity and enhanced water absorption ability, which may increase their adaptability for future culinary applications. In addition to providing insightful information about how carbonation–sonication techniques can be implemented across a variety of food matrices to improve the efficiency of processing and their nutritional value, this study offers a viable and efficient approach for managing and recycling of agri-food waste. In 2025, the potential of plastic waste-derived carbon materials (PWCMs) for ecologically friendly uses, such as carbon dioxide collection and the generation of green energy, was examined by Luu et al. (Luu et al., 2025). By incorporating PWCMs into the carbon cycle, the study highlighted their contribution to the advance of carbon neutrality. A significant strategy to help the sustainable development goals (SDGs) is the upcycling of plastic waste for the production of hydrogen. Furthermore, PWCM-based adsorbents showed remarkable efficacy in eliminating hazardous metals from wastewater. The study offered a thorough analysis of the developments and difficulties involved in turning waste plastic into useful carbon-based materials for green and sustainable energy applications.

### 3.2.3 Nano-based Applications

MXene has many advantages and unique characteristics that make it a very promising material for energy storage applications. MXene's two-dimensional form provides a large surface area, increasing its charge storage capacity and offering several active sites for chemical reactions. MXene is easily dissolved in water-based solutions and some organic solvents thanks to its surface functional groups, and its hydrophilic character improves affinity with hydrophilic electrolytes and makes ion transport easier. Furthermore, MXenes' specific structure allows them to be shaped into a variety of macroscopic structures, such as hydrogels, films, and aerogels. By allowing electrostatic self-assembly with positively charged elements, MXenes' negatively charged surfaces enable the fabrication of sophisticated composite materials. Rapid electron transport is made possible by MXene materials' high electronic conductivity (6000–8000 S/cm), which makes them appropriate for a variety of electrical appliances. Even after several charge–discharge cycles, their exceptional mechanical qualities—such as strength, bending strength, hardness, and durability against wear—guarantee energy storage systems' stability over time and longevity. Additionally, the materials can be modified to maximize particular characteristics like voltage window, capacitance, and electrolyte compatibility. Additionally, MXenes have excellent ion diffusion mobility and low ion diffusion barriers, which enhance electrochemical kinetics and make them perfect for rechargeable ion batteries. Fast ion transport is further made possible by MXenes' layered structure and interlayer gap, which advances the creation of high-powered electronics (Devi et al., 2025). Through the conversion of waste materials into valuable nanomaterials, nanotechnology offers numerous possibilities for waste management, supporting resource recovery and sustainable recycling. In their groundbreaking study, Xie et al. examined composite phase change materials (CPCMs) based on pitaya peel modified by MXene quantum dots (MQDs), which provide superior thermal characteristics for applications involving building energy-efficient structures (Xie et al., 2024). In order to develop a composite with improved thermal conductivity and stability, the polyethylene glycol (PEG) and MXene quantum dots are mixed with biomaterials, notably pitaya peel. The addition of MQDs into the porous carbon structure derived from pitaya peels enhances the composite's thermal efficiency while preserving its enthalpy across one hundred cycles, proving its recyclable stability. The prepared nanomaterial is significant promise for electronic component applications as well as building thermal management, offering an environmentally friendly approach to reduce emissions, save energy, and recycle waste materials. With a simple production way, the study demonstrated the multipurpose potential of bio-based materials and advances environmentally friendly energy options across a range of sectors.

## 4 Integration of Artificial Intelligence and Machine Learning in WtE

Recent developments in machine learning (ML) and artificial intelligence (AI) have greatly aided in the optimization of WtE operations. AI-powered models increase combustion efficiency, forecast energy yields, and improve trash sorting. Among the important uses are the following:

- Energy Yield Prediction: To predict the possible energy outputs from different waste materials, ML models use past data and external variables. In WtE facilities, this enhances the efficiency of energy conversion processes and aids in feedstock decision optimization.

- Process Control and Automation: AI technologies allow for predictive maintenance and real-time monitoring of WtE systems, spotting possible problems before they have an impact on efficiency. As a result, WtE plants operate more efficiently and have fewer interruptions.
- Combustion Efficiency: By keeping an eye on variables like temperature, pressure, and oxygen levels, AI and ML algorithms optimize the process of combustion in waste incineration systems. By modifying combustion variables in real time, these models minimize fuel use, maximize energy output, and lower pollutants.
- Waste Sorting Optimization: Compared to conventional techniques, AI algorithms—in particular, computer vision and deep learning—are utilized to recognize and classify waste products more effectively. By lowering pollution and increasing overall efficiency, this increases the precision of waste classification for the best energy conversion.
- Optimization of Biomass Conversion: AI-driven models in WtE systems that convert biomass optimize feedstock administration, moisture-level evaluation, and processing factors to increase the amount of organic waste that can be converted into biofuels like bio-oils and biogas.
- Significant gains in energy performance, affordability, and sustainable development can be made by using AI and ML into WtE technology. This will help create waste management solutions that are cleaner and more effective.

In 2025, Yan et al. addressed the worldwide energy issue and plastic waste management by combining solid oxide fuel cells (SOFCs) with plastic waste gasification in an AI-powered study to optimize a WtE system (Yan et al., 2025). The study analyzed and optimized this integrated system using ML methods based on regression. Applying nonlinear ML algorithms demonstrated a notable increase in accuracy, as seen by the electrical ML algorithm's R-squared value rising from 0.9146 (linear) and 0.9427 (quadratic) to 0.9990 (nonlinear). These settings resulted in 42% electrical efficiency, 38% exergy efficiency, and 861 kg/MWh emissions. An unparalleled electricity efficiency might be achieved by integrating plastic waste gasification with SOFCs and optimizing the process through ML, as demonstrated by this study. This strategy reduced contamination of the environment and improved energy recovery from plastic waste, establishing it as a viable and competitive energy source with enormous promise for communities and businesses around the globe.

## 5 Future Perspectives and Sustainable Energy Systems

Research on incorporating energy from clean sources including solar and wind power into the production of WtE is expanding in an effort to increase WtE efficiency and environmental sustainability. Furthermore, it is anticipated that the advancements in sophisticated material recovery approaches and nano-waste management would transform WtE strategies. By turning trash into a useful source of energy, modern WtE systems have the potential to drastically lower global carbon footprints. Future cleaner and more efficient energy options will depend on multidisciplinary collaboration between scientists, engineers, and policymakers as this sector continues to innovate.

## 6 Conclusion

Smart technologies that convert waste into energy offer a feasible and environmentally friendly way to deal with waste management and energy problems. WtE may be vital to accomplishing worldwide sustainable development goals by using AI, improving waste valorization procedures, and implementing eco-friendly processes. In order to fully realize the promise of these revolutionary energy alternatives, further developments and legislative support are essential. With the advantages offered by the evolving computer age, AI-driven models are radically transforming heat value analysis by eliminating the dependency on precise devices and previous experience required by traditional methods to determine the composition of waste. AI makes real-time predictions possible through ML and regression approaches, but traditional methods including proximate analysis, ultimate analysis, and physical composition necessitate substantial data gathering and manual handling. AI is an effective tool for optimizing incineration practices and enhancing energy recovery from municipal solid waste since it improves precision, effectiveness, and flexibility in determining the content of waste by analyzing large datasets.

## References


Abdellatief, T. M. M., Abdelkareem, M. A., & Olabi, A. G. (2024). Lemon peel oil: A groundbreaking renewable energy source for gasoline and diesel biofuel engine applications. *Science Talks, 12*, Article 100404. https://doi.org/10.1016/J.SCTALK.2024.100404

Ahmad, N., Rinaldi, A., Sidoli, M., Magnani, G., Vezzoni, V., Scaravonati, S., Pasetti, L., Fornasini, L., Gupta, H., Tamagnone, M., Ridi, F., Milanese, C., Riccò, M., & Pontiroli, D. (2024). Pretreated biomass waste melon peels for high energy density semi solid-state supercapacitors. *Journal of Power Sources, 624*, Article 235511. https://doi.org/10.1016/J.JPOWSOUR.2024.235511

Aziz, A., Mohamad Yusop, M. F., & Ahmad, M. A. (2024). Harnessing microwave energy to transform *Nephelium lappaceum* L. peel into activated carbon for chloramphenicol eradication in aqueous solutions. *Materials Chemistry and Physics, 318*, 129311. https://doi.org/10.1016/J.MATCHEMPHYS.2024.129311

Bal, B., Ghosh, S., & Das, A. P. (2018). Microbial recovery and recycling of manganese waste and their future application: A review. *Geomicrobiological Journal, 36*(1), 85–96. https://doi.org/10.1080/01490451.2018.1497731

Bala, K., Anshul, Verma, S., Dogra, A., Das, A. P., Kumar, R., & Nandini. (2024). Sustainable Management for Urban Plastic Waste Generation. *Sustainable Management of Urban Plastic Waste Through Circular Economic Approaches*, 211–236. https://doi.org/10.1201/9781003465447-13

Banerjee, S., Kundu, D., Dey, S., Kumbhakar, P., Mahapatra, P. L., Karmakar, S., Tiwari, C. S., & Banerjee, R. (2024). A cleaner and eco-friendly approach to simultaneous extraction and characterization of essential oil and pectin from Assam lemon peel and its application for energy generation through TENG devices. *Heliyon, 10*(15), Article e32999. https://doi.org/10.1016/J.HELIYON.2024.E32999

Dai, L., Hu, X., Zhou, H., Liu, X., Wu, Y., Sun, L., & Shu, X. (2025). Physicochemical, kinetic, and bond-energy analyses of the pyrolysis of carbon-fiber-reinforced polymer waste. *Journal of Analytical and Applied Pyrolysis, 186*, 106972. https://doi.org/10.1016/J.JAAP.2025.106972

Das, A., Dey, S., Bala, K., Gnanasekaran, R., & Das, A. P. (2024). Microbial bioremediation technology for sustainable treatment and management of synthetic microfiber waste. *Sustainable Microbial Technology for Synthetic and Cellulosic Microfiber Bioremediation*, 221–238. https://doi.org/10.1007/978-3-031-62600-5_12

Das, A., Dey, S., Gacem, A., Yadav, K. K., Nath, H., Ram, D. K., & Das, A. P. (2025). Revolutionizing solid hydrocarbon sourcing: Plastic's journey from waste to upcycled treasure. *Environmental Science and Engineering, Part, F218*, 331–351. https://doi.org/10.1007/978-3-031-81615-4_14

Das, A. P., Ghosh, S., Mohanty, S., & Sukla, L. B. (2015). Consequences of manganese compounds: A review. *Toxicological and Environmental Chemistry, 96*(7), 981–997. https://doi.org/10.1080/02772248.2015.1005428

Das, A. P., Sukla, L. B., Pradhan, N., & Nayak, S. (2011). Manganese biomining: A review. *Bioresource Technology, 102*(16), 7381–7387. https://doi.org/10.1016/J.BIORTECH.2011.05.018

Devi, N., Kumar, R., Singh, R. K., & Moshkalev, S. A. (2025). Recent development of MXenes and their composites in electrochemical energy storage: Current status, challenges and future prospects. *Journal of Power Sources, 636*, Article 236538. https://doi.org/10.1016/J.JPOWSOUR.2025.236538

Dey, S., Das, A., Mallick, K., Sahu, A., & Das, A. P. (2023). *Environmental Petroleum Waste: Pollution, Toxicity, Sustainable Remediation*, 159–175. https://doi.org/10.1007/978-3-031-48220-5_7

Dey, S., Dash, A., & Das, A. P. (2024). Bioremediation as a sustainable and circular solution for mining waste management and bioenergy generation. *Sustainable Management of Mining Waste and Tailings: A Circular Economy Approach*, 107–119. https://doi.org/10.1201/9781003442455-6

Ebrahimi, A., & Houshfar, E. (2025). A comprehensive exergoeconomic analysis of pyrolysis, anaerobic digestion, and integrated Py-AD plants for sustainable energy and waste management. *Fuel, 384*, Article 133928. https://doi.org/10.1016/J.FUEL.2024.133928

Fardi, Z., Shahbeik, H., Nosrati, M., Motamedian, E., Tabatabaei, M., & Aghbashlo, M. (2024). Waste-to-energy: Co-pyrolysis of potato peel and macroalgae for biofuels and biochemicals. *Environmental Research, 242*, Article 117614. https://doi.org/10.1016/J.ENVRES.2023.117614

Gaur, A., Tiwari, S., Kumar, C., & Maiti, P. (2020). Bio-waste orange peel and polymer hybrid for efficient energy harvesting. *Energy Reports, 6*, 490–496. https://doi.org/10.1016/J.EGYR.2020.02.020

Ghosh, S., Mohanty, S., Nayak, S., Sukla, L. B., & Das, A. P. (2016). Molecular identification of indigenous manganese solubilising bacterial biodiversity from manganese mining deposits. *Journal of Basic Microbiology, 56*(3), 254–262. https://doi.org/10.1002/JOBM.201500477

Ghosh, Shampa, Sinha, J. K., Ghosh, S., Vashisth, K., Han, S., & Bhaskar, R. (2023). Microplastics as an emerging threat to the global environment and human health. *Sustainability 15*(14), 10821. https://doi.org/10.3390/SU151410821

Hou, Y., Wang, Q., Zhou, K., Zhang, L., & Tan, T. (2024). Integrated machine learning methods with oversampling technique for regional suitability prediction of waste-to-energy incineration projects. *Waste Management, 174*, 251–262. https://doi.org/10.1016/J.WASMAN.2023.12.006

Hsu, H. W., Binyet, E., Nugroho, R. A. A., Wang, W. C., Srinophakun, P., Chein, R. Y., Demafelis, R., Chiarasumran, N., Saputro, H., Alhikami, A. F., Sakulshah, N., & Laemthong, T. (2024). Toward sustainability of waste-to-energy: an overview. *Energy Conversion and Management, 321*, Article 119063. https://doi.org/10.1016/J.ENCONMAN.2024.119063

Islam, N. F., Gogoi, B., Saikia, R., Yousaf, B., Narayan, M., & Sarma, H. (2024). Encouraging circular economy and sustainable environmental practices by addressing waste management and biomass energy production. *Regional Sustainability, 5*(4), Article 100174. https://doi.org/10.1016/J.REGSUS.2024.100174

Khan, M., Yan, S., Ali, M., Mahmood, F., Zheng, Y., Song, X., Li, G., Liu, J., & Wang, Y. (2024). From longan peel waste to energy storage: Porous activated carbon as a cathode matrix for advanced Li/Na-selenium batteries. *Progress in Natural Science: Materials International, 34*(2), 329–337. https://doi.org/10.1016/J.PNSC.2024.03.004

Khatun, R., Singh, S., Dubey, N. K., & Prasad Das, A. (2023a). A review on marine-based phytochemicals and their application in biomedical research. *Recent Frontiers of Phytochemicals: Applications in Food, Pharmacy, Cosmetics, and Biotechnology*, 383–395. https://doi.org/10.1016/B978-0-443-19143-5.00013-X

Khatun, R., Singh, S., Dubey, N. K., & Prasad Das, A. (2023b). A review on marine-based phytochemicals and their application in biomedical research. In *Recent Frontiers of Phytochemicals: Applications in Food, Pharmacy, Cosmetics, and Biotechnology* (pp. 383–395). Elsevier. https://doi.org/10.1016/B978-0-443-19143-5.00013-X

Lim, M., Lee, Y., Lee, Y., Yang, W., & Kim, S. (2024). Improving waste-incineration energy recovery efficiency using a reverse calculation algorithm to estimate waste composition and heating value. *Waste Management, 190*, 486–495. https://doi.org/10.1016/J.WASMAN.2024.10.014

Luo, Y., Pan, Z., Gao, R., Peng, R., Xiao, X., & He, X. (2025). Waste biomass-based jackfruit peel/n-docosane composite phase change materials for thermal energy storage and battery thermal management. *Journal of Energy Storage, 110*, Article 115277. https://doi.org/10.1016/J.EST.2024.115277

Luu, T. A., Le, V. G., Thai, V. A., Nguyen, M. K., Lin, C., Pham, M. T., Chang, S. W., & Nguyen, D. D. (2025). Plastic wastes for carbon-based materials: Investigations on recent applications towards environmentally sustainable, carbon dioxide capture and green energy. *Environmental Pollution, 370*, Article 125899. https://doi.org/10.1016/J.ENVPOL.2025.125899

Mohanty, J., Dey, S., Bala, K., & Das, A. P. (2024). Recent advances in microplastic pollution for its sustainable management. *Sustainable Management of Urban Plastic Waste Through Circular Economic Approaches*, 27–48. https://doi.org/10.1201/9781003465447-3

Mohanty, S., Ghosh, S., Nayak, S., & Das, A. P. (2016). Isolation, identification and screening of manganese solubilizing fungi from low-grade manganese ore deposits. *Geomicrobiological Journal, 34*(4), 309–316. https://doi.org/10.1080/01490451.2016.1189016

Prabina, K. E., Benya, A., Mohanty, S., Van Hullebusch, E. D., Das, A. P., & Singh, S. (2024). Recovery of multi-functional carbon fiber from COVID-19 generated waste for an effective management through circular economy. *Environmental Science and Engineering, Part, F2827*, 83–97. https://doi.org/10.1007/978-3-031-50840-0_6

Prabu, S., Vinu, M., Chiang, K. Y., Mariappan, A., Dharman, R. K., & Oh, T. H. (2025). Mango peel biomass-derive activated carbon and MnCO3 nanocomposite: The potential electrode for high-performance solid-state energy storage. *Journal of Industrial and Engineering Chemistry*. https://doi.org/10.1016/J.JIEC.2025.01.021

Priyadarshini, S., Jagatee, S., & Das, A. P. (2024). Synthetic fabrics and microfiber pollution–An assessment of their global impact. *Environmental Science and Engineering, Part, F2333*, 137–157. https://doi.org/10.1007/978-3-031-51792-1_8

Priyadarsini, S., & Das, A. P. (2024). Analytical and structural characterization of waste lithium-ion batteries for their effective recycling strategy. *Environmental Chemistry and Ecotoxicology*. https://doi.org/10.1016/j.enceco.2024.12.004

Priyadarsini, S., & Das, A. P. (2025). Lithium bioleaching: A review on microbial-assisted sustainable technology for lithium bio-circularity. *Journal of Water Process Engineering, 69*, Article 106744. https://doi.org/10.1016/j.jwpe.2024.106744

Sadasivuni, K. K., Rattan, S., Waseem, S., Brahme, S. K., Kondawar, S. B., Ghosh, S., Das, A. P., Chakraborty, P. K., Adhikari, J., Saha, P., & Mazumdar, P. (2019). Silver Nanoparticles and Its Polymer Nanocomposites—Synthesis, Optimization, Biomedical Usage, and Its Various

Applications. *Lecture Notes in Bioengineering*, 331–373. https://doi.org/10.1007/978-3-030-04741-2_11

Sahu, A., Dey, S., Mallick, K., Priyadarsini, S., Behera, I. D., & Das, A. P. (2024). A review on advanced technology for sustainable management of synthetic microplastic waste. *Sustainable Management of Urban Plastic Waste Through Circular Economic Approaches*, 262–279. https://doi.org/10.1201/9781003465447-15

Santos, N. C., Almeida, R. L. J., Monteiro, S. S., de Andrade, E. W. V., Silva, R. dos S., Albuquerque, J. C., de Figueiredo, D. V. P., Duarte, D. R., Pinheiro, L. da S. S., Martins, A. N. A., Silva, S. do N., Dias, R. A. de L., Pasquali, M. A. de B., & Rocha, A. P. T. (2024). Drying of avocado peels using carbonation-ultrasonication as pretreatment: Energy consumption, antioxidant capacity and rheological properties. *Chemical Engineering and Processing - Process Intensification, 205*, 110004. https://doi.org/10.1016/J.CEP.2024.110004

Satapathy, S., Kar, D., Priyadarsini, S., Das, A. P., & Kuanar, A. (2024). Application of microorganisms in emerging waste management strategies and remediation of synthetic and cellulosic microfibers. *Sustainable Microbial Technology for Synthetic and Cellulosic Microfiber Bioremediation*, 183–206. https://doi.org/10.1007/978-3-031-62600-5_10

Shahabuddin, S., Gaur, R., Mukherjee, N., Chandra, P., & Khanam, R. (2022). Conducting polymers-based nanocomposites: Innovative materials for waste water treatment and energy storage. *Materials Today: Proceedings, 62*(P13), 6950–6955. https://doi.org/10.1016/J.MATPR.2021.12.335

Shoeb, M., Mashkoor, F., Naved Khan, M., Kim, B. J., & Jeong, C. (2025). Waste to energy strategy: Graphene-supported Au-$Ag_2O$ polyIndole nanocomposites for antimony adsorption and their sequential utilization in supercapacitors device. *Separation and Purification Technology, 354*, Article 128656. https://doi.org/10.1016/J.SEPPUR.2024.128656

Singh Jadaun, J., Bansal, S., Sonthalia, A., Rai, A. K., & Singh, S. P. (2022). Biodegradation of plastics for sustainable environment. *Bioresource Technology, 347*, Article 126697. https://doi.org/10.1016/J.BIORTECH.2022.126697

Taer, E., Apriwandi, A., Nasution, W. M., Fudholi, A., Chitraningrum, N., & Taslim, R. (2024). Mangosteen peel waste derived Sulfur-Oxygen self-dual-doped hierarchical porous carbon nanofiber for ultrahigh energy of solid-state supercapacitor. *Bioresource Technology Reports, 28*, Article 102004. https://doi.org/10.1016/J.BITEB.2024.102004

Xie, Y., & Xi, L. (2024). MXene quantum dots modified pitaya peel-based composite phase change material with excellent thermal properties for building energy efficiency applications. *Construction and Building Materials, 449*, Article 138392. https://doi.org/10.1016/J.CONBUILDMAT.2024.138392

Yan, G., Yang, X., Shaban, M., Abed, A. M., Abdullaev, S., Alhomayani, F. M., Khan, M. N., Alkhalaf, S., Alturise, F., & Albalawi, H. (2025). Artificial intelligence-powered study of a waste-to-energy system through optimization by regression-centered machine learning algorithms. *Energy, 320*, Article 135142. https://doi.org/10.1016/J.ENERGY.2025.135142

Yao, Z., & Ma, X. (2017). A new approach to transforming PVC waste into energy via combined hydrothermal carbonization and fast pyrolysis. *Energy, 141*, 1156–1165. https://doi.org/10.1016/J.ENERGY.2017.10.008

Yatim, N. M., Shamsudin, Z., Shaaban, A., Ghafar, J. A., & Khan, M. J. H. (2024). Recovery of carbon fiber from carbon fiber reinforced polymer waste via microwave molten-carbonate pyrolysis. *Journal of Analytical and Applied Pyrolysis, 183*(1), Article 106840. https://doi.org/10.1016/J.JAAP.2024.106840

Zhu, X., Xiao, J., Chen, Y., Tang, L., Hou, H., Yao, Z., Zhang, Z., & Zhong, Q. (2022). A high-performance nano-Sn/G@C composite anode prepared by waste carbon residue from spent Lithium-ion batteries. *Chemical Engineering Journal, 450*, Article 138113. https://doi.org/10.1016/J.CEJ.2022.138113

# Nanotechnology Sustainability and Nano-waste Concerns

**Gizem Karabulut Sevk and Nuray Beköz Üllen**

## 1 Introduction to Nanotechnology and Sustainability

Nanotechnology is a multidisciplinary field of science that involves the synthesis, design, and application of various materials and devices at the nanoscale, usually ranging from 1 to 100 nm (Olawade et al., 2024). At this scale, materials display distinct physical, chemical, and biological characteristics that set them apart from their bulk forms (Das & Ansari, 2009; Wang et al., 2019). These properties enable groundbreaking innovations in many different sectors, including medicine (Karahmet Sher et al., 2024), energy (Anusha et al., 2024), agriculture (Rodríguez-Seijo et al., 2024), and environmental protection (Hasan et al., 2024). Nanotechnology offers tremendous potential to improve sustainability because it controls matter at the atomic and molecular levels (Borode et al., 2025). Innovative studies are being carried out every day to create highly efficient systems and solutions using nanotechnology products. However, the low nanoscale property also poses significant challenges, especially in terms of safety, environmental impact, and waste management, and requires the development and implementation of solutions.

Sustainability, in its most general definition, is an approach that aims to protect natural resources while meeting human needs and ensuring environmental, economic, and social balance (Kuhlman & Farrington, 2010). This concept refers to the current generation meeting its own needs while not endangering the capacity for future generations to fulfill their own requirements (Holden et al., 2014). Sustainability for nanotechnology can generally be addressed in two basic dimensions: environmental

G. Karabulut Sevk (✉) · N. Beköz Üllen
Faculty of Engineering, Department of Metallurgical and Materials Engineering, Istanbul University-Cerrahpaşa, İstanbul, Turkey
e-mail: gizem.karabulut@iuc.edu.tr

N. Beköz Üllen
e-mail: nbekoz@iuc.edu.tr



A. P. Das and S. Karakuş (eds.), *Synergizing Sustainability for Integrated Waste Management*, Sustainable Environmental Waste Management Strategies,
https://doi.org/10.1007/978-3-032-03718-3_13

and economic. Environmental sustainability can generally be defined as ensuring the renewability of natural resources, protecting ecosystems, and minimizing environmental damage (Omer, 2008). On the other hand, economic sustainability is to support long-term economic growth and prosperity and to ensure the continuity of economic systems by using resources effectively (Ekins, 2002). Thanks to this holistic approach, it is adopted as a basic guide for solving issues such as climate change, resource scarcity, biodiversity loss, and social inequality, which are global problems today.

This chapter addresses the sustainability implications of nanotechnology and the environmental and social concerns posed by nano-waste. The chapter discusses the sustainable solutions offered by nanotechnology in areas such as energy, environment, health, and agriculture while examining the challenges of managing nano-waste generated during the production and use of these technologies. The possible effects of nanomaterials on the environment and human health are evaluated in light of their recycling, reuse, and safe disposal methods. The chapter focuses on strategies for advancing nanotechnology in line with sustainable development goals.

# 2 Applications of Nanotechnology for Sustainable Development

## *2.1 Environmental Applications*

Many environmental problems that today's world is exposed to (soil pollution, water pollution, air pollution, and various wastes) are being tried to be resolved by researchers with nanotechnology products (Louis Theodore et al., 2005). In this section, nanotechnological solutions for environmental problems will be discussed.

### 2.1.1 Application of Nanotechnology in Soil Treatment

Soil has a vital function in supporting the health of ecosystems both within and on, as well as in sustaining human life (Wall et al., 2015). Healthy soils are essential for the sustainability of life due to their critical functions, including supporting agriculture, regulating the water cycle, and maintaining biodiversity (Tahat et al., 2020). However, factors such as industrial activities, pesticides, chemical wastes, and plastic pollution disrupt this natural balance by causing soil pollution (Weldeslassie et al., 2017). Soil pollution can cause the loss of productive agricultural lands, the destruction of ecosystems, and the threat to human health (Alengebawy et al., 2021). Therefore, cleaning and rehabilitating soils are of great importance. Nanotechnology offers a revolutionary tool in the solution of soil pollution. Nanoparticles (NPs) provide high efficiency in the separation of pollutants from soils and help clean heavy metals, pesticides, and organic pollutants (Dhanapal et al., 2024). In addition, biotechnological solutions developed with nanotechnology contribute to restoring soil health and increasing productivity (Rajput et al., 2022). Nanotechnology offers an innovative approach to cleaning soil pollution, enabling effective removal of pollutants (Dhanapal et al., 2024; Liu et al., 2024). Nanomaterials can interact strongly with pollutants due to their high surface area and reactivity (Zhang et al., 2020). Nanomaterials are used for various purposes in soil remediation applications and provide many environmental benefits such as reducing soil pollution, increasing fertility, or preventing erosion (Attia & Elsheery, 2020; Jatav & Rajput, 2022; Liu & Lal, 2012). The uses and purposes of nanomaterials in soil remediation applications are discussed under this title.

The most common use of nanomaterials in soil treatment is cleaning (Del Prado-Audelo et al., 2021). Nanomaterials are used for applications such as heavy metal removal (Mahmoud et al., 2017) and organic material degradation (Bakhtiari et al., 2022). Over the past few decades, various treatment techniques have been developed to mitigate heavy metal contamination. These methods fall into three main categories: physical (adsorption, coagulation, evaporation, and filtration), chemical (precipitation, oxidation, ion exchange, and electrochemical processes), and biological (biodegradation and phytoremediation) (Yu et al., 2021). Adsorption is widely preferred due to its cost-effectiveness, high removal efficiency, broad applicability, and ease of operation (Hu et al., 2015). Numerous studies in the literature focus on heavy metal removal from soil, highlighting the successful use of nanomaterials such as titanium dioxide NPs ($TiO_2$ NPs), nano-zero-valent iron (nZVI), and carbon nanotubes (CNTs). Vasarevičius et al. examined the immobilization of $Cd^{2+}$, $Cu^{2+}$, $Ni^{2+}$, and $Pb^{2+}$ using nZVI in contaminated soil (Vasarevičius et al., 2019). The results showed that immobilization efficiency was higher in soils spiked with multiple heavy metals compared to single-metal contamination, with Ni exhibiting the best immobilization. Although Cd had the lowest immobilization efficiency initially, its efficiency significantly increased with higher nZVI concentrations, reaching over 90% with 15–30% nZVI. While Cu and Pb showed the highest removal from aqueous solutions, their immobilization in single-metal polluted soils was only slightly better than Cd, likely due to high organic matter content affecting sorption. For soils rich in insoluble organic matter, using more than 5% nZVI is not recommended. Further studies are needed to assess nZVI's effectiveness in different soil types. Zarime et al. examined the efficiency of granitic residual soil-supported nano-zero-valent iron (Gr-nZVI) in removing heavy metals ($Pb^{2+}$, $Cu^{2+}$, $Co^{2+}$, $Cd^{2+}$, $Ni^{2+}$, and $Zn^{2+}$) from mixed solutions under batch equilibrium conditions (Zarime et al., 2023). In the study, adsorption tests showed rapid metal uptake within the first 5 min, particularly for $Pb^{2+}$ on nZVI ($qe = 17.89$ mg/g) and Gr-nZVI ($qe = 15.29$ mg/g). The authors stated that Gr-nZVI is an effective adsorbent for heavy metals. In another study, Matos et al. investigated the potential of CNTs for soil remediation, specifically in immobilizing heavy metal ions such as $Pb^{2+}$, $Cu^{2+}$, $Ni^{2+}$, and $Zn^{2+}$, which are commonly found in contaminated soils (Matos et al., 2017). The heavy metals exhibited different affinities for the soil, following the order Pb > Cu > Ni > Zn, with lead being the most adsorbed and zinc the least. Adsorption tests revealed that multi-walled CNTs had minimal impact on $Pb^{2+}$ and $Cu^{2+}$ immobilization, as these metals were already strongly retained by soil particles. However, the addition of multi-walled CNTs significantly enhanced the immobilization of $Ni^{2+}$ and $Zn^{2+}$ by approximately 30% compared to untreated soil. The authors stated that applicability of MWCNTs in soil stabilization and solidification techniques contributes to the long-term immobilization of heavy metals and minimizes environmental and human health risks. Zand et al. examined the co-application of $TiO_2$ NPs and plant growth-promoting rhizobacteria (PGPR) to enhance the phytoremediation of cadmium-contaminated soil using *Trifolium repens* (Zand et al., 2020). The authors declare that the co-application of $TiO_2$ NPs and PGPR, investigated in this study to enhance the phytoremediation of Cd-contaminated soil, could effectively promote the growth and performance of *Trifolium repens.* Deng et al. investigated the influence of $TiO_2$ NPs on cadmium uptake and accumulation in the hyperaccumulator plant *Boehmeria nivea* (ramie) grown in cadmium-contaminated soil (Deng et al., 2024). The results of the study highlight the potential of $TiO_2$ NPs in improving the effectiveness of phytoremediation with hyperaccumulator plants such as ramie in Cd-contaminated soils. Over the past two centuries, the synthesis of organic compounds has expanded significantly, encompassing raw materials, plastics, fuels, detergents, and various other substances. However, the environmental pollution and hazards associated with their production and persistence post-use have become a major concern. Several classes of organic compounds are considered critical environmental issues today, including polychlorinated biphenyls (PCBs), dichlorodiphenyltrichloroethane (DDT), polybrominated diphenyl ethers (PBDEs), dechlorane plus (DP), and decabromodiphenyl ethane (DBDPE) (Bakhtiari et al., 2022).

Another point of using of NPs is to increase soil fertility. At this point, nanofertilizers and nano-bioactive substances can be used (Dhoundiyal et al., 2023). Provide more efficient nutrients to plants by providing controlled release and reduce the use of chemical fertilizers (Singh et al., 2024). For example, nano-clay-based fertilizers optimize nitrogen and phosphorus release (Iqbal et al., 2019). El-Azeim et al. applied foliar application of nano fertilizers at equivalent or lower rates than the recommended non-nano-fertilizers and evaluated the yield (Abd El-Azeim et al., 2020). As a result, it produced a beneficial effect on all potato yield and yield parameters. It was observed that foliar application of nano-fertilizers at lower rates such as 25% or 50% was equal or more effective than 100% nitrogen, phosphorus, and potassium (NPK) chemical fertilizers in improving all potato yield and yield parameters under field conditions. Nano-microelements and biopolymer-based nanomaterials that promote plant growth enrich the mineral content of the soil. Additionally, the water retention capacity of the soil can be increased with nanomaterials (Abd-Elsalam & Zahid, 2024). Iron- and magnesium-based NPs were used to examine the long-term effects of NPs on soil water retention and aggregate tensile strength (Daraei et al., 2024). In their study, the authors stated that NPs should be evaluated as efficient soil enhancers to improve soil physical properties. However, they emphasized that the best NP types and optimum levels should be determined, especially for the sustainable management and prevention of potential environmental harm.

Superabsorbent polymers (SAPs) containing nano-cellulose and nano-clay create better conditions for agriculture in arid regions by trapping water (Batara et al., 2025; Zheng et al., 2023). In addition, nanomaterials can be used in areas such as supporting soil microbiota and combating erosion and desertification (Abdelhak, 2022). In conclusion, nanomaterials provide significant advantages in terms of soil sustainability. However, detailed studies should be conducted regarding the amount of use, its effects, and properties, and careful use is required.

### 2.1.2 Application of Nanotechnology in Water Treatment

Nanotechnology has emerged as a transformative tool in addressing global water challenges, offering innovative solutions for purification, filtration, and contamination removal (Ugwu et al., 2022). By leveraging materials at the nanoscale, these technologies enhance the efficiency and effectiveness of water treatment processes, making clean water more accessible and sustainable (Mauter et al., 2018).

Nanostructure-based membranes, nanomaterials used for adsorption, nanostructures that provide cleaning through catalysts, nanobiosensors that indicate water quality, and nanostructures with self-cleaning mechanisms are examples of the use of nanotechnology in water treatment (Das et al., 2015). One of the most notable applications is nanofiltration membranes, which can remove contaminants such as heavy metals, organic pollutants, and salts with high precision (Al-Rashdi et al., 2011; Covaliu-Mierlă et al., 2023; Koli & Singh, 2023). These membranes utilize nanopores to filter out even the smallest particles, including bacteria and viruses, while maintaining high water flow rates (Fahimirad et al., 2021; Goswami & Pugazhenthi, 2020). Additionally, carbon nanotubes (CNTs) and graphene-based materials have proven highly effective in adsorbing pollutants and separating oil from water, making them valuable for industrial wastewater treatment (Al-Anzi & Siang, 2017; Urfi et al., 2024; Yin et al., 2020). Bahgat et al., in their study, performed multi-walled CNTs (MWCNTs) synthesis and examined their adsorption properties. In the study, it was stated that CNTs are a good candidate as adsorbents due to their porous structure. As a result of the study, it was reported that the obtained MWCNTs played an important role in Mn removal (Bahgat et al., 2011). Nanotechnology also contributes to advanced oxidation processes (AOPs), where photocatalysts such as $TiO_2$ NPs degrade organic contaminants and neutralize harmful pathogens under ultraviolet (UV) light (Jabbar & Esmail Ebrahim, 2022). These systems are particularly beneficial in regions with limited access to traditional water treatment infrastructure. Ramli et al. studied the effect of different concentrations of $TiO_2$ nanofiller on the removal of acetaminophen from a water-based solution. The membrane containing the highest $TiO_2$ concentration (2 wt%) showed the greatest water flux and acetaminophen removal rate of 32 $L/m^2h$ and 97%, respectively. The incorporation of $TiO_2$ had a significant effect on the structure of the membrane and resulted in the formation of finger-like pores evenly distributed throughout a spongy framework. They concluded that this change resulted in increased flow rate and improved efficiency in terms of flux and rejection rate (Ramli et al., 2024). Furthermore, magnetic NPs are increasingly used to remove pollutants like arsenic and fluoride from drinking water. These particles bind to contaminants and can be easily separated using magnetic fields, offering a reusable and eco-friendly solution. Xiong et al. (2014), reported that cellulose@$\gamma$-$Fe_2O_3$ NPs synthesis via co-precipitation method and using as a adsorbent for water treatment. As a result of the study, they stated that the prepared magnetic nanocomposite showed excellent adsorption efficiency for Pb(II) and methylene blue when compared to other magnetic materials reported in the literature. In another study, Bao et al. (Bao et al., 2013) synthesized monodisperse magnetic NPs by an improved sonohydrothermal method and then applied them to remove three commonly used tetracyclines (TC), including tetracycline (TCN), oxytetracycline (OTC) and chlortetracycline (CTC) from water. As a result of the analysis, they stated that high removal efficiencies (more than 96%) were achieved in 5 min.

The integration of nanotechnology in water treatment not only addresses the immediate need for clean water but also reduces the environmental impact of treatment processes by minimizing energy consumption and waste generation. As research advances, nanotechnological innovations are expected to play a pivotal role in ensuring water security and sustainability worldwide.

### 2.1.3 Application of Nanotechnology in Air Purification

Nanotechnology offers groundbreaking solutions for air purification, addressing critical issues such as pollution, indoor air quality, and harmful emissions (Saleem et al., 2022). By utilizing materials with nanoscale properties, these technologies can efficiently capture, degrade, or neutralize air pollutants, including particulate matter, harmful gases, and biological contaminants (Yunus et al., 2012).

Nanofiltration systems are among the most effective tools for removing ultrafine particles and airborne pathogens from the air (de Almeida et al., 2022; Mallakpour et al., 2022). These systems use membranes with nanosized pores to filter out pollutants while maintaining efficient airflow (Dutta et al., 2024). Nanofiber-based filters, often made from materials such as carbon or polymer nanofibers, are particularly effective in trapping fine particulate matter (PM2.5 and PM10) and allergens (Dutta et al., 2024; Toptaş et al., 2024).

Photocatalytic NPs (such as $TiO_2$, ZnO, and $Fe_2O_3$) are widely used in air purification devices. Under exposure to UV or visible light, these NPs generate reactive oxygen species that can break down volatile organic compounds (VOCs), bacteria, and harmful gases like nitrogen oxides (NOx) into harmless by-products (Yunus et al., 2012). This technology is especially beneficial in urban environments with the high levels of vehicular and industrial emissions. Carbon-based nanomaterials, including graphene and activated carbon nanotubes, are highly efficient in adsorbing toxic gases and odors (Yin et al., 2021). These materials provide a large surface area for adsorption, making them ideal for applications in both residential and industrial air purification systems. Additionally, metal–organic frameworks (MOFs), a class of porous nanomaterials, have shown significant potential in capturing and storing greenhouse gases such as carbon dioxide and methane (Alonso et al., 2017; Roohollahi et al., 2022). MOFs can also selectively filter harmful gases, contributing to cleaner air in urban and industrial settings.

Nanotechnological solutions not only improve the efficiency and effectiveness of air purification systems but also reduce energy consumption and maintenance costs. As research continues, these innovations are poised to play a vital role in combating air pollution and improving global air quality.

## 3 Nano-waste: A New Problem in Environmental Challenges

In recent years, the rapid progress of nanotechnology has resulted in the widespread integration of nanomaterials into various sectors, including electronics, medicine, textiles, and packaging industries (Malik et al., 2023). While these materials offer innovative solutions, they also raise concerns regarding their environmental impact. Nano-waste, unlike conventional waste, poses unique challenges due to its minuscule size, making it difficult to detect and remove from ecosystems (George et al., 2022). This chapter explores the definition of nano-waste, its environmental and health implications, and potential mitigation strategies. Nano-waste consists of NPs ($\leq$100 nm) that are released into the environment following the expiration of their functional lifespan (George et al., 2022; Tiwari et al., 2024; Younis et al., 2018). Nanotechnology has led to the emergence of new unique waste streams containing residual nanomaterials that could potentially pose challenges to waste management practices and technologies at industrial scale. Classification of nano-waste is important for the implementation of correct processing processes and the establishment of appropriate laws and regulations. Classification of nano-waste allows for the effective determination of appropriate treatment processes for various types of nano-waste. It has several benefits, including reducing any adverse effects on the environment and living beings. However, there is currently no internationally accepted nano-waste classification system (Kharat et al., 2017). Musee (2011) proposed a five-stage classification for nano-waste in his study on nano-waste and the environment. Table 1 presents the classification of nano-waste by Musee.

**Table 1** Classification of nano-waste (Musee, 2011)

| Level | Nano-waste hazard | Exposure level |
|---|---|---|
| 1 | Untoxic | From less to more |
| 2 | Harmful/toxic | From less to middle |
| 3 | Toxic to very toxic | From less to middle |
| 4 | Toxic to very toxic | Middle to more |
| 5 | Very toxic to highly toxic | Middle to more |

These wastes originate from industrial production processes, automobiles, medical devices, textile applications, and cosmetic products (George et al., 2022). Common nanomaterials such as nano-silver, nano-$TiO_2$, and CNTs can contaminate soil, water, and air when they are not properly managed. The full extent of nano-waste's impact on ecosystems remains under investigation, but existing research suggests that these materials may have detrimental effects on biodiversity (Tiwari et al., 2024). When NPs enter water bodies, they can be absorbed by plankton and other aquatic organisms, thereby entering the food chain and potentially causing widespread ecological disruptions (Benson et al., 2022). Nano-waste can be produced in different categories from different types of nanomaterials used in different forms and added to different products called nanoproducts. The schematic representation of the processes from the synthesis of nanomaterials to their formation as nano-waste that negatively affects the environment is given in Fig. 1. Some examples of nano-waste are medical product, battery, sensor, electronic, textile, and industrial wastes (Zahra et al., 2022).

In terms of human health, exposure to NPs through inhalation, ingestion, or dermal contact raises significant concerns. Once inside the body, these particles can penetrate cellular structures and accumulate in organs, leading to long-term toxicity. Recent studies have indicated that nano-silver and carbon nanotubes may increase

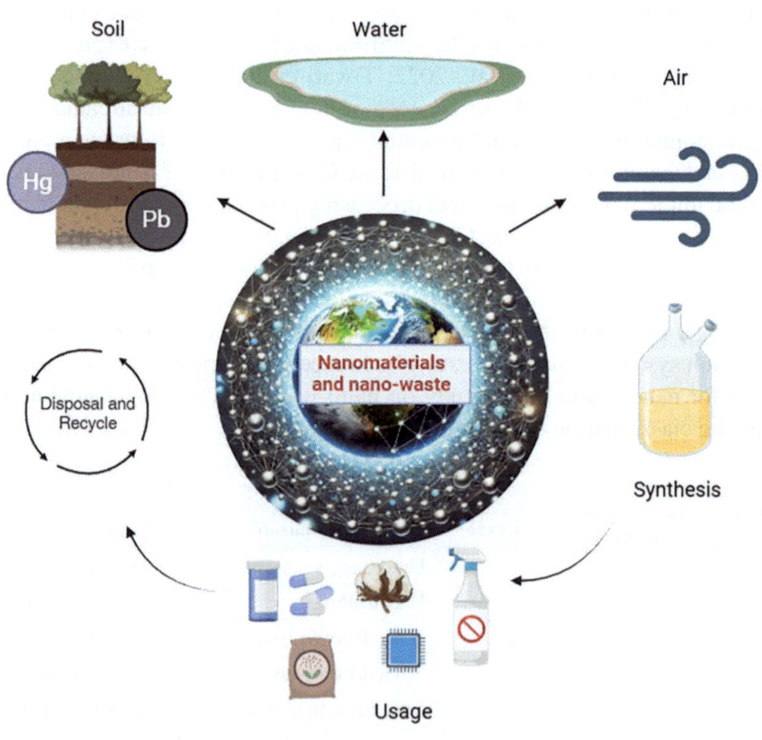


**Fig. 1** Schematic representation of the effects of nano-waste from nanoproducts in the process

the risk of respiratory diseases and carcinogenesis (Ferdous & Nemmar, 2020; Krow, 2014). Various strategies can be developed for the management of nano-waste for the sustainable protection of nature and the health of living beings.

## 4 Nano-waste Management Strategies

To overcome the global problems that we, as humanity, unfortunately, face today, one of the current problems that must be considered is the control, removal, or recycling of nano-waste (Gomez-Villalba et al., 2023). Various studies need to be carried out to eliminate and prevent the negative effects caused by the widespread use of nanotechnology (Borm et al., 2006). The most sustainable of these is the production of nanomaterials using green methods (Gupta et al., 2023). The development of eco-friendly and biodegradable nanomaterials should be provided. Green methods are widely researched by scientists today (Bai et al., 2022; Bi & Srivastava, 2024; Yuan & Golden, 2020). These methods, called green chemistry, carry out production based on 12 basic principles, and the main purpose here is to carry out production that is not harmful to the environment (Bagheri et al., 2023; Gupta et al., 2023; Nasrollahzadeh et al., 2019). Green chemistry is a rapidly growing field that focuses on designing chemical processes and products that minimize environmental impact by reducing waste, energy consumption, and hazardous substances (Beach et al., 2009; Zimmerman et al., 2020). This discipline aims to develop sustainable alternatives by applying the principles of atom economy, renewable feedstocks, and safer chemical syntheses (Chen et al., 2020). One of the most promising areas where green chemistry has found significant applications is nanotechnology. Nanotechnology, which deals with the manipulation of materials at the nanoscale level, benefits from green chemistry by promoting environmentally friendly synthesis methods (Duan et al., 2015). Traditional NPs synthesis often relies on toxic solvents and energy-intensive processes, whereas green chemistry emphasizes the use of biocompatible materials, aqueous-phase reactions, and renewable precursors (Clarke et al., 2018; Saud et al., 2024). For example, the synthesis of metal NPs using plant extracts, microbial systems, or biodegradable polymers eliminates the need for harmful chemicals while maintaining high efficiency and performance. Furthermore, nanotechnology-enabled green chemistry has led to advancements in sustainable energy, environmental remediation, and biodegradable materials (Ciambelli et al., 2020; Raja et al., 2022). By integrating green chemistry principles with nanotechnology, researchers can develop innovative and sustainable solutions for environmental challenges while enhancing material performance and functionality.

The swift progress of nanotechnology has resulted in an increased production and use of nanomaterials in various industries, including electronics, healthcare, and energy (Malik et al., 2023; Payal & Pandey, 2021). However, this progress has also resulted in a growing concern regarding nano-waste, which consists of discarded nanomaterials that may pose environmental and health risks due to their unique physicochemical properties (Dhall et al., 2024; Malode et al., 2023). Efficient recycling and recovery strategies for nano-waste are essential to minimize its impact while maximizing resource utilization. Classification of nano-waste is critical to minimizing risks to the environment and human health by increasing the effectiveness of management strategies (Zahra et al., 2022). Different types of nano-waste may have different risk levels in terms of their physicochemical properties, sources, and biological interactions (Osman, 2019; Younis et al., 2018). With correct classification, environmental pollution can be prevented by ensuring appropriate collection, processing, and disposal of waste, and nanomaterials with high recycling potential can be reused (Živković et al., 2014). Nano-waste can be classified according to their physicochemical properties, sources and environmental interactions (P. Gupta & Bhandari, 2022; Mustamam et al., 2024). According to their physical and chemical properties, they can be classified as metallic NPs, carbon-based nanomaterials, polymer-based nanostructures, and hybrid nano-wastes (P. Gupta & Bhandari, 2022). According to their sources, they can be classified as waste resulting from industrial production processes, consumer products, or laboratory studies (Chowdhury et al., 2022; Tiwari et al., 2024; Younis et al., 2018). According to their environmental and biological interactions can be classified as biodegradable or persistent, toxic or low-risk nano-wastes (Mishra et al., 2022; Pandey et al., 2024). Appropriate classification of nano-waste increases the effectiveness of management strategies and minimizes environmental and human health risks.

Several approaches have been developed for nano-waste recycling, including physical separation, chemical extraction, and biological methods (Dutta et al., 2018). For instance, nanomaterials such as carbon nanotubes, silver NPs, and quantum dots can be recovered through filtration, centrifugation, or solvent-assisted processes (Bendicho et al., 2015; Founti et al., 2024). Additionally, green chemistry principles are being integrated into recycling strategies, utilizing biodegradable agents and eco-friendly solvents to extract valuable nanomaterials (Varma, 2016). Recycled nanomaterials can be repurposed in various applications, such as catalyst regeneration, energy storage devices, and composite material production (Bassey, 2024; Faheem et al., 2025). By implementing sustainable nano-waste recycling techniques, industries can reduce environmental contamination and enhance the circular economy of nanotechnology-based products. Synthetic methods developed within the framework of green chemistry principles are of great importance in reducing the formation of nano-waste and ensuring sustainable management (Gavilan et al., 2023; Younis et al., 2018). These methods include bio-based nanomaterials, waste-reducing production, recycling and reuse (Arsenov et al., 2023). Nanomaterials obtained from biological origin or renewable sources can be more easily degraded in nature and reduce the risk of toxicity (Ahmed et al., 2022). Methods that consume low energy and minimize the use of harmful chemicals, such as sol-gel, green synthesis, biomimetic approaches (Aththanayaka et al., 2022). Recovering used nanomaterials and re-evaluating them in different industrial processes (Oldenkamp et al., 2019). These strategies contribute to sustainable nanotechnology applications by minimizing the formation of nano-waste (Zahra et al., 2022). Although the environmental impacts of nano-waste and their toxicological risks on biological systems are not yet fully understood, current studies reveal some basic risk factors such as ecotoxicity, bioaccumulation, and effects on human health (Abbas et al., 2020). Effects of nanomaterials on water resources, soil ecosystems, and air quality. The ability of some NPs to accumulate in organisms and cause long-term toxic effects (Sajid et al., 2015; Xu et al., 2022). Potential cellular damage from nano-waste taken into the body through inhalation, skin contact or ingestion (Tiwari et al., 2024). It is critical to address environmental and toxicological assessments within a regulatory framework for nano-waste management and to develop safe disposal methods (Engelmann et al., 2024). Nano-waste management strategies play an important role in ensuring the sustainability of nanotechnology applications and minimizing environmental risks (Kharat et al., 2017). In this context, both the development of waste-reducing production methods and the establishment of effective assessment mechanisms are necessary (Cainelli et al., 2015). Apart from these, considering the current status of nanomaterials and their future promises, first of all, both producers and consumers should be informed about the potential risks associated with nano-waste and encouraged to adopt sustainable disposal practices. In addition, strict regulation by governments on the production, use, and disposal of nanomaterials will be an important step in preventing significant amounts of nano-waste problems.

## 5 Conclusion

In this chapter, studies carried out for sustainability in the field of nanotechnology and the problem of nano-waste are mentioned. Although the development of nanotechnology provides many benefits, today's world is gradually facing the problem of nano-waste due to the increasing use in many areas. It is expected that this situation will increase even more in the future. Today, production with the principle of green chemistry has become quite widespread as a solution to this problem. However, more effective measures can be taken to further reduce the problems that may be encountered. Nanotechnology continues to develop day by day, and its points of use are also increasing. It is extremely important to make these processes environmentally compatible. It is very necessary to focus on studies in this field.

## References

Abbas, Q., Yousaf, B., Ullah, H., Ali, M. U., Ok, Y. S., & Rinklebe, J. (2020). Environmental transformation and nano-toxicity of engineered nano-particles (ENPs) in aquatic and terrestrial organisms. *Critical Reviews in Environmental Science and Technology, 50*(23), 2523–2581. https://doi.org/10.1080/10643389.2019.1705721

Abd-Elsalam, K. A., & Zahid, M. (2024). *Nanoscience and soil-water interaction in agroecosystem Kamel A. Abd-Elsalam, Muhammad Zahid*. CRC Press. https://www.routledge.com/Nanoscience-and-Soil-Water-Interaction-in-Agroecosystem/Abd-Elsalam-Zahid/p/book/9781032551562

Abd El-Azeim, M. M., Sherif, M. A., Hussien, M. S., Tantawy, I. A. A., & Bashandy, S. O. (2020). Impacts of nano- and non-nanofertilizers on potato quality and productivity. *Acta Ecologica Sinica, 40*(5), 388–397. https://doi.org/10.1016/J.CHNAES.2019.12.007

Abdelhak, M. (2022). Soil improvement in arid and semiarid regions for sustainable development. *Natural Resources Conservation and Advances for Sustainability*, 73–90. https://doi.org/10.1016/B978-0-12-822976-7.00026-0

Ahmed, S. F., Mofijur, M., Rafa, N., Chowdhury, A. T., Chowdhury, S., Nahrin, M., Islam, A. B. M. S., & Ong, H. C. (2022). Green approaches in synthesising nanomaterials for environmental nanobioremediation: Technological advancements, applications, benefits and challenges. *Environmental Research, 204,* 111967. https://doi.org/10.1016/J.ENVRES.2021.111967

Al-Anzi, B. S., & Siang, O. C. (2017). Recent developments of carbon based nanomaterials and membranes for oily wastewater treatment. *RSC Advances, 7*(34), 20981–20994. https://doi.org/10.1039/C7RA02501G

Al-Rashdi, B., Somerfield, C., & Hilal, N. (2011). Heavy Metals removal using adsorption and nanofiltration techniques. *Separation & Purification Reviews, 40*(3), 209–259. https://doi.org/10.1080/15422119.2011.558165

Alengebawy, A., Abdelkhalek, S. T., Qureshi, S. R., & Wang, M. Q. (2021). Heavy metals and pesticides toxicity in agricultural soil and plants: ecological risks and human health implications. *Toxics, 9*(3), 42. https://doi.org/10.3390/TOXICS9030042

Alonso, A., Moral-Vico, J., Abo Markeb, A., Busquets-Fité, M., Komilis, D., Puntes, V., Sánchez, A., & Font, X. (2017). Critical review of existing nanomaterial adsorbents to capture carbon dioxide and methane. *Science of the Total Environment, 595*, 51–62. https://doi.org/10.1016/J.SCITOTENV.2017.03.229

Anusha, J. R., Citarasu, T., Uma, G., Vimal, S., Kamaraj, C., Kumar, V., Muzammil, K., & Mani Sankar, M. (2024). Recent advances in nanotechnology-based modifications of micro/nano PET plastics for green energy applications. *Chemosphere, 352*, Article 141417. https://doi.org/10.1016/J.CHEMOSPHERE.2024.141417

Arsenov, D., Beljin, J., Jović, D., Maletić, S., Borišev, M., & Borišev, I. (2023). Nanomaterials as endorsed environmental remediation tools for the next generation: Eco-safety and sustainability. *Journal of Geochemical Exploration, 253*, 107283. https://doi.org/10.1016/J.GEXPLO.2023.107283

Aththanayaka, S., Thiripuranathar, G., & Ekanayake, S. (2022). Emerging advances in biomimetic synthesis of nanocomposites and potential applications. *Materials Today Sustainability, 20,* 100206. https://doi.org/10.1016/J.MTSUST.2022.100206

Attia, T. M. S., & Elsheery, N. I. (2020). Nanomaterials: Scope, applications, and challenges in agriculture and soil reclamation. 1–39. https://doi.org/10.1007/978-3-030-33996-8_1

Bagheri, A. R., Aramesh, N., Hasnain, M. S., Nayak, A. K., & Varma, R. S. (2023). Greener fabrication of metal nanoparticles using plant materials: A review. *Chemical Physics Impact, 7*, Article 100255. https://doi.org/10.1016/J.CHPHI.2023.100255

Bahgat, M., Farghali, A. A., El Rouby, W. M. A., & Khedr, M. H. (2011). Synthesis and modification of multi-walled carbon nano-tubes (MWCNTs) for water treatment applications. *Journal of Analytical and Applied Pyrolysis, 92*(2), 307–313. https://doi.org/10.1016/J.JAAP.2011.07.002

Bai, B., Saranya, S., Dheepaasri, V., Muniasamy, S., Alharbi, N. S., Selvaraj, B., Undal, V. S., & Gnanamangai, B. M. (2022). Biosynthesized copper oxide nanoparticles (CuO NPs) enhances the anti-biofilm efficacy against K. pneumoniae and S. aureus. *Journal of King Saud University - Science, 34*(6), 102120. https://doi.org/10.1016/J.JKSUS.2022.102120

Bakhtiari, S., Doustkhah, E., Pedram, M. Z., Yarmohammadi, M., & Seydibeyoğlu, M. Ö. (2022). Effects of nanoparticles on the biodegradation of organic materials. *Biodegradation and Biodeterioration at the Nanoscale*, 153–174

Bao, X., Qiang, Z., Ling, W., & Chang, J. H. (2013). Sonohydrothermal synthesis of $MFe_2O_4$ magnetic nanoparticles for adsorptive removal of tetracyclines from water. *Separation and Purification Technology, 117*, 104–110. https://doi.org/10.1016/J.SEPPUR.2013.03.046

Bassey, K. E. (2024). From waste to wonder: Developing engineered nanomaterials for multifaceted applications, *20*(3), 109–123. https://Gsconlinepress.Com/Journals/Gscarr/Sites/Default/Files/GSCARR-2024-0326.Pdf. https://doi.org/10.30574/GSCARR.2024.20.3.0326

Batara, B., Steven, S., Mulyana, M., Saputra, A. S., Hutahaean, A. C., Yemensia, E. V., Soekotjo, E. S. A., Abidin, A. Z., & Graha, H. P. R. (2025). Recent advances, applications, and challenges in superabsorbent polymers to support water sustainability. *Journal of Applied Polymer Science, 142*(11), Article e56588. https://doi.org/10.1002/APP.56588

Beach, E. S., Cui, Z., & Anastas, P. T. (2009). Green chemistry: A design framework for sustainability. *Energy & Environmental Science, 2*(10), 1038–1049. https://doi.org/10.1039/B904997P

Bendicho, C., Costas-Mora, I., Romero, V., & Lavilla, I. (2015). Nanoparticle-enhanced liquid-phase microextraction. *TrAC Trends in Analytical Chemistry, 68*, 78–87. https://doi.org/10.1016/J.TRAC.2015.02.007

Benson, N. U., Agboola, O. D., Fred-Ahmadu, O. H., De-la-Torre, G. E., Oluwalana, A., & Williams, A. B. (2022). Micro(nano)plastics prevalence, food web interactions, and toxicity assessment in aquatic organisms: A review. *Frontiers in Marine Science, 9*, Article 851281. https://doi.org/10.3389/FMARS.2022.851281/PDF

Bi, S., & Srivastava, R. (2024). Rosa damascena leaf extract mediated palladium nanoparticles and their anti-inflammatory and analgesic applications. *Inorganic Chemistry Communications, 162*, Article 112122. https://doi.org/10.1016/J.INOCHE.2024.112122

Borm, P. J. A., Robbins, D., Haubold, S., Kuhlbusch, T., Fissan, H., Donaldson, K., Schins, R., Stone, V., Kreyling, W., Lademann, J., Krutmann, J., Warheit, D. B., & Oberdorster, E. (2006). The potential risks of nanomaterials: A review carried out for ECETOC. *Particle and Fibre Toxicology 20063*(1), 1–35. https://doi.org/10.1186/1743-8977-3-11

Borode, A. O., Tshephe, T. S., Akinwamide, S. O., & Olubambi, P. A. (2025). Nanotechnology characterization: Emerging techniques for accurate and reliable nanostructural analysis. *Smart Nanomaterials for Environmental Applications*, 57–91. https://doi.org/10.1016/B978-0-443-21794-4.00029-6

Cainelli, G., D'Amato, A., & Mazzanti, M. (2015). Adoption of waste-reducing technology in manufacturing: Regional factors and policy issues. *Resource and Energy Economics, 39*, 53–67. https://doi.org/10.1016/J.RESENEECO.2014.11.004

Chen, T. L., Kim, H., Pan, S. Y., Tseng, P. C., Lin, Y. P., & Chiang, P. C. (2020). Implementation of green chemistry principles in circular economy system towards sustainable development goals: Challenges and perspectives. *Science of the Total Environment, 716*, Article 136998. https://doi.org/10.1016/J.SCITOTENV.2020.136998

Chowdhury, A. T., Rafa, N., Kabir, A., & Selvakumar, P. M. (2022). Consumer nanoproducts for environment. In *Handbook of consumer nanoproducts* (pp. 1169–1200). Singapore: Springer Nature Singapore. https://doi.org/10.1007/978-981-16-8698-6_67

Ciambelli, P., La Guardia, G., & Vitale, L. (2020). Nanotechnology for green materials and processes. *Studies in Surface Science and Catalysis, 179*, 97–116. https://doi.org/10.1016/B978-0-444-64337-7.00007-0

Clarke, C. J., Tu, W. C., Levers, O., Bröhl, A., & Hallett, J. P. (2018). Green and sustainable solvents in chemical processes. *Chemical Reviews, 118*(2), 747–800. https://doi.org/10.1021/ACS.CHEMREV.7B00571/ASSET/IMAGES/LARGE/CR-2017-005715_0042.JPEG

Covaliu-Mierlă, C. I., Păunescu, O., & Iovu, H. (2023). Recent advances in membranes used for nanofiltration to remove heavy metals from wastewater: A review. *Membranes, 13*(7), 643. https://doi.org/10.3390/MEMBRANES13070643

Daraei, E., Bayat, H., & Zamani, P. (2024). Effects of metal oxide nanoparticles on soil water retention curve and tensile strength. *Pedosphere, 34*(6), 1136–1145. https://doi.org/10.1016/J.PEDSPH.2023.07.017

Das, I., & Ansari, S. A. (2009). Nanomaterials in science and technology. *Journal of Scientific and Industrial Research, 68*(8), 657–667.

Das, S., Sen, B., & Debnath, N. (2015). Recent trends in nanomaterials applications in environmental monitoring and remediation. *Environmental Science and Pollution Research 22*(23), 18333–18344. https://doi.org/10.1007/S11356-015-5491-6

de Almeida, D. S., Martins, L. D., & Aguiar, M. L. (2022). Air pollution control for indoor environments using nanofiber filters: A brief review and post-pandemic perspectives. *Chemical Engineering Journal Advances, 11*, Article 100330. https://doi.org/10.1016/J.CEJA.2022.100330

Del Prado-Audelo, M. L., García Kerdan, I., Escutia-Guadarrama, L., Reyna-González, J. M., Magaña, J. J., & Leyva-Gómez, G. (2021). Nanoremediation: Nanomaterials and nanotechnologies for environmental cleanup. *Frontiers in Environmental Science, 9*, Article 793765. https://doi.org/10.3389/FENVS.2021.793765/BIBTEX

Deng, H., Lei, H., Luo, Y., Huan, C., Li, J., Li, H., He, F., Zhang, B., Yi, K., & Sun, A. (2024). The effects of titanium dioxide nanoparticles on cadmium bioaccumulation in ramie and its application in remediation of cadmium-contaminated soil. *Alexandria Engineering Journal, 86*, 663–668. https://doi.org/10.1016/J.AEJ.2023.12.019

Dhall, S., Nigam, A., Harshavardhan, M., Mukherjee, A., & Srivastava, P. (2024). A comprehensive overview of methods involved in nanomaterial production and waste disposal from research labs and industries and existing regulatory guidelines for handling engineered nanomaterials. *Environmental Chemistry and Ecotoxicology, 6*, 269–282. https://doi.org/10.1016/J.ENCECO.2024.06.002

Dhanapal, A. R., Thiruvengadam, M., Vairavanathan, J., Venkidasamy, B., Easwaran, M., & Ghorbanpour, M. (2024). Nanotechnology approaches for the remediation of agricultural polluted soils. *ACS Omega, 9*(12), 13522–13533. https://doi.org/10.1021/ACSOMEGA.3C09776/ASSET/IMAGES/LARGE/AO3C09776_0003.JPEG

Dhoundiyal, S., Kaur, A., Alam, M. A., & Sharma, A. (2023). The future of nanopesticides, nanoherbicides, and nanofertilizers. *Nanopesticides, Nanoherbicides, and Nanofertilizers*, 114–136. https://doi.org/10.1201/9781003364429-7

Duan, H., Wang, D., & Li, Y. (2015). Green chemistry for nanoparticle synthesis. *Chemical Society Reviews, 44*(16), 5778–5792. https://doi.org/10.1039/C4CS00363B

Dutta, A., Karamikamkar, S., Nofar, M., & Behzadfar, E. (2024). Nanoporous air filtering systems made from renewable sources: Benefits and challenges. *Nanoscale, 16*(32), 15059–15077. https://doi.org/10.1039/D4NR01688B

Dutta, T., Kim, K. H., Deep, A., Szulejko, J. E., Vellingiri, K., Kumar, S., Kwon, E. E., & Yun, S. T. (2018). Recovery of nanomaterials from battery and electronic wastes: A new paradigm of environmental waste management. *Renewable and Sustainable Energy Reviews, 82*, 3694–3704. https://doi.org/10.1016/J.RSER.2017.10.094

Ekins, P. (2002). Economic growth and environmental sustainability: The prospects for green growth. *Economic Growth and Environmental Sustainability*. https://doi.org/10.4324/9780203011751

Engelmann, W., Von Hohendorff, R., & da Silva Leal, D. W. (2024). A green nanotechnology approach to nanowaste management: what guarantees can the OECD protocol provide?. In *Nanotoxicology for agricultural and environmental applications* (pp. 273–280). Academic Press. https://doi.org/10.1016/B978-0-443-15570-3.00007-7

Faheem, M., Azher Hassan, M., Du, J., & Wang, B. (2025). Harnessing potential of smart and conventional spent adsorbents: Global practices and sustainable approaches through regeneration and tailored recycling. *Separation and Purification Technology, 354*, Article 128907. https://doi.org/10.1016/J.SEPPUR.2024.128907

Fahimirad, S., Fahimirad, Z., & Sillanpää, M. (2021). Efficient removal of water bacteria and viruses using electrospun nanofibers. *Science of the Total Environment, 751*, Article 141673. https://doi.org/10.1016/J.SCITOTENV.2020.141673

Ferdous, Z., & Nemmar, A. (2020). Health impact of silver nanoparticles: A review of the biodistribution and toxicity following various routes of exposure. *International Journal of Molecular Sciences 21*(7), 2375. https://doi.org/10.3390/IJMS21072375

Founti, M., Nikolaidis, G. N., Usman, M., & Cheng, S. (2024). Recent trends and advancements in green synthesis of biomass-derived carbon dots. *Eng 5*(3), 2223–2263. https://doi.org/10.3390/ENG5030116

Gavilán, H., Serrano, M. B., & Cabanelas, J. C. (2023). *Nanomaterials and their synthesis for a sustainable future* (Vol. 149, pp. 233–310). Materials Research Forum LLC.

George, B., John, A. B., Priyanila, M., & Suchithra, T. V. (2022). Two-faced nanomaterials: routes to resolve nanowaste. *International Journal of Environmental Science and Technology 20*(5), 5643–5658. https://doi.org/10.1007/S13762-022-03997-0

Gomez-Villalba, L. S., Salcines, C., & Fort, R. (2023). Application of inorganic nanomaterials in cultural heritage conservation, risk of toxicity, and preventive measures. *Nanomaterials 13*(9), 1454. https://doi.org/10.3390/NANO13091454

Goswami, K. P., & Pugazhenthi, G. (2020). Credibility of polymeric and ceramic membrane filtration in the removal of bacteria and virus from water: A review. *Journal of Environmental Management, 268*, Article 110583. https://doi.org/10.1016/J.JENVMAN.2020.110583

Gupta, D., Boora, A., Thakur, A., & Gupta, T. K. (2023). Green and sustainable synthesis of nanomaterials: Recent advancements and limitations. *Environmental Research, 231*, Article 116316. https://doi.org/10.1016/J.ENVRES.2023.116316

Gupta, P., & Bhandari, S. (2022). Classification and sources of nanowastes. In *Nanomaterials recycling* (pp. 37–60). Elsevier. https://doi.org/10.1016/B978-0-323-90982-2.00003-2

Hasan, M., Sakib, M. N., Sabroj, R. Bin, & Rahman, M. Z. (2024). Advanced nanomaterial-based biosensors for clinical diagnosis, environmental protection and industrial fermentation. *Comprehensive Materials Processing*, 30–47. https://doi.org/10.1016/B978-0-323-96020-5.00274-0

Holden, E., Linnerud, K., & Banister, D. (2014). Sustainable development: Our common future revisited. *Global Environmental Change, 26*(1), 130–139. https://doi.org/10.1016/J.GLOENVCHA.2014.04.006

Hu, R., Wang, X., Dai, S., Shao, D., Hayat, T., & Alsaedi, A. (2015). Application of graphitic carbon nitride for the removal of Pb(II) and aniline from aqueous solutions. *Chemical Engineering Journal, 260*, 469–477. https://doi.org/10.1016/J.CEJ.2014.09.013

Iqbal, M., Umar, S., & Mahmooduzzafar. (2019). Nano-fertilization to enhance nutrient use efficiency and productivity of crop plants. *Nanomaterials and Plant Potential*, 473–505. https://doi.org/10.1007/978-3-030-05569-1_19/TABLES/2

Jabbar, Z. H., & Esmail Ebrahim, S. (2022). Recent advances in nano-semiconductors photocatalysis for degrading organic contaminants and microbial disinfection in wastewater: A comprehensive review. *Environmental Nanotechnology, Monitoring & Management, 17*, Article 100666. https://doi.org/10.1016/J.ENMM.2022.100666

Jatav, H. S., & Rajput, V. D. (2022). Ecosystem Services: types, management and benefits. In *Ecosystem Services: Types, Management and Benefits*. Nova Science Publishers, Inc. https://doi.org/10.52305/PFZA6988

Karahmet Sher, E., Alebić, M., Marković Boras, M., Boškailo, E., Karahmet Farhat, E., Karahmet, A., Pavlović, B., Sher, F., & Lekić, L. (2024). Nanotechnology in medicine revolutionizing drug delivery for cancer and viral infection treatments. *International Journal of Pharmaceutics, 660*, Article 124345. https://doi.org/10.1016/J.IJPHARM.2024.124345

Kharat, M. G., Murthy, S., & Kamble, S. J. (2017). Environmental applications of nanotechnology: A review. *ADBU Journal of Engineering Technology*, *6*(3), 6031205. https://journals.dbuniversity.ac.in/ojs/index.php/AJET/article/view/408

Koli, M. M., & Singh, S. P. (2023). Surface-modified ultrafiltration and nanofiltration membranes for the selective removal of heavy metals and inorganic groundwater contaminants: A review. *Environmental Science: Water Research and Technology, 9*(11), 2803–2829. https://doi.org/10.1039/D3EW00266G

Krow, C. M. (2014). CNTs, Nano-$TiO_2$, and Nanosilver: Assessing risk and investing in safety. *Nanotechnology Law and Business, 11*. https://heinonline.org/HOL/Page?handle=hein.journals/nantechlb11&id=143&div=14&collection=journals

Kuhlman, T., & Farrington, J. (2010). What is sustainability? *Sustainability 2*(11), 3436–3448. https://doi.org/10.3390/SU2113436
Liu, R., & Lal, R. (2012). Nanoenhanced materials for reclamation of mine lands and other degraded soils: A review. *Journal of Nanotechnology, 2012*(1), Article 461468. https://doi.org/10.1155/2012/461468
Liu, X., Sathishkumar, K., Zhang, H., Saxena, K. K., Zhang, F., Naraginti, S., Anbarasu, K., Rajendiran, R., Rajasekar, A., & Guo, X. (2024). Frontiers in environmental cleanup: Recent advances in remediation of emerging pollutants from soil and water. *Journal of Hazardous Materials Advances, 16*, Article 100461. https://doi.org/10.1016/J.HAZADV.2024.100461
Louis Theodore, B., Kunz Hoboken, R. G., Wiley, J., Los, M., Gibson, S. B., Finegold, D., Bensimon, C., Daar, A., Eaton, M., Godard, B., Knoppers, B., Mackie, J., Singer, P., Friedman, M., Mottram, D., Smith, J., Beresford, N. A., Vohr, H.-W., Frumkin, H., Greenwell, A., et al. (2005). Nanotechnology: Environmental implications and solutions. *Environmental Health Perspectives, 113*(7), A488. https://pmc.ncbi.nlm.nih.gov/articles/PMC1257682/
Mahmoud, M. E., Abdou, A. E. H., Sobhy, M. E., & Fekry, N. A. (2017). Solid–solid crosslinking of carboxymethyl cellulose nanolayer on titanium oxide nanoparticles as a novel biocomposite for efficient removal of toxic heavy metals from water. *International Journal of Biological Macromolecules, 105*, 1269–1278. https://doi.org/10.1016/J.IJBIOMAC.2017.07.156
Malik, S., Muhammad, K., & Waheed, Y. (2023). Nanotechnology: A revolution in modern industry. *Molecules 28*(2), 661. https://doi.org/10.3390/MOLECULES28020661
Mallakpour, S., Azadi, E., & Hussain, C. M. (2022). Fabrication of air filters with advanced filtration performance for removal of viral aerosols and control the spread of COVID-19. *Advances in Colloid and Interface Science, 303*, Article 102653. https://doi.org/10.1016/J.CIS.2022.102653
Malode, S. J., Shanbhag, M. M., Kumari, R., Dkhar, D. S., Chandra, P., & Shetti, N. P. (2023). Biomass-derived carbon nanomaterials for sensor applications. *Journal of Pharmaceutical and Biomedical Analysis, 222*, Article 115102. https://doi.org/10.1016/J.JPBA.2022.115102
Matos, M. P. S. R., Correia, A. A. S., & Rasteiro, M. G. (2017). Application of carbon nanotubes to immobilize heavy metals in contaminated soils. *Journal of Nanoparticle Research, 19*(4), 1–11. https://doi.org/10.1007/S11051-017-3830-X/FIGURES/5
Mauter, M. S., Zucker, I., Perreault, F., Werber, J. R., Kim, J. H., & Elimelech, M. (2018). The role of nanotechnology in tackling global water challenges. *Nature Sustainability 2018 1:4, 1*(4), 166–175. https://doi.org/10.1038/s41893-018-0046-8
Mishra, A. K., Das, R., Sahoo, S., & Biswal, B. (2022). Global regulations and legislations on nanoparticles usage and application in diverse horizons. In *Comprehensive analytical chemistry* (Vol. 99, pp. 261–290). Elsevier. https://doi.org/10.1016/BS.COAC.2021.12.004
Musee, N. (2011). Nanowastes and the environment: Potential new waste management paradigm. *Environment International, 37*(1), 112–128. https://doi.org/10.1016/J.ENVINT.2010.08.005
Mustamam, M. K., Elaissari, A., Fatehah, M. O., Aziz, H. A., & Wang, M. H. S. (2024). Nano pollutant properties, occurrence and behavior in water and wastewater streams. *Industrial Waste Engineering,* 279–332. https://doi.org/10.1007/978-3-031-46747-9_7
Nasrollahzadeh, M., Sajjadi, M., Sajadi, S. M., & Issaabadi, Z. (2019). Green nanotechnology. *Interface Science and Technology, 28*, 145–198. https://doi.org/10.1016/B978-0-12-813586-0.00005-5
Olawade, D. B., Ige, A. O., Olaremu, A. G., Ijiwade, J. O., & Adeola, A. O. (2024). The synergy of artificial intelligence and nanotechnology towards advancing innovation and sustainability—A mini-review. *Nano Trends, 8*, Article 100052. https://doi.org/10.1016/J.NWNANO.2024.100052
Oldenkamp, H. F., Vela Ramirez, J. E., & Peppas, N. A. (2019). Re-evaluating the importance of carbohydrates as regenerative biomaterials. *Regenerative Biomaterials, 6*(1), 1–12. https://doi.org/10.1093/rb/rby023
Omer, A. M. (2008). Energy, environment and sustainable development. *Renewable and Sustainable Energy Reviews, 12*(9), 2265–2300. https://doi.org/10.1016/J.RSER.2007.05.001

Osman, E. M. (2019). Environmental and health safety considerations of nanotechnology, nano safety. *Biomededical Journal of Scientific & Technical Research, 19*(4), 14501–14515. https://doi.org/10.26717/BJSTR.2019.19.003346

Pandey, J. K., Bobde, P., Patel, R. K., & Manna, S. (2024). *Disposal and Recycling Strategies for Nano-engineered Materials*. Elsevier.

Payal, & Pandey, P. (2021). Role of nanotechnology in electronics: A review of recent developments and patents. *Recent Patents on Nanotechnology, 16*(1), 45–66. https://doi.org/10.2174/187221051566621012O114504/CITE/REFWORKS

Raja, R. K., Hazir, S., Balasubramani, G., Sivaprakash, G., Obeth, E. S. J., Boobalan, T., Pugazhendhi, A., Raj, R. H. K., & Arun, A. (2022). Green nanotechnology for the environment. *Handbook of Microbial Nanotechnology*, 461–478. https://doi.org/10.1016/B978-0-12-823426-6.00006-1

Rajput, V. D., Minkina, T., Upadhyay, S. K., Kumari, A., Ranjan, A., Mandzhieva, S., Sushkova, S., Singh, R. K., & Verma, K. K. (2022). Nanotechnology in the restoration of polluted soil. *Nanomaterials 12*(5), 769. https://doi.org/10.3390/NANO12050769

Ramli, W. K. W., Dollah, N., & Qing, O. L. (2024). $TiO_2$/Polysulfone composite membrane for enhanced removal of acetaminophen from water. *Journal of Physical Science, 35*(3), 1–15. https://openurl.ebsco.com/contentitem/asn:182249601?sid=ebsco:plink:crawler&id=ebsco:asn:182249601&crl=c

Rodríguez-Seijo, A., Santás-Miguel, V., Arenas-Lago, D., Arias-Estévez, M., & Pérez-Rodríguez, P. (2024). Use of nanotechnology for safety agriculture and food production: Challenges and limitations. *Pedosphere*. https://doi.org/10.1016/J.PEDSPH.2024.09.005

Roohollahi, H., Zeinalzadeh, H., & Kazemian, H. (2022). Recent advances in adsorption and separation of methane and carbon dioxide greenhouse gases using metal-organic framework-based composites. *Industrial and Engineering Chemistry Research, 61*(30), 10555–10586. https://doi.org/10.1021/ACS.IECR.2C00664/ASSET/IMAGES/LARGE/IE2C00664_0034.JPEG

Xiong, R., Wang, Y., Zhang, X., & Canhui, L. (2014). Facile synthesis of magnetic nanocomposites of cellulose@ultrasmall iron oxide nanoparticles for water treatment. *RSC Advances, 4*(43), 22632–22641. https://doi.org/10.1039/C4RA01397B

Sajid, M., Ilyas, M., Basheer, C., Tariq, M., Daud, M., Baig, N., & Shehzad, F. (2015). Impact of nanoparticles on human and environment: review of toxicity factors, exposures, control strategies, and future prospects. *Environmental Science and Pollution Research, 22*(6), 4122–4143. https://doi.org/10.1007/S11356-014-3994-1

Saleem, H., Zaidi, S. J., Ismail, A. F., & Goh, P. S. (2022). Advances of nanomaterials for air pollution remediation and their impacts on the environment. *Chemosphere, 287*, Article 132083. https://doi.org/10.1016/J.CHEMOSPHERE.2021.132083

Saud, A., Gupta, S., Allal, A., Preud'homme, H., Shomar, B., & Zaidi, S. J. (2024). Progress in the sustainable development of biobased (Nano)materials for application in water treatment technologies. *ACS Omega, 9*(27), 29088–29113. https://doi.org/10.1021/ACSOMEGA.3C08883/ASSET/IMAGES/LARGE/AO3C08883_0008.JPEG

Singh, N. B., Kumar, B., Usman, U. L., & Susan, M. A. B. H. (2024). Nano revolution: Exploring the frontiers of nanomaterials in science, technology, and society. *Nano-Structures and Nano-Objects, 39*, Article 101299. https://doi.org/10.1016/J.NANOSO.2024.101299

Tahat, M. M., Alananbeh, K. M., Othman, Y. A., & Leskovar, D. I. (2020). Soil health and sustainable agriculture. *Sustainability 12*(12), 4859. https://doi.org/10.3390/SU12124859

Tiwari, E., Khandelwal, N., & Singh, N. (2024). Nanowaste and its management: New waste management paradigm. *Solid Waste Management: Challenges and Recent Solutions*, 237–260. https://doi.org/10.1201/9781003543176-11

Toptaş, A., Nanolifli, Y., Filtreleri, H., Hava,:, Geliştirilmesi, K., & Korunması, S. (2024). Innovative nanofibrous air filters: Advancing air quality and health protection. *Sürdürülebilir Çevre Dergisi, 4*(1), 29–40. https://doi.org/10.62816/CEVDER.1491367

Ugwu, E. I., Karri, R. R., Nnaji, C. C., John, J., Padmanaban, V. C., Othmani, A., Ikechukwu, E. L., & Helal, W. M. K. (2022). Application of green nanocomposites in removal of toxic chemicals, heavy metals, radioactive materials, and pesticides from aquatic water bodies. *Sustainable Nanotechnology for Environmental Remediation*, 321–346. https://doi.org/10.1016/B978-0-12-824547-7.00018-7

Urfi, M., Bin Babar, Z., & Rizwan, K. (2024). Carbon-based nanomaterials (graphene and graphene oxide, carbon nanotubes, and carbon nanofibers) for oil-water separation. *Nanotechnology for Oil-Water Separation: From Fundamentals to Industrial Applications*, 131–151. https://doi.org/10.1016/B978-0-323-95517-1.00006-8

Varma, R. S. (2016). Greener and sustainable trends in synthesis of organics and nanomaterials. *ACS Sustainable Chemistry and Engineering, 4*(11), 5866–5878. https://doi.org/10.1021/ACS SUSCHEMENG.6B01623/ASSET/IMAGES/LARGE/SC-2016-01623M_0007.JPEG

Vasarevičius, S., Danila, V., & Paliulis, D. (2019). Application of stabilized nano zero valent iron particles for immobilization of available $Cd^{2+}$, $Cu^{2+}$, $Ni^{2+}$, and $Pb^{2+}$ ions in soil. *International Journal of Environmental Research, 13*(3), 465–474. https://doi.org/10.1007/S41742-019-00187-8/FIGURES/2

Wall, D. H., Nielsen, U. N., & Six, J. (2015). Soil biodiversity and human health. *Nature 528*(7580), 69–76. https://doi.org/10.1038/nature15744

Wang, Y., Chou, S., & Zhang, Z. (2019). Nanomaterials Innovation. *Small (Weinheim an der Bergstrasse, Germany), 15*(32), 15–17. https://doi.org/10.1002/smll.201902246

Weldeslassie, T., Naz, H., Singh, B., & Oves, M. (2017). Chemical contaminants for soil, air and aquatic ecosystem. *Modern Age Environmental Problems and Their Remediation*, 1–22. https://doi.org/10.1007/978-3-319-64501-8_1/FIGURES/1

Xu, Z., Long, X., Jia, Y., Zhao, D., & Pan, X. (2022). Occurrence, transport, and toxicity of nanomaterials in soil ecosystems: A review. *Environmental Chemistry Letters, 20*(6), 3943–3969. https://doi.org/10.1007/S10311-022-01507-9

Yin, F., Yue, W., Li, Y., Gao, S., Zhang, C., Kan, H., Niu, H., Wang, W., & Guo, Y. (2021). Carbon-based nanomaterials for the detection of volatile organic compounds: A review. *Carbon, 180*, 274–297. https://doi.org/10.1016/J.CARBON.2021.04.080

Yin, Z., Cui, C., Chen, H., Duoni, Yu, X., & Qian, W. (2020). The Application of carbon nanotube/graphene-based nanomaterials in wastewater treatment. *Small, 16*(15), 1902301. https://doi.org/10.1002/SMLL.201902301

Younis, S. A., El-Fawal, E. M., & Serp, P. (n.d.). *Nano-wastes and the environment: Potential challenges and opportunities of nano-waste management paradigm for greener nanotechnologies*. Retrieved February 13, 2018, from https://www.researchgate.net/profile/Sherif-Younis-2/publication/343976633_Nano-wastes_and_the_environment_Potential_Challenges_and_Opportunities_of_Nano-waste_Management_Paradigm_for_Greener_Nanotechnologies/links/5f4f020d299bf13a3196fadb/Nano-wastes-and-the-environment-Potential-Challenges-and-Opportunities-of-Nano-waste-Management-Paradigm-for-Greener-Nanotechnologies.pdf

Yu, G., Wang, X., Liu, J., Jiang, P., You, S., Ding, N., Guo, Q., & Lin, F. (2021). Applications of nanomaterials for heavy metal removal from water and soil: A review. *Sustainability 13*(2), 713. https://doi.org/10.3390/SU13020713

Yuan, Q., & Golden, T. D. (2020). A novel method for synthesis of clay/polymer stabilized silver nanoparticles. *Surfaces and Interfaces, 20*, Article 100620. https://doi.org/10.1016/J.SURFIN.2020.100620

Yunus, I. S., Harwin, Kurniawan, A., Adityawarman, D., & Indarto, A. (2012). Nanotechnologies in water and air pollution treatment. *Environmental Technology Reviews, 1*(1), 136–148. https://doi.org/10.1080/21622515.2012.733966

Zahra, Z., Habib, Z., Hyun, S., & Sajid, M. (2022). Nanowaste: another future waste, its sources, release mechanism, and removal strategies in the environment. *Sustainability 14*(4), 2041. https://doi.org/10.3390/SU14042041

Zand, A. D., Mikaeili Tabrizi, A., & Vaezi Heir, A. (2020). Application of titanium dioxide nanoparticles to promote phytoremediation of Cd-polluted soil: Contribution of PGPR inoculation. *Bioremediation Journal, 24*(2–3), 171–189. https://doi.org/10.1080/10889868.2020.1799929

Zarime, N. 'Aishah, Solemon, B., Wan Yaacob, W. Z., Jamil, H., Che Omar, R., & Oyekanmi, A. A. (2023). Effectiveness of artificially synthesized granitic residual soil-supported nano zero-valent Iron (Gr-nZVI) as effective heavy metal contaminant adsorbent. *Inorganics, 11*(3), 131. https://doi.org/10.3390/INORGANICS11030131/S1

Zhang, M., Tao, S., & Wang, X. (2020). Interactions between organic pollutants and carbon nanomaterials and the associated impact on microbial availability and degradation in soil: A review. *Environmental Science: Nano, 7*(9), 2486–2508. https://doi.org/10.1039/D0EN00515K

Zheng, H., Mei, P., Wang, W., Yin, Y., Li, H., Zheng, M., Ou, X., & Cui, Z. (2023). Effects of super absorbent polymer on crop yield, water productivity and soil properties: A global meta-analysis. *Agricultural Water Management, 282*, Article 108290. https://doi.org/10.1016/J.AGWAT.2023.108290

Zimmerman, J. B., Anastas, P. T., Erythropel, H. C., & Leitner, W. (2020). Designing for a green chemistry future. *Science, 367*(6476), 397–400. https://doi.org/10.1126/SCIENCE.AAY3060/ASSET/8997170C-4499-4F22-A12F-B46943929ED6/ASSETS/GRAPHIC/367_397_F3.JPEG

Živković, D., Balanović, L., Mitovski, A., Talijan, N. M., Štrbac, N., Sokić, M., Manasijević, D., Minić, D., & Ćosović, V. (2014). Nanomaterials environmental risks and recycling: Actual issues. *Recycling and Sustainable Development, 7*(1), 1–8. https://doi.org/10.5937/ROR1401001Z

# Sustainable Environmental Solutions Through Nanotechnology: Opportunities, Challenges, and Nanowaste Management

**Noor Alam Chowdhary and L. S. Songachan**

## 1 Introduction

Earth is the only known planet in our solar system that sustains life. Humans, akin to all animals, require food and water for survival. However, the natural resources essential for sustaining life are becoming progressively scarce or contaminated. Soil degradation, along with the increasing scarcity of clean water and food, presents major concerns in numerous regions globally. Waste produced by human activities has become a major environmental issue (Mishra et al., 2023). Both developed and developing nations generate vast amounts of waste, frequently facing challenges in handling it sustainably. The accumulation of nonbiodegradable and nonrecyclable materials has seriously impacted ecosystems. These pollutants are rapidly impacting water bodies, agricultural lands, wildlife, and human health (Prabina et al., 2024). A substantial quantity of this waste comes from advanced industrial processes and infrastructure development. Residual waste is often discharged into the air, water, and land, leading to severe environmental pollution (Siddiqua et al., 2022).

Environmental contamination has become a major global concern. Some problems can be handled using bioremediation, an approach that employs biological organisms to eliminate or neutralize pollutants (Bala et al., 2022). This approach has been improved through the incorporation of advanced technologies. Prominent pollutants include industrial effluents, toxic gases, fertilizers, herbicides, pesticides, heavy metals, particulate matter, oil spills, sewage, and organic compounds (Singh et al., 2020). These detrimental substances render lakes and rivers unsuitable for utilization and, in extreme situations, result in incidents such as flammable tap water

N. A. Chowdhary (✉) · L. S. Songachan
Department of Botany, Institute of Science, Banaras Hindu University, Varanasi, UP, India
e-mail: chowdharynooralam@gmail.com

L. S. Songachan
e-mail: lssongachan@gmail.com



A. P. Das and S. Karakuş (eds.), *Synergizing Sustainability for Integrated Waste Management*, Sustainable Environmental Waste Management Strategies,
https://doi.org/10.1007/978-3-032-03718-3_14

and the formation of smog in urban areas (Mair & El-Kadi, 2013). The remediation of complex contaminants is challenging because of their chemical diversity, high volatility, and low reactivity (Tratnyek & Johnson, 2006). However, recent breakthroughs in nanotechnology present promising avenues. Nanomaterials (NMs) have the unique ability to decompose complex pollutant mixtures effectively (Fig. 1). Nanotechnology is considered one of the most promising scientific disciplines for the forthcoming decades (Malik et al., 2023; Salerno et al., 2008). It has the potential to transform environmental remediation by reducing the ecological impact of existing technologies in industrialized nations while simultaneously offering novel solutions to urgent environmental challenges in developing nations (Guerra et al., 2018). The advancement of nanoscience and nanotechnology is anticipated to be crucial in handling environmental issues. Additionally, nanotechnology presents significant potential in advancing global initiatives for sustainability and ecological equilibrium.

## 2 Fundamentals of Nanotechnology and Environmental Remediation

Nanotechnology has emerged as a revolutionary scientific discipline with significant potential for addressing major environmental issues. Nanotechnology is characterized by the manipulation and application of materials at the nanoscale (1–100 nm), employing the exceptional chemical, physical, and biological properties of NMs to facilitate innovative functionalities beyond reach with bulk materials (Fig. 1) (Matteucci et al., 2018; Siddiqui et al., 2015). Nanoscale materials exhibit unique properties, including a high surface-area-to-volume ratio, increased reactivity, and tunable surface chemistry, which enable them to be highly efficient for environmental monitoring, pollution management, and remediation (Jeevanandam et al., 2018; Mondal, 2020). A diverse range of nanoparticles (NPs) has been produced through both physical and chemical approaches, which include metallic (e.g., gold [Au], silver [Ag]), metal oxide (e.g., ZnO, $TiO_2$, $Fe_3O_4$), and carbon-based NMs such as carbon nanotubes (CNTs) and graphene derivatives (Khan, 2019; Stone et al., 2010). Recent advancements extend to second-generation and third-generation NMs, characterized by complex architectures and surface functions designed for targeted interactions with environmental contaminants (Ju-Nam & Lead, 2008).

Nanotechnology's application in environmental remediation has gained significant attention for its effectiveness in detecting, transforming, and eliminating various contaminants, such as heavy metals, pesticides, organic pollutants, industrial effluents, and pathogenic microorganisms (Besha et al., 2020; Guerra et al., 2018). Contaminants frequently occur as complex mixtures, making conventional remediation methods ineffective. NMs provide improved interaction with pollutants due to their higher surface reactivity and versatile surface characteristics, enabling targeted and efficient removal (Baral et al., 2021; Masciangioli & Zhang,

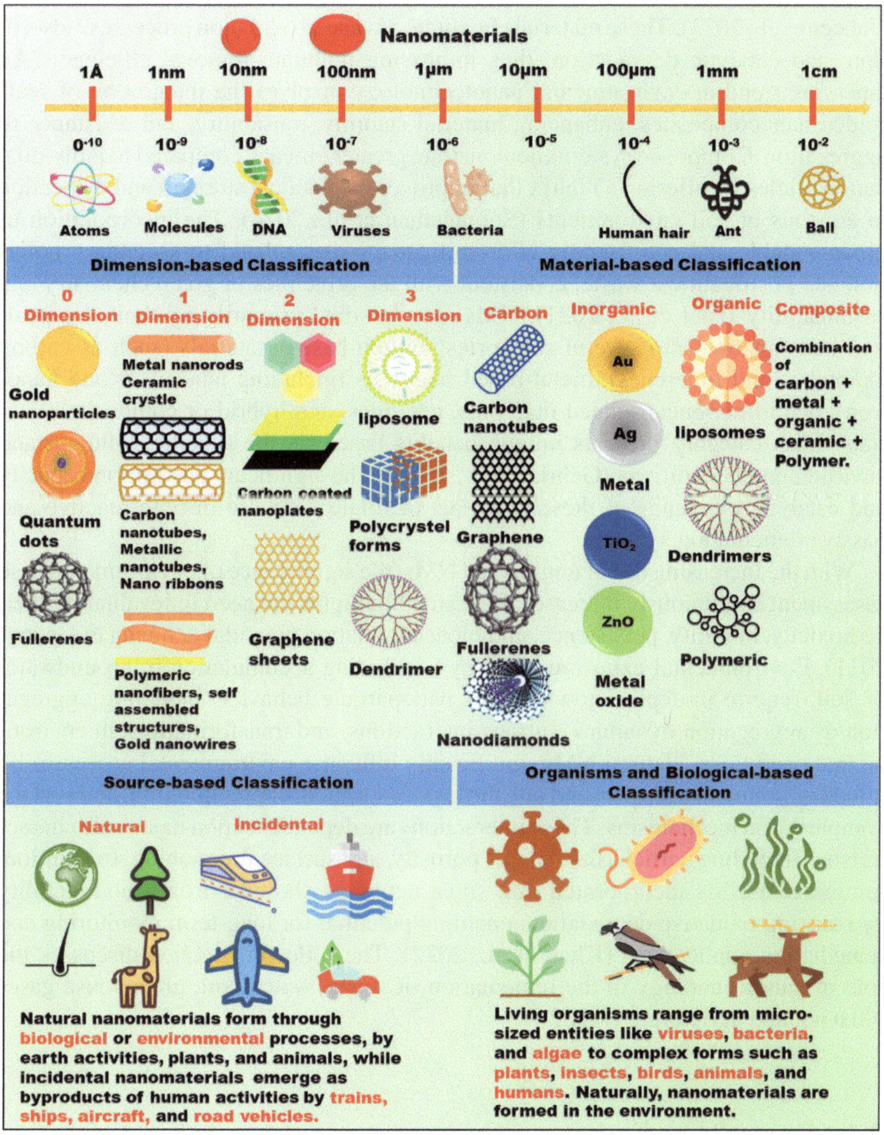


**Fig. 1** An illustration of the classification of nanomaterials based on dimensions, materials, sources, and origins

2003). Nanotechnology highlights versatility across all environmental compartments, including air, water, and soil. In water treatment, NMs including iron nanoparticles, nano-peroxides, and carbon nanotubes are utilized for the degradation of hazardous compounds in sewage, industrial wastewater, and potable water sources

(Jabeen et al., 2023). These materials facilitate advanced oxidation processes, adsorption, and catalytic degradation, thus improving pollutant removal efficiency. An emerging trend in environmental nanotechnology involves the integration of scaffolded nanocomposites, enhancing material stability, reusability, and resistance to aggregation. Composite systems demonstrate greater efficacy compared to individual nanoparticles by offering a matrix that improves mechanical strength and dispersion in aqueous or soil environments (Somanathan et al., 2024). The incorporation of biodegradable and eco-friendly NMs mitigates issues related to secondary pollution and posttreatment waste, consistent with the principles of green chemistry and sustainability (Paul et al., 2021). NMs are classified primarily by their composition and function into several categories: carbon-based materials (such as carbon nanotubes and fullerenes), metal-based materials (including nanosilver and nanoiron), dendrimer-encapsulated materials, ceramics, and hybrid or composite materials. Each category presents unique benefits based on the specific pollutant and environmental conditions (Gehrke et al., 2015). The significant adsorption capacity and catalytic efficiency of these materials facilitate their use in both reactive and passive remediation systems.

With the increasing use of engineered NMs, the significance of environmental risk assessment continuously increases. Research highlights the need for evaluating their ecotoxicity, mobility, persistence, and bioaccumulation potential (Zhuang & Gentry, 2011). Environmental exposure pathways, including accumulation in groundwater or soil, require in-depth knowledge of nanoparticle behavior, including aggregation/disaggregation dynamics, surface interactions, and transformations in environmental conditions. Natural NMs significantly influence environmental processes by affecting pollutant transport and fate through adsorption, coprecipitation, and surface complexation mechanisms. These interactions are dependent upon nanoscale characteristics including particle size, shape, porosity, and surface functioning. In addition, luminescent NMs incorporated into silica networks show environmental stability by resisting oxidative degradation, enabling potential for long-term monitoring and remediation applications (Khan et al., 2022). The following section discusses the role of nanotechnology in the remediation of waste, water, soil, greenhouse gases (GHGs), and agriculture (Fig. 2).

## *2.1 Waste Remediation*

Waste can be characterized as unwanted by-products produced mostly through human activities, resulting in undesirable impacts on both human health and the environment (Amasuomo & Baird, 2016) (Mohanty et al., 2024; Satapathy et al., 2024). Meanwhile, the United Nations (UN) has given the definition of waste as "materials that are not prime products for which the generator has no further use in terms of his/her purposes of production, transformation, or consumption, and of which he/she wants to dispose of." Although generation of waste has an extended history, significant issues arise whenever the environment's capacity to absorb waste is compromised.

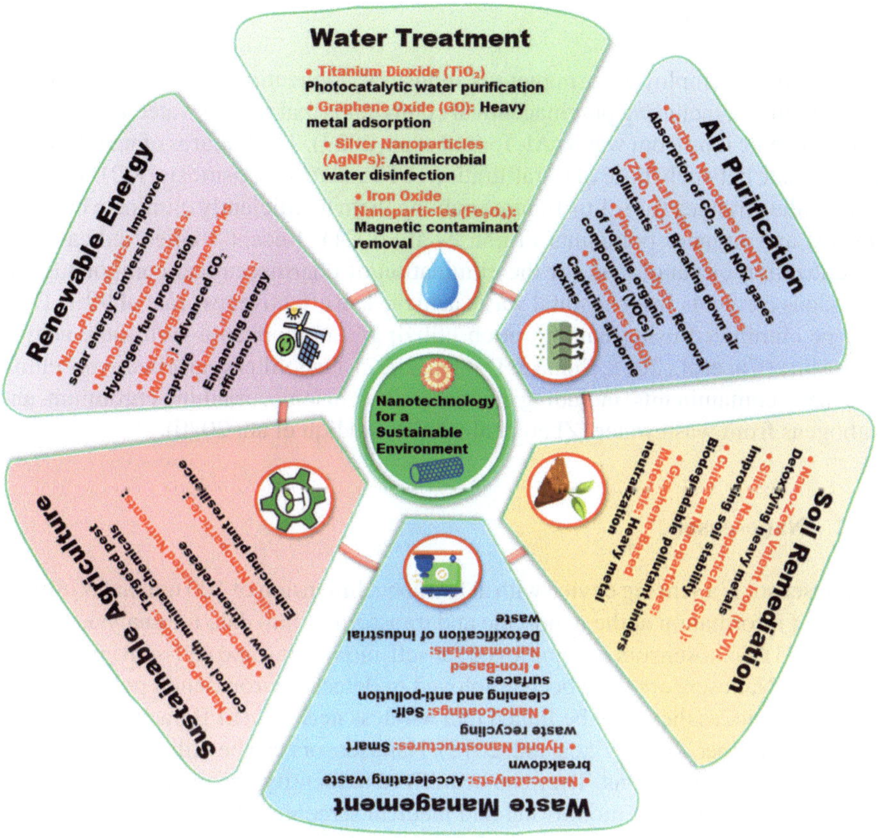


**Fig. 2** Overview of integrative role of nanotechnology in sustainable environmental remediation

Waste can be categorized into many classifications based on its physical state (gas, liquid, solid), material composition (wood, paper, glass), safety classification, and other criteria (Bal et al., 2018; Priyadarsini & Das, 2024). Among these categories, the predominant classification is the physical state of waste. Solid wastes, including metal (Das & Ghosh, 2023; Das et al., 2011; Ghosh & Das, 2018), plastic (Das et al., 2024; Mishra & Das, 2023; Mishra et al., 2019, 2021; Paul et al., 2024), food, glass, and paper, are of greater significance to the community than other materials. In recent years, more attention has been drawn toward plastic waste compared to other types of waste, due to its harmful effects on human health and the environment (Al-Mosawi et al., 2017; Sokolova et al., 2023). Consequently, the utilization of modern innovations like nanotechnology may play a key role in addressing this issue. Commonly utilized NMs for waste management include nanofilters, nanosensors, nanophotocatalysts, and nanoparticles (Taran et al., 2021).

### 2.1.1 Nanofilters

Nanofilters are employed in numerous industrial applications to eliminate various waste products, including pharmaceutical compounds that reduce water hardness and extract arsenic ions from water (Al-Toriahi et al., 2023). The features of nanofilter are intermediate between those of ultrafiltration (UF) and reverse osmosis (RO). The pore of nanofilters ranges from 1 to 10 nm, enabling them to efficiently eliminate bacteria, viruses, and organic pollutants (Taran et al., 2021). Recent studies indicate that the utilization of nanofilters for the elimination of pharmaceutical compounds and hazardous elements from hospital waste achieved an effectiveness rate of 90%. Dyes may be eliminated using nanofiltration utilizing 2-aminophenol-4-sulfonic acid or piperazine (Cai et al., 2018; Sinha & Dahiya, 2022). Nanofilters can effectively eliminate many contaminants, including anions, cations, arsenic, uranium, chromium, and pathogens from wastewater (Zhao et al., 2020; Siddiqu et al., 2020).

### 2.1.2 Nanosensor

A nanosensor is a sensing device with at least one dimension of 100 nm that is utilized to collect information at the nanoscale and transform it into data for analysis (Javaid et al., 2023). Nanosensors are employed for self-monitoring and self-sensing in smart buildings. Nanosensors have been employed to detect environmental pollutants and monitor traffic, as they may be linked to a wireless network (El Barachi et al., 2020; Patolsky & Lieber, 2005; Vikesland, 2018). Nanosensors can be utilized for chemical or mechanical applications. They can be utilized to identify chemical substances and gases for environmental monitoring and medical screening. It can monitor physical properties in a microelectromechanical system, including temperature, displacement, flow, or function as accelerometers (Su et al., 2020).

### 2.1.3 Nanophotocatalysts

Nanophotocatalysts have attracted considerable interest in the field of environmental and ecological safety, owing to their various benefits, such as affordability, remarkable stability, superior photocatalytic efficiency, and low toxicity to humans (Tahir et al., 2021). Photocatalysis, an adaptable and efficient method, has been effectively utilized in the purification of both air and water systems (Gamage & Zhang, 2010). In the fields of biomedical applications and wastewater treatment, nanophotocatalysts, especially those derived from semiconductors, are essential as they trigger advanced oxidation processes when exposed to visible or ultraviolet (UV) light (Khatun et al., 2023; Sadasivuni et al., 2019). The activation results in the production of reactive oxygen species (ROS), which are capable of effectively breaking down an extensive range of organic pollutants, including dyes (Bhattacharjee et al., 2021; Das & Swain, 2013), pharmaceuticals, and microbial contaminants (Das et al., 2015), while also facilitating the elimination of inorganic pollutants, such as heavy metals. Additionally, these materials have an important role in the photodegradation of gaseous pollutants, help with water splitting for hydrogen production, and promote the photoreduction of greenhouse gases like $CO_2$ and $N_2$. The combination of semiconductor sensitizers and quantum dots significantly boosts photocatalytic performance by expanding light absorption capabilities and enhancing charge separation efficiency, thereby advancing their potential in sustainable environmental remediation (Hooshmand et al., 2020).

## 2.2 *Water and Wastewater Remediation*

Water is an indispensable resource crucial for sustaining all living forms on Earth. In the twenty-first century, accessibility to safe and clean water is both a fundamental human right and a vital prerequisite for progress in society and economic growth. However, increasing population expansion, rapid industrialization, metropolitan development, and intensive farming methods have dramatically contributed to the production of significant quantities of wastewater (Mishra & Das, 2021, 2022; Mishra et al., 2022a, 2022b). This wastewater frequently carries numerous contaminants, making water sources both polluted and detrimental to human and environmental health (Ajith & Rajamani, 2021; Kurniawan et al., 2012). Traditional water and wastewater treatment systems have considerable constraints in dealing with the ever-strict regulatory standards for water quality (Kumar et al., 2023; Mishra et al., 2022a, 2022b). These limits necessitate the investigation and application of advanced, sustainable, and economical treatment approaches (Mishra et al., 2024). In this context, nanotechnology has emerged as a promising approach, providing enhanced treatment efficacy and cost-effectiveness. Despite ongoing problems related to scalability and environmental safety, nanotechnology-based approaches are becoming recognized for their superior efficacy and adaptability across many treatment scenarios (Khan et al., 2019). This section provides an overview of current innovations in wastewater treatment with nanotechnology. Key strategies include adsorption and biosorption using engineered NMs, nanofiltration for selective contaminant removal, photocatalysis for degradation of organic pollutants, disinfection through nanoparticle-based antimicrobial mechanisms, and deployment of nanosensors for real-time water quality monitoring (Fig. 3). These technologies combined signify a pivotal advancement in sustainable water management and environmental remediation.

### 2.2.1 Nano-based Adsorbents

Adsorption is a widely employed technique for removing contaminants from water and air, relying on the adherence of molecules to the surface of a solid material, known as the adsorbent. This process occurs through either physisorption, involving weak van der Waals forces, or chemisorption, characterized by stronger covalent or

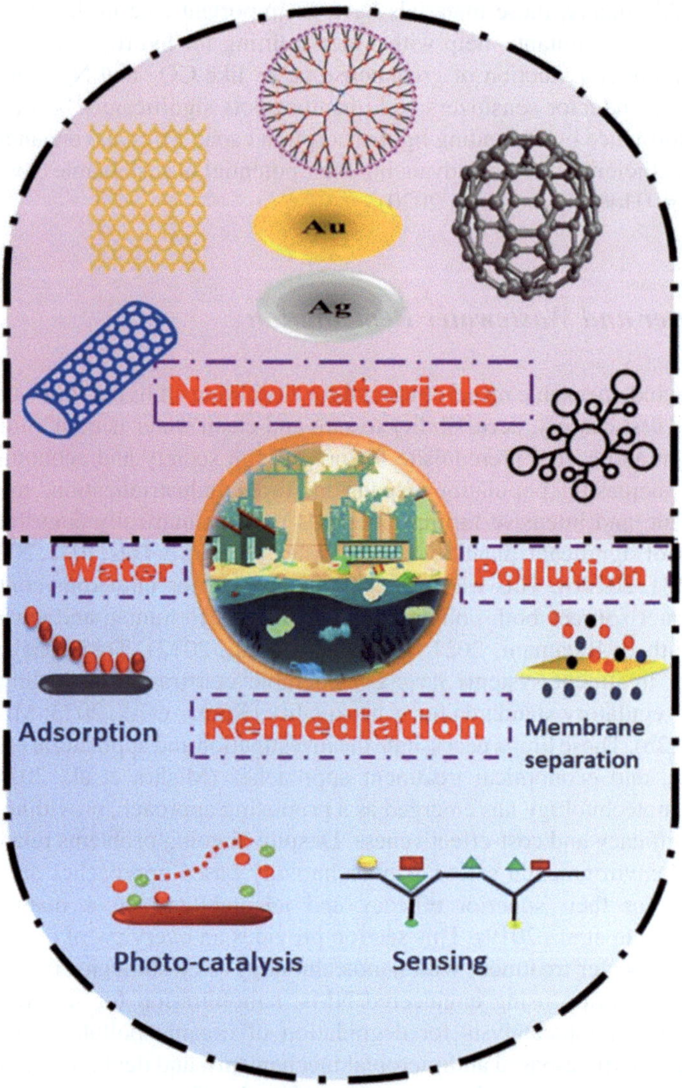


**Fig. 3** Role of carbon nanotubes, dendrimers, gold (Au) and silver (Ag) nanoparticles, liposomes, and fullerenes in water contamination remediation

ionic bonding. Nano-based adsorbents have come out as a transformative approach in treatment of wastewater because of their enhanced reactivity, high surface area, and ability to adsorb contaminants even at trace levels (Prediger et al., 2022).

### 2.2.2 Carbon-Based Nanoadsorbents

Carbon-based nanoadsorbents, including carbon nanotubes (CNTs), graphene oxide, activated carbon nanoparticles, and biochar, are highly effective for environmental remediation. These materials offer remarkable surface area, porosity, and functional group diversity. CNTs possess exceptional mechanical strength and electrical conductivity, making them ideal for adsorbing heavy metals and organic pollutants (Raninga et al., 2023). Graphene oxide, with its tunable surface chemistry, further enhances adsorption performance, while activated carbon nanoparticles excel in removing volatile organic compounds. Biochar, produced via biomass pyrolysis, is both sustainable and effective, contributing to improved water quality and soil health. These materials operate primarily through physisorption and chemisorption, targeting contaminants such as $Pb^{2+}$, $Cd^{2+}$, dyes, and pharmaceuticals. Despite their high efficiency and reusability, challenges remain in terms of production costs, nanoparticle leaching, and scalability (Pan & Xing, 2008). CNTs, in particular, outperform conventional adsorbents in removing $Cu^{2+}$, $Pb^{2+}$, $Cd^{2+}$, and $Zn^{2+}$, and their antimicrobial properties add further utility (Li et al., 2003; Liu et al., 2013).

### 2.2.3 Metal-Based Nanoadsorbents

Metal oxide nanoparticles like titanium dioxide, iron oxide, and alumina are effective for removing radionuclides and heavy metals due to their reactivity and high surface area. Adsorption occurs via surface complexation and intraparticle diffusion, with the nanoscale size significantly enhancing kinetics and capacity. For instance, nano-magnetite's arsenic adsorption increases over 100-fold when the particle size is reduced from 300 to 11 nm (Yean et al., 2005). Furthermore, superparamagnetic iron oxides such as nano-maghemite enable magnetic separation, facilitating easy recovery and reuse (Yavuz et al., 2010). These materials offer advantages over traditional adsorbents in terms of specificity, capacity, and operational simplicity.

### 2.2.4 Polymeric Nanoadsorbents

Polymeric nanoadsorbents, especially dendrimers, are notable for their highly branched architecture that supports both hydrophobic and metal-binding interactions (Shukla et al., 2017). These structures effectively adsorb organic pollutants and heavy metals. For example, dendrimer-based ultrafiltration systems have demonstrated nearly complete removal of $Cu^{2+}$, with easy regeneration by pH adjustment (Diallo & Savage, 2005). However, complex synthesis routes limit commercial viability. Chitosan-based dendrimer nanostructures have also shown promise in dye removal from textile effluents, achieving up to 99% removal efficiency and offering biodegradable, nontoxic alternatives (Sadeghi-Kiakhani et al., 2013).

### 2.2.5 Nanofiltration and Nanocomposite Membranes

Nanofiltration is a pressure-driven membrane process capable of rejecting molecules and ions smaller than 1 nm, efficiently removing hardness, color, and heavy metals from groundwater, and facilitating seawater desalination (Palit, 2017). Incorporating NMs into membranes enhances performance, with hydrophilic metal oxides (e.g., $Al_2O_3$, $TiO_2$, zeolites) improving permeability, fouling resistance, and mechanical stability (Pendergast & Hoek, 2011; Sabri et al., 2017). Antimicrobial nanoparticles like nano-Ag and CNTs reduce biofouling, although long-term efficacy and replenishment need further study (Mauter et al., 2011). Reactive membranes incorporating photocatalytic or reductive NMs (e.g., nano-$TiO_2$, nZVI) enable degradation of organic and chlorinated contaminants (Singha & Mishrab, 2020).

### 2.2.6 Thin-Film Nanocomposite (TFN) Membranes

TFN membranes integrate NMs into the active polyamide layer of conventional thin-film composite membranes to enhance water flux and selectivity. Nano-zeolites are frequently used to improve permeability and salt rejection, while also serving as transporters for antimicrobial agents such as $Ag^+$ (Lind et al., 2009). Nano-$TiO_2$ addition supports photocatalytic degradation of foulants, although high loadings may introduce defects. CNTs, when covalently bonded to membrane surfaces, offer moderate antibacterial properties while preserving membrane structure (Berekaa, 2016).

### 2.2.7 Bioinspired Membranes

Biomimetic membranes, inspired by aquaporins, offer exceptional water permeability and selectivity. Aquaporin-Z from *Escherichia coli* has been integrated into vesicle systems and membrane coatings, demonstrating high rejection of solutes and superior water transport (Kumar et al., 2007). Aligned CNT membranes exhibit ultrafast water transport due to their smooth, nano-confined channels, though achieving uniform subnanometer diameters for high selectivity remains a challenge. Functionalization at CNT openings improves ion rejection, although performance may decline with increasing ionic strength (Fornasiero, 2017).

## 2.3 *Soil Remediation*

Soil is an indispensable part of the environment, providing crucial ecosystem services for life. However, pollutants and waste have accumulated due to increasing industrialization, transportation, and urbanization, resulting in solid waste deposition in the soil. Conventional waste disposal techniques, including sanitary landfills, dumping, pyrolysis, and incineration, are laborious, costly, inefficient, and detrimental to health. Through the application of nanotechnology, science has addressed the limitations of traditional technologies. This section outlines the function of NMs in influencing the mobility, transformation, and toxicity of diverse organic and inorganic contaminants (Skinder & Hamid, 2020). They are extensively employed in several abiotic and biotic methods to eliminate contaminants from polluted soils.

### 2.3.1 Nano-assisted Abiotic Remediation of Contaminated Soils

NMs have gained significant interest for soil remediation because of their small particle size, large surface area, and known reactivity, mainly utilizing mechanisms that include sorption, reduction, and chemical oxidation (Guerra et al., 2018). The activity and toxicity of both organic and inorganic contaminants present in the soil are primarily determined by their sorption–desorption interactions with soil components (Hamid et al., 2020). Thus, the implementation of suitable amendments to polluted soils can markedly modify the mobility and toxicity of these contaminants (Hussain et al., 2020). These amendments work by multiple sorption mechanisms, such as adsorption on the surface of minerals, forming a complex with organic ligands, surface precipitation, and ion exchange (Kumpiene et al., 2008). These amendments are of two types: (i) mobilizing agents, which boost the mobility and bioavailability of soil contaminants, thus assisting in their extraction through soil washing or phytoextraction, and (ii) immobilizing agents, which sequester pollutants in the soil, thereby diminishing their mobility and bioavailability and consequently restricting their infiltration into the food chain (Robinson et al., 2009). The efficacy of soil washing in metal(loid)-contaminated soils markedly increased with the incorporation of nanoscale zerovalent iron (nZVI), which adsorbed the metal(loid)s and was later retrieved using magnetic separation (Boente et al., 2018).

The capacity to regenerate and reuse adsorbents provides a dual benefit: the possibility of pollutant recovery (for ex situ treatment or resource utilization) and reduced remediation expenses. The synthesis of stable NMs that can sustain treatment efficacy across numerous cycles is in high demand. The regeneration of adsorbents in soil remediation is less investigated than their application in wastewater treatment. Magnetic nanoparticles, due to their strong magnetism, allow for swift and effortless separation from the reaction matrix when subjected to an external magnetic field, hence promoting recycling and regeneration for reuse (Fu et al., 2018; Usman et al., 2020). The features of magnetic nanoparticles make them very promising for the adsorption and catalytic degradation of contaminants. Polyacrylamide-modified magnetite nanoparticles exhibit dual functionality, achieving a 90% reduction in soil erosion and an 82% immobilization of arsenate (Zheng et al., 2020). In situ immobilization utilizing NMs has arisen as a cost-saving and effective approach for soil remediation (Hamid et al., 2020). Nano-hydroxyapatite particles have shown efficacy in immobilizing metals in sediments (Jia et al., 2022) and soils (Dong et al., 2016) by minimizing the exchangeable metal fractions and, subsequently, their concentrations in pore water. Sodium carboxymethyl cellulose-stabilized nZVI achieved an

80% reduction of Cr(VI) from soil, with removal methods attributed to both reduction and immobilization, proven by X-ray photoelectron spectroscopy (XPS) analysis and elemental distribution data (Wang et al., 2016). In vitro studies suggested significant reductions in Cr(VI) leaching (82%) and bioaccessibility (58%) with stabilized nZVI (0.09 g $L^{-1}$), as well as a notable decrease in Cr(VI) bioaccumulation in rapeseed and Chinese cabbage by 61% and 36%, respectively, after 72 h of treatment (Wang et al., 2019).

Abiotic reductive transformations, whether independently or in conjunction with immobilization, are crucial in influencing the fate, mobility, and toxicity of redox-sensitive pollutants in soils. In pyrene-contaminated soils, the incorporation of nZVI established reducing conditions favorable for pyrene degradation, with nZVI ($k = 0.135\ min^{-1}$) exceeding micro-ZVI ($k = 0.103\ min^{-1}$) in reaction kinetics (Chang & Kang, 2009). Rajput et al. (2022) documented total Cr(VI) reduction in contaminated soils utilizing nZVI (0.10 g $L^{-1}$) under batch conditions within 2 h and further showed 99% Cr(VI) reduction in field-scale windrow tests employing 5 g $L^{-1}$ nZVI over 40 days. Despite its increased reactivity, nZVI frequently agglomerates due to its size and magnetic properties, which can hinder its mobility and efficacy in remediation. Consequently, stabilization techniques have been established, encompassing surface modification utilizing starch (Reyhanitabar et al., 2012), sodium carboxymethyl cellulose (Wang et al., 2019), green synthesis strategies (Ahmad et al., 2019), and biochar (Rajput et al., 2023).

NMs are utilized in the catalytic degradation and mineralization of organic pollutants through advanced oxidation processes (AOPs), which depend on oxidants like hydrogen peroxide ($H_2O_2$) or persulfate ($S_2O_8{}^{2-}$). These oxidants, upon activation by Fe(II), produce potent radicals (·OH or $SO_4{\cdot}^-$) for effective destruction. These processes generally need acidic conditions (pH < 4) to avoid Fe(II) precipitation, leading to problems like soil acidification, high treatment costs, and detrimental effects on soil health and microbial activity. Solid iron phases, specifically magnetite ($Fe(II)Fe(III)_2O_4$), have been suggested as catalysts for advanced oxidation processes (AOPs) at neutral pH, hence negating the necessity for acidification (Fu et al., 2018; Usman et al., 2012a, 2012b). For example, magnetite-catalyzed $H_2O_2$ and $S_2O_8{}^{2-}$ systems accomplished over 80% hydrocarbon removal from polluted soils, whereas soluble Fe(II) exhibited less than 15% effectiveness under neutral circumstances. However, the efficacy of these treatments is affected by parameters such as pollutant availability, soil organic matter concentration, and soil texture (Usman et al., 2012a, 2012b). The incorporation of chelating chemicals with iron nanoparticles can improve Fenton oxidation. Jorfi et al. (2013) established that sodium pyrophosphate was the most efficacious of various chelating agents, with a 93% elimination of pyrene. In addition, the existence of endogenous iron in soils may be adequate for catalytic oxidation, removing the need for additional iron. In soils contaminated with gasoline, extra iron was only needed when the natural Fe level was low (about 1.5 g $kg^{-1}$), but soils with higher Fe amounts (30–216 g $kg^{-1}$) did not require any added iron (Santos et al., 2018).

### 2.3.2 Nano-assisted Bioremediation of Contaminated Soils

Bioremediation is an eco-friendly, economical, flexible, and in situ approach to the elimination of organic and inorganic contaminants. Notwithstanding its benefits, this method frequently encounters limitations, including long treatment periods, restricted pollutant bioavailability, diminished efficacy in severely contaminated soils due to the pollutant toxicity to biological agents (e.g., plants, fungi, and bacteria), and the possible generation of by-products that are toxic in nature (Cecchin et al., 2017). The combination of NMs with bioremediation offers a viable resolution to address these issues. Although NMs have been utilized in chemical decontamination for more than 20 years, their use in bioremediation is very recent and still in the progressive phase. The nanobioremediation technology can be implemented either simultaneously or sequentially to improve the efficiency of contaminated soil remediation.

NMs for Bioremediation of Inorganic Pollutants

NMs have emerged as effective tools for improving the phytoremediation efficacy of heavy metals in contaminated soils. Numerous studies have indicated that NMs can boost metal absorption, reduce toxicity, and increase plant resilience to metal-induced stress. Singh and Lee (2016) investigated the influence of nano-titanium dioxide ($TiO_2$) on cadmium (Cd) accumulation in *Glycine max* and found that $TiO_2$ increased Cd uptake while simultaneously minimizing its phytotoxicity. The result was attributed to the NMs capacity to scavenge reactive oxygen species (ROS) and safeguard the plants from oxidative damage. Liang et al. (2017) conducted a field experiment to investigate the synergistic use of nano-hydroxyapatite (NHAP) and nano-carbon black (NCB) for the remediation of lead (Pb)-contaminated soils utilizing ryegrass (*Lolium temulentum* L.). Both NHAP and NCB markedly reduced Pb phytotoxicity and improved phytoextraction efficacy. Of the two, NHAP showed greater effectiveness for in situ remediation of Pb-contaminated areas.

In addition to direct interactions, NMs can mitigate heavy metal stress by regulating gene expression linked to metal stress response, oxidative stress management, water homeostasis, cell wall modification, photosynthesis, and cell division (Khodakovskaya et al., 2012; Nair & Chung, 2014). Hu et al. (2018) highlight that NMs may augment heavy metal accumulation by modifying cell wall permeability, regulating metal transporter genes, and enabling the cotransport of metals with NMs. Although beneficial, the use of NMs in bioremediation requires extra caution. Adverse effects have also been documented. Tang et al. (2015) demonstrated that the incorporation of graphene oxide (GO) alongside Cd intensified Cd toxicity in *Microcystis aeruginosa* by increasing reactive oxygen species (ROS) production. This emphasizes that the effects of NMs are notably dependent upon their type, concentration, and the biological system in which they are utilized. Gong et al., (2018a, 2018b) highlighted that various NMs can influence heavy metal adsorption and phytotoxicity differently, requiring comprehensive risk evaluations before field treatments.

### NMs for Bioremediation of Organic Pollutants

The combination of NMs with biological processes signifies a notable advancement in the treatment of persistent organic contaminants present in the soils. This nano-bio approach utilizes the combined advantages of nanotechnology and bioremediation to accomplish more efficient, quick, and precise decontamination. Recalcitrant organic contaminants can be efficiently treated using reductive procedures like dechlorination or dehalogenation utilizing extremely reactive nanoparticles, followed by biodegradation for complete remediation. Bokare et al. (2010) showed the viability of integrating palladium (Pd)/iron (Fe) nanoparticles with bioremediation for the remediation of triclosan (TCS, 2,4,4′-trichloro-2′-hydroxydiphenyl ether) in aqueous solutions ($5\ g\ L^{-1}$). The Pd/nFe-catalyzed fast reductive dechlorination of TCS in anaerobic conditions yielded 2-phenoxyphenol as the sole dechlorination product. The intermediate was subsequently oxidized by laccase enzymes produced by *Trametes versicolor* in the presence of syringaldehyde, a natural redox mediator. Similarly, Singh et al. (2013) studied the remediation of lindane (γ-hexachlorocyclohexane) in polluted soils utilizing carboxymethyl cellulose-stabilized Pd/$Fe^0$ bimetallic nanoparticles (CMC-Pd/n$Fe^0$), followed by bioremediation with *Sphingomonas* sp. strain. The combined system showed a 1.7–2.1-fold enhancement in lindane degradation efficiency and bacterial proliferation, attributed to the stimulatory influence of nanoparticles on microbial activity. Le et al. (2015) investigated the integration of chemical oxidation via Pd/nFe nanoparticles with microbial degradation for the remediation of polychlorinated biphenyls (PCBs, Aroclor 1248). During the initial phase, the bimetallic nanoparticles dechlorinated 99% of trichlorinated, 92% of tetrachlorinated, 84% of pentachlorinated, and 28% of hexachlorinated biphenyls. The residual biphenyls were later biodegraded by *Burkholderia xenovorans*, with no detected toxicity from the nanoparticles to the bacterial strain.

NMs increase the bioavailability of organic contaminants in soil, hence facilitating their accessibility to microbial and plant-based remediation agents. NMs can enhance the simultaneous uptake of contaminants and nanoparticles into plant tissues, either by direct adsorption or by altering membrane permeability due to phytotoxicity produced by nanoparticles (Gong et al., 2018a, 2018b). The effect of fullerene ($C_{60}$) exposure on the absorption of dichlorodiphenyldichloroethylene (DDE) (a metabolite of dichlorodiphenyltrichloroethane [DDT]) in *Cucurbita pepo* L., *Glycine max*, and *Solanum lycopersicum* L. cultivated in vermiculite was examined by De La Torre-Roche et al. (2012). Their findings demonstrated a 30–65% increase in DDE accumulation across all three species, presumably attributable to the cotransport of nanoparticles and contaminants. These findings collectively highlight the potential of NMs as catalysts for chemical transformation and as enhancers of pollutant uptake and degradation in integrated nanobioremediation systems.

## 2.4 Nanotechnology for the Removal of Greenhouse Gases

Greenhouse gases (GHGs), including nitrous oxide ($N_2O$), methane ($CH_4$), carbon dioxide ($CO_2$), and fluorinated gases, are the primary agents of global warming and climate change. Human activities, especially industrialization, fossil fuel consumption, and intensive agriculture, have significantly increased atmospheric greenhouse gas concentrations (Gondal, 2023). With an increase in pollution levels, there is an urgent demand for technological advancements to reduce greenhouse gas emissions. Nanotechnology offers significant potential to improve current environmental technologies and develop innovative solutions for greenhouse gas management and climate protection. NMs have a high surface-area-to-volume ratio, imparting distinctive physicochemical features including increased reactivity, selectivity, and multifunctionality. These attributes can be utilized in many environmental applications, including energy efficiency, resource conservation, waste reduction, and, particularly, the capture and elimination of greenhouse gases. A wide range of NMs such as carbon nanotubes (CNTs), zinc oxide (ZnO), titanium dioxide ($TiO_2$), iron nanoparticles, silica, dendrimers, polymers, and composite materials has been investigated for applications in air purification, water treatment, and soil remediation (Gupta et al., 2015). Wang et al. (2013) found that $TiO_2$-coated nano-calcium carbonate ($CaCO_3$) possesses excellent $CO_2$ adsorption capabilities and sintering resistance, making it effective for prolonged $CO_2$ capture. Wang et al. (2014) similarly reported that lithium orthosilicate-based adsorbents, synthesized from rice husk ash and pretreated with citric acid, retained their $CO_2$ capture capability and sintering resistance at higher temperatures (700 °C). These findings highlight the promise of NMs in carbon capture methods under industrial conditions. Wang and Lua (2012) employed metallic nickel nanoparticles as catalysts for methane decomposition and hydrogen production, showing outstanding efficiency in the thermal degradation of methane. This not only mitigates $CH_4$ emissions, a strong greenhouse gas, but also facilitates renewable energy creation via hydrogen synthesis.

Nanotechnology has been utilized to mitigate pollutants from industry, including sulfur dioxide ($SO_2$), a gas associated with acid rain, respiratory issues, and several environmental threats. Saleh et al. (2016) highlighted the efficacy of NMs in desulfurization processes, limiting $SO_2$ emissions from fossil fuel burning. Nitrogen oxides ($NO_x$), which include nitric oxide (NO) and nitrogen dioxide ($NO_2$), are significant air pollutants generated by vehicle emissions and industrial activities. The advancement of NMs-based sorbents has been a primary emphasis for $NO_x$ reduction. Low-temperature adsorbents, such as ion-exchange zeolites, activated carbon, and iron oxalate impregnated on activated charcoal fibers, show significant potential. Activated carbon can adsorb NO due to its surface functional groups, although its capacity is constrained. In contrast, carbon nanotubes exhibit enhanced adsorption properties for NO, presenting a viable alternative for efficient $NO_x$ capture. In brief, nanotechnology offers three essential functions in environmental applications: (1) remediation and purification of air, water, and soil; (2) detection and sensing of contaminants; and (3) pollution avoidance through sustainable material design and catalytic processes.

The rising demand for sustainable greenhouse gas mitigation measures highlights the significant potential of integrating nanotechnology into environmental management frameworks to reduce emissions and address climate change.

### 2.5 NMs in Agriculture

Agriculture is essential to the global economy, serving as the primary source of food and feed via the cultivation of many crops (Malhotra & Mandal, 2019). The expanding human population needs more efficient agricultural practices to guarantee global food security and nutritional sustainability (Cicek & Nadaroglu, 2015). Although conventional chemical fertilizers have historically enhanced crop growth and productivity, excessive dependence on these inputs has resulted in unforeseen environmental consequences and reduced nutrient use efficiency (Congreves & Van Eerd, 2015). The excessive and unregulated use of conventional fertilizers has increased production costs and led to water pollution, soil degradation, and greenhouse gas emissions. A significant disadvantage of chemical fertilizers is their low nutrient use efficiency, mostly attributable to their rapid nutrient release rates that frequently exceed the plant's actual absorption capacity. A considerable percentage of nutrients are lost due to leaching, volatilization, or conversion into nonbioavailable forms (Chhipa, 2017). This inefficiency hampers efforts to attain long-term agricultural sustainability. To tackle these difficulties, it is mandatory to develop innovative fertilizer systems that improve nutrient usage efficiency and reduce environmental impacts (Van Eerd et al., 2017).

Nanotechnology offers a significant opportunity for the advancement of next-generation fertilizers, including nanofertilizers, nanobiofertilizers, and nanopesticides. These nano-enabled inputs are designed to enhance the accuracy and efficacy of nutrient delivery to crops, therefore facilitating increased output while minimizing environmental impact (Feregrino-Perez et al., 2018). Nanobiomaterials in agriculture have demonstrated the capacity to significantly improve crop yield, decrease nutrient losses via leaching, and mitigate environmental hazards (Solanki et al., 2015). Due to their extensive surface area, controlled release properties, and specific targeting, nanoparticles can improve nutrient accessibility in the rhizosphere and increase the absorption efficiency of vital components in plants.

## 3 Environmental Impacts of NMs

The unique physicochemical characteristics of NMs raised early concerns about their potential impact on human and environmental health. Since risk is determined by both hazard and exposure, a life cycle assessment (LCA) strategy is crucial for the objective evaluation of the environmental concerns associated with NMs. The approach incorporates a comprehensive understanding of their mechanistic toxicity, environmental release pathways, and exposure dynamics throughout various application stages.

### *3.1 Toxicity Versus Exposure*

Toxicological research has shown that the nanoscale size of engineered NMs enhances their ability to penetrate cellular membranes and translocate to subcellular organelles (Wang et al., 2022). Such NMs may cause dose-dependent oxidative stress, modify gene expression, and impair physiological processes (Ganguly et al., 2018). The small size and high surface area of engineered NMs may impose mechanical stress on cell membranes, impacting cellular integrity. In addition, NMs consisting of reactive metals can release ions or facilitate redox processes, leading to the production of reactive oxygen species (ROS) that contribute to cytotoxicity and genotoxicity (Fu et al., 2014). Toxicities that depend on shape and structure are notably severe for one-dimensional NMs (e.g., nanotubes, nanowires), which can cause fiber-like toxicity similar to asbestos at higher concentrations (Sukhanova et al., 2018). However, initial risk evaluations predominantly focused on toxicity, frequently under controlled laboratory conditions that utilized unreasonably higher exposure levels occasionally in the milligram to sub-gram per liter range, thus amplifying perceived environmental hazards (Zhao et al., 2021).Studies on the environmental fate and transport of NMs demonstrate that they generally undergo processes like immobilization, aggregation, and passivation, significantly diminishing their bioavailability and potential for exposure (Grillo et al., 2015; Kent et al., 2014; Pourzahedi et al., 2017). Immobilization techniques are commonly employed in water treatment to reduce the unintentional release of engineered NMs and related concerns.

### *3.2 Release Pathways of NMs into the Environment*

The National Nanotechnology Initiative (NNI) of the United States emphasized the need for a life-cycle perspective in assessing the environmental risks of NMs in its 2011 environmental health and safety policy. This framework addresses both unintentional and intentional releases over the whole life cycle of nanotechnology-enabled products (NEPs) from manufacture and application to disposal (Arvidsson et al., 2018). Global production of NMs is projected to exceed $10^6$ tons per year, with silicon dioxide ($nSiO_2$), titanium dioxide ($nTiO_2$), and nanoclays being the most prevalent NMs (Janković et al., 2019; Giese et al., 2018). However, their contribution is negligible regarding naturally occurring NMs, accounting for less than 0.1% of the entire global population of nano-sized particles (Hochella et al., 2019). Silicon-based NMs represent a comparatively significant fraction (3–25%) of the worldwide silicon cycle (Dixit et al., 2023).

Quantifying NMs emissions needs awareness of their uses, transformation mechanisms (such as photodegradation, dissolution, and abrasion), and incorporation into matrices. Certain applications, like nanopesticides or in situ remediation, lead to direct environmental discharge, while others, such as textiles or packaging, experience significant physicochemical alterations before potential release. Initial research utilizing pristine NMs in controlled environments has progressed to recognize that NMs emitted from real products exhibit variations in form, chemistry, and reactivity as a result of environmental aging and interactions with matrices (Mitrano et al., 2015). In-depth investigations have examined the release of NMs from consumer and construction items, including plastics, coatings, textiles, and cement-based materials (Quadros et al., 2013; Wohlleben & Neubauer, 2016). For example, NMs utilized in coatings may fragment over time, releasing modified forms into the environment. The product's life-cycle stage significantly impacts NMs bioavailability and toxicity. A mesocosm study demonstrated that pristine copper oxide nanoparticles (nCuO) released during production exhibited higher bioavailability and toxicity to plankton compared with weathered product fragments released during usage or disposal, which predominantly affected benthic organisms due to rapid sedimentation (Gottschalk et al., 2013).

### *3.3 NMs Release in Water Treatment Applications*

In water treatment, NMs are rarely utilized in their free, suspended state. Instead, they are frequently immobilized on or integrated within substrates, such as membranes or coatings, to prevent environmental release while maintaining functionality. Applications encompass nano-zerovalent iron (nZVI) for groundwater remediation (Niewiadomski et al., 2017), quantum dots for pathogen detection (Vikesland & Wigginton, 2010), and $nTiO_2$ or hexagonal boron nitride for photocatalytic disinfection (Lu et al., 2012). Silver nanoparticles (AgNPs) are extensively researched in water treatment because of their antibacterial abilities. Studies demonstrate that silver release significantly changes based on the kind of filtration, water chemistry, and operational conditions. Under reverse osmosis membranes infused with AgNPs, silver release typically remained below the Environmental Protection Agency's (EPA) secondary maximum contaminant threshold (0.1 mg/L), especially under nonstandard batch immersion conditions (Li et al., 2021). Other investigations indicated higher Ag release in deionized water and seawater relative to surface water, due to less ionic competition (Wu et al., 2020). Additionally, dead-end filtration resulted in reduced Ag release compared to cross-flow systems, highlighting the impact of hydrodynamics on NMs behavior (Mittelman et al., 2015). These findings underscore the necessity for individualized evaluations of NMs release and associated risks under realistic environmental and operational conditions. The synthesis processes of NMs and their integration into matrices significantly impact release behavior, supporting "safety-by-design" approaches to reduce environmental risks in nanotechnology applications (Trump et al., 2024).

# 4 Management and Sustainable Use of NMs

NMs are emerging as potent agents in diverse environmental applications because of their ability to improve or catalyze chemical and biological reactions. These cover pollutant removal, microbial deactivation, and environmental monitoring. However, the characteristics that make NMs efficient can cause unanticipated and even detrimental responses in natural ecosystems or organisms exposed to them. Therefore, a "safety-by-design" strategy is essential for the development of NMs and nanomaterial-enabled technologies aimed at environmental applications (Barhoum et al., 2022; Jiménez et al., 2022). Considering that both innate hazards (e.g., nanomaterial toxicity) and exposure contribute to environmental and public health problems, safety by design must integrate strategies that reduce hazards and mitigate exposure, both of which are discussed below.

## *4.1 Hazard Mitigation*

Reducing or minimizing hazard highlights the utilization of NMs made of relatively nontoxic and abundant earth elements, like nano-titanium dioxide ($nTiO_2$), assuming inhalation risks are mitigated, and food-grade nano-magnetite. Integrating green chemistry principles diminishes toxicity, both from the NMs and their production. For instance, nano-zerovalent iron (nZVI) synthesized with neem leaf extract exhibits comparable efficacy in hexavalent chromium removal to that prepared via traditional chemical methods (Ravikumar & Prakash, 2022). Various alternative synthesis pathways utilize bio-derived reducing agents, nontoxic capping agents, and ecologically friendly solvents (Wang et al., 2015). Initiatives have been undertaken to diminish the energy consumption associated with the synthesis of NMs by low-energy approaches, including microwave and ultrasound-assisted methods (Nadagouda et al., 2011; Plata, 2009). However, these strategies may encounter barriers, including low yield and restricted control over the size and geometry of NMs. Consequently, life cycle assessment (LCA) is crucial for determining if these technologies provide genuinely reduced environmental impacts (Pati et al., 2014). The three-pillar model encompassing safe product, safe use, and safe production guides safety across the entire development process (Sánchez Jiménez et al., 2022). A new product-oriented safety-by-design model, introduced in 2021, overcomes the limitations of the linear stage-gate model by including the entire life cycle of nanotechnology-enabled products (NEPs), from manufacture to disposal (Rose et al., 2021).

This approach is exemplified by two case studies:

1. Photocatalytic paints with $nTiO_2$: An initial initiative to enhance commercial paints with $nTiO_2$ for air pollution mitigation initially met safety standards but later failed during weathering assessments. The photocatalytic breakdown of the paint matrix emitted detrimental volatile organic chemicals and reduced $nTiO_2$

content, forcing a redesign. A novel basal mineral matrix has been designed to withstand degradation while retaining active NMs (Truffier-Boutry et al., 2017).
2. Quantum dots in biosensing: The production of cadmium-free quantum dots was guided by a "Triple R" approach—Replace, Reduce, and Retain. The quantum dots, utilizing InPZn and InPZnS cores with multilayer ZnS shells, showed better optical properties and less cytotoxicity. However, aging worsened toxicity, emphasizing the necessity to enhance envelope thickness and stability (Dussert et al., 2022; Wagner & Reemtsma, 2019).

## *4.2 Minimizing Exposure*

Minimizing exposure to NMs requires employing the smallest effective quantities and optimizing their features, including size, shape, crystallinity, composition, and surface chemistry, to improve target specificity. Selectivity minimizes side reactions, resulting in less usage of NMs while retaining efficacy. Advanced techniques like nanocrystal facet engineering have shown enhanced selectivity and reduced toxicity in pollutant adsorption research (Liu et al., 2016). Another important strategy involves the immobilization of NMs on substrates such as beads, membranes, fibers, or porous structures, which reduces leaching and improves reuse. Immobilization may occur through physisorption, which includes $\pi$–$\pi$ interactions, hydrogen bonding, and electrostatic forces, or through covalent bonding, each with different stability levels (Shahriari et al., 2021). Mixed-matrix nanocomposites, formed by incorporating nanoparticles into matrices, are extensively utilized; however, they may experience compromised surface activity as a result of blockage. Core–shell electrospinning, spraying, and in situ formation of NMs are more effective in retaining the properties of NMs. Synergistic effects between NMs and substrates can enhance performance or introduce new reaction pathways (Peng et al., 2020).

NMs can be immobilized on larger particles to enhance sedimentation; however, mechanical agitation can result in their release due to erosion. Packed bed reactors utilizing nanomaterial-coated granular media present alternatives; however, they encounter limitations, including decreased surface area (Hristovski et al., 2007, 2008). Membranes and electrodes serve as effective substrates, facilitating dual functionality, such as catalysis and filtration, while also promoting selective contaminant interactions and enhancing redox processes (Chen & Ostrom, 2015; Zhang et al., 2019; Zuo et al., 2020). Emerging technologies, including single-atom catalysts, present feasible alternatives by enhancing atomic efficiency and minimizing environmental impact. These materials, while in early development, demonstrate potential for application in complex water treatment scenarios (Cheng et al., 2019). Although advancements have been made, assessments of long-term stability and release risks associated with immobilized NMs remain challenging (Froggett et al., 2014). Research indicates that the anchoring of NMs onto supports such as nanozeolites or carbon nanotubes diminishes metal ion leaching and improves antibacterial efficacy (Park et al., 2016). The release of NMs is dependent upon the properties of the material and substrate, the immobilization technique used, and the context of usage. Chemical bonding typically provides more effective anchoring compared to physical adhesion. In situ formation of NMs exhibits reduced release compared to alternative methods, such as spraying. The absence of standardized protocols for the assessment of nanomaterial release hampers comparative evaluations among various technologies.

### *4.3 Social Considerations*

In addition to technical and environmental considerations, societal acceptance is crucial for the sustainable execution of NMs in environmental applications. Obstacles include higher expenses, safety worries, and regulatory ambiguity. In water treatment, pricing frequently represents the primary obstacle in comparison to conventional procedures. Mitigation options mean using less of certain additives, using recycled materials, producing more to lower costs, and allowing for lower purity levels that are acceptable for commercial use (Brame et al., 2011). To address the public's concern and regulatory complexities, early stakeholder participation is crucial. Surveys and interviews can reveal and reduce concerns or misconceptions, improving technology legitimization, which is defined as the perceived appropriateness of a technology within its societal context (Harris-Lovett et al., 2015).

Surveys in Northwestern Europe suggested that public approval depends on the acknowledged need, utility, and accessibility of the technology to end users (Gupta et al., 2022). Raised public awareness of NMs and their uses markedly fosters acceptance, underscoring the need for education and outreach (Kidd et al., 2020). From an industrial perspective, regulatory issues were subordinate to public opinion and financial considerations. Companies preferred clear regulation and independent certification to mitigate safety concerns and enhance long-term market survival. To ensure the appropriate and sustainable use of NMs in environmental systems, it is crucial to incorporate safety by design, green production, and efficient immobilization into the development process. These efforts must include stakeholder participation, public education, and certification of nanomaterial-enabled products to ensure safety standards and build societal confidence.

## 5 Conclusion and Future Recommendations

Nanotechnology presents transformative opportunities for sustainable environmental solutions, offering advanced methods for pollution control, water purification, and soil remediation. Its integration with existing technologies and green synthesis approaches can significantly reduce ecological footprints. However, real-world applications remain limited due to concerns over safety, regulatory gaps, and lack of large-scale validation. To ensure safe and effective deployment, a multidisciplinary approach involving policy frameworks, industry partnerships, and societal engagement is essential. Several key points are outlined below as future recommendations to support the effective and responsible application of nanotechnology in environmental remediation.

- Promote real-field validations and scale-up studies to bridge the gap between laboratory findings and practical implementation
- Establish robust regulatory frameworks and standardized protocols for safe nanoparticle production, usage, and disposal
- Encourage green and cost-effective synthesis methods to minimize environmental risks
- Expand ecotoxicological studies across a broader range of organisms and ecosystems, including long-term and cumulative exposure effects
- Foster interdisciplinary collaboration among scientists, industries, policymakers, and communities to ensure inclusive and responsible nanotechnology development
- Develop effective nanowaste management systems to address end-of-life nanoparticle disposal and reduce secondary pollution risks

## References

Ahmad, S., Munir, S., Zeb, N., Ullah, A., Khan, B., Ali, J., et al. (2019). Green nanotechnology: A review on green synthesis of silver nanoparticles—An ecofriendly approach. *International Journal of Nanomedicine*, 5087–5107.

Ajith, M. P., & Rajamani, P. (2021). Nanotechnology for water purification–current trends and challenges. *Journal of Nanotechnology and Nanomaterials, 2*(2), 88–91.

Al-Mosawi, A. I., Abdulsada, S. A., & Al-Maamori, M. H. (2017). Mechanical properties of tri sustainable NBR/cement waste/rice husks silica blend. *Journal of Polymer & Composites, 5*(5), 29–31.

Al-Toriahi, A. K. M., Azooz, E. A., & Al-Mulla, E. A. J. (2023). Metal nanoparticles and nano-filters for the disposal of hospital waste: A review. *Nano Biomedicine & Engineering, 15*(2).

Amasuomo, E., & Baird, J. (2016). The concept of waste and waste management. *Journal of Management & Sustainability, 6*, 88.

Arvidsson, R., Baun, A., Furberg, A., Hansen, S. F., & Molander, S. (2018). Proxy measures for simplified environmental assessment of manufactured nanomaterials. *Environmental Science & Technology, 52*(23), 13670–13680.

Bal, B., Ghosh, S., & Das, A. P. (2018). Microbial recovery and recycling of manganese waste and their future application: A review. *Geomicrobiological Journal, 36*(1), 85–96. https://doi.org/10.1080/01490451.2018.1497731

Bala, S., Garg, D., Thirumalesh, B. V., Sharma, M., Sridhar, K., Inbaraj, B. S., & Tripathi, M. (2022). Recent strategies for bioremediation of emerging pollutants: A review for a green and sustainable environment. *Toxics, 10*(8), 484.

El Barachi, M., Rahman, S. A., Mourad, A., & Al Orabi, W. (2020). Nanosensors for traffic condition monitoring. In *Nanosensors for smart cities* (pp. 187–208). Elsevier.

Baral, A., Satish, L., Zhang, G., Ju, S., & Ghosh, M. K. (2021). A review of recent progress on nano $MnO_2$: Synthesis, surface modification and applications. *Journal of Inorganic and Organometallic Polymers and Materials, 31*, 899–922.

Barhoum, A., García-Betancourt, M. L., Jeevanandam, J., Hussien, E. A., Mekkawy, S. A., Mostafa, M., et al. (2022). Review on natural, incidental, bioinspired, and engineered nanomaterials: history, definitions, classifications, synthesis, properties, market, toxicities, risks, and regulations. *Nanomaterials, 12*(2), 177.

Berekaa, M. M. (2016). Nanotechnology in wastewater treatment; influence of nanomaterials on microbial systems. *International Journal of Current Microbiology and Applied Sciences, 5*(1), 713–726.

Besha, A. T., Liu, Y., Fang, C., Bekele, D. N., & Naidu, R. (2020). Assessing the interactions between micropollutants and nanoparticles in engineered and natural aquatic environments. *Critical Reviews in Environmental Science and Technology, 50*(2), 135–215.

Bhattacharjee, J., Mishra, S., & Das, A. P. (2021). Recent advances in sensor-based detection of toxic dyes for bioremediation application: A review. *Applied Biochemistry and Biotechnology, 2021*, 1–20. https://doi.org/10.1007/S12010-021-03767-7

Boente, C., Sierra, C., Martínez-Blanco, D., Menéndez-Aguado, J. M., & Gallego, J. R. (2018). Nanoscale zero-valent iron-assisted soil washing for the removal of potentially toxic elements. *Journal of Hazardous Materials, 350*, 55–65.

Bokare, V., Murugesan, K., Kim, Y. M., Jeon, J. R., Kim, E. J., & Chang, Y. S. (2010). Degradation of triclosan by an integrated nano-bio redox process. *Bioresource Technology, 101*(16), 6354–6360.

Brame, J., Li, Q., & Alvarez, P. J. (2011). Nanotechnology-enabled water treatment and reuse: Emerging opportunities and challenges for developing countries. *Trends in Food Science & Technology, 22*(11), 618–624.

Cai, Z., Dwivedi, A. D., Lee, W. N., Zhao, X., Liu, W., Sillanpää, M., et al. (2018). Application of nanotechnologies for removing pharmaceutically active compounds from water: Development and future trends. *Environmental Science: Nano, 5*(1), 27–47.

Cecchin, I., Reddy, K. R., Thomé, A., Tessaro, E. F., & Schnaid, F. (2017). Nanobioremediation: Integration of nanoparticles and bioremediation for sustainable remediation of chlorinated organic contaminants in soils. *International Biodeterioration & Biodegradation, 119*, 419–428.

Chang, M. C., & Kang, H. Y. (2009). Remediation of pyrene-contaminated soil by synthesized nanoscale zero-valent iron particles. *Journal of Environmental Science and Health Part A, 44*(6), 576–582.

Chen, A., & Ostrom, C. (2015). Palladium-based nanomaterials: Synthesis and electrochemical applications. *Chemical Reviews, 115*(21), 11999–12044.

Cheng, Y., Liu, Y., Fan, W., Yan, Z., & Ye, X. (2019). Triple helix on globalization: A case study of the China International Nanotech Innovation Cluster. *Information Development, 35*(2), 272–289.

Chhipa, H. (2017). Nanofertilizers and nanopesticides for agriculture. *Environmental Chemistry Letters, 15*, 15–22.

Cicek, S., & Nadaroglu, H. (2015). The use of nanotechnology in the agriculture. *Advances in Nano Research, 3*(4), 207.

Congreves, K. A., & Van Eerd, L. L. (2015). Nitrogen cycling and management in intensive horticultural systems. *Nutrient Cycling in Agroecosystems, 102*, 299–318.

Das, A. P., & Swain, S. (2013). Algal biosorption of toxic dye methylene blue. *A Potential Source of Food, Feed, Biochemicals, Biofuels and Biofertilizers, International Conference on Algal Biorefinery, Indian Institute of Technology, 13.*

Das, A. P., Ghosh, S., Mohanty, S., & Sukla, L. B. (2015). Advances in manganese pollution and its bioremediation. In L. B. Sukla, N. Pradhan, S. Panda, & B. K. Mishra (Eds.), *Environmental microbe biotechnology* (1st ed., pp. 313–328). Springer. https://doi.org/10.1007/978-3-319-19018-1_16

Das, A. P., & Ghosh, S. (2023). Microorganisms in manganese biomining processes. *Manganese Mining Microorganisms*, 55–61. https://doi.org/10.1016/B978-0-12-822147-1.00013-3

Das, A., Dey, S., & Das, A. P. (2024). Ecotoxicological impacts of synthetic microfiber pollutants and development of sustainable mitigation strategies. *Environmental Chemistry and Ecotoxicology.* https://doi.org/10.1016/J.ENCECO.2024.12.003

Das, A. P., Sukla, L. B., Pradhan, N., & Nayak, S. (2011). Manganese biomining: A review. *Bioresource Technology, 102*(16), 7381–7387. https://doi.org/10.1016/J.BIORTECH.2011.05.018

De La Torre-Roche, R., Hawthorne, J., Deng, Y., Xing, B., Cai, W., Newman, L. A., et al. (2012). Fullerene-enhanced accumulation of p, p′-DDE in agricultural crop species. *Environmental Science & Technology, 46*(17), 9315–9323.

Diallo, M. S., & Savage, N. (2005). Nanoparticles and water quality. *Journal of Nanoparticle Research, 7*(4), 325–330.

Dixit, R., Arpita, K., & Chand, V. (2023). 1 The current scenario in nanocomposite and nanohybrid materials. *Nanocomposite and Nanohybrid Materials: Processing and Applications, 17.*

Dong, A., Ye, X., Li, H., Zhang, Y., & Wang, G. (2016). Micro/nanostructured hydroxyapatite structurally enhances the immobilization for Cu and Cd in contaminated soil. *Journal of Soils and Sediments, 16*, 2030–2040.

Dussert, F., Sarret, G., Wegner, K. D., Proux, O., Landrot, G., Jouneau, P. H., et al. (2022). Physicochemical transformation and toxicity of multi-shell InP quantum dots under simulated sunlight irradiation, in an environmentally realistic scenario. *Nanomaterials, 12*(20), 3703.

Feregrino-Perez, A. A., Magaña-López, E., Guzmán, C., & Esquivel, K. (2018). A general overview of the benefits and possible negative effects of the nanotechnology in horticulture. *Scientia Horticulturae, 238*, 126–137.

Fornasiero, F. (2017). Water vapor transport in carbon nanotube membranes and application in breathable and protective fabrics. *Current Opinion in Chemical Engineering, 16*, 1–8.

Froggett, S. J., Clancy, S. F., Boverhof, D. R., & Canady, R. A. (2014). A review and perspective of existing research on the release of nanomaterials from solid nanocomposites. *Particle and Fibre Toxicology, 11*, 1–28.

Fu, H., Yang, Y., Zhu, R., Liu, J., Usman, M., Chen, Q., & He, H. (2018). Superior adsorption of phosphate by ferrihydrite-coated and lanthanum-decorated magnetite. *Journal of Colloid and Interface Science, 530*, 704–713.

Fu, P. P., Xia, Q., Hwang, H. M., Ray, P. C., & Yu, H. (2014). Mechanisms of nanotoxicity: Generation of reactive oxygen species. *Journal of Food and Drug Analysis, 22*(1), 64–75.

Gamage, J., & Zhang, Z. (2010). Applications of photocatalytic disinfection. *International Journal of Photoenergy, 2010*(1), 764870.

Ganguly, P., Breen, A., & Pillai, S. C. (2018). Toxicity of nanomaterials: Exposure, pathways, assessment, and recent advances. *ACS Biomaterials Science & Engineering, 4*(7), 2237–2275.

Gehrke, I., Geiser, A., & Somborn-Schulz, A. (2015). Innovations in nanotechnology for water treatment. *Nanotechnology, Science and Applications*, 1–17.

Ghosh, S., & Das, A. P. (2018). Metagenomic insights into the microbial diversity in manganese-contaminated mine tailings and their role in biogeochemical cycling of manganese. *Scientific Reports, 8*(1), 1–12. https://doi.org/10.1038/s41598-018-26311-w

Giese, B., Klaessig, F., Park, B., Kaegi, R., Steinfeldt, M., Wigger, H., et al. (2018). Risks, release and concentrations of engineered nanomaterial in the environment. *Scientific Reports, 8*(1), 1565.

Gondal, A. H. (2023). Nanotechnology advancement in the elimination of chemical toxins from air spectrums. *International Journal of Environmental Science and Technology, 20*(11), 12775–12792.

Gong, X., Huang, D., Liu, Y., Peng, Z., Zeng, G., Xu, P., et al. (2018a). Remediation of contaminated soils by biotechnology with nanomaterials: Bio-behavior, applications, and perspectives. *Critical Reviews in Biotechnology, 38*(3), 455–468.

Gong, Y., Li, X., Liao, G., Ding, Y., Li, J., & Cao, Y. (2018b). Cytotoxicity and ER stress–apoptosis gene expression in ZnO nanoparticle exposed THP-1 macrophages: Influence of pre-incubation with BSA or palmitic acids complexed to BSA. *RSC Advances, 8*(28), 15380–15388.

Gottschalk, F., Sun, T., & Nowack, B. (2013). Environmental concentrations of engineered nanomaterials: Review of modeling and analytical studies. *Environmental Pollution, 181*, 287–300.

Grillo, R., Clemente, Z., de Oliveira, J. L., Campos, E. V. R., Chalupe, V. C., Jonsson, C. M., et al. (2015). Chitosan nanoparticles loaded the herbicide paraquat: The influence of the aquatic humic substances on the colloidal stability and toxicity. *Journal of Hazardous Materials, 286*, 562–572.

Guerra, F. D., Attia, M. F., Whitehead, D. C., & Alexis, F. (2018). Nanotechnology for environmental remediation: Materials and applications. *Molecules, 23*(7), 1760.

Gupta, A., Gupta, S., & Mukherjee, G. (2022). Nanotechnology and regulatory issues. In *Handbook of nanotechnology in nutraceuticals* (pp. 441–460). CRC Press.

Gupta, N., Fischer, A. R., & Frewer, L. J. (2015). Ethics, risk and benefits associated with different applications of nanotechnology: A comparison of expert and consumer perceptions of drivers of societal acceptance. *NanoEthics, 9*, 93–108.

Hamid, Y., Tang, L., Hussain, B., Usman, M., Lin, Q., Rashid, M. S., et al. (2020). Organic soil additives for the remediation of cadmium contaminated soils and their impact on the soil-plant system: A review. *Science of the Total Environment, 707*, 136121.

Harris-Lovett, S. R., Binz, C., Sedlak, D. L., Kiparsky, M., & Truffer, B. (2015). Beyond user acceptance: A legitimacy framework for potable water reuse in California. *Environmental Science & Technology, 49*(13), 7552–7561.

Hochella Jr, M. F., Mogk, D. W., Ranville, J., Allen, I. C., Luther, G. W., Marr, L. C., et al. (2019). Natural, incidental, and engineered nanomaterials and their impacts on the Earth system. *Science, 363*(6434), eaau8299.

Hooshmand, S., Kargozar, S., Ghorbani, A., Darroudi, M., Keshavarz, M., Baino, F., & Kim, H. W. (2020). Biomedical waste management by using nanophotocatalysts: The need for new options. *Materials, 13*(16), 3511.

Hristovski, K., Baumgardner, A., & Westerhoff, P. (2007). Selecting metal oxide nanomaterials for arsenic removal in fixed bed columns: From nanopowders to aggregated nanoparticle media. *Journal of Hazardous Materials, 147*(1–2), 265–274.

Hristovski, K., Westerhoff, P., & Crittenden, J. (2008). An approach for evaluating nanomaterials for use as packed bed adsorber media: A case study of arsenate removal by titanate nanofibers. *Journal of Hazardous Materials, 156*(1–3), 604–611.

Hu, T., Li, H., Li, J., Zhao, G., Wu, W., Liu, L., et al. (2018). Absorption and bio-transformation of selenium nanoparticles by wheat seedlings (*Triticum aestivum* L.). *Frontiers in Plant Science, 9*, 597.

Hussain, B., Lin, Q., Hamid, Y., Sanaullah, M., Di, L., Khan, M. B., et al. (2020). Foliage application of selenium and silicon nanoparticles alleviates Cd and Pb toxicity in rice (*Oryza sativa* L.). *Science of the Total Environment, 712*, 136497.

Jabeen, S., Ahmad, N., Bala, S., Bano, D., & Khan, T. (2023). Nanotechnology in environmental sustainability and performance of nanomaterials in recalcitrant removal from contaminated water: A review. *International Journal of Nano Dimension, 14*(1).

Janković, N. Z., & Plata, D. L. (2019). Engineered nanomaterials in the context of global element cycles. *Environmental Science: Nano, 6*(9), 2697–2711.

Javed, R., Bilal, M., Ali, J. S., Khan, S., & Cheema, M. (2023). Nanotechnology: A tool for the development of sustainable agroindustry. In *Agricultural and environmental nanotechnology: Novel technologies and their ecological impact* (pp. 317–339). Springer Nature Singapore.

Jeevanandam, J., Barhoum, A., Chan, Y. S., Dufresne, A., & Danquah, M. K. (2018). Review on nanoparticles and nanostructured materials: History, sources, toxicity and regulations. *Beilstein Journal of Nanotechnology, 9*(1), 1050–1074.

Jia, X., Shi, N., Tang, W., Su, Z., Chen, H., Tang, Y., et al. (2022). Nano-hydroxyapatite increased soil quality and boosted beneficial soil microbes. *Plant Nano Biology, 1*, 100002.

Jiménez, A. S., Puelles, R., Perez-Fernandez, M., Barruetabeña, L., Jacobsen, N. R., Suarez-Merino, B., et al. (2022). Safe(r) by design guidelines for the nanotechnology industry. *NanoImpact, 25*, 100385.

Jorfi, S., Rezaee, A., Moheb-Ali, G. A., & Jaafarzadeh, N. A. (2013). Pyrene removal from contaminated soils by modified Fenton oxidation using iron nano particles. *Journal of Environmental Health Science and Engineering, 11*, 1–8.

Ju-Nam, Y., & Lead, J. R. (2008). Manufactured nanoparticles: An overview of their chemistry, interactions and potential environmental implications. *Science of the Total Environment, 400*(1–3), 396–414.

Kent, R. D., Oser, J. G., & Vikesland, P. J. (2014). Controlled evaluation of silver nanoparticle sulfidation in a full-scale wastewater treatment plant. *Environmental Science & Technology, 48*(15), 8564–8572.

Khan, S. H. (2019). Green nanotechnology for the environment and sustainable development. In *Green materials for wastewater treatment* (pp. 13-46). Springer International Publishing.

Khan, N. A., Khan, S. U., Ahmed, S., Farooqi, I. H., Dhingra, A., Hussain, A., & Changani, F. (2019). Applications of nanotechnology in water and wastewater treatment: A review. *Asian Journal of Water, Environment and Pollution, 16*(4), 81–86.

Khan, Y., Sadia, H., Ali Shah, S. Z., Khan, M. N., Shah, A. A., Ullah, N., et al. (2022). Classification, synthetic, and characterization approaches to nanoparticles, and their applications in various fields of nanotechnology: A review. *Catalysts, 12*(11), 1386.

Khatun, R., Singh, S., Dubey, N. K., & Prasad Das, A. (2023). A review on marine-based phytochemicals and their application in biomedical research. In *Recent frontiers of phytochemicals: Applications in food, pharmacy, cosmetics, and biotechnology* (pp. 383–395). Elsevier. https://doi.org/10.1016/B978-0-443-19143-5.00013-X

Khodakovskaya, M. V., De Silva, K., Biris, A. S., Dervishi, E., & Villagarcia, H. (2012). Carbon nanotubes induce growth enhancement of tobacco cells. *ACS Nano, 6*(3), 2128–2135.

Kidd, J., Westerhoff, P., & Maynard, A. D. (2020). Public perceptions for the use of nanomaterials for in-home drinking water purification devices. *NanoImpact, 18*, 100220.

Kumar, M., Pandit, S., Patel, V., Khanna, N., Nag, M., Lahiri, D., Ray, R. R., Das, A., & Das, D. (2023). Maximization of energy recovery from starch processing wastewater by thermophilic dark fermentation coupled with microbial fuel cell technology. *Geomicrobiology Journal*, 1–12. https://doi.org/10.1080/01490451.2023.2209555

Kumar, M., Grzelakowski, M., Zilles, J., Clark, M., & Meier, W. (2007). Highly permeable polymeric membranes based on the incorporation of the functional water channel protein Aquaporin Z. *Proceedings of the National Academy of Sciences, 104*(52), 20719–20724.

Kumpiene, J., Lagerkvist, A., & Maurice, C. (2008). Stabilization of As, Cr, Cu, Pb and Zn in soil using amendments—A review. *Waste Management, 28*(1), 215–225.

Kurniawan, T. A., Sillanpää, M. E., & Sillanpää, M. (2012). Nanoadsorbents for remediation of aquatic environment: Local and practical solutions for global water pollution problems. *Critical Reviews in Environmental Science and Technology, 42*(12), 1233–1295.

Le, T. T., Nguyen, K. H., Jeon, J. R., Francis, A. J., & Chang, Y. S. (2015). Nano/bio treatment of polychlorinated biphenyls with evaluation of comparative toxicity. *Journal of Hazardous Materials, 287*, 335–341.

Li, L., Bi, Z., Hu, Y., Sun, L., Song, Y., Chen, S., et al. (2021). Silver nanoparticles and silver ions cause inflammatory response through induction of cell necrosis and the release of mitochondria in vivo and in vitro. *Cell Biology and Toxicology, 37*(2), 177–191.

Li, Y. H., Ding, J., Luan, Z., Di, Z., Zhu, Y., Xu, C., et al. (2003). Competitive adsorption of $Pb^{2+}$, $Cu^{2+}$ and $Cd^{2+}$ ions from aqueous solutions by multiwalled carbon nanotubes. *Carbon, 41*(14), 2787–2792.

Liang, L., Lan, F., Ge, S., Yu, J., Ren, N., & Yan, M. (2017). Metal-enhanced ratiometric fluorescence/naked eye bimodal biosensor for lead ions analysis with bifunctional nanocomposite probes. *Analytical Chemistry, 89*(6), 3597–3605.

Lind, M. L., Jeong, B. H., Subramani, A., Huang, X., & Hoek, E. M. (2009). Effect of mobile cation on zeolite-polyamide thin film nanocomposite membranes. *Journal of Materials Research, 24*(5), 1624–1631.

Liu, X., Zhang, H., Ma, Y., Wu, X., Meng, L., Guo, Y., et al. (2013). Graphene-coated silica as a highly efficient sorbent for residual organophosphorus pesticides in water. *Journal of Materials Chemistry A, 1*(5), 1875–1884.

Liu, Y., Nam Ong, C., & Xie, J. (2016). Emerging nanotechnology for environmental applications. *Nanotechnology Reviews, 5*(1), 1–2.

Lu, F., Wang, F., Cao, L., Kong, C. Y., & Huang, X. (2012). Hexagonal boron nitride nanomaterials: Advances towards bioapplications. *Nanoscience and Nanotechnology Letters, 4*(10), 949–961.

Mair, A., & El-Kadi, A. I. (2013). Logistic regression modeling to assess groundwater vulnerability to contamination in Hawaii, USA. *Journal of Contaminant Hydrology, 153*, 1–23.

Malhotra, S. P. K., & Mandal, T. (2019). Zinc oxide nanostructure and its application as agricultural and industrial material. *Contaminants in Agriculture and Environment: Health Risks and Remediation, 1*, 216.

Malik, S., Muhammad, K., & Waheed, Y. (2023). Nanotechnology: A revolution in modern industry. *Molecules, 28*(2), 661.

Masciangioli, T., & Zhang, W. X. (2003). Peer reviewed: Environmental technologies at the nanoscale.

Matteucci, F., Giannantonio, R., Calabi, F., Agostiano, A., Gigli, G., & Rossi, M. (2018). Deployment and exploitation of nanotechnology nanomaterials and nanomedicine. In *AIP conference proceedings* (Vol. 1990, No. 1). AIP Publishing.

Mauter, M. S., Wang, Y., Okemgbo, K. C., Osuji, C. O., Giannelis, E. P., & Elimelech, M. (2011). Antifouling ultrafiltration membranes via post-fabrication grafting of biocidal nanomaterials. *ACS Applied Materials & Interfaces, 3*(8), 2861–2868.

Mishra, S., Rath, C. Charan, & Das, A. P. (2019). Marine microfiber pollution: A review on present status and future challenges. *Marine Pollution Bulletin, 140*, 188–197. https://doi.org/10.1016/J.MARPOLBUL.2019.01.039

Mishra, S., Rout, P. K., & Das, A. P. (2021). Emerging microfiber pollution and its remediation. In R. Prasad (Ed.), *Environmental and microbial biotechnology* (pp. 247–266). Springer. https://doi.org/10.1007/978-981-15-5499-5_9

Mishra, S., & Das, A. P. (2021). Current treatment technologies for removal of microplastic and microfiber pollutants from wastewater. *Wastewater Treatment: Cutting-Edge Molecular Tools, Techniques and Applied Aspects*, 237–251. https://doi.org/10.1016/B978-0-12-821881-5.00011-8

Mishra, S., Singh, R. P., Rout, P. K., & Das, A. P. (2022a). Membrane bioreactor (MBR) as an advanced wastewater treatment technology for removal of synthetic microplastics. *Development in Wastewater Treatment Research and Processes*, 45–60. https://doi.org/10.1016/B978-0-323-85583-9.00022-3

Mishra, S., & Das, A. P. (2022). Treatment of the wastewater polluted with synthetic microfiber released from washing machine. In B. B. Das, H. Hettiarachchi, P. K. Sahu, & S. Nanda (Eds.), *Lecture notes in civil engineering* (Vol. 207, pp. 109–117). Springer Science and Business Media Deutschland GmbH. https://doi.org/10.1007/978-981-16-7509-6_9/COVER

Mishra, S., & Das, A. P. (2023). Microbial remediation of synthetic microfiber contaminated wastewater. In M. P. Shah (Ed.), *Microbial technologies in industrial wastewater treatment* (pp. 337–350). Springer. https://doi.org/10.1007/978-981-99-2435-6_17

Mishra, S., Dash, D., & Das, A. P. (2022b). Detection, characterization and possible biofragmentation of synthetic microfibers released from domestic laundering wastewater as an emerging source of marine pollution. *Marine Pollution Bulletin, 185*, 114254. https://doi.org/10.1016/J.MARPOLBUL.2022.114254

Mishra, S., Dash, D., & Das, A. P. (2024). Isolation and molecular characterization of polyester-degrading bacterial strains from municipal wastewater and their application in microplastic degradation in contaminated wastewater. *Geomicrobiology Journal*. https://doi.org/10.1080/01490451.2024.2376101

Mishra, S., Ghosh, S., van Hullebusch, E. D., Singh, S., & Das, A. P. (2023). A critical review on the recovery of base and critical elements from electronic waste-contaminated streams using microbial biotechnology. *Applied Biochemistry and Biotechnology, 2023*, 1–30. https://doi.org/10.1007/S12010-023-04440-X

Mitrano, D. M., Motellier, S., Clavaguera, S., & Nowack, B. (2015). Review of nanomaterial aging and transformations through the life cycle of nano-enhanced products. *Environment International, 77*, 132–147.

Mittelman, A. M., Fortner, J. D., & Pennell, K. D. (2015). Effects of ultraviolet light on silver nanoparticle mobility and dissolution. *Environmental Science: Nano, 2*(6), 683–691.

Mohanty, J., Dey, S., Bala, K., & Das, A. P. (2024). Recent advances in microplastic pollution for its sustainable management. *Sustainable Management of Urban Plastic Waste Through Circular Economic Approaches*, 27–48. https://doi.org/10.1201/9781003465447-3

Mondal, S. (2020). Potential of nanotechnology for rural applications. *Arabian Journal for Science and Engineering, 45*(7), 5011–5042.

Nadagouda, M., Speth, T. F., Impellitteri, C., & Zhao, Y. (2011). Nanomaterials synthesis, applications, and toxicity. *Journal of Nanotechnology, 2011*(1), 218562.

Nair, P. M. G., & Chung, I. M. (2014). A mechanistic study on the toxic effect of copper oxide nanoparticles in soybean (*Glycine max* L.) root development and lignification of root cells. *Biological Trace Element Research, 162*, 342–352.

Niewiadomski, P., Stefaniuk, D., & Hoła, J. (2017). Microstructural analysis of self-compacting concrete modified with the addition of nanoparticles. *Procedia Engineering, 172*, 776–783.

Palit, S. (2017). Application of nanotechnology, nanofiltration, and drinking and wastewater treatment—A vision for the future. In *Water purification* (pp. 587–620). Academic Press.

Pan, B. O., & Xing, B. (2008). Adsorption mechanisms of organic chemicals on carbon nanotubes. *Environmental Science & Technology, 42*(24), 9005–9013.

Park, T. J., Lee, K. G., & Lee, S. Y. (2016). Advances in microbial biosynthesis of metal nanoparticles. *Applied Microbiology and Biotechnology, 100*, 521–534.

Pati, P., McGinnis, S., & Vikesland, P. J. (2014). Life cycle assessment of "green" nanoparticle synthesis methods. *Environmental Engineering Science, 31*(7), 410–420.

Patolsky, F., & Lieber, C. M. (2005). Nanowire nanosensors. *Materials Today, 8*(4), 20–28.

Paul, A., Priyadarsini, S., Karakuş, S., Behera, J. K., Gnanasekaran, R., & Das, A. P. (2024). Fate and accumulation of synthetic microfibers in the environment. *Sustainable Microbial Technology for Synthetic and Cellulosic Microfiber Bioremediation*, 61–76. https://doi.org/10.1007/978-3-031-62600-5_4

Paul, A., Choudhary, B. L., Majumdar, S., & Rani, K. (2021). Nanotechnology application in soil and crop management. *Crop Management under Stress Conditions, 167*, 223.

Pendergast, M. M., & Hoek, E. M. (2011). A review of water treatment membrane nanotechnologies. *Energy & Environmental Science, 4*(6), 1946–1971.

Peng, Z., Liu, X., Zhang, W., Zeng, Z., Liu, Z., Zhang, C., et al. (2020). Advances in the application, toxicity and degradation of carbon nanomaterials in environment: A review. *Environment International, 134*, 105298.

Plata, D. L. (2009). *Carbon nanotube synthesis and detection: Limiting the environmental impact of novel technologies*. Massachusetts Institute of Technology.

Pourzahedi, L., Vance, M., & Eckelman, M. J. (2017). Life cycle assessment and release studies for 15 nanosilver-enabled consumer products: Investigating hotspots and patterns of contribution. *Environmental Science & Technology, 51*(12), 7148–7158.

Prabina, K. E., Benya, A., Mohanty, S., Van Hullebusch, E. D., Das, A. P., & Singh, S. (2024). Recovery of multi-functional carbon fiber from COVID-19 generated waste for an effective management through circular economy. *Environmental Science and Engineering, Part, F2827*, 83–97. https://doi.org/10.1007/978-3-031-50840-0_6

Prediger, P., Vieira, M. G. A., Camparotto, N. G., de Figueiredo Neves, T., da Silva, P. M. M., & de Vargas Brião, G. (2022). Adsorption and desorption aspects of carbon-based nanomaterials: Recent applications for water treatments and toxic effects. *Liquid and Crystal Nanomaterials for Water Pollutants Remediation*, 59–87.

Priyadarsini, S., & Das, A. P. (2024). Analytical and structural characterization of waste lithium-ion batteries for their effective recycling strategy. *Environmental Chemistry and Ecotoxicology*. https://doi.org/10.1016/j.enceco.2024.12.004

Quadros, M. E., Pierson, R., IV., Tulve, N. S., Willis, R., Rogers, K., Thomas, T. A., & Marr, L. C. (2013). Release of silver from nanotechnology-based consumer products for children. *Environmental Science & Technology, 47*(15), 8894–8901.

Rajput, V. D., Kumari, A., Minkina, T., Barakhov, A., Singh, S., Mandzhieva, S. S., et al. (2023). A practical evaluation on integrated role of biochar and nanomaterials in soil remediation processes. *Environmental Geochemistry and Health, 45*(12), 9435–9449.

Rajput, V. D., Minkina, T., Upadhyay, S. K., Kumari, A., Ranjan, A., Mandzhieva, S., et al. (2022). Nanotechnology in the restoration of polluted soil. *Nanomaterials, 12*(5), 769.

Raninga, M., Mudgal, A., Patel, V. K., Patel, J., & Sinha, M. K. (2023). Modification of activated carbon-based adsorbent for removal of industrial dyes and heavy metals: A review. *Materials Today: Proceedings, 77*, 286–294.

Ravikumar, A., & Prakash, K. S. (2022). Green and sustainable approaches of nanoparticles. In *Handbook of consumer nanoproducts* (pp. 1433–1453). Springer Nature Singapore.

Reyhanitabar, A., Alidokht, L., Khataee, A. R., & Oustan, S. (2012). Application of stabilized FeO nanoparticles for remediation of Cr (VI)-spiked soil. *European Journal of Soil Science, 63*(5), 724–732.

Robinson, B. H., Bañuelos, G., Conesa, H. M., Evangelou, M. W., & Schulin, R. (2009). The phytomanagement of trace elements in soil. *Critical Reviews in Plant Sciences, 28*(4), 240–266.

Rose, J., Auffan, M., de Garidel-Thoron, C., Artous, S., Auplat, C., Brochard, G., et al. (2021). The SERENADE project; a step forward in the safe by design process of nanomaterials: The benefits of a diverse and interdisciplinary approach. *Nano Today, 37*, 101065.

Sabri, N. S. M., Hasbullah, H., Said, N., Ibrahim, N., Kasmani, R. M., Ali, R. R., & Rahman, S. A. (2017). Hydrophilicity enhancement of metal oxide nanoparticles incorporated polysulfone ultrafiltration membrane. *Journal of Applied Membrane Science & Technology, 20*(1).

Sadasivuni, K. K., Rattan, S., Waseem, S., Brahme, S. K., Kondawar, S. B., Ghosh, S., Das, A. P., Chakraborty, P. K., Adhikari, J., Saha, P., & Mazumdar, P. (2019). Silver nanoparticles and its polymer nanocomposites—Synthesis, optimization, biomedical usage, and its various applications. In *Lecture notes in bioengineering* (pp. 331–373). https://doi.org/10.1007/978-3-030-04741-2_11

Sadeghi-Kiakhani, M., Arami, M., & Gharanjig, K. (2013). Application of a biopolymer chitosan-poly (propylene) imine dendrimer hybrid as an antimicrobial agent on the wool fabrics. *Iranian Polymer Journal, 22*, 931–940.

Saleh, T. A., Shuaib, T. D., Danmaliki, G. I., & Al-Daous, M. A. (2016). Carbon-based nanomaterials for desulfurization: classification, preparation, and evaluation. In *Applying nanotechnology to the desulfurization process in petroleum engineering* (pp. 154–179). IGI Global Scientific Publishing.

Salerno, M., Landoni, P., & Verganti, R. (2008). Funded research projects as a tool for the analysis of emerging fields: The case of nanotechnology. In *SPRU 40th anniversary conference–the future of science, technology and innovation policy* (Vol. 6).

Santos, A., Firak, D. S., Emmel, A., Siedlecki, K., Lopes, A., & Peralta-Zamora, P. (2018). Evaluation of the Fenton process effectiveness in the remediation of soils contaminated by gasoline: Effect of soil physicochemical properties. *Chemosphere, 207*, 154–161.

Satapathy, S., Kar, D., Priyadarsini, S., Das, A. P., & Kuanar, A. (2024). Application of microorganisms in emerging waste management strategies and remediation of synthetic and cellulosic microfibers. *Sustainable Microbial Technology for Synthetic and Cellulosic Microfiber Bioremediation*, 183–206. https://doi.org/10.1007/978-3-031-62600-5_10

Shahriari, S., Sastry, M., Panjikar, S., & Singh Raman, R. K. (2021). Graphene and graphene oxide as a support for biomolecules in the development of biosensors. *Nanotechnology, Science and Applications*, 197–220.

Shukla, G., Dixit, R., Kumar, A., Singh, R., Rani, A., & Kumar, P. (2017). Nanotechnology: An innovative approach for waste water treatment. *Applied Nanotechnology: An Introduction.*

Siddiqua, A., Hahladakis, J. N., & Al-Attiya, W. A. K. (2022). An overview of the environmental pollution and health effects associated with waste landfilling and open dumping. *Environmental Science and Pollution Research, 29*(39), 58514–58536.

Siddique, T. A., Dutta, N. K., & Roy Choudhury, N. (2020). Nanofiltration for arsenic removal: Challenges, recent developments, and perspectives. *Nanomaterials, 10*(7), 1323.

Siddiqui, M. H., Al-Whaibi, M. H., & Mohammad, F. (2015). *Nanotechnology and plant sciences* (303 pp). Springer International Publishing. https://doi.org/10.1007/978-3-319-14502-0

Singh, P., Singh, V. K., Singh, R., Borthakur, A., Madhav, S., Ahamad, A., et al. (2020). Bioremediation: a sustainable approach for management of environmental contaminants. In *Abatement of environmental pollutants* (pp. 1–23). Elsevier.

Singh, J., & Lee, B. K. (2016). Influence of nano-$TiO_2$ particles on the bioaccumulation of Cd in soybean plants (*Glycine max*): A possible mechanism for the removal of Cd from the contaminated soil. *Journal of Environmental Management, 170*, 88–96.

Singh, R., Manickam, N., Mudiam, M. K. R., Murthy, R. C., & Misra, V. (2013). An integrated (nano-bio) technique for degradation of γ-HCH contaminated soil. *Journal of Hazardous Materials, 258*, 35–41.

Singha, I., & Mishrab, P. K. (2020). Nano-membrane filtration a novel application of nanotechnology for waste water treatment. *Materials Today: Proceedings, 29*, 327–332.

Sinha, N., & Dahiya, P. (2022). Removal of emerging contaminants from pharmaceutical wastewater through application of bionanotechnology. In *Development in wastewater treatment research and processes* (pp. 247–264).

Skinder, B. M., & Hamid, S. (2020). Nanotechnology: a modern technique for pollution abatement. In *Bioremediation and biotechnology: Techniques for noxious substances remediation* (Vol. 4, pp. 295–311).

Sokolova, T., Krishna, A., & Döring, T. (2023). Paper meets plastic: The perceived environmental friendliness of product packaging. *Journal of Consumer Research, 50*(3), 468–491.

Solanki, P., Bhargava, A., Chhipa, H., Jain, N., & Panwar, J. (2015). *Nano-fertilizers and their smart delivery system. Nanotechnologies in food and agriculture* (pp. 81–101).

Somanathan, A., Mathew, N., & Arfin, T. (2024). Environmental impacts and developments in waste-derived nanoparticles for air pollution control. *Waste-Derived Nanoparticles*, 281–318.

Stone, D., Harper, B. J., Lynch, I., Dawson, K., & Harper, S. L. (2010). Exposure assessment: Recommendations for nanotechnology-based pesticides. *International Journal of Occupational and Environmental Health, 16*(4), 467–474.

Su, X., Sutarlie, L., & Loh, X. J. (2020). Sensors, biosensors, and analytical technologies for aquaculture water quality. *Research.*

Sukhanova, A., Bozrova, S., Sokolov, P., Berestovoy, M., Karaulov, A., & Nabiev, I. (2018). Dependence of nanoparticle toxicity on their physical and chemical properties. *Nanoscale Research Letters, 13*, 1–21.

Tahir, M. B., Asiri, A. M., Malik, M. F., Khalid, N. R., Iqbal, T., Rafique, M., et al. (2021). Role of nano-photocatalysts in detoxification of toxic heavy metals. *Current Analytical Chemistry, 17*(2), 126–137.

Tang, Y., Tian, J., Li, S., Xue, C., Xue, Z., Yin, D., & Yu, S. (2015). Combined effects of graphene oxide and Cd on the photosynthetic capacity and survival of Microcystis aeruginosa. *Science of the Total Environment, 532*, 154–161.

Taran, M., Safaei, M., Karimi, N., & Almasi, A. (2021). Benefits and application of nanotechnology in environmental science: An overview. *Biointerface Research in Applied Chemistry, 11*(1), 7860–7870.

Tratnyek, P. G., & Johnson, R. L. (2006). Nanotechnologies for environmental cleanup. *Nano Today, 1*(2), 44–48.

Truffier-Boutry, D., Fiorentino, B., Bartolomei, V., Soulas, R., Sicardy, O., Benayad, A., et al. (2017). Characterization of photocatalytic paints: A relationship between the photocatalytic properties–release of nanoparticles and volatile organic compounds. *Environmental Science: Nano, 4*(10), 1998–2009.

Trump, B. D., Antunes, D., Palma-Oliveira, J., Nelson, A., Hudecova, A. M., Rundén-Pran, E., et al. (2024). Safety-by-design and engineered nanomaterials: The need to move from theory to practice. *Environment Systems and Decisions, 44*(1), 177–188.

Usman, M., Farooq, M., Wakeel, A., Nawaz, A., Cheema, S. A., ur Rehman, H., et al. (2020). Nanotechnology in agriculture: Current status, challenges and future opportunities. *Science of the total environment, 721*, 137778.

Usman, M., Faure, P., Ruby, C., & Hanna, K. (2012a). Application of magnetite-activated persulfate oxidation for the degradation of PAHs in contaminated soils. *Chemosphere, 87*(3), 234–240.

Usman, M., Faure, P., Ruby, C., & Hanna, K. (2012b). Remediation of PAH-contaminated soils by magnetite catalyzed Fenton-like oxidation. *Applied Catalysis b: Environmental, 117*, 10–17.

Van Eerd, L. L., Turnbull, J. J. D., Bakker, C. J., Vyn, R. J., McKeown, A. W., & Westerveld, S. M. (2017). Comparing soluble to controlled-release nitrogen fertilizers: Storage cabbage yield, profit margins, and N use efficiency. *Canadian Journal of Plant Science, 98*(4), 815–829.

Vikesland, P. J. (2018). Nanosensors for water quality monitoring. *Nature Nanotechnology, 13*(8), 651–660.

Vikesland, P. J., & Wigginton, K. R. (2010). Nanomaterial enabled biosensors for pathogen monitoring-a review. *Environmental Science & Technology, 44*(10), 3656–3669.

Wagner, S., & Reemtsma, T. (2019). Things we know and don't know about nanoplastic in the environment. *Nature Nanotechnology, 14*(4), 300–301.

Wang, Y., Dimkpa, C., Deng, C., Elmer, W. H., Gardea-Torresdey, J., & White, J. C. (2022). Impact of engineered nanomaterials on rice (*Oryza sativa* L.): A critical review of current knowledge. *Environmental Pollution, 297*, 118738.

Wang, H. Y., & Lua, A. C. (2012). Development of metallic nickel nanoparticle catalyst for the decomposition of methane into hydrogen and carbon nanofibers. *The Journal of Physical Chemistry C, 116*(51), 26765–26775.

Wang, J., Fang, Z., Cheng, W., Yan, X., Tsang, P. E., & Zhao, D. (2016). Higher concentrations of nanoscale zero-valent iron (nZVI) in soil induced rice chlorosis due to inhibited active iron transportation. *Environmental Pollution, 210*, 338–345.

Wang, K., Zhao, P., Guo, X., Li, Y., Han, D., & Chao, Y. (2014). Enhancement of reactivity in $Li_4SiO_4$-based sorbents from the nano-sized rice husk ash for high-temperature $CO_2$ capture. *Energy Conversion and Management, 81*, 447–454.

Wang, T., Liu, Y., Wang, J., Wang, X., Liu, B., & Wang, Y. (2019). In-situ remediation of hexavalent chromium contaminated groundwater and saturated soil using stabilized iron sulfide nanoparticles. *Journal of Environmental Management, 231*, 679–686.

Wang, Y., Sun, H., Duan, X., Ang, H. M., Tadé, M. O., & Wang, S. (2015). A new magnetic nano zero-valent iron encapsulated in carbon spheres for oxidative degradation of phenol. *Applied Catalysis b: Environmental, 172*, 73–81.

Wang, Y., Zhu, Y., & Wu, S. (2013). A new nano CaO-based $CO_2$ adsorbent prepared using an adsorption phase technique. *Chemical Engineering Journal, 218*, 39–45.

Wohlleben, W., & Neubauer, N. (2016). Quantitative rates of release from weathered nanocomposites are determined across 5 orders of magnitude by the matrix, modulated by the embedded nanomaterial. *NanoImpact, 1*, 39–45.

Wu, Q., Miao, W. S., Zhang, Y. D., Gao, H. J., & Hui, D. (2020). Mechanical properties of nanomaterials: A review. *Nanotechnology Reviews, 9*(1), 259–273.

Yavuz, C. T., Mayo, J. T., Suchecki, C., Wang, J., Ellsworth, A. Z., D'Couto, H., et al. (2010). Pollution magnet: Nano-magnetite for arsenic removal from drinking water. *Environmental Geochemistry and Health, 32*, 327–334.

Yean, S., Cong, L., Yavuz, C. T., Mayo, J. T., Yu, W. W., Kan, A. T., et al. (2005). Effect of magnetite particle size on adsorption and desorption of arsenite and arsenate. *Journal of Materials Research, 20*(12), 3255–3264.

Zhang, C., Wang, W., Duan, A., Zeng, G., Huang, D., Lai, C., et al. (2019). Adsorption behavior of engineered carbons and carbon nanomaterials for metal endocrine disruptors: Experiments and theoretical calculation. *Chemosphere, 222*, 184–194.

Zhang, L., Liu, T., Xie, Y., Zeng, Z., & Chen, J. (2020). A new classification method of nanotechnology for design integration in biomaterials. *Nanotechnology Reviews, 9*(1), 820–832.

Zhao, J., Lin, M., Wang, Z., Cao, X., & Xing, B. (2021). Engineered nanomaterials in the environment: Are they safe? *Critical Reviews in Environmental Science and Technology, 51*(14), 1443–1478.

Zhao, L., Lu, L., Wang, A., Zhang, H., Huang, M., Wu, H., et al. (2020). Nano-biotechnology in agriculture: Use of nanomaterials to promote plant growth and stress tolerance. *Journal of Agricultural and Food Chemistry, 68*(7), 1935–1947.

Zhuang, J., & Gentry, R. W. (2011). Environmental application and risks of nanotechnology: a balanced view. In *Biotechnology and nanotechnology risk assessment: minding and managing the potential threats around us* (pp. 41–67). American Chemical Society.

Zuo, K., Huang, X., Liu, X., Gil Garcia, E. M., Kim, J., Jain, A., et al. (2020). A hybrid metal–organic framework–reduced graphene oxide nanomaterial for selective removal of chromate from water in an electrochemical process. *Environmental Science & Technology, 54*(20), 13322–13332.

# Application of Nanotechnology for Sustainable Agricultural Waste Management and Energy Recycling

**Soumya Hota, Sudeshna Dey, and Alok Prasad Das**

## 1 Introduction

The global population has increased over the years, from 4 billion in the year 1970 to 8 billion by 2021, with projections suggesting it will rise to 10 billion by 2050 and 12 billion by 2100 (do Espirito Santo Pereira et al., 2021). In other words, due to the degenerating conditions of human civilization, there have been, and still exist, significantly large numbers of both livestock and croplands, which have also exacerbated the issue of agricultural waste (AW) generation. Besides, in addition to the significant economic and demographic growth experienced over the last century, India, China, and Africa have also expanded their capacity to produce AW. One of the burdens faced in the twenty-first century is the proper management of agricultural waste (Ghosh and Das 2017). Waste in agriculture management in this context deals with the wastes that are generated due to agricultural practices and their management, which includes aspects like processes dealing with treatment and disposal (Ghosh and Das 2020a, 2020b). Waste of this kind can comprise different materials such as manure, leftover crops, and various other materials utilized in the process of farming. The Ministry of New and Renewable Energy (MNRE) is a ministerial department within the Government of India that has estimated that India generates more than 500 million tons (Mt) of crop waste every year, of which some portions are used for animal feed and other uses, while the majority is still being burnt in agricultural plots around the country (Das et al., 2023). The practice of burning crop residues was regarded as one of the contributors that can adversely affect health, while waste management summarizes the consequences of actions or inactions (Mallick et al., 2023). If waste is not managed properly, there is a risk of pollution of surface and groundwater, air, and land, as well as climate change, and depletion of beneficial organic matter in the

S. Hota · S. Dey · A. P. Das (✉)
Department of Life Sciences, Rama Devi Women's University, Bhubaneswar, Odisha, India
e-mail: alok1503@gmail.com

 
A. P. Das and S. Karakuş (eds.), *Synergizing Sustainability for Integrated Waste Management*, Sustainable Environmental Waste Management Strategies,
https://doi.org/10.1007/978-3-032-03718-3_15

soil. Moreover, it can cause great harm to the well-being of individuals. Agricultural residues, when burned, tend to release greenhouse gases: carbon dioxide ($CO_2$), nitrous oxide ($N_2O$), and methane among others.

The major categories of agricultural wastes (AWs) that have elicited public debate and jeopardized agricultural systems' sustainability include crop residues, livestock wastes, poultry waste, slaughterhouse waste, including excreta, hair, bones, and meat, aquaculture waste, and many others. According to the classification of catering wastes, all the generated agricultural waste can be divided into two major components: organic waste and inorganic waste. Organic waste consists of agricultural industrial waste which comprises crop residues such as straw and rice husks, animal dung, and processing wastes such as peel, seeds, and pulp. Inorganic waste, including plastics, pesticides, and fertilizers, proves difficult to find a good method of disposal (Mishra, et al., 2019). The agriculture sector is where most of the agricultural solid waste is produced. For example, in animal production systems, animal rearing contributes to solid waste through livestock production (Adejumo et al., 2021). This waste also includes bedding/litter, animal carcasses, impaired feed and water troughs, and equipment such as unexpected damage caused by animals (Mishra and Das 2021, 2023). Horticultural production solid wastes refer to the solid wastes produced during the growing and maintenance of ornamental horticultural plants and ornamental landscapes (Peng et al., 2023). Such wastes include grass cuttings and trimmings. There are solid wastes of agricultural origin derived from wood processing and cuttings, as well as from the cultivation of other crops to make paper, where agricultural products are used as raw materials (Mishra et al., 2022a, 2022b). Despite generating chemical wastes from pesticides, insecticides, and herbicides, agricultural practices still heavily rely on these products. It has been found that approximately 2% of pesticides remain in containers after usage, which some ill-informed and ill-mannered users may dispose of in lakes or even on open ground, leading to contamination of food and pollution of the environment and water sources, claiming many lives. Cooking gardening organic rubbish solid wastes are created for these purposes (Mohanty et al., 2016, 2017). For example, in many homes, agricultural households consider orange peel and banana peel as agricultural solid waste and thus throw them away in the house (El-Ramady et al., 2020).

The improper management and disposal of waste, including organic and inorganic materials, can have serious consequences (Dey et al., 2023a, 2023b). Its impact on people's health, global warming, well-being of other creatures, and the surroundings is difficult to overstate. Agricultural activities tend to produce wastes that are rich in nutrients like nitrates, phosphates, and other minerals that are likely to be washed away into water bodies. This causes eutrophication (Yadav et al., 2023) resulting in a lack of sufficient oxygen in the water; hence, the aquatic life is affected, making the water unfit for use (Das et al., 2024). Moreover, excessive agricultural practices, where chemicals such as pesticides and fertilizers are used, may lead to the pollution of local rivers or lakes where such practices are carried out (Das et al., 2015). The careless segregation and dumping of solid agricultural wastes may also restrict the free flow of water in outlets, and this greatly contributes to incessant floods. The absence of long-standing agricultural by-product management results in the accumulation of agricultural wastes that provide breeding grounds for insects, larvae, and mosquitoes, disturbing organisms and contributing to environmental degradation and health risks (Mishra et al., 2021, 2023, 2024). Improper disposal of animal waste can bring about both direct and indirect effects on human health and environmental quality (Ghosh et al., 2018). For instance, most soils commonly found in animal production systems have high salt levels due to the application of condensed forms of animal excreta or high rates of manure application without management practices which detrimentally alters the organisms and plants occupying the soil. Moreover, diseases can be transmitted from animals to humans via feces (Paul et al., 2023, 2024). The increasing levels of agro-waste may cause pollution of soils by heavy metals and other toxicants. Such contaminants may interfere with the soil microbiome leading to reduced fertility which in turn affects the growth of crops. Additionally, the open burning of field residues adds to the problem of air pollution through the emission of noxious gases including $CO_2$ and methane which are greenhouse gases inducing global warming. The large-scale and frequent burning of agricultural waste has detrimental effects on people's health since the flames expose them to harmful pollution. Smog is a common consequence of the extensive burning of agricultural wastes across the world, especially in the early fall. Animal husbandry is said to be accountable for 37% of methane emissions globally and 65% of more harmful emissions like nitrous oxides compared to carbon dioxides (Prabina et al., 2024). It is wrong to assume that burning agricultural solid waste does not contribute to emissions that cause climate change. Burning activities cause air pollution and exposure to particulates that are known to be harmful to the respiratory organs of human beings. Gas concentrations that are generally dispersed in low concentrations in the environment like ammonia and hydrogen sulfide from biodegradable waste are toxic and cause headaches, shortness of breath, and fatigue even at very low concentrations (Priyadarshini and Das 2024). There is frequent exposure to organic dust from crop and/or livestock waste among farmers which is associated with chronic obstructive pulmonary disease (COPD). Endotoxins present in organic dust, which agricultural laborers are in contact with, are associated with pulmonary diseases. In women, it has been shown that the exposure to adverse environmental factors, particularly those in agricultural waste, caused disturbances such as prolonged menstrual periods and early menopause. Certain classes of pollutants such as dioxins which can be produced when agricultural waste is burned are known to cause adverse health effects including congenital malformations in babies and a heightened risk of certain cancers (Das and Mishra, 2008). Water pollution due to improper sanitary disposal of animal wastes can result in enteric diseases such as cholera and dysentery (Das and Mishra, 2009). The runoff due to agriculture also contains antibiotic-resistant bacteria which poses health threats to the population.

Agricultural waste management is important to ensure sustainable farming and protect the environment (Das et al., 2024). Conventional means have always been adopted for recycling and managing agricultural wastes effectively. Composting is among the earliest and most practical techniques for transforming biodegradable waste (Aswathi et al., 2023) into compost, which improves the quality and productivity of soils. Anaerobic digestion is a method of biological treatment in which organic substrates are degraded by microorganisms in a gas-tight chamber where biogas (mainly methane and carbon dioxide) is formed and can be put into use (Reed et al., 2012). Biochar is a carbon-rich material, heated or pyrolyzed under controlled industrial conditions of oxygen deprivation. When applied as an animal feed, it reduces methane emissions from livestock, enhances agricultural outcomes, and improves soil quality. The eradication of hunger and poverty largely depends on the growth of agriculture. Thus, to advance agricultural growth, we must make courageous steps. There is a need to embrace innovations that deal with agricultural productivity enhancement (Periakaruppan et al., 2023). Agricultural growth cannot just be recorded centrally. It has to be planned and carried out toward achieving specific objectives and goals, with attention to factors like social exclusion, health, climate change, energy, ecological processes, natural resources, and good governance among others. Certainly, advanced technologies such as nanotechnology are crucial for sustainable agricultural development (Fatima et al., 2021; Raj and Das, 2023). Nanotechnology encourages a greener economy by promoting sustainable development and the sustainability of industries (Ghosh et al., 2016).

There is a desperate need to prefer nanotechnology over standard waste management systems since they have been proven useful in meeting food security, environmental sustainability, and resource efficiency (Neme et al., 2021). Conventional methods are often responsible for excessive application of fertilizers and pesticides which may lead to soil degradation and water contamination. Whereas, in an advanced way, nanotechnology brings more complex and sophisticated methodologies in the mode of improved nutrient dispersion and enhanced crop management, as well as better resource management using precision farming techniques. In particular, agricultural waste can be more effectively harnessed with the aid of nanotechnology. Nanotechnology-based materials such as slow-release fertilizers can provide long-term nutrients to the plants when applied, ensuring that important nutrients are available as required, minimizing nutrient runoff, and enhancing soil quality. Farmers may increase crop yields while minimizing waste and environmental effects by applying agrochemicals with the help of nanoparticles (NPs). Additionally, nanotechnology facilitates advancements in pathogen detection and soil health management, providing tools that are faster, more sensitive, and cost-effective compared to conventional methods (Prasad et al., 2017). It can assist in the rehabilitation of polluted terrestrial composition by employing nanomaterials to eliminate contaminants from agricultural runoff or effluents which also enhances the water quality and lessens the environmental issues (Balusamy et al., 2023). Controlled bioenergy, eco-farm, smart farms, and even nanofuels greatly improve the effectiveness of the photosynthesis process and allow the generation of bioenergy from agricultural waste with a lesser amount of energy spent altogether. The age-old techniques employed in handling agricultural wastes have been very useful over the years; however, the introduction of nanotechnology in agriculture has provided new ways of improving efficiency, sustainability, and environmental conservation (Mgadi et al., 2024) (Fig. 1).

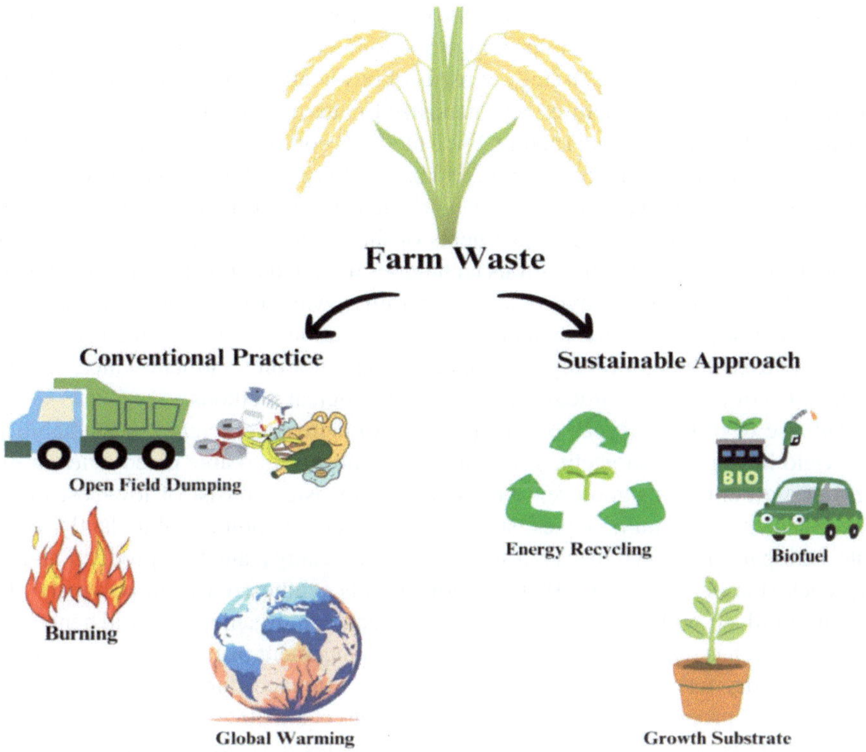


**Fig. 1** Difference between conventional practice and sustainable practice

## 2 Nanoparticles (NPs) and Their Types

Approximately 800 million individuals all around the world suffer from hunger, and the greatest reliance on conventional agricultural inputs has resulted in environmental degradation and depletion of soil health. Agriculture is also receiving a possible boost with nanotechnology by managing resources more efficiently, increasing crop productivity, and even reducing the pollution caused by seed coats, precision farming, and food processing among others. Nanotechnology is an interdisciplinary field that deals with the manipulation of matter at the level of the nanoscale (1–100 nm). At this scale, materials possess distinctive chemical, physical, and biological properties that are appreciably different from their bulk forms. The agents that participate in the process of integration of nanotechnology are the nanomaterials, which are quite small in shape and size and are therefore invisible to the naked eye. Due to the remarkable features and properties of nanotechnology, it has evolved to advance significantly in many aspects of human life, agribusiness being one of the fields that has witnessed massive development because of this technology. Nanomaterials can be fabricated using top-down (reducing larger structures to nanoscale) or bottom-up (assembling from atoms/molecules) approaches with techniques like vapor deposition, laser ablation, and nanoimprinting (Selva Preetha et al., 2017; Abd Elkodous et al., 2022). The top-down strategy consists of physically carving or milling already existing macroscopic structures into nanosized patterns. This procedure can be of physical or chemical (or both) nature. Such methods comprise nanolithography, electrospinning, sputtering, laser ablation, and ball milling. As for bottom-up strategies, already existing nanoparticles can be fabricated by constructing elements known as atoms or molecules, using adjustments of factors like pressure, temperature, or element addition. Among the various techniques included in this broad grouping are coprecipitation, pyrolysis, hydrothermal and microwave irradiation processes, and sol–gel methods. More often than not, the above treatments are performed before the nanosynthesis, which normally involves carrying out a pretreatment which may be in the form of physicochemical, chemical, or biological methods.

The use of different types of nanoparticles to increase efficiency, sustainability, and yield has made nanotechnology one of the most remarkable discoveries in the field of agriculture for good reasons. Since such wastes consist of low-cost materials, such as seeds, pulp, leaves, stems, bagasse, etc., which are abundantly available, nanoparticle biosynthesis is a sustainable, low-energy, and low-production-cost approach (Paul et al., 2023, 2024). Nanomaterials synthesized from polymers like pectin, cellulose, and silica or pure metals are organic, inorganic, or hybrid nanocomposites. This has many advantages including enhancement of food preservation due to their ability to contain bioactive agents with extended durability and versatile antimicrobial properties (Gupta and Xie, 2018; Das et al., 2024). Depending on their content and function in agriculture, these nanoparticles can also be classified in many ways. Metallic nanoparticles AgNPs, for instance, which possess antibacterial properties, help shield plants and aid in their enhancement. Copper, on the other hand, promotes growth and disease resistance in plants while acting as a fungicide, and this is used in the form of copper nanoparticles (CuNPs). In contrast, gold nanoparticles (AuNPs) are applied primarily in biosensing and as vehicles of nutrients or agrochemicals. Zinc oxide nanoparticles (ZnONPs) increase the capacity of plants to uptake nutrients and contribute to their well-being as well as support seed development. There are also nanoparticles and other natural polymers based on carbons like graphene (Mahmoud et al., 2021). Fertilizer use is improved by fertilizer based on carbon nanotubes, and vertical growth is improved by nutrient delivery (Singh et al., 2024). In particular, silica nanoparticles (SiNPs) improve water and nutrient retention in soils (Sarkar et al., 2021). In addition, polymeric nanoparticles, particularly those that have the capability of biodegradation, extend the controlled release of fertilizers to plants over time (Vejan et al., 2021). Hybrid nanomaterials are multifunctional fertilizers that enhance soil and nutrient supply by incorporating organic and inorganic materials. Finally, environmental stimulus-dependent fertilization is incorporated into nanofertilizers. A detailed description of such systems shows that they are integrated with other systems that are responsive to the demand of the plants or the environment so that the use of fertilizers is efficient. In addition to that, a decrease in nutrient waste and an increase in efficiency are attained by engineering delivery methods like nano-aptamers or nanocapsules that enhance nutrient uptake at the roots or leaves of plants. For instance, in one research, it was suggested that nanofertilizers could improve ammunition usage efficiency and hence crop yields with lesser overall nutrients thanks to a period of use elongation from three days to above thirty days. This approach promotes the reduction of nutrient leaching and loss, therefore supporting more eco-friendly farming practices (Das and Ghosh, 2018). Nanobiosensors, which are composite of nanoparticles, deploy monitoring of the soil and atmospheric parameters in a continuous mode and allow implementation of resource use efficiency maximization techniques—precision farming. The various nanoparticles incorporated in agriculture are also geared at increasing crop productivity and quality but at the same time encouraging environmentally friendly farming practices. With more research, the adoption of these new materials within agricultural practices will be more and more observed, bringing development in line with the goals of sustainable development (Das et al., 2024).

## 3 Synthesis of Nanoparticles from Agricultural Waste

Banana husk, coconut fibers, eggshell membrane, monkey-nut rind, onion peel, mango peel, sapota peel, pomegranate peel, rice husk, watermelon husk, human hair, orange peel, tamarind seed, algal extract, tea waste, marine waste, and slaughterhouse waste have been synthesized into nanoparticles (Das and Swain, 2013; Das et al., 2023; El-Ramady et al., 2020). Silver nanoparticles are synthesized using the plasma chemical extraction approach, followed by the degradation of methylene blue under sun irradiation, and orange peel extract is also used (Das and Swain, 2013). The waste residue from various nut parts such as shells, kernels, and extracts is rich in hemicellulose, lignin, and cellulose. Such material may be employed to prepare bionanomaterials as a reducing agent or nanocatalyst, which is quite advantageous because the material is natural, eco-friendly, economical, and reduces waste generation (Khan, 2020). Recently, eggshells have also been recognized as valuable agricultural waste. The primary constituent of eggshells, inorganic porous calcium carbonate, can lend itself to conversion to profit-making products. Employing the sol–gel approach, eggshells may be processed to yield nano-calcium oxide. To design magnetic $CuFe_2O_4$ nanoparticles for the dual purpose of water treatment as well as possessing catalytic and antimicrobial properties, the unique hierarchical porous structure of the eggshell membrane is employed. Human hair is another biowaste as it constitutes pigments, water, proteins, and lipids in a fibrous structure. Some cultural practices, like the burning of human hair, pose dangers to the ecosystem. Human hair keratin has been used to synthesize gold and silver nanoparticles. The relative antibacterial properties of silver and gold NPs were demonstrated against several strains of bacteria, including *Klebsiella pneumoniae*, *Staphylococcus aureus*, *Escherichia coli*, and *Pseudomonas aeruginosa*. The aqueous extract of marine algae *Sargassum myriocystum* helps in the synthesis of silver nanoparticles that prevent *Aedes aegypti* and *Culex quinquefasciatus* mosquito vectors, show anticancer activity against HeLa cells, inhibit clinical human pathogens, and also act as drug for cancer and bacterial infections. Silver nanoparticles have been fabricated using an aqueous extract of tea waste. The tea waste extract can also be utilized to prepare iron nanoparticles. Pomegranate peel extract, which is rich in phenolic compounds, gallic acid, some fatty acids, and anthocyanidins, has also been used for the preparation of silver nanoparticles. In addition, an inexpensive, nontoxic, and eco-friendly approach has been successfully used for the preparation of silver NPs from *Tamarindus indica* fruit shells. The silver NPs isolated from the fruit shell extract also possess medicinal applications for the treatment of breast cancer in humans. The banana peel extract, which acts as a reductant in the synthesis of silver nanoparticles, consists of complex organic polymers like lignin, cellulose, hemicellulose, and pectin. Zinc oxide NPs can also be synthesized from the slaughtering of goats and extraction of waste. Table 1 lists various types of nanomaterials with their respective applications.

Wastes from agricultural practices can be used as a potential precursor for the formation of nanoparticles. In agricultural field, weeds can be used to synthesize various types of metal and metal oxide nanoparticles. For instance, CuNPs may be extracted from *Lantana camara*, silver NPs from *Ipomoea carnea*, and copper oxide NPs from *Gloriosa superba* L. According to Skiba et al., AgNPs have been prepared from orange peels. Pomegranate peel extracts act as reducing and stabilizing agents in the biosynthesis of AuNPs, which are among the most stable nanoparticles. The process of converting organics, especially the red and green husks of watermelon, into nanogold is simple. According to Bankar et al., banana peel extracts that are biodegradable and eco-friendly have been utilized in the synthesis of green palladium nanoparticles. Sharon et al. found that copper nanomaterials synthesized from jackfruit (*Artocarpus heterophyllus*) killed all *Aedes aegypti* larvae. Phang et al. demonstrated that CuO NPs, which are synthesized from waste papaya peel extract, are potent photocatalysts because they can degrade palm oil mill effluent (POME) showing greater than average photocatalytic activity. Pioneers have synthesized zinc oxide nanoparticles from the waste peel of *Musa acuminata* and *Punica granatum* as antibacterial agents. Zinc oxide nanoparticles have been synthesized using the biowaste of dragon fruit peel (*Hylocereus polyrhizus*), which acts both as a reducing agent and stabilizer. Titanium dioxide nanoparticles are derived from agricultural wastes using bacterial cellulose (BC). Bacterial cellulose (BC), also known as biotype, is prepared using sugarcane molasses and rotten apple waste. The results of the work of Lei et al. have drawn a lot of attention since they revealed the synthesis of cellulose nanocrystals from waste paper for various purposes. As a significant proportion of office waste paper ends up in garbage each year, it not only gives rise to waste but also contributes to environmental pollution.

## 4 Effects of Agricultural Waste on the Environment

Almost every country depends on agriculture for its economic position relative to other countries. While this is evident in many places, primary agricultural production contributes directly to economic growth but tends to harm the environment

**Table 1** Nanomaterials with their sources, synthesis methods, and applications

| Nanomaterial | Agro-waste source | Synthesis method | Application |
|---|---|---|---|
| Zinc oxide nanoparticles | Wheat straw residue | Precipitation method | • Water purification<br>• Antimicrobial agent in agricultural fields<br>• Pest and disease control |
| Silver nanoparticles | Sugarcane bagasse/fiber | Biological synthesis using plant extract | • Antimicrobial applications<br>• Pathogen detection in crops<br>• Pest control |
| Biochar nanoparticles | Coconut shells, coconut husks | Pyrolysis followed by grinding | • Soil amendment<br>• Enhanced water retention<br>• Carbon sequestration<br>• Nutrient recycling |
| Titanium dioxide nanoparticles | Coffee ground waste | Sol–gel method | • Photocatalytic degradation of pesticides<br>• Pollutant decomposition |
| Iron oxide nanoparticles | Banana peel waste | Coprecipitation | • Microbial fuel cells for energy recovery<br>• Heavy metal removal in soil and water |
| Carbon nanotubes | Corn stalks and leaves | Chemical vapor deposition (CVD) | • Improved soil structure and enhanced microbial growth |
| Copper oxide nanoparticles | Orange peels or citrus peel waste | Hydrothermal synthesis | • Antifungal and antibacterial agents<br>• Pest and disease control<br>• Water purification |
| Graphene oxide nanoparticles | Soybean processing waste | Oxidation and exfoliation | • Biosensors for pathogen detection<br>• Soil amendment<br>• Pollutant adsorption |
| Tin oxide nanoparticles | Mango processing waste | Hydrothermal method | • Biosensors for pollutant detection<br>• Heavy metal removal<br>• Soil conditioning |
| Starch nanoparticles | Potato processing waste (peels) | Acid hydrolysis | • Biodegradable films for agricultural use<br>• Controlled release of fertilizers |

(continued)

**Table 1** (continued)

| Nanomaterial | Agro-waste source | Synthesis method | Application |
|---|---|---|---|
| Copper nanoparticles | Tea leaves or used tea waste | Green synthesis using plant extracts | • Pesticide and herbicide activity<br>• Antimicrobial coatings<br>• Plant growth enhancement |
| Calcium oxide nanoparticles | Poultry industry waste | Calcination | • Soil pH adjustment<br>• Heavy metal remediation<br>• Nutrient enhancement |
| Chitosan nanoparticles | Seafood waste | Ionic gelation | • Biofertilizer<br>• Biodegradable packaging<br>• Water purification |
| Vanadium oxide nanoparticles | Wine and juice industry waste | Sol–gel synthesis | • Catalysts in waste-to-energy systems<br>• Environmental remediation<br>• Enhanced plant nutrient uptake |
| Magnesium oxide nanoparticles | Pineapple processing waste | Sol–gel method | • Soil nutrient enhancement<br>• Water purification<br>• pH control |

through agricultural waste (Kolawole et al., 2024). However, with the rise in agricultural output, there has also been a noticeable rise in agricultural waste. More often than not, agriculture-related solid and organic waste is carelessly disposed of or left unattended which is detrimental to the environment. The wastes from agriculture include groundnut shells, palm kernel shells, sugarcane bagasse, cereal husks, urine, litter, animal feces, wasted feed, waste milk, beddings, animal carcasses, dairy parlor washings, paunch waste, abattoir wastewater, animal viscera, horns, bones, feathers, fur, blood, placenta, fetal membranes, birth tissue, aborted fetuses, waste wood, spoiled packaging products such as bags, boxes, and plastics, and worn out machinery and equipment components like tires and batteries (Das et al., 2023, 2024a, 2024b). All these wastes pose serious health and environmental hazards, particularly to farmworkers. Farm animals produce about 250 billion gallons of waste yearly from excreta and urine. This waste is stored in lagoons or "cesspools" which are open pits for holding excreta and urine. During rain, the capacity of the cesspool increases, which may wash out the fecal slurry onto nearby lands and water, contaminating the water and degrading the aesthetics of the surrounding area. Some hazardous gases emitted by cesspools include carbon dioxide, hydrogen sulfide, and ammonia. Even at low concentrations, hydrogen sulfide, the most dangerous substance, causes severe adverse health effects that include tonsillitis, runny nose, burning eyes, sore head, wheezing, shortness of breath, excessive coughing, diarrhea, nausea, fatigue, tension, depression, asthma and bronchitis in humans. Nitrates that leak from lagoons and spray fields into the ground increase the risk of blue baby syndrome, which can cause death in infants and are related to spontaneous abortions in females.

The Department for Environment, Food, and Rural Affairs, USA, 2012, states that exposure to fetuses that have been aborted, birthing fluids, and postpartum materials—which are parts of animal wastes—poses a danger to the health of the pregnant women and their fetuses that will finally result in zoonotic diseases, characterized by feverish and influenza-like symptoms. Other than posing human health hazards, animal waste is an environmental concern, which affects the healthiness and aesthetics of the environment. It was also observed that chicken production emits airborne contaminants such as dust, odor, and ammonia that are largely released due to manure denitrification, which can vent directly into the environment. The various health and environmental issues that agricultural wastes inflict on rural women include dioxins and air contamination due to burning of agricultural wastes, the existence of cesspools, dirty environments from heaps of agricultural wastes, poor visibility due to smoke, nitrate seepage, stench/odor from decaying wastes, environmental defacement, eutrophication of water bodies, destruction of aquatic life, spontaneous abortion, blue baby syndrome, prolonged menstruation, and early menopause. Agricultural wastes are very rich in poisonous by-products that include excreta, urine, blood, and animal by-products, like slurry, straw, and shells, among others, which are not ecologically friendly. Emitted gases include methane, carbon dioxide, nitrous oxide, ammonia, and hydrogen sulfide. Sludge also pollutes the landscape and makes it unstable. The health-related issues associated with environmental pollution include headaches, cholera, dysentery, diarrhea, and malaria, among many. It is already known that burning biomass is the source of dioxins, which causes health and environmental issues because they persist in the environment, remain for long periods before decomposing into other chemical forms, accumulate, and deposit in the fatty tissues of humans and animals. They can also travel many miles in the atmosphere. Thus, they are sometimes produced in one location, but they end up drifting into another far-off location. Exposure to dioxins has numerous negative health effects on humans, especially for women, which include neurological damage in the fetus due to exposure during pregnancy, reproductive problems such as premature menopause and delayed menstruation, changes in thyroid hormone levels, various types of cancer in humans, and chloracne, which is an acne-like eruption of blackheads, cysts, and pustules found on the cheeks, behind the ears, in the armpits, and in the groin.

## 5 Nanotechnology in Agricultural Waste Management

By processing, producing, and collecting grain goods, organic products, vegetables, and trees, as well as livestock cultivation, a variety of methods have been put forth to repurpose waste that ends up on the planet. Modern goods including natural acids, biofuels, protein-rich feed, fragrance compounds, and bioactive metabolites are made from these wastes. Enhancing the efficiency of photosynthetic process will result in notable increases in agricultural output. Since plant photosynthesis is essential to food security, increasing the efficiency of the photosynthetic process can result in significant increases in crop yield. C3 plants, which constitute 85% of existing plant species, are considered superior to others in photosynthetic efficiency under cool, wet conditions. Enhancing the process of photosynthesis may include improving the catalytic mechanism of an enzyme present along with other enzymes, particularly Rubisco, which adds $CO_2$ to ribulose bisphosphate and also takes part in photorespiration as an oxygen-binding enzyme. Nonetheless, this approach is considered ineffective because during deglaciation hunting Rubisco in C3 plants, in deciduous tropical forests as much as 30% of the fixed carbon committed annually is lost, the plant process most limiting in growth to the effluent of photosynthesis. Phytomass may be looked at again as a biophysical process, i.e., when sunlight impinges on the leaf, photosynthetic pigments chlorophyll-a and chlorophyll-b absorb photons along with their absorption spectra, and the energy flows into the biochemical pathways of photosynthesis. The efficiency of this process is very low such that the efficiency of solar energy conversion, i.e., the percentage of energy contained in the plant to the energy of light absorbed, is 2.4% and 3.7% for healthy C3 and C4 crops, respectively (Kreslavski et al., 2023). In the face of great challenges, there are ongoing efforts to overcome this restriction of photosynthesis. Novel technological advancements have at our disposal tools that facilitate modifications to the process of photosynthesis that were unachievable before. However, no doubt that nanotechnology belongs to these new approaches. In the early work, titanium oxide nanoparticles ($nTiO_2$) were considered to have a photocatalytic activity due to its high activity of the crystalline anatase phase; therefore, $nTiO_2$ was expected to enhance light absorption of the plant leaves, thus increasing photosynthetic activity. Specifically, $nTiO_2$ was shown to prevent chloroplasts from photochemical stress-induced senescence, to stimulate Rubisco carboxylation, which causes an increase in the photosynthesis rate, and to be beneficial to biophysical parameters of photosynthesis like electron transport chain and Chl-photophosphorylation (Sukirtha et al., 2020). Finally, $nTiO_2$ improves leaf water conductance and transpiration rate in addition to photosynthesis.

The initial plan of integrating nanomaterials with living plants to improve their natural functions has been accurately targeted. This is referred to as plant nanobionics. It is likely to allow the building of plants that grow quickly and be major players in the construction of artificial systems of photosynthesis, maybe a source of clean energy. The first reported application of nanobionics in plant systems was from a research group at MIT. They used perfusion to deliver a suspension of single-walled carbon nanotubes (SWCNTs) (Zulfiqar et al., 2019) to the leaves of *Arabidopsis thaliana* and isolated chloroplasts of *Spinacia oleracea.* In both cases, there were no observable signs of stress, and the SWCNTs could be detected within the thylakoids. The treatment extended the shelf life of isolated chloroplasts by about 2 h and enhanced the rate of electron transport relative to the control. As stated by the authors, to absorb solar energy at wavelengths to which chloroplasts only incidentally absorb, their high electrical conductance could enable semiconductor SWCNTs to absorb. The authors discovered that under their experimental setup, SWCNT-chloroplast assemblies increased the rate of electron transport and encouraged photosynthetic activity higher than the control. Macronutrients (Mg, K, N, Ca, P, S) and micronutrients (B, Cl, Zn, Fe, Mo, Cu, Mn) greatly depend on crop yield in agricultural lands. Nutrient use efficiency (NUE) measures the efficiency of plants in utilizing available nutrients. In all agroecosystems, NUE is below 50% (Zhang et al., 2021) because of leaching, soil properties, fertilizer characteristics, and gaseous losses. Urea is a leading source of N-fertilizer; it does not assimilate in the soil immediately as hydrolysis and volatilization of it take place at a higher rate while other enzymatic urease within the soil absorbs considerable loss of nitrogen. Various micronutrients have defensive mechanisms within plants to counter disease and other abiotic stresses. These vital micronutrients are also acquired by both animals and humans from plants. Soil micronutrient deficiencies or limited availability of micronutrients in soils are associated with lower crop productivity as well as the nutritional quality of food.

## *5.1 Nanoproducts*

### 5.1.1 Nanofertilizers

Best nanofertilizers are an integral part of any agricultural practice, as they outperform chemical fertilizers in terms of their effectiveness and environmental safety, while also contributing to the reduction of urea use by improving nitrogen use efficiency. Therefore, they improve fertility, productivity, and the agricultural output value of soil. The use of nanofertilizers also aims to make possible controlled release of nutrients concerning several triggering factors such as temperature, moisture, and other abiotic factors. Due to certain characteristics such as larger surface area as well as particle sizes smaller than the leaf and root pores of plants, nanofertilizers can penetrate plants more deeply, enhance nutrient absorption, and optimize the use of available resources. Extensive research conducted has concluded that these nanofertilizers accelerate the growth, quality, and yield of crops, which in turn leads to improved food production and better-quality crops for livestock and human use.

### 5.1.2 Nanopesticides and Nanofungicides

Nanopesticides have supplanted traditional pesticides. The good side of incorporating pesticides in nanoform is that the active agents can be released in controlled ways and times by controlling the membrane of the nanocapsule. The unwanted overkill is also reduced by dispensing a small quantity over large intervals. There has also been considerable development in the area of controlling diseases and pests by application of nanopesticides (Kah et al., 2018). Nanopesticides are designed for specific targets, therefore minimizing environmental pollution as well as the impact on nontarget flora. Using pesticides is one of the effective means of controlling weeds, fungi, and insects that pose a risk to growing plants. Copper is coated with either a nanofungicide or a nanopesticide. This is an important aspect in the control of infections in regular and organic farming systems. It is demonstrated that copper nanoparticles have both insecticidal and fungicidal effects on the pests disturbing the crops. Hence, copper-based nanomaterials like nanopesticides, nanofungicides, nanofertilizers, and nanoherbicides are available for application (Khan, 2020).

### 5.1.3 Nanoinsecticides and Nanoherbicides

Food is an essential element in the everyday life of the ever-growing population of the world. Consequently, the need to grow more and more food is persistent, which, in turn, leads to better measures aimed at crop protection from pests (Ghosh and Das,). Out of all the technology involved in the management of pests, controlled-release insecticide microencapsulated nanocapsules remain the most effective in the management of pests. This is regarding the current shift toward protecting the environment and the successful application of biodegradable, rather than synthetic, polymers for encapsulating agrochemicals (Das et al., 2024). Heavy dependence on synthetic pesticide counterparts is not without repercussions; for example, it leads to excessive residual toxicity, which contaminates the environment after a few applications (Dey et al., 2023a, 2023b). Since the majority of the insect pest populations are prone to insecticide overuse, an alternative pest control strategy is also warranted.

In the past, weed control was mainly achieved through the application of herbicides. Weeds that are grown together with crops are usually known to suppress crop growth. The use of herbicide may influence growth and other developmental processes in plants. An herbicide is a kind of pesticide that aims to destroy all unwanted vegetations including weeds. Herbicides, when nanocapsules are used to inject them into tissues and cuticles, are released over time and in a sustained manner. More soil-resilient herbicides are capable of decreasing the rate of germination and the fresh weight. Therefore, preemergence herbicide with vegetative inhibition for the purpose of suppressing weed emergence is expected.

### 5.1.4 Nanobarcode Technology

Barcoding plays a crucial part in managing livestock and agriculture in daily life. It has been extensively used as an identifying tag that can be spotted by optical microscopes and ultraviolet (UV) light because of its precise efficiency. The agricultural micro barcodes are made using a special type of soil that is doped with multiple fluorescent materials and inked within micro-barreled glass nanostructures. A machine-readable, durable, and simple encoding element would be porous nanowells in which the application of nanobarcode technology can be extended for multipurpose identification tags for genetically modified plants that are resistant to drought, insects, pests, salt, and other stresses at a reasonable cost. The use of barcodes has recently become one of the important practices in the management of livestock and farming in general. It has gained popularity as an identification mark that can be seen under an optical microscope and ultraviolet rays and is upheld because of its considerable efficiency. They are developed by pharmacologically embedding special rare earth elements in micrometer-sized glass nanobarcodes, which have been widely adopted in the area of agriculture. The nanoparticles must be able to be produced by machine, have to be durable and quick to bear inscriptions, and are more miniaturized than taggant particles that are utilized in a semiautomated process, besides having high capabilities for the manufacturing of nanobarcodes. The uses of nanobarcode technology in multiplexed analysis can be profitably incorporated in the ID tags for genetic enhancement of plants for resistance to drought, insects, pests, and salinity. Nanotechnology stands out in terms of tagging agro-food products and their distribution. Nanobioprocessing and nanobarcoding are also useful in assessing the quality of crop yield besides diagnosing diseases. Barcoding methods were invented to exclusively measure the interactions of a particular label. The DNA microarray is one such example of this kind of approach, and it performs positional encoding to identify a particular set of responses in a mixed reaction. In contrast, the radioisotope which is used as a marker is the one that labels the disease-causing organism or the fertilizer element. A fluorimeter is used to detect tagged plant diseases by attaching microscopic probes to them. At present, battery-operated nanobarcodes are being manufactured, and once completed, they will be integrated with the Global Positioning System (GPS), facilitating the identification of any type of pesticides and viruses. In gene-expression-quantification techniques aimed at the detection of phytopathogens, quantum dots have also been used as nanobarcodes, as invented by a group of scientists. Other works include a discussion of applications of barcode technologies which are somewhat peripheral, and more insistently that such a, as it were, the leitmotif of enormity, actually that this will be a massive step forward toward improving industrious development of the nano barcode.

### 5.1.5 Nanosensors in Agriculture for the Detection of Heavy Metals, Pathogens, Pesticide Residues, and Nutrient Deficiency

Developing silver nanoparticles for use in agricultural problems has primarily focused on their antibacterial effects and wastewater restoration capabilities (Mishra and Das, 2022; Das et al., 2024a, 2024b). Silver nanoparticles first found application in the food packaging industry in 2008, where they were utilized to eliminate food-borne viruses from products stored for extended periods. Early experiments suggested that the concentration of 100 ppm of silver nanoparticles was effective against powdery mildew by killing the fungal hyphae and swollen conidia in members of the Cucurbitaceae family. It has been found that layers of engineered silver nanoparticles improve the sensitivity of the detection of methyl parathion and Raman detection. By incorporating electrical nanosensing, the silver nanoparticles (NPs) present in plants and processed food have also been detected and quantified. Heavy metals such as lead, mercury, chromium, arsenic, cadmium, nickel, and copper affect human health as well as the environment (Priyadarsini et al., 2025). The use of toxic substances is also associated with the presence of some agricultural activities (Priyadarsini & Das, 2024, 2025). The nitrocellulose (NC) membrane was coated with a cadmium (Cd) ion chelating protein known as cadmium(II)-EDTA-BSA as well as goat anti-mouse immunoglobulin G (IgG). This membrane, along with a concentrated colloidal gold probe, was then applied onto a glass fiber membrane to create the highly sensitive, nanosensors in the type of Immunochromatographic strip kind of cadmium sensors anti sensors (Zamani et al., 2019). Nanosensors that detect a variety of pathogens have ultimately been developed to assist in agriculture. The detection of bean pod mottle virus (BPMV), arabis mosaic virus (ArMV), and tomato ringspot virus (ToRSV) was possible using MNPs, where MNPs were sonicated for amino functionalization and synthesis of $NH_2$–$Fe_3O_4/SiO_2$ MNPs. Antibodies were covalently immobilized in these MNPs. Gold NPs of different sizes were created using sodium citrate dehydrate and gold(III) chloride trihydrate ($HAuCl_4$), and the electrode used in electrochemical sensors to detect *Ganoderma boninense* was changed layer by layer. Because of their ability to detect a wide range of objects, nanosensors are crucial in the agricultural sector. Furthermore, several nanosensors have been created that can eradicate the risk that pesticides pose to both crops and consumers. The imazapyr pesticide was detected using fluorescent nanosensors by employing $Yb_2O_3$ doped with 3-aminopropyl-triethoxysilane and hydrothermal approaches. Precipitation of the copper nitrate using an aqueous NaOH solution determines the glyphosate detection with the resultant copper oxide/multiwalled copper nanotube (CuO/MWCNT) fluorescence sensor.

# 6 Energy Recycling from Agricultural Waste

With each passing year, the world's demand for energy grows, and the development of new, effective, and environmentally friendly technologies for energy generation and storage becomes increasingly urgent. As pointed out by Dr. Wade Adams of Rice University, energy in the next 50 years will be the central problem for humanity, and nanotechnology shall make it possible to solve this problem. In this way, nanotechnology implemented in energy recycling processes facilitates creative waste-to-energy generating activities which encourages the establishment of a circular economy. One of the key elements of sustainable agriculture is energy recycling, which aims to minimize waste and make the most out of available resources. Agricultural waste like crop residues and animal manures or by-products can turn into potential biofuels by biomass conversion technologies. Organic waste can also break down under anaerobic conditions by microorganisms, which is known as anaerobic digestion, and can provide mainly methane as biogas for electricity and heat. Nanoparticles can be used to optimize microbial activity and enhance biogas production rates. Pyrolysis is the process of heating biomass without oxygen to create char, syngas, and bio-oil. Higher yields of energy-rich products can be obtained by the use of nanomaterials to enhance heat transport and catalyze reactions during pyrolysis. Gasification is turning organic compounds into synthetic gas (syngas) via partial oxidation at elevated temperatures. Nanocatalysts may enhance this gasification process, reduce emissions, and improve efficiency. Nanoparticles can therefore enhance the combustion performance of standard biological fuels leading to reduced emissions and higher energy release. For instance, metallic nanoparticles enhance the stability and ignition properties of the fuel.

When it comes to the production of biodiesel, nanocatalysts are effectively utilized. The purpose of nanocatalysts is to speed up the process of transesterification, which is crucial in biodiesel production from oils of plant or animal origin. Such types of nanocatalysts are able to work even with less stringent requirements leading to less energy and monetary expenditure. Supercapacitors and high-performance batteries capable of storing power from renewable sources, i.e., wind and solar, are often embedded with nanomaterials such as graphene and carbon nanotubes. These technologies allow for the saving of additional energy for future use, making farmers less reliant on external energy resources. Energy resources can also be used more efficiently and effectively by fitting nanotechnology into the smart energy grid. Nanosensors and nanodevices help farmers monitor energy usage in real time, reducing operational energy waste and facilitating energy conservation. The incorporation of nanotechnology-enabled energy recycling systems not only tackles the issue of waste generation but also has positive effects on the environment. Emissions normally caused by the use of fossil fuel-based sources are cut down by turning agricultural biomass waste into biofuels and other energy forms. This step is critical to mitigate the effects of climate change. Resource conservation is a technique for management that focuses on the primary causes of resource crises and on those resources that are very crucial for survival. Energy recycling reverses the process of extraction for new useful resources by repurposing currently underutilized agricultural waste. This model endorses such land-use patterns, which are ecologically sound as well as beneficial to the ecosystem. Energy recycling technologies based on nanotechnology bear the potent seeds of a paradigm shift and evolution in the agricultural industry, allowing farmers to turn trash into treasure and increase environmental stewardship. Nanotechnology is essential to the development of a circular economy in agriculture because it enhances biomass conversion processes, produces advanced biofuels, and incorporates effective energy storage solutions (Dey et al., 2024). The potential for creative applications, as research in this field develops, will help promote sustainable farming methods and advance global environmental objectives (Ghosh et al., 2023a, 2023b).

## 7 Applications of Nanotechnology

Numerous novel materials and devices with a wide range of uses can be created using nanotechnology. In recent years, nanotechnology has made good strides in the field of agriculture. Agricultural productivity is enhanced through the application of nanotechnology in delivery systems such as nanopesticides, nanofertilizers, nanofungicides, nanoherbicides, and nanosensors for crop disease diagnosis, genetic modification, animal and plant health assessment, postharvest management, and so on. Furthermore, it contributes to improving the health of cultivated plants while adding no detrimental effects to the soil. Moreover, it reduces the nitrogen loss caused by soil microorganisms and leaching (Dimka et al., 2020). Nanocapsules have spatially targeted the distribution of herbicides throughout plant tissues and cuticles and the delivery of macromolecules to enhance disease resistance in plants and improve general growth. Agronomic natural resources such as water, pesticides, and fertilizers can be effectively used using nanosensors for accurate agricultural techniques. For instance, by using nanomaterials, farmers might be able to determine the presence of infecting insects or even information on stress conditions such as drought using just satellite photography of the field. An irrigation point scan or the automatic adjustment of pesticide treatments can be done if it is found that the crop is infested by pests or the soil is dry. Nanosensors also senses the amount of nutrients in the soil and the presence of plant diseases. The application of fertilizers in nanocapsules is aimed at preventing the depletion of soil fertility. An increase in agricultural productivity is a natural phenomenon that is aimed at alleviating poverty and hunger. In this regard, we may say that nanotechnology is making more sustainable competitiveness and advancement—hence assisting in the transition to a green economy (Sonu et al., 2023; Pradhan et al., 2023).

Nanotechnology has various applications such as the synthesis and characterization of food, preservation of different food products for longer periods, and also transmission of genes using nanoparticles for developing transgenic resistance in plants against certain insects. Accurate agricultural methods have increased crop yields without causing land or water damage. The primary goal of using nanomaterials in agriculture is to increase plant yields and decrease the amount of plant protection products sprayed. There is no way to discuss about farming without recognizing the huge benefits that nanotechnologies can offer to agricultural and food-producing activities. Moreover, some nanotechnology-based applications are helpful in various agricultural interventions such as improving fertilizer use efficiency, protecting plants from diseases and pests, etc. The impacts of various nanoparticles on phytotoxicity and plant development such as magnetite ($Fe_3O_4$), alumina, zinc, and zinc oxide nanoparticles on seed germination and root elongation of five flowering plants: maize, cucumbers, lettuce, radishes, and green beans; zinc oxide and sulfur nanoparticles in tomatoes; silver nanoparticles on seedling growth in wheat; and MnO, FeO, ZnO, NiO, CuO, and AlO nanoparticles in plant roots have been presented. Zinc is a plant microelement necessary in small quantities for successful mediated processes. Plants need reactive oxygen species and protection of their cells from oxidative stresses and therefore require zinc only in very minute concentrations. The deficiency of zinc translates to lower yields and poorer quality of food crops. Effective synthesis of MgOH, MgO, CuO, and ZnO nanoparticles has been achieved via aqueous extracts of *Chamaemelum nobile* flowers, *Olea europaea* leaves, and *Punica granatum* peels. This method also reduces the quantity of hazardous substances that are contaminating the environment. Nanobiofertilizer is a combination of nanoparticles with biofertilizers as one product that is cheaper, more potent, and more environmentally friendly than nanoparticles or biofertilizers alone. The nanobiofertilizers represent the encapsulation of the biofertilizers within nanoparticles. The green synthesis of these nanoparticles will further enhance their performance and qualities. Nanobiofertilizers are comparatively better than other traditional approaches. They also play their part much better than those ordinary salts that have been used so far in agricultural conditions to enhance the yield of crops. Nanobiofertilizers provide excellent and long-lasting effects compared with traditional chemical fertilizers. It enhances the strength and activity of soil along with the morphological, physiological, biochemical, and yield properties of plants (Akhtar et al., 2022). Nanoparticles like ZnO NPs are found in products like paint, cosmetics, animal feeds, and fertilizers. The $TiO_2$ nanoparticles are commonly used in sunscreens for protection against cell damage by blocking UV light (Baker et al., 2014).

Nanoparticle biosynthesis is becoming increasingly popular as a new biotechnology and nanotechnology focus because of the growing need to reduce the usage of chemicals that harm the environment, such as pesticides. Silver nanoparticles are of particular relevance due to their extensive use in food packaging, where they are used as antibacterial agents along with other nanomaterials. Gold, silver, magnesium oxide (MgO), silicon oxide ($SiO_2$), zinc oxide (ZnO), and titanium dioxide ($TiO_2$) nanoparticles are presently in use. The food sector is at the forefront of developing its capabilities as well as the provision of high-value nutritious food. Preservatives, such as UV radiation, help protect the food from the external environments. Additionally, extreme water–vapor-resistant active packaging nanomaterials are employed, for instance, to help incorporate the product within the suitable environmental conditions for a prolonged duration. Nanotechnology is applied in food for the detection of microorganisms, gases, and chemicals using nanosensors. Recent development of nanostructures with the use of nanotechnology plays a great role in the food industry as they protect from microorganisms. The above properties are also helpful in prolonging the storage of food materials without compromising on the quality of the food because of the prevention of spoilage caused by several disease-causing microorganisms (Das et al., 2023b; Das and Ghosh 2023c, 2023d). To extend the longevity of chicken and meat products and protect them, a biodegradable nanosystem consisting of polyvinyl alcohol (PVA) and montmorillonite K10 clay and silver NPs was created, demonstrating water-resistant, light-protecting, and antimicrobial properties against *Staphylococcus aureus* and *Salmonella typhimurium* in contaminated samples of these coatings (Duhan et al., 2017; Das & Ghosh, 2023a, 2023b).

## 8 Conclusion

To summarize, the utilization of nanotechnology for better agricultural waste management and energy recycling is an area that holds great promise in increasing the productivity and sustainability of farming activities. In this way, the function of nanotechnology goes beyond just the environmental concerns of waste disposal as it mobilizes renewable resources from waste. The application of nanoparticles in lower biomass conversion, biofuel, and energy storage, for example, points to sustainable agricultural practices where wastes are reduced and all sources are of resources. In the future, as research progresses, it is equally critical to consider the potential negative impacts of nanomaterials and their health and environmental consequences over time. In conclusion, integrating nontoxic technology within agricultural systems is a promising approach to enhance food production methods that meet the sustainability agenda of the world so that the environment is healthy for future generations.

## References

Abd Elkodous, M., Hamad, H., Abdel Maksoud, M., Ali, G., El Abboubi, M., Bedir, A., Eldeeb, A., Ayed, A., Gargar, Z., Zaki, F., Farage, D., Matsuda, A., Abdelnour, M., Sabra, B., Elsayed, A., Abdelrazek, T., Abdelhameed, S., Gharieb, M., Rabee, M., … Kawamura, G. (2022). Cutting-edge development in waste-recycled nanomaterials for energy storage and conversion applications. *Nanotechnology Reviews, 11*(1), 2215–2294. https://doi.org/10.1515/ntrev-2022-0129

Akhtar, N., Ilyas, N., Meraj, T. A., Pour-Aboughadareh, A., Sayyed, R. Z., Mashwani, Z. U., Poczai, P. (2022). Improvement of plant responses by nanobiofertilizer: A step towards sustainable agriculture. *Nanomaterials (Basel), 12*(6), 965. https://doi.org/10.3390/nano12060965. PMID: 35335778; PMCID: PMC8949119.

Aswathi, V. P., Meera, S., Maria, C. G. A., et al. (2023). Green synthesis of nanoparticles from biodegradable waste extracts and their applications: A critical review. *Nanotechnology for Environmental Engineering, 8*, 377–397. https://doi.org/10.1007/s41204-022-00276-8

Baker, T. J., Tyler, C. R., & Galloway, T. S. (2014). Impacts of metal and metal oxide nanoparticles on marine organisms. *Environmental Pollution, 186*, 257–271. https://doi.org/10.1016/J.ENVPOL.2013.11.014

Balusamy, S. R., Joshi, A. S., Perumalsamy, H., et al. (2023). Advancing sustainable agriculture: A critical review of smart and eco-friendly nanomaterial applications. *Journal of Nanbiotechnology, 21*, 372. https://doi.org/10.1186/s12951-023-02135-3

Das, A. P., & Ghosh, S. (2018). Bioleaching of Manganese from mining waste materials. *Materials Today: Proceedings, 5*(1), 2381–2390. https://doi.org/10.1016/J.MATPR.2017.11.459

Das, A. P., & Ghosh, S. (2023a). Introduction. *Manganese Mining Microorganisms*, 1–4. https://doi.org/10.1016/B978-0-12-822147-1.00005-4

Das, A. P., & Ghosh, S. (2023b). Mechanism of manganese bioleaching. *Manganese Mining Microorganisms*, 87–98. https://doi.org/10.1016/B978-0-12-822147-1.00014

Das, A. P., & Ghosh, S. (2023c). Microorganisms in manganese biomining processes. *Manganese Mining Microorganisms*, 55–61. https://doi.org/10.1016/B978-0-12-822147-1.00013-3

Das, A. P., & Ghosh, S. (2023d). Mechanism of manganese bioleaching. *Manganese Mining Microorganisms*, 87–98. https://doi.org/10.1016/B978-0-12-822147-1.00014-5

Das, A. P., & Mishra, S. (2008). Hexavalent Chromium (VI): Health hazards and environmental pollutant. *The Journal of Environment & Development, 2*(3), 386–392.

Das, A. P., & Mishra, S. (2009). Hexavalent chromium [Cr (VI)]: Yellow water pollution and its remediation. *Sarovar Saurabh ENVIS Newsl Wetl Ecosyst, 5*(2).

Das, A. P., & Swain, S. (2013). Algal biosorption of toxic dye methylene blue. *A Potential Source of Food, Feed, Biochemicals, Biofuels and Biofertilizers, International Conference on Algal Biorefinery, Indian Institute of Technology, 13.*

Das, A. P., Ghosh, S., Mohanty, S., & Sukla, L. B. (2015). Advances in manganese pollution and its bioremediation. In L. B. Sukla, N. Pradhan, S. Panda, & B. K. Mishra (Eds.), *Environmental microbe biotechnology* (1st ed., pp. 313–328). Springer. https://doi.org/10.1007/978-3-319-19018-1_16

Das, A. P., Mohanty, A. K., van Hullebusch, E. D., Figueiredo, G., Das, A. P., Mohanty, A. K., van Hullebusch, E. D., & Figueiredo, G. (2023). Bioremediation as an emerging technology for the removal of synthetic microplastic pollutants from marine ecosystem. *MarPB, 194*, 115297. https://doi.org/10.1016/J.MARPOLBUL.2023.115297

Das, A., Dey, S., Bala, K., Gnanasekaran, R., & Das, A. P. (2024a). Microbial bioremediation technology for sustainable treatment and management of synthetic microfiber waste. *Sustainable Microbial Technology for Synthetic and Cellulosic Microfiber Bioremediation*, 221–238. https://doi.org/10.1007/978-3-031-62600-5_12

Dey, S., Bissoyi, A., Koteswara Reddy, G., & Das, A. P. (2024a). Synthetic microfibers: An urban emerging pollutant demanding for sustainable environmental solutions. *Sustainable Microbial Technology for Synthetic and Cellulosic Microfiber Bioremediation*, 107–121. https://doi.org/10.1007/978-3-031-62600-5_7

Dey, S., Das, A., Mallick, K., Sahu, A., & Das, A. P. (2023a). *Environmental petroleum waste: Pollution, toxicity, sustainable remediation* (pp. 159–175). https://doi.org/10.1007/978-3-031-48220-5_7

Dey, S., Dash, A., & Das, A. P. (2024b). Bioremediation as a sustainable and circular solution for mining waste management and bioenergy generation. *Sustainable Management of Mining Waste and Tailings: A Circular Economy Approach*, 107–119. https://doi.org/10.1201/978 1003442455-6/BIOREMEDIATION-SUSTAINABLE-CIRCULAR-SOLUTION-MINING-WASTE-MANAGEMENT-BIOENERGY-GENERATION-SUDESHNA-DEY-AKANSHYA-DASH-ALOK-PRASAD-DAS

Dey, S., Tripathy, B., Kumar, M. S., & Das, A. P. (2023b). Ecotoxicological consequences of manganese mining pollutants and their biological remediation. *Environmental Chemistry and Ecotoxicology*. https://doi.org/10.1016/J.ENCECO.2023.01.001

Dimkpa, C. O., Fugice, J., Singh, U., & Lewis, T. D. (2020). Development of fertilizers for enhanced nitrogen use efficiency - Trends and perspectives. *Science of the Total Environment, 731*, 139113. https://doi.org/10.1016/j.scitotenv.2020.139113. Epub 2020 May 5. PMID: 32438083.
do Espirito Santo Pereira, A., Caixeta Oliveira, H., Fernandes Fraceto, L., & Santaella, C. (2021). Nanotechnology potential in seed priming for sustainable agriculture. *Nanomaterials (Basel)*, 11(2), 267. https://doi.org/10.3390/nano11020267. PMID: 33498531; PMCID: PMC7909549.
Duhan, J. S., Kumar, R., Kumar, N., Kaur, P., Nehra, K., & Duhan, S. (2017). Nanotechnology: The new perspective in precision agriculture. *Biotechnology Reports (Amsterdam, Netherlands), 24*(15), 11–23. https://doi.org/10.1016/j.btre.2017.03.002. PMID:28603692; PMCID:PMC5454086.
El-Ramady, H., El-Henawy, A., Amer, M., Omara, A. E.-D., Elsakhawy, T., Elbasiouny, H., Elbehiry, F., Abou Elyazid, D., & El-Mahrouk, M. (2020). Agricultural waste and its nano-management: Mini review. *Egyptian Journal of Soil Science, 61*. https://doi.org/10.21608/EJSS.2020.46807.1397
Fatima, F., Hashim, A., & Anees, S. (2021). Efficacy of nanoparticles as nanofertilizer production: A review. *Environmental Science and Pollution Research International, 28*(2), 1292–1303. https://doi.org/10.1007/s11356-020-11218-9. Epub 2020 Oct 17 PMID: 33070292.
Flores-Contreras, E. A., González-González, R. B., Pablo Pizaña-Aranda, J. J., Parra-Arroyo, L., Rodríguez-Aguayo, A. A., Iñiguez-Moreno, M., González-Meza, G. M., Araújo, R. G., Ramírez-Gamboa, D., Parra-Saldívar, R., & Melchor-Martínez, E. M. (2024). Agricultural waste as a sustainable source for nanoparticle synthesis and their antimicrobial properties for food preservation. *Frontiers in Nanotechnology, 6*, 1346069. https://doi.org/10.3389/fnano.2024.1346069
Fraceto, L. F., Grillo, R., de Medeiros, G. A., Scognamiglio, V., Rea, G., & Bartolucci, C. (2016). Nanotechnology in agriculture: Which innovation potential does it have? *Frontiers in Environmental Science, 4*, 20. https://doi.org/10.3389/fenvs.2016.00020
Ghosh, S., & Das, A. P. (2017). Bioleaching of manganese from mining waste residues using *Acinetobacter* sp. *Geology, Ecology, and Landscapes, 1*(2), 77–83. https://doi.org/10.1080/24749508.2017.1332847
Ghosh, S., & Das, A. P. (2020a). Microbial metagenomics: Current advances in investigating microbial ecology and population dynamics. In S. K. Nayak & B. B. Mishra (Eds.), *Frontiers in soil and environmental microbiology* (1st ed., pp. 247–254). CRC Press.
Ghosh, S., & Das, A. P. (2020b). Microbial metagenomics: Current advances in investigating microbial ecology and population dynamics. In S. K. Nayak & B. B. Mishra (Eds.), *Frontiers in soil and environmental microbiology* (1st ed., pp. 247–254). CRC Press.
Ghosh, S., Kumar, M. S., Bal, B., & Das, A. P. (2018). Application of bioengineering in revamping human health. *Synthetic Biology: Omics Tools and Their Applications*, 21–37. https://doi.org/10.1007/978-981-10-8693-9_2
Ghosh, S., Mohanty, S., Akcil, A., Sukla, L. B., & Das, A. P. (2016). A greener approach for resource recycling: Manganese bioleaching. *Chemosphere, 154*, 628–639. https://doi.org/10.1016/J.CHEMOSPHERE.2016.04.028
Ghosh, S., Sinha, J. K., Ghosh, S., Vashisth, K., Han, S., & Bhaskar, R. (2023a). Microplastics as an emerging threat to the global environment and human health. *Sustainability, 15*(14), 10821. https://doi.org/10.3390/SU151410821
Ghosh, S., Sinha, J. K., Ghosh, S., Vashisth, K., Han, S., & Bhaskar, R. (2023b). Microplastics as an emerging threat to the global environment and human health. *Sustainability, 15*(14), 10821. https://doi.org/10.3390/SU151410821
Gupta, R., & Xie, H. (2018). Nanoparticles in daily life: applications, toxicity and regulations. *Journal of Environmental Pathology, Toxicology and Oncology, 37*(3), 209–230. https://doi.org/10.1615/JEnvironPatholToxicolOncol.2018026009. PMID:30317972; PMCID:PMC6192267.
Kah, M., Kookana, R. S., Gogos, A., & Bucheli, T. D. (2018). A critical evaluation of nanopesticides and nanofertilizers against their conventional analogues. *Nature Nanotechnology, 13*(8), 677–684. https://doi.org/10.1038/s41565-018-0131-1. Epub 2018 May 7 PMID: 29736032.

Khan, S. H. (2020). Green nanotechnology for the environment and sustainable development. In M. Naushad & E. Lichtfouse (Eds.), *Green materials for wastewater treatment. environmental chemistry for a sustainable world* (Vol. 38). Springer. https://doi.org/10.1007/978-3-030-17724-9_2

Kolawole, I. D., Kolawole, G. O., Sanni-manuel, B. A., et al. (2024). Economic impact of waste from food, water, and agriculture in Nigeria: Challenges, implications, and applications—A review. *Discover Environment, 2*, 51. https://doi.org/10.1007/s44274-024-00086-6

Kreslavski, V. D., Shmarev, A. N., Ivanov, A. A., Zharmukhamedov, S. K., Strokina, V., Kosobryukhov, A., Yu, M., Allakhverdiev, S. I., & Shabala, S. (2023). Effects of iron oxide nanoparticles ($Fe_3O_4$) and salinity on growth, photosynthesis, antioxidant activity and distribution of mineral elements in wheat (*Triticum aestivum*). *Functional Plant Biology, 50*(11), 932–940. https://doi.org/10.1071/FP23085. PMID: 37573788.

Mahmoud, T., Sayed, M. A., Ragab, A. A., & Mohamed, E. A. (2021). Advanced waste recycling technologies for manufacturing of nanomaterials for green energy applications. In A. S. H. Makhlouf & G. A. M. Ali (Eds.), *Waste recycling technologies for nanomaterials manufacturing. Topics in mining, metallurgy and materials engineering*. Springer. https://doi.org/10.1007/978-3-030-68031-2_21

Mallick, K., Sahu, A., Dubey, N. K., & Das, A. P. (2023). Harvesting marine plastic pollutants-derived renewable energy: A comprehensive review on applied energy and sustainable approach. *Journal of Environmental Management, 348*, 119371. https://doi.org/10.1016/J.JENVMAN.2023.119371

Mgadi, K., Ndaba, B., Roopnarain, A., Rama, H., & Adeleke, R. (2024). Nanoparticle applications in agriculture: Overview and response of plant-associated microorganisms. *Frontiers in Microbiology, 15*, 1354440. https://doi.org/10.3389/fmicb.2024.1354440

Mishra, S., & Das, A. P. (2021). Current treatment technologies for removal of microplastic and microfiber pollutants from wastewater. *Wastewater Treatment: Cutting-Edge Molecular Tools, Techniques and Applied Aspects*, 237–251. https://doi.org/10.1016/B978-0-12-821881-5.00011-8

Mishra, S., & Das, A. P. (2022). Treatment of the wastewater polluted with synthetic microfiber released from washing machine. In B. B. Das, H. Hettiarachchi, P. K. Sahu, & S. Nanda (Eds.), *Lecture notes in civil engineering* (Vol. 207, pp. 109–117). Springer Science and Business Media Deutschland GmbH. https://doi.org/10.1007/978-981-16-7509-6_9/COVER

Mishra, S., & Das, A. P. (2023). Microbial remediation of synthetic microfiber contaminated wastewater. In M. P. Shah (Ed.), *Microbial technologies in industrial wastewater treatment* (pp. 337–350). Springer. https://doi.org/10.1007/978-981-99-2435-6_17

Mishra, S., Dash, D., & Das, A. P. (2022a). Detection, characterization and possible biofragmentation of synthetic microfibers released from domestic laundering wastewater as an emerging source of marine pollution. *Marine Pollution Bulletin, 185*, 114254. https://doi.org/10.1016/J.MARPOLBUL.2022.114254

Mishra, S., Dash, D., & Das, A. P. (2024). Isolation and molecular characterization of polyester-degrading bacterial strains from municipal wastewater and their application in microplastic degradation in contaminated wastewater. *Geomicrobiology Journal*. https://doi.org/10.1080/01490451.2024.2376101

Mishra, S., Dash, D., Al-Tawaha, A. R. M. S., & Das, A. P. (2022b). A review on heavy metal ion adsorption on synthetic microfiber surface in aquatic environments. *Applied Biochemistry and Biotechnology, 194*(10), 4639–4654. https://doi.org/10.1007/S12010-022-04029-W/METRICS

Mishra, S., Ghosh, S., van Hullebusch, E. D., Singh, S., & Das, A. P. (2023). A critical review on the recovery of base and critical elements from electronic waste-contaminated streams using microbial biotechnology. *Applied Biochemistry and Biotechnology, 2023*, 1–30. https://doi.org/10.1007/S12010-023-04440-X

Mishra, S., Rath, C. C., & Das, A. P. (2019). Marine microfiber pollution: A review on present status and future challenges. *Marine Pollution Bulletin, 140*, 188–197. https://doi.org/10.1016/J.MARPOLBUL.2019.01.039

Mishra, S., Swain, S., Sahoo, M., Mishra, S., & Das, A. P. (2021). Microbial colonization and degradation of microplastics in aquatic ecosystem: A review. *Geomicrobiology Journal*. https://doi.org/10.1080/01490451.2021.1983670

Mohanty, S., Ghosh, S., Nayak, S., & Das, A. P. (2016). Isolation, identification and screening of manganese solubilizing fungi from low-grade manganese ore deposits. *Geomicrobiological Journal, 34*(4), 309–316. https://doi.org/10.1080/01490451.2016.1189016

Mohanty, S., Ghosh, S., Nayak, S., & Das, A. P. (2017). Bioleaching of manganese by *Aspergillus* sp. isolated from mining deposits. *Chemosphere, 172*, 302–309. https://doi.org/10.1016/J.CHEMOSPHERE.2016.12.136

Neme, K., Nafady, A., Uddin, S., & Tola, Y. B. (2021). Application of nanotechnology in agriculture, postharvest loss reduction and food processing: Food security implication and challenges. *Heliyon., 7*(12), e08539. https://doi.org/10.1016/j.heliyon.2021.e08539. PMID:34934845; PMCID:PMC8661015.

Oluseun Adejumo, I., & Adebukola Adebiyi, O. (2021). Agricultural solid wastes: Causes, effects, and effective management. *IntechOpen*. https://doi.org/10.5772/intechopen.93601

Paul, A., Dey, S., Ram, D. K., & Das, A. P. (2023). Hexavalent chromium pollution and its sustainable management through bioremediation. *Geomicrobiology Journal*, 1–11. https://doi.org/10.1080/01490451.2023.2218377

Paul, A., Priyadarsini, S., Karakuş, S., Behera, J. K., Gnanasekaran, R., & Das, A. P. (2024). Fate and accumulation of synthetic microfibers in the environment. *Sustainable Microbial Technology for Synthetic and Cellulosic Microfiber Bioremediation*, 61–76. https://doi.org/10.1007/978-3-031-62600-5_4

Peng, X., Jiang, Y., Chen, Z., et al. (2023). Recycling municipal, agricultural and industrial waste into energy, fertilizers, food, and construction materials, and economic feasibility: A review. *Environmental Chemistry Letters, 21*, 765–801. https://doi.org/10.1007/s10311-022-01551-5

Periakaruppan, R., Romanovski, V., Thirumalaisamy, S. K., Palanimuthu, V., Sampath, M. P., Anilkumar, A., Sivaraj, D. K., Ahamed, N. A. N., Murugesan, S., Chandrasekar, D., et al. (2023). Innovations in modern nanotechnology for the sustainable production of agriculture. *ChemEngineering, 7*(4), 61. https://doi.org/10.3390/chemengineering7040061

Prabina, K. E., Benya, A., Mohanty, S., Van Hullebusch, E. D., Das, A. P., & Singh, S. (2024). Recovery of multi-functional carbon fiber from COVID-19 generated waste for an effective management through circular economy. *Environmental Science and Engineering, Part, F2827*, 83–97. https://doi.org/10.1007/978-3-031-50840-0_6

Pradhan, G., Tripathy, B., Ram, D. K., Digal, A. K., & Das, A. P. (2023). Bauxite mining waste pollution and its sustainable management through bioremediation. *Geomicrobiology Journal*, 1–10. https://doi.org/10.1080/01490451.2023.2235353

Prasad, R., Bhattacharyya, A., & Nguyen, Q. D. (2017). Nanotechnology in sustainable agriculture: Recent developments, challenges, and perspectives. *Frontiers in Microbiology, 8*, 1014. https://doi.org/10.3389/fmicb.2017.01014

Priyadarshini, B., & Das, A. P. (2024). The adverse impact of synthetic microplastics released from disposable face masks pollutants on aquatic microbiota, aquatic species and humans: A review. *Environmental Science and Engineering, Part, F2827*, 67–81. https://doi.org/10.1007/978-3-031-50840-0_5

Priyadarsini, S., & Das, A. P. (2024). Analytical and structural characterization of waste lithium-ion batteries for their effective recycling strategy. *Environmental Chemistry and Ecotoxicology, 7*, 182–191. https://doi.org/10.1016/j.enceco.2024.12.004

Priyadarsini, S., & Das, A. P. (2025). Isolation and identification of native lithium-resistant bacteria from electronic waste contaminated sites: A focused investigation on tolerance assessment with potential lithium bio-recovery application. *Geomicrobiology Journal, 42*(10), 898–912. https://doi.org/10.1080/01490451.2025.2539509

Priyadarsini, S., & Das, A. P. (2025). Lithium bioleaching: A review on microbial-assisted sustainable technology for lithium bio-circularity. *Journal of Water Process Engineering, 69*, 106744. https://doi.org/10.1016/j.jwpe.2024.106744

Raj, K., & Das, A. P. (2023). Lead pollution: Impact on environment and human health and approach for a sustainable solution. *Environmental Chemistry and Ecotoxicology, 5*, 79–85. https://doi.org/10.1016/J.ENCECO.2023.02.001

Reed, R. B., Ladner, D. A., Higgins, C. P., Westerhoff, P., & Ranville, J. F. (2012). Solubility of nano-zinc oxide in environmentally and biologically important matrices. *Environmental Toxicology and Chemistry, 31*(1), 93–99. https://doi.org/10.1002/etc.708. PMID:21994124; PMCID:PMC4713012.

Sarkar, J., Mridha, D., Sarkar, J., Orasugh, J. T., Gangopadhyay, B., Chattopadhyay, D., Roychowdhury, T., & Acharya, K. (2021). Synthesis of nanosilica from agricultural wastes and its multifaceted applications: A review. *Biocatalysis and Agricultural Biotechnology, 37*, 102175. https://doi.org/10.1016/J.BCAB.2021.102175

Selva Preetha, P., & Balakrishnan, N. (2017). A review of nano fertilizers and their use and functions in soil. *International Journal of Current Microbiology and Applied Sciences, 6*(12), 3117–3133. https://doi.org/10.20546/ijcmas.2017.612.364

Sonu, R. G. M., Pathania, D., Abhimanyu, U. R., Rustagi, S., Huh, Y. S., Gupta, V. K., Kaushik, A., & Chaudhary, V. (2023). Agro-waste to sustainable energy: A green strategy of converting agricultural waste to nano-enabled energy applications. *Science of the Total Environment, 875*, 162667. https://doi.org/10.1016/j.scitotenv.2023.162667. Epub 2023 Mar 7. PMID: 36894105.

Sukirtha, T., & Saranya, N. (2020). Applications of nanotechnology in agricultural waste—Review paper. *Scholars Academic Journal of Biosciences, 8*. 259–268. https://doi.org/10.36347/sajb.2020.v08i09.001

Vejan, P., Khadiran, T., Abdullah, R., & Ahmad, N. (2021). Controlled release fertilizer: A review on developments, applications and potential in agriculture. *Journal of Controlled Release, 10*(339), 321–334. https://doi.org/10.1016/j.jconrel.2021.10.003. Epub 2021 Oct 6 PMID: 34626724.

Yadav, A., Yadav, K., & Abd-Elsalam, K. A. (2023). Nanofertilizers: Types, delivery and advantages in agricultural sustainability. *Agrochemicals., 2*(2), 296–336. https://doi.org/10.3390/agrochemicals2020019

Zamani, A., Marjani, A. P., & Mousavi, Z. (2019). Agricultural waste biomass-assisted nanostructures: Synthesis and application. *Green Processing and Synthesis, 8*(1), 421–429. https://doi.org/10.1515/gps-2019-0010

Zhang, P., Guo, Z., Ullah, S., Melagraki, G., Afantitis, A., & Lynch, I. (2021). Nanotechnology and artificial intelligence to enable sustainable and precision agriculture. *Nature Plants, 7*(7), 864–876. https://doi.org/10.1038/s41477-021-00946-6

Zulfiqar, F., Navarro, M., Ashraf, M. Y., Akram, N. A., & Munné-Bosch, S. (2019). Nanofertilizer use for sustainable agriculture: Advantages and limitations. *Plant Science: An International Journal of Experimental Plant Biology, 289*, 110270.

# Recent Advances in Valorizing Spent Coffee Grounds for Biodiesel Production with Sustainability Considerations

**Merve Ceylan, Ibrahim Metin Hasdemir, Ekin Selvi, and Dilek Özmen**

## 1 Introduction

Global waste generation from urban, agricultural, and industrial sources is increasing at an accelerated pace. Improper disposal or lack of recycling of these wastes—particularly when they are directly deposited into landfills—can result in severe environmental challenges (Mishra et al., 2019, 2023). Transforming waste into energy and high-value products offers a promising solution for addressing the escalating challenges associated with waste generation (Das & Ghosh, 2022; Das et al., 2024). Beyond its economic and environmental benefits, waste valorization enables the production of biofuels and value-added products without the use of edible crops, preventing competition with the food supply chain. Among domestic organic wastes, spent coffee grounds (SCG) stand out due to their significant potential for various valorization applications (Mishra & Das, 2023; Mohanty et al., 2018).

Coffee, a widely consumed beverage since the seventeenth century, is an agricultural commodity second only to oil and its derivatives in global trade volume (Atabani et al., 2019). According to the International Coffee Organization (ICO) (2023), 10.38 million tons of green coffee were consumed in October 2022–September 2023,

M. Ceylan · E. Selvi · D. Özmen (✉)
Department of Chemical Engineering, Faculty of Engineering, Yalova University, Yalova, Turkey
e-mail: dilek.ozmen@yalova.edu.tr

M. Ceylan
e-mail: merve.ceylan@yalova.edu.tr

E. Selvi
e-mail: ekin.selvi@yalova.edu.tr

I. M. Hasdemir
Department of Chemical Engineering, Faculty of Engineering, Istanbul University—Cerrahpasa, Istanbul, Turkey
e-mail: hasdemir@iuc.edu.tr



A. P. Das and S. Karakuş (eds.), *Synergizing Sustainability for Integrated Waste Management*, Sustainable Environmental Waste Management Strategies,
https://doi.org/10.1007/978-3-032-03718-3_16

resulting in approximately 6.92 million tons of dry biomass, i.e., SCG after brewing. SCG are often disposed of in landfills through domestic, commercial, and industrial waste streams or incinerated; however, being rich in organic compounds, SCG release greenhouse gases when decomposed or incinerated, contributing to global warming. By converting this biomass into valuable products, the coffee sector can significantly reduce its environmental impact and create a more sustainable, resilient global coffee value chain (ICO, 2023; Atabani et al., 2019).

SCG are natural sources of insoluble and soluble fiber, antioxidants, essential amino acids, and low-glycemic sugars lipids as detailed in Table 1. Research demonstrates that SCG can serve as a source of various valuable products, including biofuels, oil, lignin, cellulose and hemicellulose, antioxidants, caffeine, polyphenols, pigments, flavonoids, fertilizers, and bio-based adsorbent. The spent coffee valorization products are depicted in Fig. 1. SCG are rich in fatty acids. The oil content of SCG varies depending on the coffee varieties (such as *Arabica, Robusta*), processing, brewing, and cultivation methods, with values reported to range from 6.3 to 28.3% (Caetano et al., 2012) and 7 to 21.5% (Atabani et al., 2019). The SCG oil contains long-chain fatty acids, such as palmitic, oleic, and linoleic, making it suitable for biodiesel production via transesterification.

Biodiesel, also known as FAME (Fatty Acid Methyl Esters), is a sustainable, environmentally friendly, and non-toxic alternative that can be used in diesel engines with little or no modifications (Caetano et al., 2012). It comprises monoalkyl esters of long-chain fatty acids derived from renewable sources such as vegetable oils or animal fats, meeting the specifications outlined in ASTM D6751. Vegetable oils and animal fats primarily consist of triacylglycerols (TAGs), which are composed of long-chain fatty acids esterified to a glycerol backbone (1,2,3-propanetriol). The production of biodiesel is achieved through a chemical process known as transesterification (Fig. 2). This reaction involves the conversion of TAGs into fatty acid alkyl esters (FAAEs) and

**Table 1** Chemical composition of SCG according to different literature sources

| Component | Composition (w %) | | | |
|---|---|---|---|---|
| Hemicellulose | 36.70 | 32–42 | | |
| Cellulose | 8.60 | | | 8.60–24.30 |
| Lignin | | | | |
| Lipids | 6.30–28.30 | 9.42–17.96 | 7.50–13.10 | 7.0–21.50 |
| Proteins | 13.60 | 11–16 | 14.80 | 8.97–17.44 |
| Ash | 1.60 | | 1.59 | 1.30–2.77 |
| Caffeine | | 0–0.40 | 0.02–0.08 | |
| Chlorogenic acids | | 8.20 | | |
| Moisture | 12.10 | 80 | 7.47 | |
| References | Caetano et al. (2012) | Le et al. (2017) | Somnuk et al. (2017) | Atabani et al. (2019) |

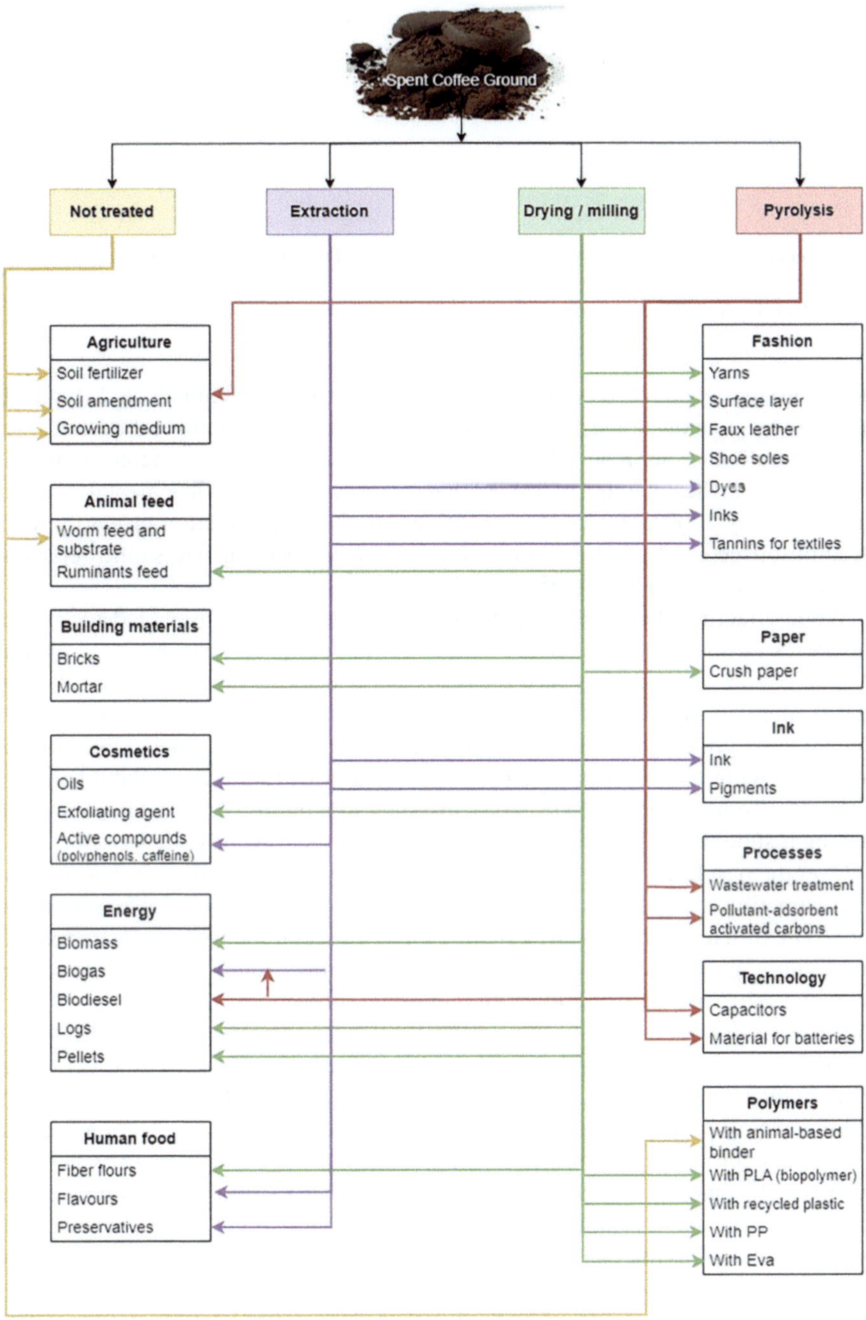


**Fig. 1** Sustainable valorization products from SCG (adapted from ICO-coffee development report 2022–23)

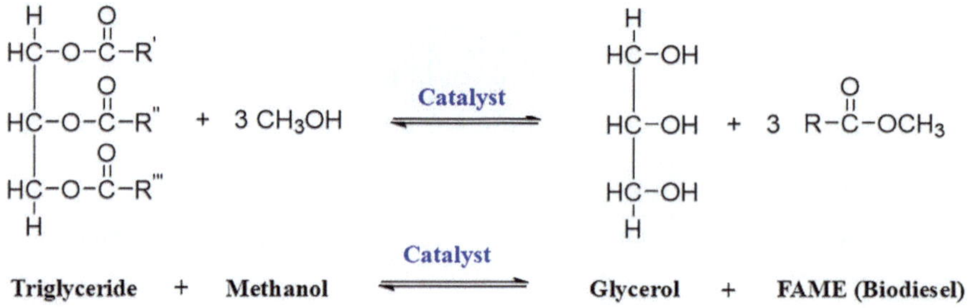


**Fig. 2** Transesterification reaction

the by-product glycerol, by reacting the TAGs with a short-chain alcohol (typically methanol or ethanol) in the presence of a catalyst at elevated temperatures.

Together with other renewable energy sources, biodiesel has garnered significant attention due to the depletion of fossil fuel reserves and rising environmental concerns since the early 2000s. However, conventional biodiesel production primarily relies on edible oilseeds, i.e., first-generation biodiesel feedstocks such as sunflower, palm, rapeseed, and soybean (up to 95%).

While effective, their use competes with food supply, increases feedstock costs, and raises sustainability concerns. Therefore, efforts have been made to explore alternative sources, including inedible seeds, waste cooking oils, animal fats, sewage sludge, algal oil, and SCG (Karmee, 2018). As a low-cost, non-food feedstock, SCG has attracted growing attention from the scientific community, policymakers, and industry, owing to its substantial lipid content and widespread availability. SCG-extracted oil contains long-chain fatty acids, such as palmitic, oleic, and linoleic acids, making it suitable for biodiesel production via transesterification.

Over the past few years, substantial efforts have been directed toward developing both extraction technologies for obtaining oil from SCG and processes for converting the SCG-extracted oil into biodiesel (Fig. 3). The most commonly investigated extraction technologies can be categorized into the following groups: (a) classical solvent extraction; (b) Soxhlet extraction with traditional solvents and their blends; (c) innovative extraction methods, such as supercritical fluid extraction (SFE), ultrasound-assisted extraction (UAE), and pressurized liquid extraction (PLE). These techniques result in different oil yields and compositions, governed by the operating conditions and the overall process choice, including factors such as solvent selection. Various successful approaches spanning from pre-treatment processes applied to SCG before extraction to innovative extraction methods have been explored to improve these processes. Still, there remains substantial potential for future enhancements, with examples including the use of cutting-edge solvent systems such as deep eutectic solvents (DES) and ionic liquids (IL).

The use of SCG-extracted oils in biodiesel production has been reported in several studies in the last decade (Kamil et al., 2019). Al-Hamamre et al. (2012) estimated that utilizing SCG, with an average fatty acid content of 14%, could supply approximately 1.123 million tons of biodiesel to the global coffee market. In 2017, the UK

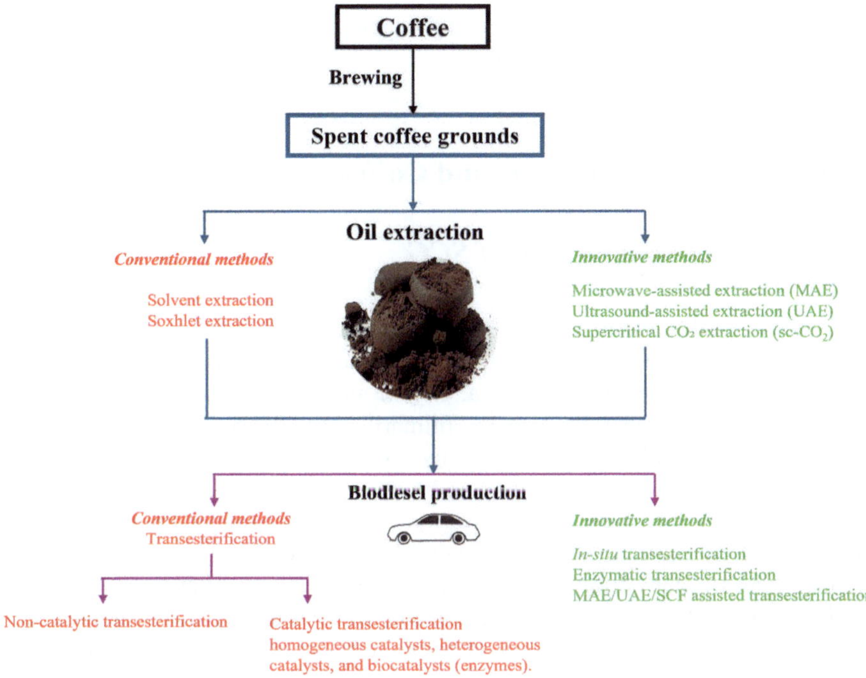


**Fig. 3** Biodiesel production from SCG

made a pioneering move by using biofuel from SCG in public buses in London, as reported by BBC News (2017). Biodiesel extracted from SCG supplied by local coffee companies is blended with diesel and used to power public buses without the need for engine modification. SCG-derived biodiesel offers numerous benefits, such as reduced raw material costs, waste elimination, similar operational efficiency to petro-diesel, lower tailpipe emissions (Kamil et al., 2019). Parallel to the development in extraction methods, considerable focus has been placed on the transesterification techniques for SCG. SCG-transesterification methods include conventional catalytic transesterification as well as enzymatic transesterification that employs biocatalysts for more effective and sustainable reactions. Innovative approaches such as ultrasound-assisted transesterification (UAT), supercritical fluid transesterification (SFE), and reactive extraction/in situ transesterification have been explored to improve SCG-based biodiesel production.

In this chapter, the physical and chemical characteristics of SCG are outlined, followed by an overview of both conventional and innovative coffee oil extraction methods. The transesterification processes of SCG are then described, highlighting the performance and feasibility of various proven and emerging technologies. Furthermore, the chapter presents an overview of life cycle assessment (LCA) and techno-economic analysis conducted on SCG-biodiesel, evaluating its industrial potential. Finally, the chapter identifies areas for future research and highlights the limitations of existing literature that could enable SCG-based biodiesel production to become a standard industry practice.

## 2 Characterization of SCG and Coffee Oil

### *2.1 Chemical Composition of SCG*

Coffee was first discovered in Ethiopia approximately 1000 years ago. The coffee tree belongs to the family *Rubiaceae*, and the genus *Coffea* comprises 103 species, which are divided into two subgenera: *Coffea* and *Baracoffea.* Among these, the subgenus *Coffea* includes the species primarily used in coffee production. The two most economically and commercially significant species are *Coffea arabica* (commonly known as Arabica coffee) and *Coffea canephora* (commonly known as *Robusta* coffee) (Atabani et al., 2019). Various methods are used to brew coffee, each producing different caffeine content (Atabani et al., 2019).

Since there are many different coffee varieties, SCG may also have varying properties depending on factors such as the region where the coffee is grown, climate conditions, and brewing methods. The main components of SCG are carbon, hydrogen, oxygen, nitrogen, sulfur, and ash, with carbon content ranging from 46 to 69%. Table 1 presents the chemical composition of SCG as reported in various literature sources. SCG generally has a moisture content typically ranging from 12 to 80% (Table 1).

### *2.2 Physical Properties and Chemical Composition of SCG Oil*

According to the literature, the fatty acids derived from the extraction of SCG encompass a broad spectrum, comprising both saturated and unsaturated groups. Table 2 provides an overview of the fatty acid composition of SCG oil, based on findings from various studies. These studies highlight notable differences in the oil compositions obtained. Palmitic and linoleic acids were the most abundant in all studies, though their percentages varied based on factors such as the type of coffee beans, the extraction method, and the solvent. For instance, Araujo et al. (2019) reported 31.31% palmitic acid and 44.18% linoleic acid in their SCG oil extracts, while Tarigan et al. (2019) found 32.90% palmitic acid and 41.68% linoleic acid.

**Table 2** Fatty acid composition of SCG oil as reported in various studies

| Fatty acids* | | | | | | References |
|---|---|---|---|---|---|---|
| C16:0 | C18:0 | C18:1 | C18:2 | C18:3 | C20:0 | |
| 34.44 | 0.00 | 7.74 | 43.12 | 1.18 | 2.83 | Somnuk et al. (2017) |
| 31.31 | 8.60 | 8.65 | 45.43 | 1.36 | ND | Araujo et al. (2019) |
| 32.90 | 7.50 | 9.02 | 44.18 | 1.40 | 2.80 | Tarigan et al. (2019) |
| 33.21 | 7.51 | 11.22 | 42.50 | 0.91 | 3.14 | Coelho et al. (2020) |
| 34.26 | 5.84 | 16.03 | 41.68 | 1.13 | 0.97 | Wang (2024) |

* C16:0 palmitic acid, C18:0 stearic acid, C18:1 oleic acid, C18:2 linoleic acid, C18:3 linolenic acid, C20:0 arachidic acid, ND: not detected

# 3 Oil Extraction Methods from SCG

## 3.1 Conventional Technologies

The literature outlines various methods for extracting oil from SCG. The oil yield reported in the studies in this section is expressed as the percentage of oil per 100 g of dry coffee waste.

### 3.1.1 Solvent Extraction

This widely practiced method involves combining a designated quantity of SCG with a suitable organic solvent, followed by vigorous shaking for a set period. The resulting oil-rich solvent mixture is then filtered to remove solid residues, and the solvent is recovered through vacuum evaporation, leaving the fatty acids isolated. The extraction process is generally performed at room temperature with n-hexane being the solvent most frequently used. Solvent extraction is also carried out with heating (without agitation), a process known as maceration. Several attempts have been made to explore the use of traditional solvent extraction for oil recovery from SCG. A recent study tested a range of solvents, including hexane, aqueous and anhydrous ethanol, and methanol (Somnuk et al., 2017). Hexane provided the highest yield of 14.7%. Further, an attempt was made to scale up the process in a prototype-scale continuous extraction unit using hexane, where a slightly lower yield of 11.8% was achieved. Compared to modern techniques such as Soxhlet and supercritical fluid extraction (SFE), traditional methods result in significantly lower oil yields, making them less favorable for efficient extraction processes.

### 3.1.2 Soxhlet Extraction

This most widely employed method for oil recovery from SCG operates on the same principle as solvent extraction. The method is named after the Soxhlet apparatus, which ensures continuous contact between the solvent and the sample, allowing the solvent to be recycled for repeated use. First, the SCG is weighed in the Soxhlet filter, which is positioned in the extraction chamber of the Soxhlet apparatus. The organic solvent, contained in the flask below, is heated, and upon reaching its boiling point, evaporates. It then condenses in the reflux condenser at the top of the Soxhlet apparatus and drips onto the SCG in the filter. The dissolved fatty acids transfer into the liquid phase and accumulate in the extraction chamber. When the chamber is full, the fatty acid-rich solvent flows back into the flask below through the connecting pipe. This cyclic process continues until extraction is complete. In the literature, most studies set a predetermined extraction time (Caetano et al., 2012; Efthymiopoulos et al., 2018), while others monitor the process until the solvent becomes colorless (Couto et al., 2009; Le et al., 2017). In the end, the fatty acid-rich solvent is evaporated for recovery and reuse.

Table 3 outlines the findings of studies on oil recovery from SCG using Soxhlet extraction. The highest oil yields obtained with each solvent, along with the respective experimental conditions are provided. n-hexane stands out as the most commonly used solvent, though alternative solvents, e.g., ethanol, heptane, isopropanol, and toluene, have also been explored. As seen in Table 3, oil yields mostly ranged from 8 to 22%, with the highest yield reaching 24.3% (Efthymiopoulos et al., 2018). A decrease in the solid-to-solvent ratio leads to an increase in oil yield. The larger solvent volume at lower ratios promotes the diffusion of oil molecules, resulting in a higher yield (Efthymiopoulos et al., 2018). Extraction times are typically between 4 and 8 h, with higher yields often achieved when the process extends for at least 6 h. The particle size and moisture content of SCG are identfied as the most significant factors affecting the Soxhlet extraction yield in these studies (Cante et al., 2021).

### 3.1.3 Supercritical Fluid Extraction (SFE)

The use of supercritical fluids is also widely adopted for oil recovery, with extensive applications in industries such as food, pharmaceuticals, and cosmetics. In this method, a fluid—typically $CO_2$—is brought to a supercritical state and introduced to the SCG inside a pressure tank, where it dissolves the fatty acids. Once the extraction is complete, the pressure is reduced, enabling the evaporation of liquid $CO_2$ and its separation from the extracted fatty acids. The evaporated solvent is then returned to its supercritical state and reintroduced into the pressure vessel. This process is repeated until the desired yield of fatty acids is achieved. The key advantages of SFE include the absence of toxic organic solvents and the ability to operate under moderate pressure and temperature, attributed to the thermophysical properties of $CO_2$. Notably, it offers an eco-friendly alternative to traditional methods that use toxic and flammable solvents like n-hexane.

**Table 3** Oil recovery from SCG using different solvents in Soxhlet extraction: Extraction conditions and yields

| Solvent | Solid-solvent ratio (g $mL^{-1}$) | Time (min.) | Yield (%) | Reference |
|---|---|---|---|---|
| Isopropanol | 1:4.2 | 70 | 11.43 | Al-Hamamre et al. (2012) |
| | 1:20 | 150–570 | 21.0 | Caetano et al. (2012) |
| Ethanol | 1:4.2 | 40 | 11.9 | Al-Hamamre et al. (2012) |
| | 1:9 | 480 | 20.9 | Efthymiopoulos et al. (2018) |
| | 1:20 | 150–570 | 16.0 | Caetano et al. (2012) |
| | 1:30 | – | 15.64 | Araujo et al. (2019) |
| Pentane | 1:4.2 | 30 | 15.18 | Al-Hamamre et al. (2012) |
| | 1:9 | 480 | 18.61 | Efthymiopoulos et al. (2018) |
| n-hexane | 1:4.2 | 30 | 15.28 | Al-Hamamre et al. (2012) |
| | 1:9 | 480 | 24.3–16.3 | Efthymiopoulos et al. (2018) |
| | 1:20 | 150–570 | 16.0 | Caetano et al. (2012) |
| | 1:30 | – | 14.52 | Araujo et al. (2019) |
| | – | 180 | 18.30 | Couto et al. (2009) |
| | 1:15 | 900–960 | 16.0 | Le et al. (2017) |
| | 1:43 | 240 | 8.6 | Hibbert et al. (2019) |
| | – | 300 | 16.6 | Cante et al. (2021) |
| Toluene | 1:4.2 | 30 | 14.32 | Al-Hamamre et al. (2012) |
| | 1:9 | 480 | 22.51 | Efthymiopoulos et al. (2018) |
| Heptane | 1:9 | 480 | 14.49 | Efthymiopoulos et al. (2018) |
| | 1:20 | 150–570 | 18.0 | Caetano et al. (2012) |
| Dichloromethane | 1:9 | 480 | 23.83 | Efthymiopoulos et al. (2018) |
| Acetone | 1:4.2 | 40 | 12.92 | Al-Hamamre et al. (2012) |
| Ethyl acetate | 1:30 | | 15.04 | Araujo et al. (2019) |
| Chloroform | 1:4.2 | 40 | 11.15 | Al-Hamamre et al. (2012) |

Few studies in the open literature focus on the recovery of valuable components from SCG using SFE. Although several studies focus on the recovery of phenols and triglycerides, the majority investigates the efficiency of oil recovery. SFE studies in literature exclusively use $CO_2$ as the supercritical fluid often combined with co-solvents such as ethanol and isopropanol. Couto et al. (2009) were the first to attempt the use of supercritical $CO_2$ for oil recovery from SCG. Extraction tests showed that with SFE, a maximum yield of 15.4% can be achieved within the first 3 h; however, this yield is approximately 85% of that of n-hexane Soxhlet extraction.

In a more recent study, Coelho et al. (2020) explored both sc-$CO_2$ and co-solvent-assisted sc-$CO_2$ extraction, assessing the impact of various co-solvents (isopropanol, ethanol, and ethyl lactate) on yield. Surpassing the Soxhlet extraction yield of 10.4%, the co-solvent-assisted sc-$CO_2$ method achieved a maximum yield of 11–12%. A notable advantage of adding a co-solvent was the halving of the time required to reach this maximum yield. Likewise, Couto et al. (2009) found that adding ethanol as a co-solvent boosted the yield by 6.5% and reduced the time required to reach the maximum yield to roughly one-third of that needed for the sc-$CO_2$ method.

### 3.1.4 Other Methods

Ultrasound-assisted extraction (UAE) and microwave-assisted extraction (MAE) are among the most widely applied methods for oil extraction from SCG, following Soxhlet and sc-$CO_2$ extraction. UAE leverages the cavitation effect generated by high-frequency sound waves. Gas bubbles generated by the sound waves collapse, leading to the breakdown of cell walls and enabling the solvent to penetrate the SCG particles more efficiently, thereby enhancing the extraction rate and reducing solvent usage. MAE is a technique that utilizes microwave energy to heat and rupture cell walls, enabling the rapid and efficient release of valuable compounds such as fatty acids into the solvent. The microwave-induced heating of the solvent further enhances the extraction yield; MAE is advantageous due to its short processing times (Hibbert et al., 2019) and low energy consumption (Coelho et al., 2020). The open literature offers limited research on oil recovery from SCG using UAE and MAE processes.

Hibbert et al. (2019) examine the impact of factors such as extraction time, solvent volume, and temperature on oil recovery from SCG using MAE. The maximum oil yield was 11.54%, with optimal conditions being 95 °C, 160 mL of n-hexane, and 10 min. MAE required 24 times less time and less solvent compared to Soxhlet extraction for the same yield. Le et al. (2017) compared UAE, MAE, Soxhlet, and maceration methods, finding that MAE achieved the highest yield of 17.96% with a 1:15 solid-to-solvent ratio and hexane. Soxhlet extraction provided a similar yield to MAE, while maceration had the lowest yield. In terms of extraction time, the ranking was: MAE < UAE < Soxhlet << maceration, emphasizing the efficiency of MAE and UAE for oil recovery from SCG.

Pressurized liquid extraction (PLE), also known as accelerated solvent extraction (ASE), has emerged in recent years as an efficient and environmentally friendly method for extracting oil. PLE uses high pressure and temperature for extraction, resulting in a faster process compared to atmospheric conditions, with a reduced need for solvent (Hibbert et al., 2019). A detailed study by Efthymiopoulos et al. (2018) on SCG valorization via PLE examined the influence of pressure, temperature, time, and solvent type on both extraction efficiency and the composition of the extracted oil. The study findings revealed that the oil yield improved with rising temperature when polar solvents were used. At 165 °C, ethanol yielded the highest oil yield of 28.8%, surpassing the maximum yield obtained with Soxhlet extraction (n-hexane) by 20%. The composition of the obtained oil varied depending on the solvent used (Efthymiopoulos et al., 2018). Moreover, it was observed that extraction pressure had a significant effect on oil yield. Ultimately, compared to Soxhlet extraction, the PLE method, which operates at higher temperatures, enables the extraction of more oil in a shorter time and with lower SCG-to-solvent ratios.

## *3.2 Innovative Technologies for SCG Extraction*

As detailed in earlier sections, traditional and modern methods, including microwave-assisted and supercritical fluid extraction, enable oil extraction from SCG; however, the necessity for greener and more efficient approaches persists. Open literature highlights a limited number of innovative studies focused on improving existing techniques. This chapter evaluates these promising technologies for recovering oil from SCG with a focus on sustainability. The improvements include (i) various pre-treatment processes prior to extraction, (ii) innovative extraction methods, and (iii) the use of more effective, green solvents. These improvements are designed to enhance the yield and quality of the extracted oil, minimize the use of toxic solvents, and shorten processing time.

The incorporation of co-solvents is the primary improvement applied in sc-$CO_2$ extraction of oil from SCG. Co-solvents significantly enhance the solubility of the target compounds, increasing the yield. Various co-solvents, such as ethanol, isopropanol, and ethyl acetate, have been used (Coelho et al., 2020; Couto et al., 2009). These studies demonstrated that co-solvents notably shorten the extraction time and enhances yield. For instance, Couto et al. (2009) found that ethanol as a co-solvent in sc-$CO_2$ at 50 °C and 20.0 MPa increased yield by 6.5% and reduced extraction time by 60%, improving solvent density, and creating new chemical interactions between the co-solvent and oil molecules.

Another innovative extraction method, similar to co-solvent sc-$CO_2$ extraction, is gas-expanded liquid (GXL) extraction. GXL involves adding compressed $CO_2$ to a liquid solvent, thereby increasing solubility and enhancing mass transfer. GXL can operate at lower pressures than sc-$CO_2$ extraction, while requiring less solvent compared to pressurized liquid extraction (PLE). Araujo et al. (2019) conducted the first study in the literature on oil extraction from SCG using the gas-expanded liquid extraction method, with ethanol as the solvent. The GXL extraction was compared with Soxhlet, sc-$CO_2$, and PLE, with the highest yield (15.9%, 80 °C, 20 MPa,

25 min.) achieved using GXL extraction (sc-$CO_2$+ ethanol). A comparable yield was achieved with PLE using ethanol alone; however, the use of co-solvent significantly reduced extraction time and the amount of organic solvent required.

Cante et al. (2021) proposed a novel subcritical liquefied gas extraction approach using the hydrofluorocarbon solvent Norflurane (R134a). Norflurane (R134a), with its fluorinated hydrocarbon structure, low toxicity, and ability to liquefy at low pressures, provides an eco-friendly and cost-effective alternative to commercial solvents and supercritical $CO_2$. This method operates at milder pressures and temperatures, offering an energy-efficient option compared to sc-$CO_2$ extraction. Using recirculating solvent system at 0.5–1.1 MPa, the process achieved 92% of the Soxhlet extraction yield (16.6%) in 75, 90, and 285 min with wet, moist, and dried SCG, respectively, without the need for pre-drying.

In recent years, bio-based solvents have been extensively studied as a sustainable alternative to hexane for oil recovery from oilseeds. One study has explored the use of bio-based solvent alternatives for valorization of SCG. Chemat et al. (2022) demonstrated that Soxhlet extraction with bio-based solvent 2-methyloxolane (2-MeOx) can achieve high oil extraction yields, showing that a bio-based solvent can effectively substitute the toxic, petroleum-drived hexane. The oil yield was 12.5% using hexane, whereas with anhydrous 2-MeOx and aqueous 2-MeOx (95.5%), it reached 13.7% and 15.8%, respectively.

One improvement in the Soxhlet extraction for oil recovery from SCG involves using solvent blends. Combining solvents with different structures creates a synergistic effect, enhancing oil dissolution, reducing extraction times, and lowering energy consumption. Caetano et al. (2012) studied the efficiency of a hexane–isopropanol mixture in various ratios. They found that after 3 h, a 50:50 blend yielded 21.5% oil, while isopropanol alone produced 21% after 6.8 h, demonstrating the advantage of solvent mixtures in improving efficiency.

## 4 Methods for Biodiesel Production from SCG

Biodiesel production from SCG has garnered significant attention as a sustainable and renewable energy solution. This process includes stages of collecting, transporting, drying, oil extraction, and the conversion of SCG into biodiesel. SCG stands out due to its higher heating value (5960 kcal/kg) and energy density compared to conventional feedstocks such as rice husk (3805.3 kcal/kg) and wheat straw (4185.0 kcal/kg) (Karmee, 2018). Ongoing research is exploring both traditional and novel methods for converting SCG into biodiesel as detailed in the following section.

## 4.1 Conventional Technologies

Conventional methods for SCG-to-biodiesel production primarily rely on transesterification, which can be performed via catalytic or non-catalytic routes, each with distinct operational requirements. Non-catalytic transesterification is generally slow, necessitating high pressure and temperature to reach completion. In contrast, catalytic transesterification is the most widely used method due to its advantages, including lower pressure and temperature requirements, higher product yield, the production of biodiesel without significant by-products, and the avoidance of toxic substances. Catalysts used in biodiesel production are categorized into three main groups: homogeneous catalysts, heterogeneous catalysts, and biocatalysts (enzymes). A key drawback of heterogeneous catalysts is their reaction time, which is significantly longer than that of homogeneous catalyst (Wang, 2024).

Table 4 presents a comparative summary of selected recent studies on biodiesel production from SCG, highlighting differences in catalysts used, oil-to-alcohol ratio, and reaction conditions. Among conventional methods, Uddin et al. (2019) utilized a two-step approach, applying sulfuric acid in the first stage and KOH in the second. Among studies using enzymatic catalysts, Karmee (2018) achieved a 96% biodiesel yield using the lipase enzyme Novozyme-435 over a 15-h reaction time. Despite the success of conventional methods in converting SCG into biodiesel, research is ongoing to explore more efficient and sustainable approaches, which will be discussed in detail in the following section.

## 4.2 Innovative Technologies for Biodiesel Production from SCG

Various emerging technologies have been explored to enhance biodiesel yield, reduce reaction times, and improve the sustainability of biodiesel production from SCG. Selected studies on different novel approaches can be seen in Table 4. Among these, Wang, (2024) investigated the transesterification of SCG in an ultrasonic reactor using a green nanocatalyst derived from pomegranate and orange peel activated carbon (POPAC) at varying weight percentages (wt.%). Similarly, another innovative method includes electrolytic transesterification (Annal et al., 2024), which focuses on optimizing parameters such as voltage, process time, and catalyst concentration (wt.%) to achieve maximum efficiency. In addition, supercritical ethanol transesterification has been investigated under different temperatures and reaction times in both batch and continuous reactors to increase biodiesel yield from SCG (Supang et al., 2024).

Furthermore, Nguyen et al. (2020) explored the use of 1,8-diazabicyclo [5.4.0] undec-7-ene (DBU) as the catalyst, directing optimization efforts toward methanol amount, DBU concentration, and reaction temperature. In another study, Amberlyst ion-exchange resin catalyst was used but the biodiesel yield was lower (89.62%)

**Table 4** Various studies investigating biodiesel production from SCG*

| Methods | Catalysis | Yield (%) | Reaction Conditions | | | References |
|---|---|---|---|---|---|---|
| | | | Oil-to-alcohol molar ratio | Temp. (°C) | Time (h) | |
| *Conventional methods* | | | | | | |
| Enzymatic TE | Lipase enzyme (Novozyme-435) | 96 | 1:5 | 40 | 15 | Karmee (2018) |
| Two-step TE | Acid ($H_2SO_4$) + base (KOH) | High conversion rate | 1:12 | 60 | 1<br>2 | Uddin et al. (2019) |
| *Innovative methods* | | | | | | |
| TE with enzyme immobilization | *Thermomyces lanuginosus* on citric acid modified $Fe_3O_4$/Au magnetic nanoparticles (solvent-free) | 98.4 | 1:6 | 50 | 24 | Sarno and Iuliano (2018) |
| Reactive extraction Soxhlet (RES) with wet SCG | Methanol/ co-solvent 1:1 and 0.25 M NaOH in methanol as the catalyst | 97 | – | 65 | 0.5 | Tarigan et al. (2019) |
| Direct esterification | DBU (1,8-diazabicyclo [5.4.0] undec-7-ene) | 97.18 | 6.25 mL methanol/g wet SCG | 60 | 0.48 | Nguyen et al. (2020) |
| In situ solvo-thermal co-TE | Wet SCG/ microalgae mixture | 79.9 | – | 198 | 2.2 | Abomohra et al. (2021) |
| Ultrasonication | POPAC green nanocatalyst | 95.24 | 1:10 | 60 | 0.5 | Wang (2024) |
| Electrolysis | NaOH | 98.32 | 0.75 wt.% catalyst | 40 Volt | 2 | Annal et al. (2024) |
| sc-ethanol TE | Ethanol (SCF) | 88.37 | 1:30 | 275 | 0.67 | Supang et al. (2024) |

* TE: transesterification, SCF: supercritical fluid

compared to the conventional two-step transesterification process. In a different approach, Sarno and Iuliano (2018) investigated the use of *Thermomyces lanuginosus* lipase immobilized on citric acid (CA)-functionalized $Fe_3O_4$/Au magnetic nanoparticles for transesterification reactions. They introduced an innovative and efficient method for CA ligand exchange, which significantly enhanced enzyme stability and loading capacity. The immobilized lipase was then applied to produce biodiesel from SCG in a solvent-free environment. Notably, the enzyme maintained more than 90%

of its initial catalytic activity after 60 days of storage. The process achieved a biodiesel yield of 51.7% within just 3 h, with the yield approaching nearly 100% after 24 h, highlighting the system's rapid enzymatic kinetics. The resulting biodiesel met the EN14214 standard, with an ester content of 98.4 $\pm$ 0.23%. Furthermore, the iodine value of 76.67 g iodine/100 g complied with European biodiesel regulations.

Lastly, Abomohra et al. (2021) evaluated the in *situ* co-transesterification of wet SCG/microalgae mixture for enhanced biodiesel production. They showed that the SCG-algae mixture has dual effects in increasing biodiesel yield and quality compared to individual transesterification. In the studies examined, the biodiesel standards align closely biodiesel produced from other feedstocks, such as rice husk and soybean oil, and comply with established specifications, such as EN 14214 and ASTM D6751 standards for biodiesel quality. Overall, these developments demonstrate the potential of SCG as a valuable feedstock for biodiesel production and highlight ongoing efforts to optimize and innovate production methodologies.

## 5 Life Cycle Assessment and Techno-economic Analysis of Biodiesel Production from SCG

A realistic assessment of the potential of secondary biodiesel feedstocks, such as SCG, can only be made by considering environmental impacts and economic indicators. Subsequent research has provided detailed insights into the environmental and techno-economic aspects of SCG valorization routes.

Studies examining the environmental impact of SCG valorization, including biodiesel production, can be classified into four groups: (i) Comparing SCG with other feedstocks and petro-diesel (Kookos, 2018; Kamil et al., 2019); (ii) comparing different biodiesel production processes for SCG (Tuntiwiwattanapun, 2017); (iii) evaluating biodiesel production against alternative disposal and reuse methods such as composting, incineration, brick-making, landfilling, and anaerobic digestion (Rivera et al., 2020); and (iv) comparing different biomass-to-fuel and biomass-to-energy routes (Gu et al., 2023; Kisiga et al., 2024). Comparisons are made using standard life cycle assessment procedures; however, performance metrics vary across studies.

Kookos (2018) conducted a simulation study applying a gate-to-gate LCA, focusing exclusively on emissions from the transesterification process, excluding other life cycle stages. The study utilized conventional extraction followed by two-stage transesterification. Greenhouse gas (GHG) emissions were 0.73 $CO_2$-eq/kg biodiesel. The study concluded that the environmental impact of SCG-biodiesel was comparable to conventional feedstocks, such as soybean oil, and considerably lower than conventional diesel production. Kamil et al. (2019) expanded the LCA system boundaries from waste collection to distribution stations, covering combustion in engines and transportation. The environmental performance was evaluated using experimental data on biodiesel production via n-hexane extraction and transesterification, followed by real-time engine combustion tests. The results substantiated the potential of SCG-derived biodiesel over petro-diesel, showing an 80.5% reduction in $CO_2$ emissions and a 10% higher conversion efficiency. Emissions from extraction (0.252 kg $CO_2$-eq/kg biodiesel) and transesterification (0.154 kg $CO_2$-eq/kg biodiesel) stages were significantly lower than those reported by Kookos (2018).

Tuntiwiwattanapun et al. (2017) conducted a theoretical study comparing the environmental impact of an on-site in situ transesterification process—using methanol for both extraction and transesterification—to the conventional n-hexane process. LCA showed that n-hexane process was more environmentally sustainable than in situ transesterification, with lower environmental impacts in 13 out of 15 categories. An important outcome of the LCA was conventional transesterification was associated with three times the respiratory risk of in situ transesterification, due to the use of hazardous n-hexane.

Two comprehensive LCA studies were applied to the biodiesel production process proposed by Tuntiwiwattanapun et al. (2017). Rivera et al. (2020) conducted a comparative analysis of biodiesel production with five alternative reuse/disposal routes using a cradle-to-grave approach. The LCA indicated that biodiesel production had the highest impact in 11 out of 16 categories, making it the least sustainable option, followed by composting. Still, it has a negative climate change impact, saving −4.3 kg $CO_2$-eq/ton of SCG-treated and $CO_2$ and $CH_4$ emissions associated with methanol recovery units were identified as the primary contributors. Expressed per MJ of energy produced, the carbon impact of SCG-derived biodiesel is estimated at approximately −1.65 g $CO_2$-eq/MJ, significantly lower than other waste-based biodiesels, typically ranging from −88 to +80 g $CO_2$-eq/MJ, and much lower than fossil diesel emissions (83.8 g $CO_2$-eq/MJ), thus meeting the EU Renewable Energy Directive's 60% reduction target for new production plants (Rivera et al., 2020).

Tuntiwiwattanapun et al. (2017) stressed the critical role of feedstock transport distance when assessing the environmental benefits of a low-carbon production pathway over landfill disposal, as long transport distances may render the alternative unsustainable due to transport-associated emissions.

A few simulation studies assessing the environmental sustainability of SCG conversion routes to fuel and energy are available in the open literature (Gu et al., 2023; Kisiga et al., 2024). Considering a LCA boundary from SCG collection from shops to fuel and energy generation and use, Kisiga et al. (2024) found that transesterification is a relatively unsustainable option compared to other reuse strategies. The ranking in terms of environmental impact was anaerobic digestion < pyrolysis < transesterification < gasification < fermentation < hydrothermal liquefaction. Aiming to compare SCG-biooil and SCG-biodiesel production, Gu et al. (2023) conducted a more comprehensive cradle-to-gate LCA, including the cultivation, milling, transport, brewing, and reuse stages of the coffee life cycle. An interesting result, countering the findings of previous studies (Kisiga et al., 2024), was conversion via the transesterification process (0.0140 kg $CO_2$-eq/MJ) produced lower $CO_2$ emissions than conversion via the pyrolysis process (0.0192 kg $CO_2$-eq/MJ). However, the emissions associated with the cultivation and transport of coffee required for transesterification per 1 MJ of biofuel produced were considerably larger, rendering the total $CO_2$ emissions from this technology 3.49 times larger than those from pyrolysis. Overall, transesterification was found to be a less sustainable option for SCG valorization, ranking below pyrolysis but still outperforming certain reuse strategies such as gasification and hydrothermal liquefaction in terms of overall sustainability.

In terms of process economics, the techno-economic feasibility of SCG-based biodiesel production has been examined in a limited number of studies. The main conclusions are as follows: First, SCG-based biodiesel production is economically viable only at large-scale operations (Kookos, 2018; Yeoh & Ng, 2022). Capacity estimations indicate that production becomes cost-effective at scales exceeding 42 kton/year, where economies of scale significantly reduce unit production costs (Kookos, 2018). Second, SCG-based biodiesel production is not economically feasible unless integrated with other value-added products, including co-generated electricity and glycerol (Kookos, 2018), defatted SCG pellets and glycerol (Kamil et al., 2019), activated carbon and glycerol or high-value biochemicals (Yeoh & Ng, 2022). Third, logistical challenges in the collection and transportation of SCG significantly impact the economic viability, especially at large scale (Kookos, 2018). On-site collection strategies, such as sourcing SCG directly from coffee production facilities (Yeoh & Ng, 2022) or central facilities (Kookos, 2018), could mitigate transportation costs. Fourth, the major contributors to energy and capital costs are hexane extraction and solvent recovery (Gu et al., 2023), and transesterification (Kamil et al., 2019). Lastly, advancements in extraction, transesterification processes, and integration strategies are expected to enhance the economic feasibility of SCG-based biodiesel, making it a more competitive alternative in the biofuel market (Kamil et al., 2019; Rivera et al., 2020; Tuntiwiwattanapun et al., 2017). However, although few studies compare the economic performance of SCG-to-biodiesel with other fuel alternatives, available results offer valuable insights. A comparison of two distinct transesterification methods, based on energy consumption—rather than economic factors—demonstrates that conventional transesterification (6.49 MJ/kg) demands significantly less energy than in situ transesterification (11.38 MJ/kg) (Tuntiwiwattanapun et al., 2017). This difference is primarily due to the high energy consumption associated with methanol recovery, which accounts for 73% of the total energy used in in situ transesterification. Furthermore, in evaluating the economic viability of alternative waste-to-fuel conversion pathways, pyrolysis emerges as a more cost-effective option, requiring 3.8 times less production cost per unit of energy (1 MJ) compared to transesterification (Gu et al., 2023).

## 6 Emerging Challenges and Opportunities in Valorizing SCG

Building on the improvements in Sect. 3.2, promising research areas remain for SCG oil recovery. Ionic liquids (ILs) and deep eutectic solvents (DES) have gained attention as alternatives to traditional solvents for recovering valuable components from biomass. Recent research has focused on ILs for recovering components like caffeine and polysaccharides from SCG and husks, but oil recovery from SCG using ILs remains unexplored. Similarly, DESs, known for their eco-friendly properties, have been tested for extracting other coffee waste by-products like nutraceuticals and bioactive phenolics, but not for oil recovery. Both ILs and DESs hold significant potential for oil recovery from SCG, and various IL and DES structures are expected to be further explored for this application in the near future.

Although significant progress has been made in optimizing transesterification processes, ongoing research in areas such as alternative catalysts and reaction conditions are still essential. Recent innovations, such as the use of immobilized enzyme systems and reactive Soxhlet extraction, are still in their early stages and present promising areas for further exploration. For instance, combining extraction and transesterification in a single step within a Soxhlet apparatus has been shown to reduce the reaction time to 0.5 h while achieving a yield of 97% (Table 4) (Tarigan et al., 2019). These emerging techniques require deeper investigation and development to optimize their efficiency and enable scale-up.

Based on recent studies, future directions in sustainability and techno-economics for SCG-based biodiesel production should explore the recovery of additional products, including bioactive compounds and biofuels, through a cascade biorefinery approach. Furthermore, scenarios involving the use of renewable energy sources to meet the steam and electricity demands of SCG-based biodiesel conversion processes should be developed and tested (Tuntiwiwattanapun et al., 2017). A key gap in the current literature is the replacement of petrochemical-based n-hexane and methanol with bio-based alternatives, as this holds significant potential for improving sustainability in SCG-based biodiesel production (Kamil et al., 2019; Rivera et al., 2020). Moreover, more comprehensive modeling studies are needed to assess the process viability and sustainability of these approaches.

## 7 Conclusions

Growing concerns regarding the environmental impact and limited availability of diesel fuel have intensified research into alternative and sustainable energy sources such as biodiesel. Among various potential feedstocks, spent coffee grounds (SCG) have emerged as a promising candidate due to their abundance and valuable chemical composition. The production process involves oil extraction from SCG, followed by transesterification to convert the extracted oil into biodiesel. A growing body of research emphasizes key methods for biodiesel production from SCG, including traditional Soxhlet extraction, innovative approaches like supercritical fluid extraction (SFE), and cutting-edge techniques such as gas-expanded liquid (GXL) extraction. Among these, Soxhlet extraction, being the most widely used method, achieves an oil yield ranging from 8 to 22%. The yield is influenced by factors such as coffee variety, processing method, drying time, solvent type, and applied extraction technique. Innovative approaches, such as ultrasound-assisted extraction (UAE) and microwave-assisted extraction (MAE), show promise in reducing extraction time and solvent usage, offering environmental benefits, though further optimization is needed for enhanced oil yields for large-scale application. Solvent selection is a critical factor influencing both yield and sustainability. Future research should focus on minimizing the environmental impact through the use of advanced solvent systems (e.g., ILs and DESs), co-solvent applications and hybrid or innovative extraction methods to enhance both economic feasibility and sustainability. Both traditional and innovative methods of biodiesel production face challenges such as the high water content of spent coffee grounds and the presence of free fatty acids. Excess water can negatively impact the transesterification reaction, reducing biodiesel yield, making drying a crucial preprocessing step. Additionally, free fatty acids can lead to saponification, particularly in the presence of base catalysts, which decreases biodiesel yield and complicates purification. Traditional methods typically address these challenges using a two-stage acid–base catalysis approach, whereas innovative methods focus on enhancing efficiency and sustainability through enzymatic catalysts, nanocatalysts, and supercritical fluid technologies. Sustainability benefits of biodiesel production from SCGs are evident, with key factors such as collection and transportation, the chemicals used in extraction and transesterification, and the chosen biodiesel production process shaping the overall carbon footprint and environmental impact. Key findings indicate that biodiesel (1.65 g $CO_2$-eq/MJ) from SCG has a lower environmental impact than petroleum-based diesel (83.8 g $CO_2$-eq/MJ) (Rivera et al., 2020) and is comparable to conventional feedstocks like soybean oil. However, LCA results vary depending on system boundaries and SCG-to-biodiesel conversion routes. The highest contributors to the carbon footprint of SCG-biodiesel are the emissions from n-hexane used in extraction and methanol used in transesterification. Conventional transesterification outperforms in situ transesterification, showing three times less environmental impact. However, alternative valorization methods, such as anaerobic digestion and pyrolysis, could be more sustainable than biodiesel production due to their lower overall emissions. Bio-based alternatives for n-hexane and methanol, along with green energy integration, are crucial for enhancing sustainability. Future research should focus on process configurations using bio-based versions of these chemicals and other green processes and chemical alternatives. Lastly, from an economic perspective, biodiesel production from SCG is feasible only in large-scale operations and when integrated with the production of other value-added products. Logistical challenges in SCG collection, transportation, and high energy and capital costs limit widespread implementation. Advancements in waste-to-biodiesel conversion technologies are likely to position SCG-derived biodiesel as a more competitive and sustainable alternative in the biofuel market in the near future.

**Acknowledgements** The authors would like to acknowledge the support of Yalova University Scientific Research Projects Coordination Unit for funding this study under project number 2023/AP/0013. Additionally, the authors also acknowledge the use of ChatGPT (OpenAI) as an assistive tool for grammatical editing and proofreading of this section.

## References

Abomohra, A. E., Zheng, X., Wang, Q., Huang, J., & Ebaid, R. (2021). Enhancement of biodiesel yield and characteristics through in-situ solvo-thermal co-transesterification of wet microalgae with spent coffee grounds. *Bioresource Technology, 323*, 124640. https://doi.org/10.1016/j.biortech.2020.124640

Al-Hamamre, Z., Foerster, S., Hartmann, F., Kröger, M., & Kaltschmitt, M. (2012). Oil extracted from spent coffee grounds as a renewable source for fatty acid methyl ester manufacturing. *Fuel, 96*, 70–76. https://doi.org/10.1016/j.fuel.2012.01.023

Annal, U. N., Vaithiyanathan, R., Natarajan, A., Rajadurai, V., Kumar, P. S. M., & Li, Y. Y. (2024). Electrolytic biodiesel production from spent coffee grounds: Optimization through response surface methodology and artificial neural network. *Journal of the Taiwan Institute of Chemical Engineers, 165*, 105697. https://doi.org/10.1016/j.jtice.2024.105697

Araujo, M. N., Azevedo, A. Q. P. L., Hamerski, F., Voll, F. A. P., & Corazza, M. L. (2019). Enhanced extraction of spent coffee grounds oil using high-pressure $CO_2$ plus ethanol solvents. *Industrial Crops and Products, 141*, 111723. https://doi.org/10.1016/j.indcrop.2019.111723

Atabani, A. E., Al-Muhtaseb, A. H., Kumar, G., Dattatraya, G., Saratale, G. D., Aslam, M., Khan, H. A., Said, Z., & Mahmoud, E. (2019). Valorization of spent coffee grounds into biofuels and value-added products: Pathway towards integrated bio-refinery. *Fuel, 254*, 115640. https://doi.org/10.1016/j.fuel.2019.115640

BBC News. (2017, Nov 21). *London buses to be powered by coffees.* https://www.bbc.com/news/uk-england-london-42044852

Caetano, N. S., Silva, V. F., & Mata, T. M. (2012). Valorization of coffee grounds for biodiesel production. *Chemical Engineering Transactions, 26*, 267–272. https://doi.org/10.3303/CET1226045

Cante, R. C., Garella, I., Gallo, M., & Nigro, R. (2021). Effect of moisture content on the extraction rate of coffee oil from spent coffee grounds using Norflurane as solvent. *Chemical Engineering Research and Design, 165*, 172–179. https://doi.org/10.1016/j.cherd.2020.11.002

Chemat, A., Ravi, H. K., Hostequin, A. C., Burney, H., Tomao, V., & Fabiano-Tixier, A. S. (2022). Valorization of spent coffee grounds by 2-methyloxolane as bio-based solvent extraction. Viable pathway towards bioeconomy for lipids and biomaterials. *OCL, 29*, 7. https://doi.org/10.1051/ocl/2021052

Coelho, J. P., Filipe, R. M., Robalo, M. P., Boyadzhieva, S., Cholakov, G. S., & Stateva, R. P. (2020). Supercritical $CO_2$ extraction of spent coffee grounds. Influence of co-solvents and characterization of the extracts. *The Journal of Supercritical Fluids, 161*, 104825. https://doi.org/10.1016/j.supflu.2020.104825

Couto, R. M., Fernandes, J., Da Silva, M. G., & Simoes, P. C. (2009). Supercritical fluid extraction of lipids from spent coffee grounds. *The Journal of Supercritical Fluids, 51*(2), 159–166. https://doi.org/10.1016/j.supflu.2009.09.009

Das, A., & Ghosh, S. (2022). Role of microorganisms in extenuation of mining and industrial wastes. *Geomicrobiology Journal, 39*, 173–175. https://doi.org/10.1080/01490451.2022.2038953

Das, A., Ray, S. K., Mohanty, M., Mohanty, J., Dey, S., & Das, A. P. (2024). Ecotoxicity of microplastic wastes and their sustainable management: A review. *Environmental Chemistry and Ecotoxicology, 6*, 144–152. https://doi.org/10.1016/J.ENCECO.2024.05.003

Efthymiopoulos, I., Hellier, P., Ladommatos, N., Russo-Profili, A., Eveleigh, A., Aliev, A., Kay, A., & Mills-Lamptey, B. (2018). Influence of solvent selection and extraction temperature on yield and composition of lipids extracted from spent coffee grounds. *Industrial Crops and Products, 119*, 49–56. https://doi.org/10.1016/j.indcrop.2018.04.008

Gu, J., Lee, A., Choe, C., & Lim, H. (2023). Comparative study of biofuel production based on spent coffee grounds transesterification and pyrolysis: Process simulation, techno-economic, and life cycle assessment. *Journal of Cleaner Production, 428*, 139308. https://doi.org/10.1016/j.jclepro.2023.139308

Hibbert, S., Welham, K., & Zein, S. H. (2019). An innovative method of extraction of coffee oil using an advanced microwave system: In comparison with conventional Soxhlet extraction method. *SN Applied Sciences, 1*, 1–9. https://doi.org/10.1007/s42452-019-1457-5

International Coffee Organization. (2023). *Coffee development report 2022–23*. https://ico.org/documents/cy2024-25/coffee-development-report-2022-23.pdf

Kamil, M., Ramadan, K. M., Olabi, A. G., Shanableh, A., Ghenai, C., Al Naqbi, A. K., Awad, O. I., & Ma, X. (2019). Comprehensive evaluation of the life cycle of liquid and solid fuels derived from recycled coffee waste. *Resources, Conservation and Recycling, 150*, 104446. https://doi.org/10.1016/j.resconrec.2019.104446

Karmee, S. K. (2018). A spent coffee grounds based biorefinery for the production of biofuels, biopolymers, antioxidants and biocomposites. *Waste Management, 72*, 240–254. https://doi.org/10.1016/j.wasman.2017.10.042

Kisiga, W., Chetty, M., & Rathilal, S. (2024). Environmental impact assessment of alternative technologies for production of biofuels from spent coffee grounds. *Energy Science & Engineering, 12*(11), 4823–4842. https://doi.org/10.1002/ese3.1933

Kookos, I. K. (2018). Technoeconomic and environmental assessment of a process for biodiesel production from spent coffee grounds (SCGs). *Resources, Conservation and Recycling, 134*, 156–164. https://doi.org/10.1016/j.resconrec.2018.02.002

Le, P. T. K., Vu, Q. T. H., Nguyen, Q. T. V., Tran, K. A., & Le, K. A. (2017). Extraction and evaluation the biological activities of oil from spent coffee grounds. *Chemical Engineering Transactions, 56*, 1729–1734. https://doi.org/10.3303/CET1756289

Mishra, S., & Das, A. P. (2023). Microbial Remediation of Synthetic Microfiber Contaminated Wastewater. In M. P. Shah (Ed.), *Microbial Technologies in Industrial Wastewater Treatment* (pp. 337–350). Singapore: Springer. https://doi.org/10.1007/978-981-99-2435-6_17

Mishra, S., Ghosh, S., van Hullebusch, E. D., Singh, S., & Das, A. P. (2023). A critical review on the recovery of base and critical elements from electronic waste-contaminated streams using microbial biotechnology. *Applied Biochemistry and Biotechnology, 2023*, 1–30. https://doi.org/10.1007/S12010-023-04440-X

Mishra, S., Rout, P. K., & Das, A. P. (2019). Solar photovoltaic panels as next generation waste: a review. *Biointerface Research in Applied Chemistry, 9*(6).

Mohanty, S., Ghosh, S., Bal, B., & Das, A. P. (2018). A review of biotechnology processes applied for manganese recovery from wastes. *Reviews in Environmental Science and Bio/Technology 17*(4), 791–811. https://doi.org/10.1007/S11157-018-9482-1

Nguyen, H. C., Nguyen, M. L., Wang, F. M., Juanb, H. Y., & Su, C. H. (2020). Biodiesel production by direct transesterification of wet spent coffee grounds using switchable solvent as a catalyst and solvent. *Bioresource Technology, 296*, 122334. https://doi.org/10.1016/j.biortech.2019.122334

Rivera, X. C. S., Gallego-Schmid, A., Najdanovic-Visak, V., & Azapagic, A. (2020). Life cycle environmental sustainability of valorisation routes for spent coffee grounds: From waste to resources. *Resources, Conservation and Recycling, 157*, 104751. https://doi.org/10.1016/j.resconrec.2020.104751

Sarno, M., & Iuliano, M. (2018). Active biocatalyst for biodiesel production from spent coffee ground. *Bioresource Technology, 266*, 431–438. https://doi.org/10.1016/j.biortech.2018.06.108

Somnuk, K., Eawlex, P., & Prateepchaikul, G. (2017). Optimization of coffee oil extraction from spent coffee grounds using four solvents and prototype-scale extraction using circulation process. *Agriculture and Natural Resources, 51*, 181–189. https://doi.org/10.1016/j.anres.2017.01.003

Supang, W., Ngamprasertsith, S., Sakdasri, W., & Sawangkeaw, R. (2024). Biodiesel production from spent coffee grounds by using ethanolic extraction and supercritical transesterification. *BioEnergy Research, 17*(4), 2429–2439. https://doi.org/10.1007/s12155-024-10782-z

Tarigan, J. B., Ginting, M., Mubarokah, S. N., Sebayang, F., Karo-karo, J., Nguyen, T. T., Ginting, J., & Stepu, E. K. (2019). Direct biodiesel production from wet spent coffee grounds. *RSC Advances, 9*, 35109–35116. https://doi.org/10.1039/c9ra08038d

Tuntiwiwattanapun, N., Usapein, P., & Tongcumpou, C. (2017). The energy usage and environmental impact assessment of spent coffee grounds biodiesel production by an in-situ transesterification process. *Energy for Sustainable Development, 40*, 50–58. https://doi.org/10.1016/j.esd.2017.07.002

Uddin, M. N., Techato, K., Rasul, M. G., Hassan, N. M. S., & Mofijur, M. (2019). Waste coffee oil: A promising source for biodiesel production. *Energy Procedia, 160*, 677–682. https://doi.org/10.1016/j.egypro.2019.02.221

Wang, J. (2024). Biodiesel production from spent coffee grounds utilizing pomegranate/orange peel biochar as a green and renewable nanocatalyst: Compression ignition engine performance and emission. *Industrial Crops and Products, 218*, 118890. https://doi.org/10.1016/j.indcrop.2024.118890

Yeoh, L., & Ng, K. S. (2022). Future prospects of spent coffee ground valorisation using a biorefinery approach. *Resources, Conservation and Recycling, 179*, 106123. https://doi.org/10.1016/j.resconrec.2021.106123

# Economic Potential of Recyclable Wastes in Municipal Solid Waste: A Case Study of the Marmara Region

**Mirac Nur Ciner, Serdar Aydın, Emine Elmaslar Özbaş, H. Kurtulus Ozcan, Sinan Güneysu, and Atakan Öngen**

## 1 Introduction

Solid waste is defined as used and residual materials comprising municipal solid waste (MSW) (solid waste generated by households), industrial and commercial solid waste (solid waste produced by shops, hotels, offices, and hospitals), construction and demolition waste, and marine debris (solid waste generated from coastal areas and the sea) (Wang et al. 2006; Mishra & Das, 2023; Mohanty et al., 2018). According to recent sources, solid waste is generally defined as materials that are unused, discarded, or nonrecyclable (Mishra et al., 2023). These wastes originate from various sources, including household, commercial, industrial, and construction activities (Mishra et al., 2024; Tripathy et al., 2024). In particular, MSW consists of waste collected from households, commercial establishments, and small-scale industrial activities in urban areas. Globally, the amount of solid waste was estimated to have reached approximately 2.1 billion tons in 2023, with projections indicating an increase to 3.8 billion tons by 2050 (UNEP, 2024). Ineffective solid waste management can lead to significant environmental and economic challenges (Osra et al., 2024). An effective waste management system aims to consider waste as a resource and integrate it into recycling and reuse processes (Mishra & Das, 2021).

M. N. Ciner (✉) · S. Aydın (✉) · E. Elmaslar Özbaş · H. K. Ozcan · S. Güneysu · A. Öngen
Engineering Faculty, Environmental Engineering Department, Istanbul University-Cerrahpaşa, Avcılar, Istanbul, Türkiye
e-mail: mirac.ciner@iuc.edu.tr

S. Aydın
e-mail: mirac.ciner@iuc.edu.tr

S. Aydın · E. Elmaslar Özbaş · H. K. Ozcan · A. Öngen
Environmental and Earth Sciences Application and Research Center, Istanbul University-Cerrahpaşa, Avcılar, İstanbul, Türkiye

 
A. P. Das and S. Karakuş (eds.), *Synergizing Sustainability for Integrated Waste Management*, Sustainable Environmental Waste Management Strategies,
https://doi.org/10.1007/978-3-032-03718-3_17

Additionally, the United Nations' 2024 Global Waste Management Outlook report emphasizes the critical importance of waste reduction and management through circular economy strategies. By implementing these approaches, the link between economic growth and waste generation can be decoupled, facilitating the achievement of sustainability goals (UNEP, 2024).

Waste management can be defined as the set of activities required to manage waste from its generation to its final disposal (Kumar et al., 2023; Mishra & Das, 2022). These activities primarily include waste collection, transportation, processing, and disposal (Das et al., 2024; Das & Ghosh, 2022; Mishra et al., 2019). Additionally, it encompasses the legal and regulatory framework related to waste management and monitoring activities. The hierarchy presented in Fig. 1 illustrates the most common solid waste management practices prioritizing resource efficiency and sustainability (MoUD, 2016).

Recycling is a crucial practice that enables the recovery of valuable materials from post-consumption waste streams, diverting them from final disposal and reintegrating them into the consumer market (Priyadarsini & Das, 2024, 2025). Additionally, new products manufactured from recycled raw materials significantly contribute to energy savings, thereby supporting sustainable production processes (Cognet et al., 2020).

Solid wastes are categorized based on their components. According to the classification by Zheng et al. (2009), these wastes consist of nonbiodegradable inorganic materials, synthetic organic compounds resistant to biodegradation, biodegradable natural organic materials, unidentified and environmentally degraded chemicals,

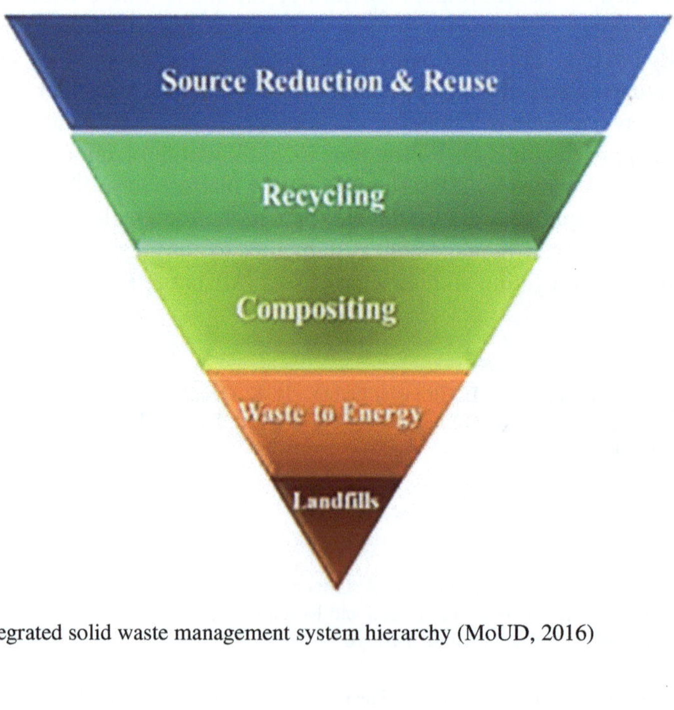


**Fig. 1** Integrated solid waste management system hierarchy (MoUD, 2016)

toxic organic compounds, metals and their derivatives, and partially biodegradable natural organic substances (Dey et al., 2023a, 2023b).

Biodegradability varies depending on the composition of the waste, the types of microorganisms present, and environmental conditions (Bala et al., 2024; Mishra et al., 2022). In another classification by Gündüzalp and Güven (2016), solid wastes are categorized into MSW, construction and demolition waste, waste from energy and fuel production, agricultural and food industry waste, forestry and forest product waste, recreational and garden waste, livestock waste, slaughterhouse waste, wastewater treatment sludge, septic waste, and hospital waste (Das & Ghosh, 2018; Ray et al., 2024). These classifications contribute to the more effective planning of waste management processes and the optimization of recycling strategies.

Contemporary research emphasizes the necessity of implementing the 4R principles—reduce, reuse, recycle, and recover—for the sustainable management of solid waste. Numerous studies highlight the importance of establishing a circular economy during the processing of MSW. When waste is perceived as a resource and appropriately managed, it can generate significant economic benefits (Silva et al., 2017). For instance, vermicomposting is an ancient process in which organic waste is converted into valuable compost through the synergistic activity of worms and microbes. This method has also been utilized for detoxifying industrial waste to produce fertilizers for agriculture (Bhat et al., 2018). Microbial anaerobic digestion is another approach that transforms animal waste into renewable energy (biogas) and nutrient-rich biofertilizer (Manyi-Loh et al., 2013). Other examples of value-added products derived from waste include lactic acid production from food waste (Li et al., 2018), activated carbon production from solid waste (Adelopo et al., 2018), and the recovery of cement, salts, metals, and rare earth elements from MSW incineration ash (Kahvecioğlu et al., 2003). In summary, a societal shift towards perceiving waste as a resource for the production of valuable products will foster the development of a circular economy while simultaneously ensuring the sustainable management of solid waste.

Solid waste management is one of the foremost global environmental challenges. The amount of solid waste generated worldwide is rapidly increasing, primarily driven by population growth, urbanization, and changes in consumption patterns (Voukkali et al., 2024).

When waste management is inadequate, it leads to severe consequences such as environmental pollution, public health issues, and economic losses. For instance, in 2020, the direct global cost of waste management was estimated at 252 billion USD. However, when accounting for the environmental damage and health-related impacts caused by improper waste disposal, this cost increased to 361 billion USD (UNEP, 2024).

Although a significant portion of waste is recyclable, global recycling rates remain low, highlighting the importance of transitioning to a circular economy. Recovering recyclable waste reduces the demand for raw materials and contributes to energy savings (U.S. EPA, 1998). Globally, approximately 2.01 billion tons of solid waste are generated annually, with at least 33% of this waste not being managed in an environmentally safe manner (World Bank, 2025).

Global solid waste recycling rates vary significantly across countries and regions. As of 2024, it is estimated that approximately 13.5% of solid waste worldwide is recycled. When examining recycling rates by country, Slovenia has the highest rate at 68.9%. Other notable countries include Germany (66.5%), Switzerland (55.9%), Norway (38.5%), the United States (44%), and Turkey (12.3%) (World Population Review, 2025). Recycling rates also differ based on income levels. High-income countries generally achieve higher recycling rates compared to low- and middle-income countries. For instance, in high-income nations, the average per capita waste generation is approximately 1.5 kg per day, whereas in low-income countries, this figure is around 0.5 kg per day (World Population Review, 2025).

Many countries worldwide are developing various strategies for sustainable waste management. One of these strategies is the circular economy model, which envisions a system where waste is considered a resource. This approach aims to reduce waste generation while ensuring sustainable economic growth through recycling and reuse. According to the United Nations' 2024 Global Waste Management Outlook report, the adoption of a circular economy could generate an estimated net annual gain of 108.5 billion USD (UNEP, 2024).

The Zero Waste Regulation in Türkiye was officially published in the Official Gazette on July 12, 2019, and subsequently came into force. This regulation adopts a circular economy-based approach to waste management, aiming to promote source-segregated waste collection, recycling, and waste prevention. As part of its recovery targets, the regulation sets a goal of increasing the recycling rate to 60% by 2035 (Zero Waste Regulation, 2019).

The aim of this study is to analyze the composition of solid waste generated in the Marmara Region, with a particular focus on assessing the recycling potential and economic value of packaging waste. The region's high industrial production and dense population necessitate the effective planning of waste management. In this context, the study seeks to compare waste types across different provinces and develop strategic recommendations to optimize the recycling process. Additionally, by projecting market values for 2026, 2031, and beyond, the study highlights the economic contributions of packaging waste recycling. Based on these evaluations, the research aims to establish a scientific foundation for the development of sustainable waste management policies and the enhancement of recycling processes in the Marmara Region.

## 2 Materials and Methods

### *2.1 Site Description*

The Marmara Region, situated in northwestern Türkiye, stands out as one of the country's most crucial areas in terms of economy, trade, and population dynamics (Fig. 2). Encompassing roughly 8.5% of Türkiye's total land area, the region comprises 11 provinces, including Istanbul, Bursa, Kocaeli, Balıkesir, Tekirdağ, Sakarya, Edirne, Çanakkale, Kırklareli, Bilecik, and Yalova (RoT, 2015). The region serves as Turkey's economic powerhouse, with Istanbul at its core acting as the nation's financial and commercial center. Other major cities, such as Bursa, Kocaeli, and Tekirdağ, contribute significantly to Turkey's industrial output. As the most densely inhabited part of the country, the Marmara Region accommodates over 30% of Türkiye's total population (MMU, 2024).

Positioned at the intersection of Asia and Europe, the Marmara Region holds a strategically vital location, incorporating key maritime passages such as the Bosphorus and the Dardanelles. This geographical advantage significantly strengthens the region's role in both domestic and global trade, as well as logistics networks. Due to its economic prominence, the region serves as a major hub for commercial activities, facilitating the movement of goods and services between continents. Notably, Istanbul, which lies within the Marmara Region, contributes approximately 30.4% of Türkiye's gross domestic product (GDP), underscoring its pivotal influence on the national economy (TURKSTAT, 2022).

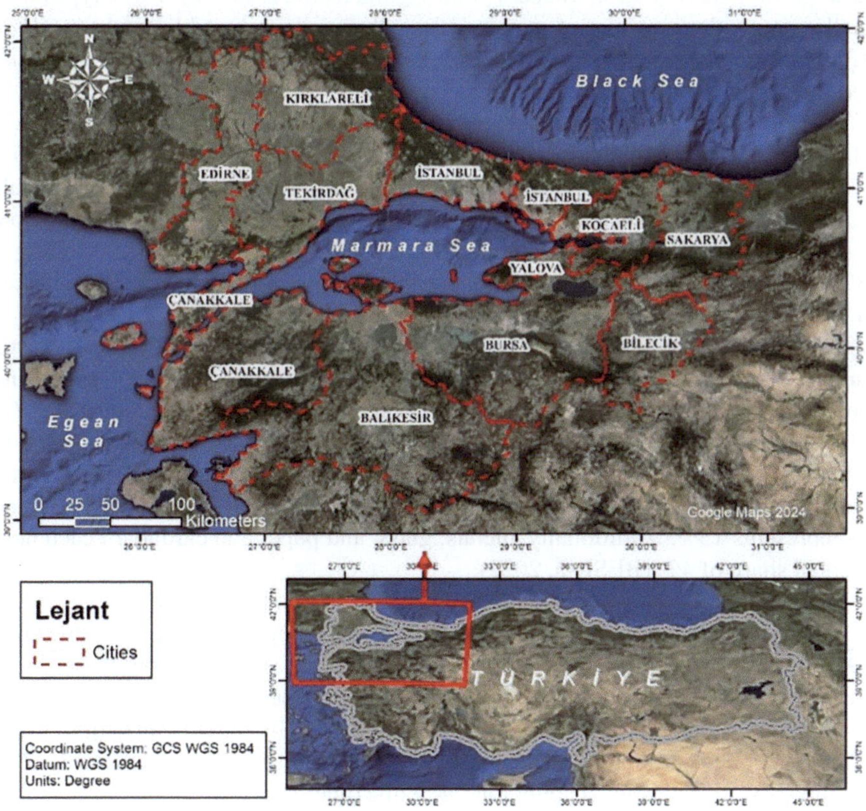


**Fig. 2** Marmara region

However, the Marmara Region faces considerable challenges in environmental sustainability due to extensive industrial activities and rapid urbanization. A crucial dimension of this challenge lies in the economic potential of recyclable waste, particularly in the effective management of packaging waste. Efficient recycling strategies not only mitigate environmental impacts but also contribute to the circular economy by creating economic value from waste materials. Given its high industrial output and dense population, the Marmara Region serves as a critical area for developing and implementing sustainable waste management strategies to maximize the economic benefits of recyclable materials.

## 2.2 *Solid Waste Characterization*

In Istanbul's waste composition, biodegradable waste accounts for the largest portion, making up 48.17% of the total waste. This is followed by paper and cardboard at 10.38% and plastic bags at 9.57%. Other notable categories include other combustible waste (8.9%), glass (4.93%), diapers (4.86%), and textiles (4.69%). Waste types with smaller proportions consist of metals (1%), other noncombustible waste (1.71%), PET plastics (1.48%), composite waste (0.89%), hazardous waste (0.86%), and electronic waste (0.18%) (IESR, 2022).

In Tekirdağ Province, biodegradable waste represents the highest share at 46%. It is followed by other combustible waste (14%), paper and cardboard (13%), and glass (9%). Plastics and other noncombustible waste each contribute 8%, while metals make up 2% of the total waste (TESR, 2022). In Edirne Province, biodegradable waste dominates the total waste composition at 61%. Paper and cardboard follow at 18%, with ash accounting for 14%. Additionally, glass waste makes up 5%, whereas metals and plastics each hold a 1% share (EESR, 2021).

In Kırklareli Province, biodegradable waste constitutes 39% of the total waste, ranking first. This is followed by other waste (16%), plastics (14%), paper and cardboard (9%), glass (8%), and ash (7%). Park and garden waste contribute 5%, while metals and other combustible waste each account for 1% (KESR-a, 2022). In Balıkesir Province, biodegradable waste holds the largest share at 43% of the total waste. Other combustible waste follows at 18%, while plastics make up 14%, and ash represents 13%. Additionally, metals, glass, and paper and cardboard each hold an equal share of 4% (BESR-a, 2022).

In Çanakkale Province, biodegradable waste accounts for 40% of the total waste, making it the most prevalent category. This is followed by paper and cardboard (19%), other waste (19%), plastics (15%), and glass (5%). Metals contribute 2% of the total waste (CESR, 2011). In Bursa Province, biodegradable waste represents the largest portion at 54% of total waste. It is followed by other combustible waste (13%), paper and cardboard (10%), plastic bags (7%), glass (7%), and plastics (6%). Additionally, PET plastics account for 2%, while other noncombustible waste makes up 1% (Ünlü et al., 2022).

In Bilecik Province, biodegradable waste holds the highest proportion at 48%. Other waste follows at 15%, with paper and cardboard at 14%, plastics at 10%, and glass at 8%. Ash waste accounts for 4%, while metals constitute 1% (BESR-b, 2022). In Sakarya Province, biodegradable waste represents 46.65% of total waste. Other combustible waste follows at 25.15%, while plastics contribute 8.84%, paper and cardboard 7.25%, and glass 4.82%. Smaller proportions include metals (1.04%), ash (2.2%), park and garden waste (2.82%), other waste (0.36%), other noncombustible waste (0.49%), hazardous waste (0.32%), and electronic waste (0.05%) (SESR, 2022).

In Kocaeli Province, biodegradable waste comprises 54.2% of total waste, making it the largest category. It is followed by other combustible waste (15.15%), plastics (14.58%), paper and cardboard (5.92%), and glass (4.88%). Lower proportions include metals (1.51%), ash (1.87%), park and garden waste (0.64%), hazardous waste (0.91%), electronic waste (0.15%), and other noncombustible waste (0.19%) (KESR-b, 2022). In Yalova Province, biodegradable waste accounts for 49.5% of total waste, the highest share. Ash follows at 15%, with paper and cardboard at 11%, plastics at 9.7%, metals at 5%, other waste at 5%, and glass at 4.8% (YESR, 2023). The waste composition of all provinces in the Marmara Region is illustrated in Fig. 3.

## 3 Results and Discussion

The annual market values of packaging waste recovered from MSW in the Marmara Region, based on its recycling potential, can be summarized as follows. These assessments vary depending on the type of packaging material and the time frame and are projected for the years 2026, 2031, and beyond. These projections have been determined in accordance with the Regulation on the Control of Packaging Waste (Packaging Waste Control Regulation, 2021).

The annual market value of recycled paper and cardboard waste is estimated to reach $49,013,221 by 2026. In accordance with regulatory requirements, this value is expected to increase to $55,548,317 by 2031 and remain stable thereafter. Plastic waste, which has the highest recycling potential among packaging materials, is projected to maintain a constant value of $105,952,305 across all scenarios. For glass waste, the annual market value from recycling is set at $31,731,607 in 2026, rising to $33,998,150 by 2031, with expectations that it will remain at this level beyond 2031. Although glass packaging waste has a relatively lower market value, it still provides economic contributions to sustainable waste management. For metal packaging waste, both minimum and maximum market values have been calculated. The minimum market value is projected to reach $11,124,884 by 2026, increasing to $12,979,031 by 2031, with stabilization at that level. The maximum market value is estimated at $71,663,360 in 2026, rising to $83,607,254 by 2031, with expectations to remain at or above this level beyond 2031 (Fig. 4). These regulation-based projections illustrate the expected annual revenue from recycling packaging waste in Istanbul by

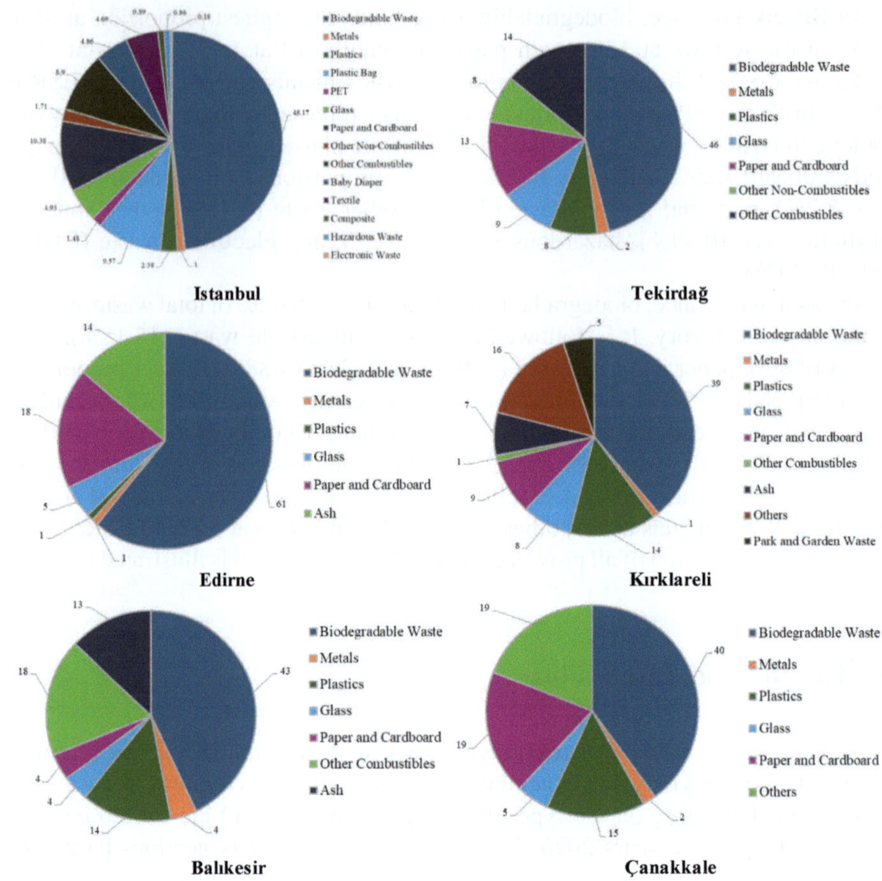


**Fig. 3** Waste characterization of provinces in the Marmara region

material type, emphasizing the significant economic potential of plastic and metal waste recycling.

In Tekirdağ Province, the annual revenue from the recycling of paper and cardboard packaging waste is projected to reach \$867,609 by 2026 and increase to \$983,291 by 2031, remaining stable thereafter. The annual revenue from plastic packaging waste is expected to remain constant at \$892,050 across all three scenarios. For glass waste, the estimated annual revenue from recycling is projected to increase from \$818,753 in 2026 to \$877,235 in 2031, with expectations of stabilization beyond 2031.

For metal packaging waste, both minimum and maximum market values have been considered. The minimum revenue is projected to reach \$314,478 by 2026, increasing to \$366,891 by 2031, and stabilizing at this level thereafter. The maximum revenue is estimated at \$2,025,779 by 2026, rising to \$2,363,409 by 2031, and remaining constant beyond 2031 (Fig. 5).

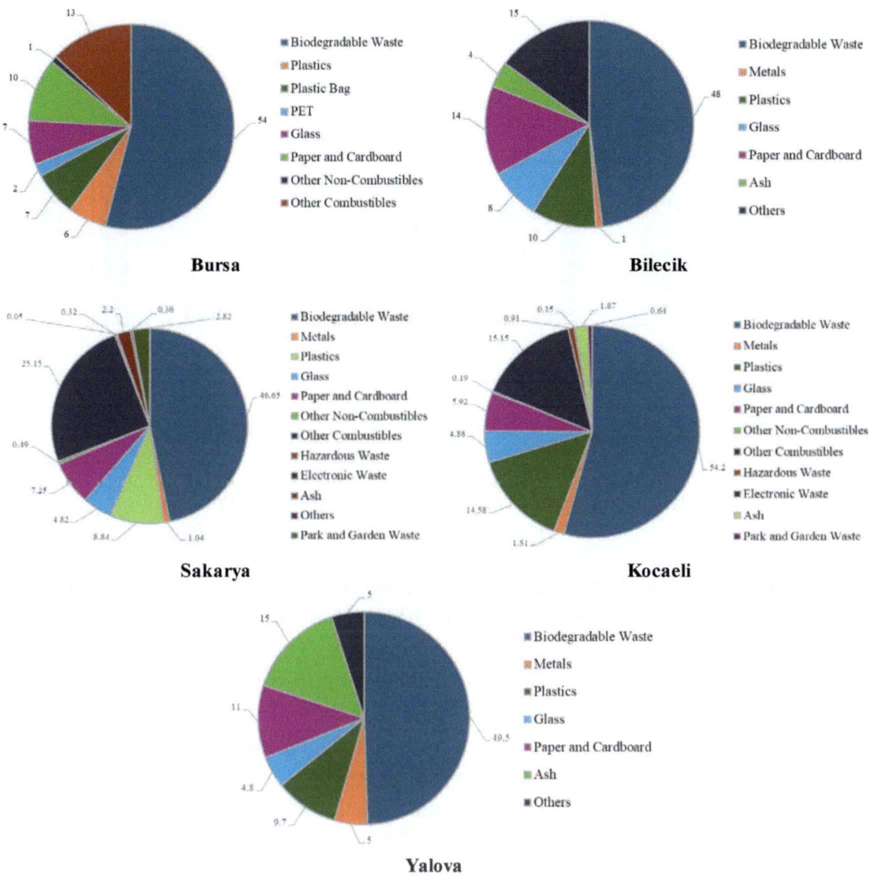


**Fig. 3** (continued)

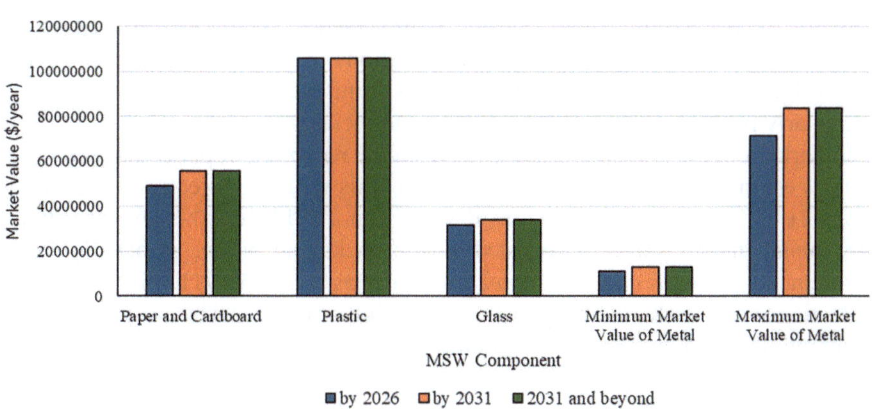


**Fig. 4** Annual market values of recycled packaging waste in Istanbul

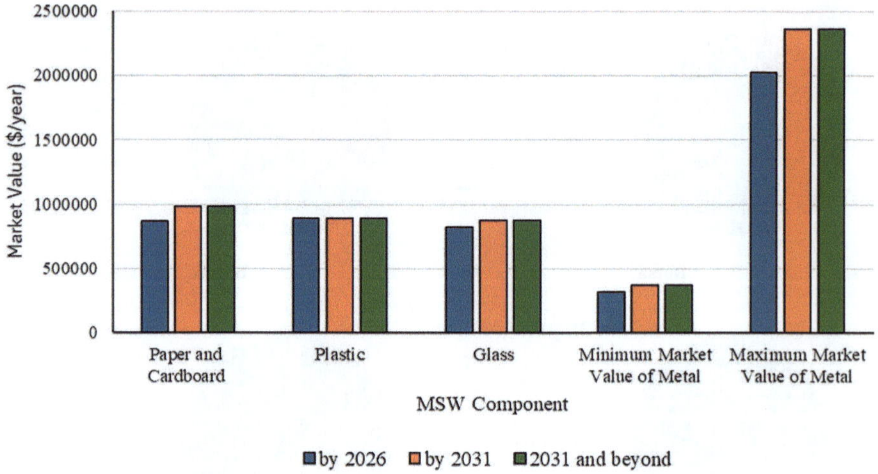


**Fig. 5** Annual market values of recycled packaging waste in Tekirdağ Province

In Edirne, the annual market value from the recycling of paper and cardboard waste is estimated to reach $2,431,844 by 2026. This value is projected to increase to $2,756,090 by 2031 and remain stable thereafter. The market value of recycled plastic packaging waste, in accordance with regulatory requirements, is estimated at a minimum of $225,726 across all scenarios, representing at least 55% recovery. Although plastic has a relatively lower market value in this context, its recycling potential is crucial for sustainable waste management.

The annual market value of recycled glass waste in Edirne is projected to reach $920,794 by 2026, increasing to $986,565 by 2031, with expectations to remain at this level beyond 2031. Glass packaging waste represents a significant recyclable material in terms of annual revenue potential. For metal packaging waste, the minimum market value is estimated at $318,305 in 2026, rising to $371,355 by 2031, and stabilizing thereafter. The maximum market value is projected to increase from $2,050,428 in 2026 to $2,392,166 by 2031, with expectations to remain at this level beyond 2031 (Fig. 6). In Edirne, the economic gains from packaging waste recycling are particularly prominent in paper, cardboard, and metal waste categories.

In Kırklareli, the annual market value from the recycling of paper and cardboard waste is estimated to reach $866,842 by 2026. This value is projected to increase to $982,421 by 2031, with expectations that it will not fall below this level thereafter. The annual market value of recycled plastic packaging waste, as per regulatory requirements, remains constant across all scenarios at $2,252,911 (Fig. 7). Among packaging waste in Kırklareli, plastic stands out as the material with the highest economic potential, making it a key component in recycling efforts.

The annual market value of recycled glass waste in Kırklareli is projected to reach $1,050,308 by 2026, increasing to $1,125,330 by 2031, with expectations to remain at or above this level beyond 2031. For metal packaging waste, both minimum and maximum market values have been calculated. The minimum market value is

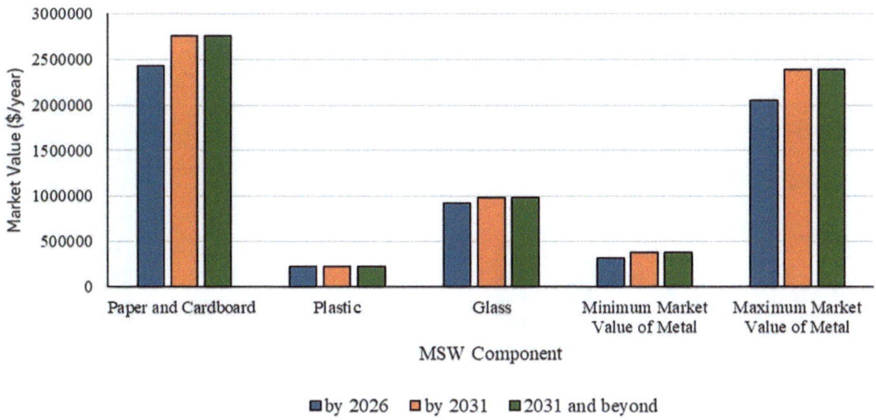


**Fig. 6** Annual market values of recycled packaging waste in Edirne Province

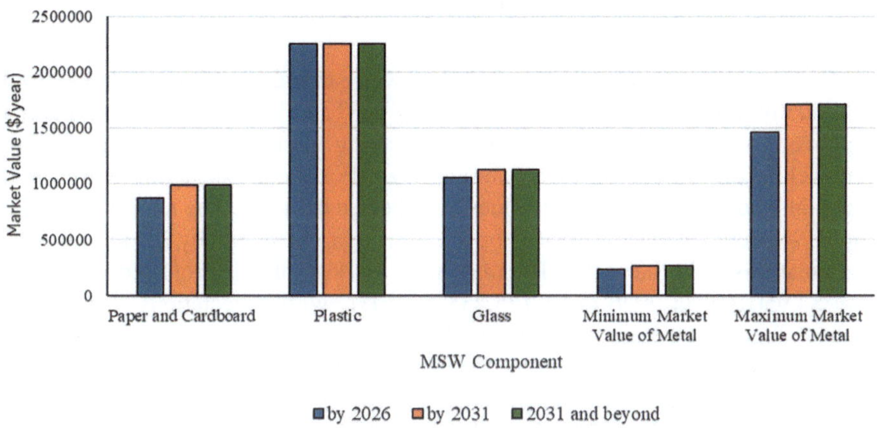


**Fig. 7** Annual market values of recycled packaging waste in Kırklareli Province

expected to reach $226,922 by 2026, rising to $264,743 by 2031, with stabilization at this level thereafter. The maximum market value is projected to increase from $1,461,769 in 2026 to $1,705,397 by 2031, with expectations to remain at or above this level beyond 2031. In Kırklareli, the economic returns from recycling are particularly high for plastic and glass waste, while metal waste also presents a noteworthy revenue potential.

In Balıkesir, the annual market value from the recycling of paper and cardboard waste is estimated to reach $326,169 by 2026. This value is projected to increase to $369,659 by 2031 and remain at or above this level thereafter (Fig. 8). The annual market value of recycled plastic packaging waste is expected to remain constant across all scenarios at $1,907,347. Among recyclable materials in Balıkesir, plastic stands out as one of the packaging materials with the highest economic potential.

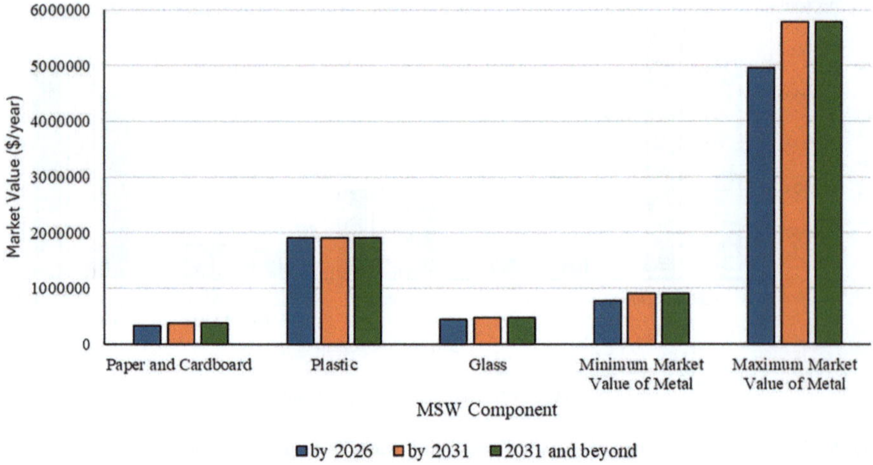


**Fig. 8** Annual market values of recycled packaging waste in Balıkesir Province

The annual market value of recycled glass waste in Balıkesir is projected to reach $444,603 by 2026, increasing to $476,360 by 2031, with expectations to remain at or above this level beyond 2031. Glass packaging waste contributes significantly to recycling efforts due to its economic value. For metal packaging waste, both minimum and maximum market values have been calculated. The minimum market value is expected to reach $768,463 by 2026, rising to $896,540 by 2031, with stabilization at this level thereafter. The maximum market value is projected to increase from $4,950,219 in 2026 to $5,775,255 by 2031, with expectations to remain at or above this level beyond 2031. In Balıkesir, the economic potential of packaging waste recycling is particularly high for plastic and metal waste, highlighting their significance in sustainable waste management strategies.

In Çanakkale, the annual market value from the recycling of paper and cardboard waste is estimated to reach $4,770,759 by 2026. This value is projected to increase to $5,406,860 by 2031 and remain at or above this level beyond 2031. The annual market value of recycled plastic packaging waste is expected to remain constant across all scenarios at $6,292,795 (Fig. 9). Among recyclable materials in Çanakkale, plastic waste stands out as one of the materials with the highest economic potential in recycling.

The annual market value of recycled glass waste in Çanakkale is projected to reach $1,711,327 by 2026, increasing to $1,833,565 by 2031, with expectations to remain at or above this level beyond 2031. For metal packaging waste, both minimum and maximum market values have been calculated. The minimum market value is estimated at $1,183,160 in 2026, rising to $1,380,353 by 2031, with expectations to stabilize at this level thereafter. The maximum market value is projected to increase from $7,621,582 in 2026 to $8,891,845 by 2031, with expectations to remain at or above this level beyond 2031. In Çanakkale, the economic potential of packaging

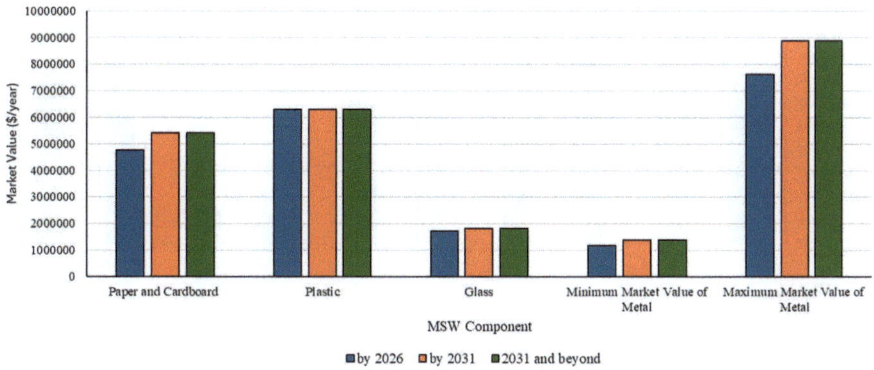


**Fig. 9** Annual market values of recycled packaging waste in Çanakkale Province

waste recycling is particularly high for plastic and metal waste, highlighting their significance in sustainable waste management efforts.

In Bursa, the annual market value from the recycling of paper and cardboard waste is estimated to reach $7,636,103 by 2026. This value is projected to increase to $8,654,250 by 2031 and remain at or above this level beyond 2031 (Fig. 10). The annual market value of recycled plastic packaging waste is expected to remain constant across all scenarios at $19,137,338. Among recyclable materials in Bursa, plastic waste stands out as the packaging material with the highest economic potential.

The annual market value of recycled glass waste in Bursa is projected to reach $7,286,166 by 2026, increasing to $7,806,607 by 2031, with expectations to remain at or above this level beyond 2031. In Bursa, the economic potential of packaging waste recycling is particularly prominent in plastic waste, highlighting its significance in sustainable waste management.

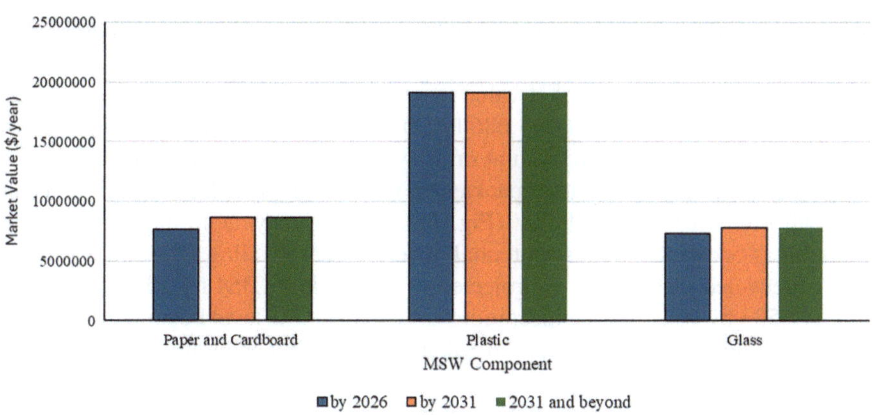


**Fig. 10** Annual market values of recycled packaging waste in Bursa Province

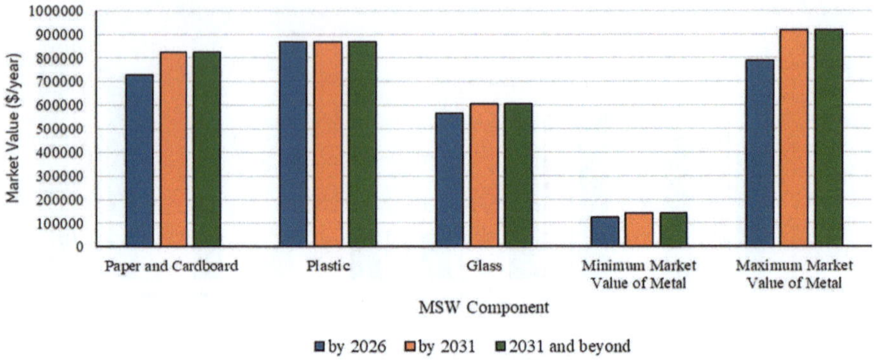


**Fig. 11** Annual market values of recycled packaging waste in Bilecik Province

In Bilecik, the annual market value from the recycling of paper and cardboard waste is estimated to reach $725,231 by 2026. This value is projected to increase to $821,928 by 2031 and remain at or above this level beyond 2031. The annual market value of recycled plastic packaging waste is expected to remain constant across all scenarios at $865,499 (Fig. 11). In Bilecik, plastic waste represents a significant recycling potential, highlighting its importance in sustainable waste management efforts.

The annual market value of recycled glass waste in Bilecik is projected to reach $564,895 by 2026, increasing to $605,244 by 2031, with expectations to remain at this level beyond 2031. Glass is recognized as a significant component contributing to the recycling economy. For metal packaging waste, both minimum and maximum market values have been calculated. The minimum market value is estimated at $122,047 in 2026, rising to $142,388 by 2031, with expectations to stabilize at this level thereafter. The maximum market value is projected to increase from $786,194 in 2026 to $917,226 by 2031, remaining at or above this level beyond 2031. In Bilecik, the economic potential of packaging waste recycling is particularly high for plastic, metal, paper, and cardboard waste, highlighting their importance in sustainable waste management efforts.

In Sakarya, the annual market value from the recycling of paper and cardboard waste is estimated to reach $1,933,686 by 2026. This value is projected to increase to $2,191,511 by 2031 and remain at or above this level beyond 2031. The annual market value of recycled plastic packaging waste is expected to remain constant across all scenarios at $3,939,297 (Fig. 12). Among recyclable materials, plastic stands out as one of the highest-potential materials for recycling. The annual market value of recycled glass waste is projected to reach $1,752,364 by 2026, increasing to $1,877,533 by 2031, with expectations to remain at this level beyond 2031. Glass represents a significant component in Sakarya's recycling economy, contributing to sustainable waste management.

For metal packaging waste, both minimum and maximum market values have been determined. The minimum market value is estimated to reach $653,523 in

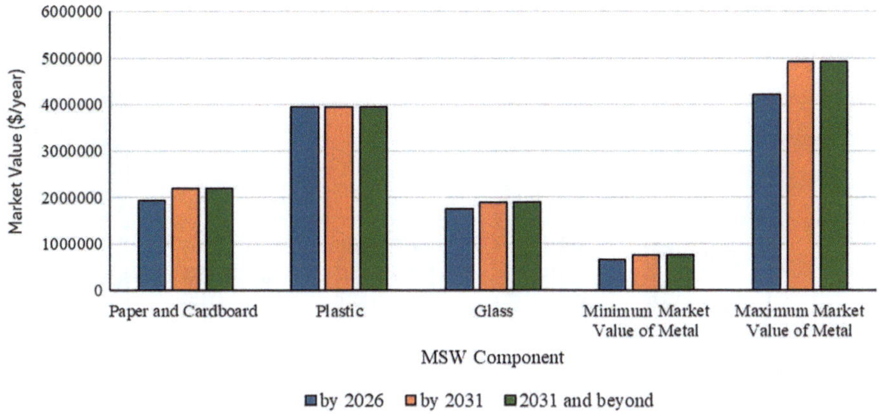


**Fig. 12** Annual market values of recycled packaging waste in Sakarya Province

2026, rising to $762,444 by 2031, with expectations to remain at this level beyond 2031. The maximum market value is projected to increase from $4,209,811 in 2026 to $4,911,446 by 2031, stabilizing at this level thereafter. In Sakarya, the economic potential of packaging waste recycling is particularly high for plastic and metal waste, emphasizing their significant role in sustainable waste management strategies.

In Kocaeli, the annual market value from the recycling of paper and cardboard waste is estimated to reach $2,724,673 by 2026. This value is projected to increase to $3,087,963 by 2031 and remain at or above this level beyond 2031. The annual market value of recycled plastic packaging waste is expected to remain constant across all scenarios at $11,211,628 (Fig. 13). Among recyclable materials, plastic stands out as the packaging material with the highest economic return potential in Kocaeli. The annual market value of recycled glass waste is projected to reach $3,061,553 by 2026, increasing to $3,280,235 by 2031, with expectations to remain at this level beyond 2031. Glass plays a significant role in the recycling economy, contributing to sustainable waste management efforts.

The minimum market value of recycled metal packaging waste in Kocaeli is estimated to reach $1,637,379 by 2026, increasing to $1,910,275 by 2031, with expectations to remain at this level beyond 2031. The maximum market value is projected to rise from $10,547,533 in 2026 to $12,305,455 by 2031, stabilizing at or above this level thereafter. In Kocaeli, the economic potential of packaging waste recycling is particularly significant for plastic and metal waste, highlighting their crucial role in sustainable waste management and economic recovery efforts.

In Yalova, the annual market value from the recycling of paper and cardboard waste is estimated to reach $1,091,686 by 2026. This value is projected to increase to $1,237,244 by 2031 and remain stable beyond 2031 (Fig. 14). The annual market value of recycled plastic packaging waste is expected to remain constant across all scenarios at $1,608,404. Plastic is one of the key recyclable materials in Yalova, offering significant recycling potential. The annual market value of recycled glass

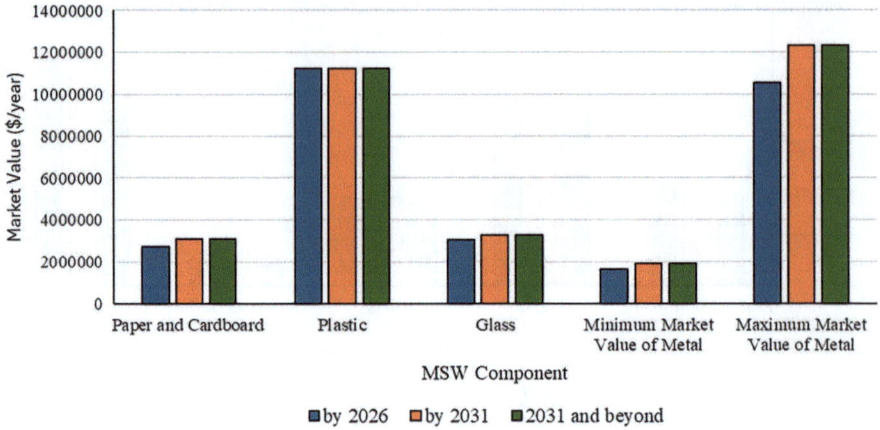


**Fig. 13** Annual market values of recycled packaging waste in Kocaeli Province

waste is projected to reach $649,345 by 2026, increasing to $695,727 by 2031, with expectations to remain at or above this level beyond 2031.

The minimum market value of recycled metal packaging waste in Yalova is estimated to reach $1,169,108 by 2026, increasing to $1,363,959 by 2031, with expectations to remain at this level beyond 2031. The maximum market value is projected to rise from $7,531,064 in 2026 to $8,786,241 by 2031, stabilizing at or above this level thereafter. In Yalova, the economic potential of packaging waste recycling is particularly significant for metal and plastic waste, highlighting their crucial role in sustainable waste management and economic recovery efforts.

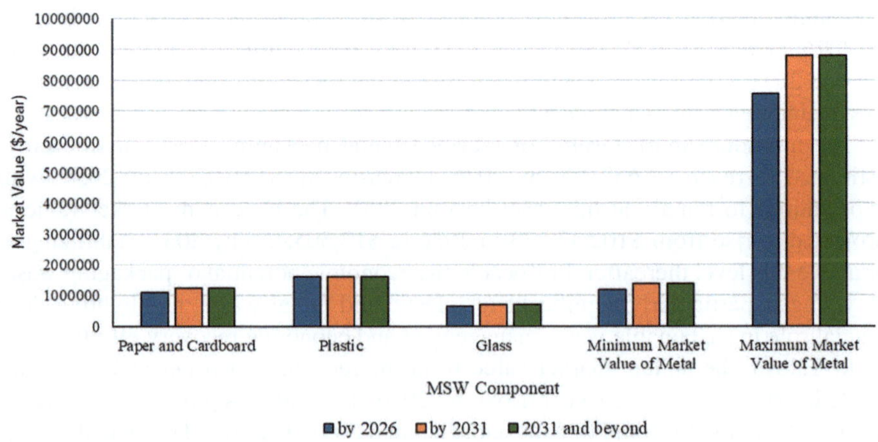


**Fig. 14** Annual market values of recycled packaging waste in Yalova Province

## 4 Conclusions

The Marmara Region is not only one of Türkiye's most significant economic, commercial, and demographic hubs but also holds great importance in terms of sustainable waste management and recycling potential. The region's high industrial production and dense population contribute to increased waste generation while simultaneously enhancing the economic potential of recycling activities.

In this study, the composition of solid waste in the Marmara Region has been analyzed, with a particular focus on the recycling potential of packaging waste. The findings indicate that the annual market value of recycled packaging waste in the region presents significant economic opportunities. Notably, plastic, metal, glass, and paper/cardboard packaging waste stand out in terms of economic value creation. In industrially intensive cities such as Istanbul, Kocaeli, Bursa, and Tekirdağ, the recycling potential of plastic and metal packaging waste is higher compared to other provinces. However, expanding recycling practices and adopting effective waste management strategies across the entire Marmara Region are essential requirements for achieving sustainable development goals.

The recycling of packaging waste presents a significant opportunity for both ensuring environmental sustainability and generating economic value. Market value projections for 2026, 2031, and beyond highlight the necessity of long-term planning in waste management policies. Strategies such as implementing incentives for packaging waste recycling, improving waste collection infrastructure, and increasing consumer awareness will play a crucial role in enhancing the recycling capacity of the region.

In conclusion, the opportunities presented by the Marmara Region in terms of sustainable waste management should be assessed through a holistic approach that considers both economic and environmental aspects. The implementation of effective waste management policies and the optimization of recycling processes will not only contribute to the region's economic growth but also yield positive outcomes for environmental sustainability.

## References

Adelopo, A. O., Haris, P. I., Alo, B. I., Huddersman, K., & Jenkins, R. O. (2018). Multivariate analysis of the effects of age, particle size and landfill depth on heavy metals pollution content of closed and active landfill precursors. *Waste Management, 78*, 227–237.

Bala, K., Anshul, Verma, S., Dogra, A., Das, A. P., Kumar, R., & Nandini. (2024). Sustainable management for urban plastic waste generation. In *Sustainable management of urban plastic waste through circular economic approaches* (pp. 211–236). https://doi.org/10.1201/9781003465447-13

Bhat, S. A., Singh, S., Singh, J., Kumar, S., & Vig, A. P. (2018). Bioremediation and detoxification of industrial wastes by earthworms: Vermicompost as powerful crop nutrient in sustainable agriculture. *Bioresource Technology, 252*, 172–179.

Cognet, M., Cambedouzou, J., Madhavi, S., Carboni, M., & Meyer, D. (2020). Targeted removal of aluminium and copper in Li-ion battery waste solutions by selective precipitation as valuable porous materials. *Materials Letters, 268*, Article 127564.

Das, A. P., & Ghosh, S. (2018). Bioleaching of Manganese from mining waste materials. *Materials Today: Proceedings, 5*(1), 2381–2390. https://doi.org/10.1016/J.MATPR.2017.11.459

Das, A., & Ghosh, S. (2022). Role of microorganisms in extenuation of mining and industrial wastes. *Geomicrobiology Journal, 39*, 173–175. https://doi.org/10.1080/01490451.2022.2038953

Das, A., Ray, S. K., Mohanty, M., Mohanty, J., Dey, S., & Das, A. P. (2024). Ecotoxicity of microplastic wastes and their sustainable management: A review. *Environmental Chemistry and Ecotoxicology, 6*, 144–152. https://doi.org/10.1016/J.ENCECO.2024.05.003

Dey, S., Das, A., Mallick, K., Sahu, A., & Das, A. P. (2023a). *Environmental petroleum waste: Pollution, toxicity, sustainable remediation* (pp. 159–175). https://doi.org/10.1007/978-3-031-48220-5_7

Dey, S., Das, A., Mallick, K., Sahu, A., & Das, A. P. (2023b). *Environmental petroleum waste: Pollution, toxicity, sustainable remediation* (pp. 159–175). https://doi.org/10.1007/978-3-031-48220-5_7

Gündüzalp, A. A., & Güven, S. (2016). Atık, çeşitleri, atık yönetimi, geri dönüşüm ve tüketici: Çankaya belediyesi ve semt tüketicileri örneği. *Hacettepe Üniversitesi Sosyolojik Araştırmalar E-Dergisi, 9*(1), 1–19. (in Turkish).

Kahvecioğlu, Ö., Kartal, G., Güven, A., & Timur, S. (2003). Metallerin çevresel etkileri-I. *Metalurji dergisi, 136*, 47–53 (in Turkish)

Kumar, M., Pandit, S., Patel, V., Khanna, N., Nag, M., Lahiri, D., Ray, R. R., Das, A., & Das, D. (2023). Maximization of energy recovery from starch processing wastewater by thermophilic dark fermentation coupled with microbial fuel cell technology. *Geomicrobiology Journal*, 1–12. https://doi.org/10.1080/01490451.2023.2209555

Manyi-Loh, C. E., Mamphweli, S. N., Meyer, E. L., Okoh, A. I., Makaka, G., & Simon, M. (2013). Microbial anaerobic digestion (bio-digesters) as an approach to the decontamination of animal wastes in pollution control and the generation of renewable energy. *International Journal of Environmental Research and Public Health, 10*(9), 4390–4417.

Marmara Municipalities Union (MMU). (2024). *Marmara Sea.* Retrieved January 17, 2025, from https://www.marmara.gov.tr/tr/marmara-denizi

Ministry of Urban Development (MoUD). (2016). *Municipal solid waste management manual—Part I: An overview* (Vol. 1, p. 62). Ministry of Urban Development. Retrieved January 6, 2025, from http://cpheeo.gov.in/upload/uploadfiles/files/Part1(1).pdf

Mishra, S., & Das, A. P. (2021). Current treatment technologies for removal of microplastic and microfiber pollutants from wastewater. In *Wastewater treatment: Cutting-edge molecular tools, techniques and applied aspects* (pp. 237–251). https://doi.org/10.1016/B978-0-12-821881-5.00011-8

Mishra, S., & Das, A. P. (2022). Treatment of the wastewater polluted with synthetic microfiber released from washing machine. In B. B. Das, H. Hettiarachchi, P. K. Sahu, & S. Nanda (Eds.), *Lecture Notes in Civil Engineering* (Vol. 207, pp. 109–117). Springer Science and Business Media Deutschland GmbH. https://doi.org/10.1007/978-981-16-7509-6_9/COVER

Mishra, S., & Das, A. P. (2023). Microbial remediation of synthetic microfiber contaminated wastewater. In M. P. Shah (Ed.), *Microbial technologies in industrial wastewater treatment* (pp. 337–350). Springer. https://doi.org/10.1007/978-981-99-2435-6_17

Mishra, S., Dash, D., & Das, A. P. (2022). Detection, characterization and possible biofragmentation of synthetic microfibers released from domestic laundering wastewater as an emerging source of marine pollution. *Marine Pollution Bulletin, 185*, 114254. https://doi.org/10.1016/J.MARPOLBUL.2022.114254

Mishra, S., Dash, D., & Das, A. P. (2024). Isolation and molecular characterization of polyester-degrading bacterial strains from municipal wastewater and their application in microplastic degradation in contaminated wastewater. *Geomicrobiology Journal*. https://doi.org/10.1080/01490451.2024.2376101

Mishra, S., Ghosh, S., van Hullebusch, E. D., Singh, S., & Das, A. P. (2023). A critical review on the recovery of base and critical elements from electronic waste-contaminated streams using microbial biotechnology. *Applied Biochemistry and Biotechnology, 2023*, 1–30. https://doi.org/10.1007/S12010-023-04440-X

Mishra, S., Rout, P. K., & Das, A. P. (2019). Solar photovoltaic panels as next generation waste: A review. *Biointerface Research in Applied Chemistry*, *9*(6).

Mohanty, S., Ghosh, S., Bal, B., & Das, A. P. (2018). A review of biotechnology processes applied for manganese recovery from wastes. *Reviews in Environmental Science and Bio/Technology 17*(4), 791–811. https://doi.org/10.1007/S11157-018-9482-1

Official Gazette. (2019, July 12). *Zero Waste Regulation* (Issue No. 30829). Ministry of Environment and Urbanization. Retrieved January 2, 2025, from https://www.resmigazete.gov.tr/eskiler/2019/07/20190712-9.htm

Official Gazette. (2021, June 26). *Regulation on the control of packaging waste (Issue No. 31523).* Ministry of Environment and Urbanization. Retrieved January 2, 2025, from https://www.resmigazete.gov.tr/eskiler/2021/06/20210626-18.htm.

Osra, F. A., Elbisy, M. S., Mosaıbah, H. A., Osra, K., Ciner, M. N., & Ozcan, H. K. (2024). Environmental impact assessment of a dumping site: A case study of Kakia Dumping Site. *Sustainability, 16*(10), 3882.

Priyadarsini, S., & Das, A. P. (2024). Analytical and structural characterization of waste lithium-ion batteries for their effective recycling strategy. *Environmental Chemistry and Ecotoxicology*. https://doi.org/10.1016/j.enceco.2024.12.004

Priyadarsini, S., & Das, A. P. (2025). Lithium bioleaching: A review on microbial-assisted sustainable technology for lithium bio-circularity. *Journal of Water Process Engineering, 69*, Article 106744. https://doi.org/10.1016/j.jwpe.2024.106744

Ray, S. K., Dey, S., Das, A. P., & John, A. (2024). Microplastic waste management through greener technologies: A review. *Sustainable Management of Urban Plastic Waste Through Circular Economic Approaches*, 237–261. https://doi.org/10.1201/9781003465447-14

Regions of Türkiye (RoT). (2015). *Türkiye's Seven Regions.* Retrieved January 17, 2025, from https://turkiyeninbolgelri.weebly.com/tuumlrk304yen304n-boumllgeler304/turkiyenin-yedi-bolgesi

Silva, A., Rosano, M., Stocker, L., & Gorissen, L. (2017). From waste to sustainable materials management: Three case studies of the transition journey. *Waste Management, 61*, 547–557.

T.R. Ministry of Environment and Urbanization. (2011). Çanakkale Environmental Status Report (CESR) 2011.

T.R. Ministry of Environment, Urbanization, and Climate Change. (2022). *Istanbul Environmental Status Report (IESR) 2022.* Retrieved January 20, 2025, from https://webdosya.csb.gov.tr/db/ced/icerikler/istanbul_-cdr2022-20230914131022.pdf

T.R. Ministry of Environment, Urbanization, and Climate Change. (2022). *Tekirdağ Environmental Status Report (TESR) 2022.* Retrieved January 20, 2025, from https://webdosya.csb.gov.tr/db/ced/icerikler/tek-rdag_-cdr2022-20231109081520.pdf

T.R. Ministry of Environment, Urbanization, and Climate Change. (2021). *Edirne Environmental Status Report (EESR) 2021.* Retrieved January 20, 2025, from https://webdosya.csb.gov.tr/db/ced/icerikler/ed-rne_-cdr2021-20220914161705.pdf

T.R. Ministry of Environment, Urbanization, and Climate Change. (2022). *Kırklareli Environmental Status Report (KESR-a) 2022.* Retrieved January 20, 2025, from https://webdosya.csb.gov.tr/db/ced/icerikler/k-rklarel-_-cdr2022-20231115150809.pdf

T.R. Ministry of Environment, Urbanization, and Climate Change. (2022). *Balıkesir Environmental Status Report (BESR-a) 2022.* Retrieved January 20, 2025, from https://webdosya.csb.gov.tr/db/ced/icerikler/balikes-r_-cdr2022-20240207092126.pdf

T.R. Ministry of Environment, Urbanization, and Climate Change. (2022). *Bilecik Environmental Status Report (BESR-b) 2022.* Retrieved January 20, 2025, from https://webdosya.csb.gov.tr/db/ced/icerikler/b-lec-k_-cdr2022-20231218100136.pdf

T.R. Ministry of Environment, Urbanization, and Climate Change. (2022). *Sakarya Environmental Status Report (SESR) 2022*. Retrieved January 20, 2025, from https://webdosya.csb.gov.tr/db/ced/icerikler/sakarya-ilcdr-2022-20230828130907.pdf

T.R. Ministry of Environment, Urbanization, and Climate Change. (2022). *Kocaeli Environmental Status Report (KESR-b) 2022*. Retrieved January 20, 2025, from https://webdosya.csb.gov.tr/db/ced/icerikler/kocaeli-ilcdr-2022-20230706104825.pdf

T.R. Ministry of Environment, Urbanization, and Climate Change. (2023). *Yalova Environmental Status Report (YESR) 2023*. Retrieved January 20, 2025, from https://webdosya.csb.gov.tr/db/ced/icerikler/yalova-icdr_2023-1-20240712081326.pdf

Tripathy, B., Dey, S., Dubey, N. K., Lo, W.-C., & Das, A. P. (2024). *An omics-based approach for managing the risk of COVID-19-generated wastes* (pp. 51–66). https://doi.org/10.1007/978-3-031-50840-0_4

Turkish Statistical Institute (TURKSTAT). (2022). *Gross Domestic Product by Province 2022*. Retrieved January 17, 2025, from https://data.tuik.gov.tr/Bulten/Index?p=Il-Bazinda-Gayrisafi-Yurt-Ici-Hasila-2022-45867

U.S. Environmental Protection Agency. (1998). *Solid waste definition and solid and hazardous waste exclusions guidance document*. Retrieved January 11, 2025, from https://semspub.epa.gov/work/03/2079159.pdf

United Nations Environment Programme. (2024). *Global waste management outlook 2024*. UNEP. Retrieved January 4, 2025, from https://www.unep.org/resources/global-waste-management-outlook-2024

Ünlü, H., Gür Parlayıcı, S., & Topçu, T. (2022, September). *Bursa Integrated Solid Waste Management Plan (BISWMP)*. Bursa Metropolitan Municipality. Retrieved January 20, 2025, from https://www.bursa.bel.tr/dosyalar/galeri_x/bzhd37rcoa8s4wg.pdf

Voukkali, I., Papamichael, I., Loizia, P., & Zorpas, A. A. (2024). Urbanization and solid waste production: Prospects and challenges. *Environmental Science and Pollution Research, 31*(12), 17678–17689.

Wang, J. J., Li, J. Z., Wu, X. R., & Shim, Y. K. (2006). Synthesis of Styryl-substituted Z/E-Chlorin derivatives with Chlorophyll-a skeleton. *Chinese Journal of Chemistry, 24*(7), 933–938.

World Bank. (2025). *Trends in solid waste management*. Retrieved January 12, 2025, from https://datatopics.worldbank.org/what-a-waste/trends_in_solid_waste_management.html

World Population Review. (2025). *Recycling rates by country 2025*. World Population Review. Retrieved January 14, 2025, from https://worldpopulationreview.com/country-rankings/recycling-rates-by-country

Zheng, J., Jin, Y. Q., Chi, Y., Wen, J. M., Jiang, X. G., & Ni, M. J. (2009). Pyrolysis characteristics of organic components of municipal solid waste at high heating rates. *Waste Management, 29*(3), 1089–1094.

# Waste Water Treatment Using the Constructed Wetland: Recent and Future Prospects

**Vivek Raj, Santosh Kumar Karn, Rakesh Kumar, Aman Prakash, Kanchan Kumari, Alok Prasad Das, and Neeraj Choudhary**

## 1 Introduction

Over the last decade, worldwide governments have started making managing pollution and contamination a top priority due to increasing environmental awareness. The effectiveness of deprivation practice and the approach's cost-effectiveness are typically the deciding factors when selecting the environmental remediation method for kind of waste (Omondi & Navalia, 2020). A remediation method's effect on the environment is important because the degradation process might occasionally result in byproducts that are more harmful than the initial contamination. Multiple approaches must often be integrated for an effective strategy because no single remediation technique is uniformly effective for all toxins and sources, as researchers have highlighted. Constructed wetlands (CWs) have gained recognition as a promising

V. Raj (✉) · K. Kumari
Department of Biotechnology, Patna Women's College, (Autonomous), Patna University, Patna, Bihar, India
e-mail: dr.vivekraj07bt@gmail.com

R. Kumar
School of Pharmaceutical Sciences, Lovely Professional University, Phagwara, Punjab, India

S. K. Karn
Department of Biochemistry & Biotechnology, Sardar Bhagwan Singh University, Balawala, Dehradun, Uttarakhand, India

A. Prakash
Department of Marine Biology, Leon H. Charney School of Marine Sciences, University of Haifa, Haifa, Israel

A. P. Das
Department of Life Sciences, Rama Devi Women University, Bhubaneswar, Odisha, India

N. Choudhary
Department of Pharmacognosy, GNA University, Phagwara, Punjab, India

 
A. P. Das and S. Karakuş (eds.), *Synergizing Sustainability for Integrated Waste Management*, Sustainable Environmental Waste Management Strategies,
https://doi.org/10.1007/978-3-032-03718-3_18

technology for wastewater remediation. Natural wetlands utilize processes such as biodegradation, sorption, phytostabilization, phytoextraction, and rhizofiltration to effectively remove contaminants from water resources (Omondi & Navalia, 2020).

Wastewater treatment systems are known as constructed wetlands. This is engineered to mimic natural processes for breaking down and eliminating contaminants (Brovelli et al., 2011) (Mishra & Das, 2023; Mishra et al., 2024a). Wetlands can effectively treat a variety of wastewater types, including municipal sewage, etc. While the industrial effluent from sources such as petroleum refineries and sour-water treatment, agricultural runoff, stormwater, textile wastewater, landfill leachates, mining drainage, and many other types (Vymazal et al., 2021). Constructed wetlands utilize natural processes to break down pollutants, thereby providing a highly environmentally friendly means of remediation with minimal adverse environmental consequences, as noted by Dan et al. (2011). Soil saturation in wetlands typically persists for a considerable duration, leading to the development of anaerobic environments. Wetland's variations include both natural fresh water and saltwater types as well as artificially created CWs (Omondi & Navalia, 2020). Wetlands constructed for water treatment involve a variety of intricate and interconnected processes involving water, soil, plants, animals, microorganisms, and environmental factors (Mishra & Das, 2021; Mishra et al., 2022a, 2022b). The authors use a variety of remediation techniques, like biodegradation, phytoremediation, and natural attenuation (Truu et al., 2015). Wetlands act multiple processes, including physical practices like sedimentation and filtration; chemical processes such as adsorption and precipitation; and biological processes like decomposition and plant assimilation. These ecosystems are typically marked by extensive numbers of vascular plants, which aid in slowing down water movement, fostering diverse microenvironments, and offering surfaces for microorganisms to attach and sites for contaminants to bind (Thorslund et al., 2017). Dead plants in wetlands that have decomposed underwater offer further sites for sorption and exchange once their above-water parts are exposed. The plant debris produced has a high organic carbon and nutrient content, specifically nitrogen and phosphorus, that facilitate microbial metabolic processes. Low oxygen levels in wetlands are typically found where the soil is saturated. As a result, vegetation in these ecosystems consists of vascular plant species that are adapted to low-oxygen conditions (Thorslund et al., 2017). Germs play a crucial role in deprivation of contaminants and makeover of pollutants. Each microorganism possesses specific degradation pathways tailored to the breakdown of particular contaminants. The ability of the contaminant degradation process to be effective relies on the presence of required microorganisms and suitable environmental circumstances being available (Bruch et al., 2011). Wastewater treatment using constructed wetlands has been implemented worldwide (Kumar et al., 2023; Mishra & Das, 2022). In Europe, CWs have been utilized since the past half of the twentieth century, with Germany serving as country in adopting this technology on the continent (Barik, 2018). Many countries have built constructed wetlands. CWs have been adopted by several African countries, including Kenya, South Africa, and Seychelles, for wastewater treatment purposes. Wastewater treatment using subsurface-constructed wetlands in Africa has a cost of roughly $5 per person. This compares to the expense of mechanical systems like activated sludge systems, which is approximately $50 per person (Barik, 2018; Azubuike et al., 2016). A recent study indicated that total cost of wastewater treatment, which encompasses operation, maintenance, and sludge disposal, varies between $0.30 and $0.88 per cubic meter ($m^3$) (Moral Pajares et al., 2019). CWs have been widely adopted worldwide for wastewater treatment, offering several advantages. These include low construction costs, environmentally sustainable practices, relatively low-maintenance and operational expenses, significant flexibility in landscape design, and enhanced opportunities for the reuse of treated water. Despite these benefits, constructed wetlands present certain limitations, such as the need for extensive land areas, variability in treatment effectiveness, and their unsuitability for applications requiring compliance with specific discharge standards. Furthermore, constructed wetlands demonstrate a low tolerance for conditions of near-complete drying. Local regulations and bylaws also play a crucial role for selection of flow type in constructed wetlands.

## 2 Wetland Systems: Natural and CWs

Wetland environments like swamps, lakes, and river floodplains are vital to ecological well-being, as they efficiently filter out erosion, store carbon, and provide necessary habitats and food for a broad spectrum of organisms (Stefanakis, 2018). Natural wetlands are often insufficient for effectively removing water pollutants caused by human activities, as these pollutants can interfere with ecosystem processes. Innovative wetland systems have been developed to address this constraint, with notable progress achieved in Germany throughout the 1970s. Engineered systems using plants are able to not only remove contaminants and treat leftover water but also fulfill various roles, including flood management, aquaculture to produce food and fiber, and providing habitats to mitigate the loss of natural areas caused by agricultural and urban development (Hua, 2003). Constructed wetlands are designed with specific sizes, locations, substrate types, hydraulic conditions, and controlled retention times. These systems are characterized by low-maintenance costs, reliance on renewable energy sources, and the integration of natural elements, allowing them to effectively handle significant volumes of water with diverse contaminants. Furthermore, constructed wetlands can serve as destinations for public tours and also host environmental education (Hua, 2003). Wetlands with limited scope are marked by their susceptibility to high levels of ammonia, the need for large land areas, uncertainties about treatment efficacy, and the potential to attract insect pests, including mosquitoes (Luederitz et al., 2001). These systems necessitate longer holding periods in comparison to traditional technical systems. Findings over the past 25 years have shown positive results in building wetlands for treating household and city sewage. Their application in industrial waste management is still problematic because of the higher levels of contaminants found in these environments.

# 3 Wetland Treatment Systems

Wetlands may categorized by water movement, types of aquatic plants like free-floating, emergent, and submerged, and the direction of flow (horizontal or vertical) (Hua, 2003). The several classifications of constructed wetlands, encompassing surface flow (SF) wetlands, subsurface flow (SSF) wetlands, and hybrid systems. Multistage hybrid systems are designed to achieve specific goals, including wastewater treatment under both aerobic and anaerobic conditions. Constructed wetlands with aeration systems are equipped with air pumps that link with systems of underground air supply pipes, which enable oxygenation within the system (Marín-Muñiz et al., 2020). The air pump's introduction of air bubbles boosts $O_2$ transfer rates in both horizontal flow (HF) and vertical flow (VF) constructed wetlands, thereby fostering aerobic environments. These systems are assessed to consume approximately 250 g of oxygen per square meter per day, with air flow rates surpassing 0.6 cubic meters per square meter per hour and a distribution pattern of 30 cm by 30 cm. Routinely aerated wetlands can substantially boost $O_2$ transfer rates to over 1 $m^3/m^2/h$, thus promoting nitrification and denitrification processes, along with the biogeochemical conditions in both subsurface and plant communities (Marín-Muñiz et al., 2020). In literature, wetlands are often referred to by various terms, such as free-surface flow, vertical subsurface flow (VSSF), surface flow (SF), subsurface flow (SSF), horizontal subsurface flow (HSSF), and subsurface vertical flow (SSVF).

## *3.1 Surface Flow (SF)*

SF wetlands are used for wastewater treatment and as wildlife habitats through keeping the wastewater surface level above the soil beneath it. These wetlands are made up of a base layer, usually clay and sand, which holds plant roots, as well as water management systems that help keep a shallow layer of water above ground (Fig. 1) (Gogoi et al., 2018). Oxygen levels in surface flow wetlands fluctuate with depth, with upper sections exhibiting higher concentrations and lower concentrations being present in the deeper sections. These wetlands are able to treat stormwater, agricultural runoff, etc. Subsurface flow wetlands require less land area than SF wetlands; however, they have higher construction and operational costs, and they utilize more complex technologies. One drawback of these wetlands is the potential for unpleasant odors and a need for more space related to SSF constructed wetlands and other remediation techniques (Stefanakis, 2018).

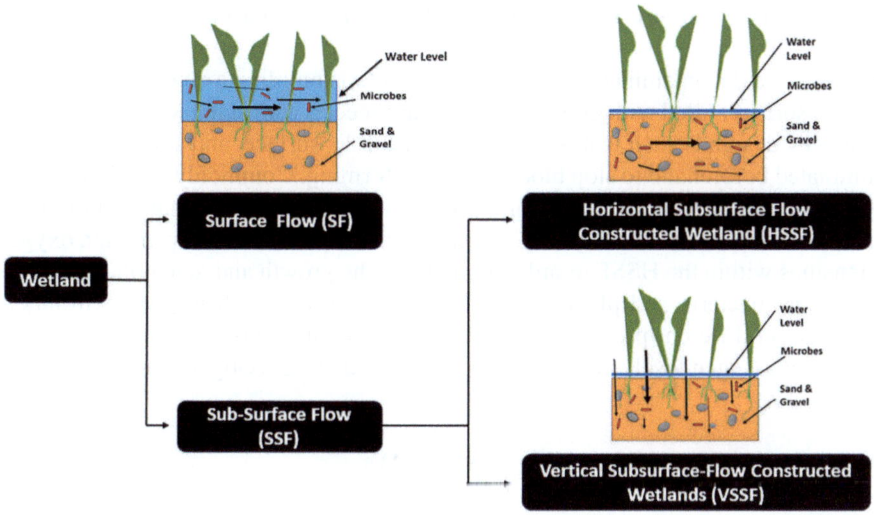


**Fig. 1** Types of wetland systems

## 3.2 *Subsurface Flow (SSF)*

Wetlands with subsurface flow are built using substrates made of porous materials, including rock and gravel, with sealed bottoms that keep water levels consistently below the soil's surface. These systems are especially effective for wastewater treatment in cases where solid levels are low, whereas sedimentation tanks or basins are utilized to handle wastewater with greater solid concentrations. SSF wetlands present several advantages, including enhanced unkind weather tolerance, condensed odor issues, increased sorption and exchange sites, and more efficient land use compared to surface flow wetlands. Notably, SSF wetlands can be integrated into public parks due to their subsurface water levels. However, they also face limitations, including higher construction costs, suitability primarily for smaller flows, and the potential for pore clogging (Marín-Muñiz et al., 2020). Wetland purification encompasses a range of processes, including sorption, filtration, sedimentation, settling, ion exchange, plant uptake, and microbial oxidation (Gogoi et al., 2018). Constructed wetlands featuring subsurface flow have effectively treated heavy oil-produced water containing a variety of contaminants, such as asphalt, gelatinoid, wax, polymers, surfactants, and other synthetic compounds. Despite their success, there is currently no established design guideline for optimal filtration rates; some studies have suggested a rate of 0.52 m/day (Guittonny-Philippe et al., 2015). The efficiency of removing chemical oxygen demand is notably affected by the concentration of dissolved oxygen (DO), with low oxygen levels and high organic waste inputs resulting in lower efficiency. In subsurface flow systems, two main types of flow have been identified: HSSF and VSSF.

### 3.2.1 Horizontal Subsurface Flow (HSSF) Constructed Wetland

A HSSF is a decontamination technique that entails building a channel beneath the Earth's surface, filled with gravel and sand, and seeded with suitable vegetation. The channel's design is such that water flows beneath its surface, and solid particles are eliminated beforehand to stop blockages from forming. Commonly, a sedimentation tank is utilized to remove these solid pollutants. The cross-sectional area of HSSF has crucial impact on its efficiency optimization. It is crucial to keep enough oxygen intensities within the HSSF in order to facilitate the growth and metabolic processes of aerobic bacteria and plants. Oxygen transfer into the subsurface environment happens via different processes, with root systems of plants being a key factor in increasing oxygen levels in their immediate surroundings (Gogoi et al., 2018).

### 3.2.2 Vertical Subsurface Flow (VSSF) CWs

VSSF is designed to allow water to flow downward, passing over filter layers before entering the channels. VSSF systems typically rely on gravity or pumping systems and require determining the time it takes for wastewater to trickle over the vertical channels. Increased oxygen levels in vertical subsurface flow (VSSF) systems support the proliferation of aerobic bacteria, thereby making them highly effective for nitrification and creating a more aerobic environment than horizontal subsurface flow (HSSF) systems. VSSF enables a range of degradation and decontamination processes, such as biodegradation, adsorption, and precipitation (Herath & Vithanage, 2015). VSSF is also effective in eliminating organic substance and deferred solids. Additionally, the rate of VSSF is lower than that of HSSF, owing to its smaller land area requirements (Hua, 2003). VSSF is more aerobic than HSSF and is more suitable for nitrification.

## *3.3 Hybrid CWs*

A hybrid CW system is designed to integrate the benefits of both hybrid and vertical constructed wetland (VCW) configurations, thus resulting in higher treatment efficiency (Fernandez-Fernandez et al., 2020). The system consists of parallel vertical constructed wetlands (VCWs) that function in sequence, taking advantage of the low energy demands and cost-effectiveness of both configurations. In this setup, the VCW segment is tasked with removing organic matter and suspended solids and concurrently enables nitrification within its aerobic environment. Following this, the horizontal constructed wetland (HCW) facilitates the processes of denitrification and the removal of additional organic matter and suspended solids. In combination, this hybrid wetland system is more effective in treating a broad spectrum of contaminants present in wastewater, including those generated by winery operations, pharmaceuticals, and oil field activities (Almuktar et al., 2018). The hybrid system's incorporation of diverse wetland setups facilitates both aerobic and anaerobic processes, thereby maximizing organic breakdown and improving nitrogen elimination through nitrification, denitrification, and ammonification.

## 4 Categories of Contaminants Treated in Constructed Wetlands

In the last decades, CWs have proven to be a highly effective method of bioremediation for wastewater treatment, showcasing their high efficiency as a natural treatment process. Conventional water treatments have been successfully applied to treat residential, municipal, and agricultural wastewater. CWs are also a promising green technology capable of treating organic-rich wastewater, removing contaminants like heavy metals (for example, Cr (VI) and Se) and pharmaceuticals and personal care products (PPCPs) as described by Nguyen et al. (2019).

### *4.1 National and Municipal Wastewater Contaminants*

The urban population growth has enlarged the impact of household water supply and sewage systems, resulting in increased amounts of wastewater. Municipal wastewater treatment that is properly managed can decrease surface water pollution issues and save valuable water supplies while using sewage nutrients for crop cultivation (Vymazal et al., 2021) (Mishra et al., 2022a, 2022b). Wastewater is often regarded as a potential source of high-quality water, but high operating costs for wastewater treatment facilities pose a significant concern in numerous regions, particularly in developing nations (Das & Mishra, 2009; Nawaz et al., 2023). Rapid expansion of constructed wetlands globally can be attributed to several key benefits, including substantially reduced operational and maintenance expenditures. CW treatment methods can remove various contaminants from wastewater, including nitrogen, phosphorus, organics, solids, metals, and coliforms. These contaminants can be removed through a complex interconnected system of plants, media, bulk water, and biomass population. Nitrogen, a primary pollutant in wastewater, is present in both organic and inorganic forms. Its removal in constructed wetlands can be accomplished through nitrification and denitrification, processes influenced by factors such as pH, temperature, dissolved oxygen levels, and ammonia concentration. Additionally, the efficiency of nitrification and denitrification depends on factors like plant species and the specific type of wetland used (Zhi et al., 2015). Nitrogen in organic compounds is present in several forms: amino acids, which are peptide chains that form proteins; urea, a product of mammalian metabolism that removes excess ammonia when the body uses amino acids for energy; uric acid, a compound produced by birds and insects; and the nucleic acid components purines and pyrimidines, which are involved in DNA synthesis (Saeed and Sun, 2015). Nitrogen compounds such as ammonium ($NH_{4+}$), nitrite ($NO_2$-), nitrate ($NO_{3-}$), nitrous oxide ($N_2O$), free ammonia ($NH_3$), and dissolved nitrogen ($N_2$) can be efficiently eliminated through the use of SSF wetland networks. The removal processes encompass both biological processes like ammonification, nitrification, denitrification, plant uptake, biomass assimilation, and dissimilatory nitrate reduction and physicochemical processes like ammonia volatilization and adsorption (Saeed & Sun, 2015).

The nitrogen and organic matter elimination in CW is heavily dependent on both environmental factors, such as pH, temperature, and the availability of oxygen, as well as operational considerations like organic carbon, hydraulic loading rates, feeding methods, retention times, pollutant loads, recycling processes, and plant harvesting techniques. Removal of phosphorus in constructed wetland systems is accomplished through immobilization or elimination via chemical precipitation, microbial action, plant absorption, and other physical–chemical processes (Mishra et al., 2024b; Priyadarsini & Das, 2025). Recent studies have revealed removal efficiencies of 77% for biochemical oxygen demand (BOD), 78% for COD, 99% or NH4-N, and a complete removal of 100% for total nitrogen (TN) (Tee et al., 2012). A range of substrate materials such as rice husk, organic wood mulch, zeolite, aggregates, alum sludge, slag, peat, malt, compost, and shale have been used to improve contaminant removal performance in different constructed wetlands systems. Commonly used media substrates include fine to coarse gravel, limestone, zeolite, basic $O_2$ furnace slag, activated carbon, and lightweight aggregates. Additionally, free-floating, emergent plants, and microalgae can achieve phosphorus removal efficiencies exceeding 90% in CWs. The performance of constructed wetlands in removing phosphorus is impacted by several key factors, including the plant's species, hydraulic retention time, temperature, CW parts, influent concentration, and seasonal changes (Tee et al., 2012).

### 4.2 *Toxic Metals/Metalloids Contaminated Water*

Engineered wetlands are a successful method for eliminating heavy metals like copper, cadmium, zinc, lead, and cobalt, with their effectiveness dependent on both the type of medium employed and the type of wastewater being treated. The performance of elimination processes can depend on the form of the metal in the incoming water and the physical and chemical properties of the water, encompassing redox potential, pH, and wetland layout. Treatment of industrial wastewater containing metals using wetlands has become a major focal point due to the substantial potential it offers for environmental restoration (Liu et al., 2020). Soils in wetland areas, characterized by their low oxygen levels and high levels of organic material, are prone to accumulating heavy metals. Wetlands' metal removal process hinges significantly on soil, as well as water and vegetation above it. Metals can penetrate wetlands in dissolved or particulate states via wastewater, with the free metal ions being the most readily available form for biological uptake. These systems contain sediments that serve as both repositories and sinks for the accumulation of metals. Heavy metals in industrial wastewater are a priority for treatment, requiring cost-effective and efficient methods to address them (Liu et al., 2020). Studies have demonstrated that employing constructed wetlands for elimination of metals is a highly effective method, providing promising results. The effectiveness of these wetland treatment systems in removing metals hinges on the quality of the incoming water and the specific metal removal processes used. The wetland's efficiency in removing metal mass ranges from 27 to 99%. According to research, an average of 7.6 kg of metals per hectare is stored annually within the internal compartments of a wetland, amounting to approximately 54% of the incoming metal load (Yeh, 2008). Table 1 illustrates the efficiency of a range of plant species in eliminating heavy metals from various types of wastewater.

Prior research has assessed how well certain heavy metals, such as copper (Cu) and nickel (Ni), are removed from artificial wetlands (Lim et al., 2003). For example, a study examined the effectiveness of a laboratory-scale wetland system intended to treat stormwater runoff in terms of heavy metal removal. According to the findings, heavy metal removal efficiency in artificial wetlands varied between 81.7% and 91.8%. For copper (Cu), the clearance rates were 36.6–372.7 mg/m$^2$/day; for lead (Pb), they were 75.8%–95.3% and 30.8–387 mg/m$^2$/day, while for zinc (Zn), it is 82.8–90.4% and 33.6–362.1 mg/m$^2$/day (Lim et al., 2003).

**Table 1** Heavy metals elimination using CW treatment

| Metals | Types of waste | CW (media used) | Plants | Types of CW | Removal efficiency (%) | Sources |
|---|---|---|---|---|---|---|
| Zn | Industrial wastewater | Sediments | Macrophyte | Free water surface wetland | 55 | Hadad et al. (2006) |
| Ni | Industrial wastewater | Sediments | Macrophyte | Free water surface wetland | 69 | Hadad et al. (2006) |
| Fe | Industrial wastewater | Compost-based and gravel-based | Macrophyte | Free water surface wetland | 83 | Hadad et al. (2006) |
| Pb | Domestic wastewater | Gravel-filled subsurface- flow | *Typha latifolia* | Surface flow | > 75 | Lim et al. (2003) |
| Cd | Domestic wastewater | Gravel-filled subsurface- flow | *Typha latifolia* | Surface flow | > 80 | Lim et al. (2003) |
| Co | Domestic wastewater | Gravel-filled subsurface- flow | *Typha latifolia* | Surface flow | > 70 | Lim et al. (2003) |

CWs are successfully eliminated, with elements such as copper, cadmium, zinc, lead, nickel, and cobalt being removed, primarily due to factors including the choice of media and the type of wastewater being treated. The effectiveness of various removal methods depends on the chemical form of metals entering the system and the physical and chemical properties of the water, which encompass redox potential, pH, and the type of wetland.

### *4.3 Industrial Wastewater*

Discharging industrial wastewater into aquatic environments poses a major threat to ecosystems. Effective industrial wastewater management encounters substantial technical challenges in achieving efficient treatment and producing high-quality treated effluent (Stefanakis, 2018). CWs are recommended as a low-cost, low-maintenance solution for industrial wastewater treatment, especially in developing countries. A study in Argentina explored the feasibility of treating industrial wastewater using a pilot-scale wetland system, achieving average metal removal efficiencies of 83% for Fe, 82% for Cr, 69% for Ni, and 55% for Zn (Lim et al., 2003). Previous studies have shown that Duckweed-microalgae constructed wetlands (DM-CWs) effectively remove various chemical and microbiological contaminants, achieving removal rates of 68% for COD, 71% for BOD, 66% for ammoniacal nitrogen, 21.5% for total phosphorus, 68.5% for total nitrogen, and 47.2% for fecal coliforms. Constructed wetlands (CWs) utilizing *Scirpus grossus* plants also present a potential method for extracting fuel from wastewater (Al-Baldawi et al., 2013).

### *4.4 Oilfield Formed Wastewater*

It is a process in oil and gas industry that has a considerable environmental impact, often leading to the contamination of natural ecosystems, including surface water (Stefanakis, 2018). Oil exploration and making generate substantial volumes of more contaminated water, often containing harmful organic and inorganic pollutants. Oilfield wastewater is often held in large ponds situated in isolated areas, and it can comprise a combination of groundwater, produced water, and chemical additives used in extraction methods. Studies indicate that choosing the right kind of CW can substantially improve the quality of oilfield wastewater (Hassan et al., 2021). Species like *Acacia nilotica*, *Brassica napus*, and *Ricinus communis* are effective in detoxifying oilfield wastewater because of their ability to withstand high saline conditions. A current study showed that CWs can effectively clean and remediate oilfield wastewater, reducing pollutant concentrations, 500–30 mg/L (Stefanakis et al., 2014). The data also indicated a substantial water loss resulting from evapotranspiration, which spans a range of 30%–50%. Studies have demonstrated that constructed wetlands can effectively eliminate nitrogen and phosphorus from petroleum-polluted wastewater.

Treatment of petroleum wastewater generally leads to a 10–12% rise in the pH of the effluent. Constructed wetland systems with a vertical surface flow, planted with water hyacinth (*Eichhornia crassipes*), are capable of effectively treating petroleum refinery or oilfield wastewater, thus meeting discharge standards established by both the Federal Environmental Protection Agency (FEPA) and the WHO for surface water quality (Adewuyi & Olowu, 2012). In a vertical surface flow constructed wetland, biotreatment of petroleum wastewater resulted in high levels of pollutant removal: 91.5% for turbidity, 94.6% for BOD, 80.2% for COD, 92.6% for total petroleum hydrocarbons (TPH), 9 mg (0.4%) for oil and grease, 94.8% for iron, 94% for cadmium, 93% for chromium, 92.5% for lead, 92.2% for nickel, and 57.7% for chloride (Adewuyi & Olowu, 2012).

## 4.5 *Oil Refinery Wastewater*

Petroleum production, storage, distribution, and processing generate substantial volumes of wastewater throughout the extraction process (Jain et al., 2020). Accidental spills of water and fuel mixtures, along with leaks from underground storage tanks, also contribute to wastewater production in oil refineries. The process of oil exploration and extraction produces toxic wastewater, which is comprised of many organic compounds, inorganic substances, and heavy metals, resulting in pollution to both land- and water-based ecosystems (Mustapha, 2018). Industrial operations at oil refineries produce a range of organic pollutants. These are including BTEX compounds—benzene, toluene, and ethylbenzene; xylene; polycyclic aromatic hydrocarbons (PAHs), linear alkylbenzene sulfonates; chlorinated hydrocarbons; and methyl tert-butyl ether (MTBE). A variety of aliphatic and aromatic petroleum hydrocarbons is synthesized, encompassing methyl tert-butyl ether; phenol; 2,3,5,6-tetramethylphenol; naphthalene; xylene; tetradecane; 4-chloro-3-methylphenol; and 3-tert-butylphenol, with each compound existing at different concentration levels (Saien & Shahrezaei, 2012). Due to metal addition and speciation during the refining process, a recent study discovered that refined effluent is more harmful than crude oil (Adewuyi & Olowu, 2012). The high levels of heavy metals from oil exploration and extraction, like iron (Fe), chromium (Cr), cadmium (Cd), lead (Pb), mercury (Hg), silver (Ag), molybdenum (Mo), copper (Cu), zinc (Zn), and arsenic (As), are characteristics of petroleum wastewater (Mustapha, 2018). *Eichhornia crassipes*-planted vertical surface flow constructed wetland (VSFCW) can be used to remediate secondary effluent from petroleum refineries (Adewuyi & Olowu, 2012). Using a vertical surface flow constructed wetland (VSFCW) to treat wastewater with more turbidity (18.30 NTU), BOD (20.40 mg/L), COD (86 mg/L), total petroleum hydrocarbons (TPH) (16.6 mg/L), oil and grease (18.4 mg/L), and heavy metals (cadmium [0.034 mg/L], lead [0.12 mg/L], chromium [0.47 mg/L], iron [1.54 mg/L], nickel [0.09 mg/L], and chloride [1412 mg/L]) can achieve removal efficiencies of 90–95%. The inherent ability of plants is to absorb metals, which is aided by microbial processes (Haberl et al., 2004), plant absorption, and adsorption onto media and sediments in the system (Aslam et al., 2007). This improves the removal of heavy metals from these systems. In built wetlands (CWs), heavy metals can be eliminated from wastewater via a number of processes, such as precipitation as insoluble salts, redox transformation, biosorption, bioaccumulation, and dissimilatory sulfate reduction (Lim et al., 2003). Since calcined sludge was shown to be effective in extracting copper in a prior study, treating various wastes at CW can increase removal efficiency (Tony et al., 2012).

## *4.6 Pharmaceuticals and Personal Care Products (PPCPs)*

Significant developments in the pharmaceutical and cosmetics industries have revolutionized healthcare and beauty services, leading to a global increase in the manufacture of a variety of pharmaceuticals and personal care products (PPCPs) in recent decades (Carvalho et al., 2014). The biotechnology has become a popular and economical approach for treating wastewater by using aquatic macrophytes and microalgae. Pharmaceuticals and personal care products (PPCPs) can be effectively removed from the environment using a proven method called phytoremediation in built wetlands (CWs). Numerous nations' engineers and researchers have concentrated on aquatic macrophytes' capacity to clean and recycle contaminants from industrial and municipal wastewater (Nguyen et al., 2019; Verlicchi & Zambello, 2014). When used to remove pharmaceuticals and personal care products (PPCPs), the phytoremediation process in built wetlands (CWs) can serve as a primary, secondary, or tertiary treatment stage (Verlicchi & Zambello, 2014). According to a prior study, absorption rates of over 99.9% for caffeine, 99.98% for acetaminophen, 99.6% for ibuprofen (IBU), 99.4% for naproxen, and 98% for triclosan were attained using constructed wetland (CW) treatment (Nguyen et al., 2019). Another study showed elimination efficiencies of 70%–90% for naproxen, diclofenac, galaxolide, and tonalide and at above 95% for caffeine, carboxy-ibuprofen, hydrocinnamic acid, hydroxyibuprofen, methyl dihydrojasmonate, oxybenzone, salicylic acid, and ibuprofen (IBU). Using horizontal subsurface flow (HSSF) constructed wetlands planted with *Phragmites australis* to treat water contaminated with pharmaceuticals and personal care products (PPCPs) like diclofenac, IBU, naproxen, tonalide, and bisphenol showed encouraging results, with removal rates of approximately 90% (Ávila et al., 2010). Biodegradation and biosorption are both responsible for the elimination. According to a number of studies, plant species like *Typha latifolia, Phragmites australis, Juncus effusus, Iris pseudacorus,* and *Berula erecta* may efficiently remove or break down up to 90% of ibuprofen (IBU) and iohexol from wastewater. On the other hand, it has been shown that *Scirpus validus*-planted constructed wetlands (CWs) can reduce wastewater under hydroponic conditions by 74% for carbamazepine and 98% for naproxen (Zhang et al., 2013). In horizontal subsurface flow (HSSF), vertical flow (VF), and surface flow (SSF) constructed wetlands (CWs), a variety of plant species, like *Phragmites australis, Typha angustifolia, Spathiphyllum, Oryza sativa, Capsicum annuum, Scirpus validus, Populus nigra*, and *Cucumis sativus*, have been used to remove drugs and personal care products (PPCPs) like caffeine, ibuprofen, naproxen, chlortetracycline, diclofenac, carbamazepine, tonalide, and ketoprofen. It was found that the elimination effectiveness for these PPCPs ranged from 70 to 95% (Schmidt et al., 2002; Zhi et al., 2015). A recent study used horizontal flow (HF) and vertical flow (VF) designed wetland systems to remove five endocrine-altering substances (EDCs): triclosan (TCS), bisphenol A (BPA), nonylphenol (NP), nonylphenol monoethoxylate (NP1EO), and diethoxylate (NP2EO). From 48 to 99%, the removal efficiencies varied. Both aerobic and anaerobic biodegradation and adsorption onto the organic matter in the wetland substrate were the methods used to remove these EDCs. The removal efficiencies of the HF CW system were greater for NP1EO and NP2EO, but lower for BPA when compared to the VF CW system (Papaevangelou et al., 2016).

Persistent organic pollutants (POPs) include all pharmaceuticals and personal care products (PPCPs). SSF and FSF CW systems have also been used to treat other chemicals, like PPCPs, aldrin, chlordane, dieldrin, endrin, heptachlor, mirex, toxaphene, hexachlorobenzene (HCB), polychlorinated biphenyls (PCBs), DDT, polycyclic aromatic hydrocarbons (PAHs), including polychlorinated dibenzo-p-dioxins, and dibenzofurans (Polińska et al., 2021; Wojciechowska, 2013). Most persistent organic pollutants (POPs) can be successfully removed by the phytoremediation process in built wetlands (CWs), with the possibility of almost 100% organic compound removal from wastewater. *Acorus calamus*, *Carex rostrata*, *Phragmites australis*, *Schoenoplectus lacustris* (L.) Palla, *Typha* spp., *Nymphaea alba*, *Nuphar lutea*, *Myriophyllum spicatum*, *Lemna minor*, *Spirodela polyrhiza*, and *Eichhornia crassipes* are among the plant species used in CWs to treat other POPs (Polińska et al., 2021). In constructed wetlands (CWs), the removal effectiveness of the phytoremediation process was shown to vary between 66 and 100% depending on whether free water surface (FWS), subsurface flow (SSF), or vertical subsurface flow (VFS) constructed wetlands were used (Polińska et al., 2021).

## 4.7 Agricultural Wastewater

Discharging wastewater into natural networks such as rivers, lakes, and groundwater can adversely affect the future usability and functionality of these aquatic ecosystems. Rainfall, floods, and runoff can transport wastewater from livestock, leading to contamination (Nan et al., 2020). Surface flow (SF) constructed wetlands help lower nutrient concentrations, organic matter, and suspended solids while supporting wildlife habitats. They also help control mosquito populations, mitigate unpleasant odors associated with constructed wetlands, and minimize human contact (Wang et al., 2018). CWs can effectively treat biochemical oxygen demand (BOD), total suspended solids (TSS), and fecal coliforms in dairy parlor wastewater. CWs planted with various macrophytes demonstrate removal efficiencies of 75–85% for TSS from agricultural wastewater. The efficiency of dissolved humic substance removal is directly related to the retention time; a 7-day retention period resulted in nitrogen removal rates ranging from 47 to 75%, phosphorus removal rates between 37 and 74%, and a 99% reduction in fecal coliforms. In order to achieve removal rates of 91%–99% for BOD, TSS, NH3-N, total Kjeldahl nitrogen (TKN), and fecal coliforms, respectively, CWs embedded with Juncus, Scirpus, Typha, and mixed species are essential for the degradation process (Cronk, 1996). Pesticides from wastewater combined with agricultural runoff can also be efficiently removed by constructed wetlands. The elimination of pesticides from this wastewater is greatly influenced by the sediment load, according to recent studies (Ignatowicz et al., 2020). The effectiveness of removing boscalid from washing water produced during the cleaning of pesticide spraying equipment was investigated in a recent study using two pilot-scale horizontal subsurface flow (HSSF) built wetlands (CWs) (Papaevangelou et al., 2017). The findings demonstrated the efficacy of horizontal subsurface flow (HSSF) systems in treating boscalid-contaminated washing water from agricultural equipment, with removal efficiencies ranging from 49 to 100%. Furthermore, a study discovered that HSSF CW systems are capable of efficiently eliminating pesticides; systems that contain *Typha latifolia* and *Phragmites australis* have been shown to achieve clearance rates of up to 58.4% and 73.7%, respectively. Both plant species' roots, leaves, and shoots contained high levels of terbuthylazine (Gikas et al., 2018). In a different study, washing water from pesticide spraying equipment was treated using three HSSF-built wetlands, with a focus on fluopyram. Due to microbial biodegradation, adsorption onto porous media, and bioaccumulation processes, the HSSF system showed a fluopyram elimination efficacy of more than 96% (Parlakidis et al., 2021).

## 4.8 Fertilizer and Landfill Leachates

Landfill leachate can be dangerous and poses a serious risk to the environment. One of several techniques for handling landfill leachate is the use of constructed wetlands (CWs). Consider a HSSF CW that was planted with leachate from a composting facility in Isfahan, Iran, that had been treated with vetiver (Bakhshoodeh et al., 2017). The results of leachate treated by built wetlands (CWs) and untreated leachate were compared during the five-month trial. The 24 L/d was the stocking rate for the horizontal subsurface flow (HSSF) CW. The findings showed that 75% of the BOD had been eliminated. This was an improvement above the 26% removal rate without CWs, even though the COD removal rate was only about 53%. $NH_3$-N, $NO_3$-N, and total nitrogen (TN) had removal rates of about 70%, 74%, and 74%, respectively (Bakhshoodeh et al., 2017).

## 4.9 *Mine Drainage*

Acid mine drainage (AMD) is a highly hazardous byproduct of mining operations that use a lot of water to extract precious minerals from ores. AMD is carcinogenic and dangerous for the environment. To treat AMD, however, built wetlands (CWs) can be used, which lowers the effluent's toxicity. Utilizing biological, chemical, biogeochemical, geochemical, and physical processes, CWs are an environmentally beneficial method of reducing wastewater toxins. According to Pat-Espadas et al. (2018), AMD has a low pH and is mostly composed of heavy metal compounds.

Sedimentation, sorption, plant accumulation, and precipitation are some of the techniques used to remove heavy metals from acid mine drainage (AMD). Although accumulation in plants is responsible for a component of the removal of heavy metals, this only makes up a minor percentage of the elimination. Conversely, it has been demonstrated that sedimentation considerably lowers the levels of heavy metals in AMD in artificial wetlands (CWs) (Pat-Espadas et al., 2018).

### 4.9.1 Stormwater Runoff

Stormwater runoff is an unpredictable, inconsistent feedstock, which can be treated using constructed wetlands. Interest in using constructed wetlands to clean up stormwater runoff is growing, particularly since the runoff frequently contains high levels of metals including copper (Cu), cadmium (Cd), zinc (Zn), and lead (Pb) (Liu et al., 2020; Vymazal, 2011; Zhang et al., 2014). Researchers have previously examined the efficiency of a laboratory-scale wetland system in eliminating heavy metals from stormwater runoff. These studies found removal efficiencies for the studied metals spanned 81.7–91.8%, while removal rates varied between 36.6 and 372.7 mg/m$^2$/day for copper (Cu), 75.8%–95.3% and 30.8–387 mg/m$^2$/day for lead (Pb), and 82.8%–90.4% and 33.6–362.1 mg/m$^2$/day for zinc (Zn). Constructed wetland systems are capable of treating stormwater runoff, particularly in settings with low hydraulic and pollutant loads, effectively eliminating pollutants, nutrients, and sediments from the runoff (Lim et al., 2003; Qasaimeh et al., 2015).

### 4.9.2 Microplastic on CWs

Synthetic organic polymers, which are commonly known as plastics, play a significant role in various sectors, encompassing household and personal care items, medical devices, textiles, farming, construction materials, infrastructure development, and land management practices. Individuals can come into contact with micro and nanoplastics through personal care products that contain these imperceptible particles, which are too small to be seen without a microscope (Duis & Coors, 2016; Helcoski et al., 2020). Humans and the environment are at risk due to the widespread usage of plastic products, which are incredibly slow biodegradable materials. Microplastic (MP) movement and distribution in wetlands are influenced by vegetation cover, soil texture, substrate apertures, and morphological characteristics. (Chen et al., 2021). In wetlands created by horizontal subsurface flow, macroinvertebrates have been employed as a tertiary treatment to study the distribution and elimination of MPs. (Wang et al., 2020). Species including *Sphagnum palustre*, *Vigna radiata*, *Spirodela polyrhiza*, and *Allium cepa* are used to reduce or control microplastic particles in multistage wetlands, which are crucial for treating microplastics (Mateos-Cárdenas et al., 2021). Strategies for controlling MPs's movement and distribution in water include converting hazardous MP compounds into less toxic or harmless forms through root accumulation or immobilization within plant species (Mateos-Cárdenas et al., 2021). Mangroves possess the ability to absorb considerable amounts of microplastics, with their capacity to do so determined by factors such as density, pigmentation patterns, root systems, decomposed plant matter, and branch structures (Kumar et al., 2021). Microplastic particles attract soil organic carbon, increasing the stability of microplastic in soil sediments (Li et al., 2020). A recent study found an average concentration of 4.9 items/individual in freshwater reeds (Yin et al., 2021). Microplastic fibers contribute positively to plant community growth by reducing soil bulk density and increasing aeration. Plastic particles can also increase plant biomass (Mateos-Cárdenas et al., 2021). Some studies indicate that plastic particles can have a detrimental impact on plant germination and root growth. Although findings on their influence on plant photosynthesis have been inconsistent (Mateos-Cárdenas et al., 2021).

## 5 Microorganisms Used in the Wetland

Wetlands consist of three main components: vegetation, water management systems, and the underlying soil composition, as well as diverse invertebrate and microbial communities. According to Schmidt et al. (2020), bacteria are crucial in the decomposition of pollutants and in facilitating the movement of pollutants from the environment to plants. Heterotrophic bacteria break down organic contaminants such as petroleum hydrocarbons through metabolic processes that utilize organic carbon. In contrast, autotrophic bacteria develop their cells by utilizing carbon dioxide. During aerobic absorption, the organic pollutant functions as an electron donor and oxygen as an electron acceptor, resulting in the breakdown of the pollutant into $CO_2$ through a process that generates secondary products from the original contamination (Nguyen et al., 2019). The organic pollutant acts as an electron giver and oxygen as an electron acceptor during aerobic digestion, breaking the pollutant down into $CO_2$ via a process that creates daughter products of the original contamination (Nguyen et al., 2019). It's essential to note that the final outcomes of these processes can be more hazardous than the pollutants they were designed to eliminate. It is essential that processes deplete at the end and emit $CO_2$ and $H_2O$. Surface flow constructed wetlands' shallow surface and subsurface flow constructed wetlands' uppermost layers offer ideal conditions (i.e., oxygen content) for the aerobic breakdown of pollutants (Nguyen et al., 2019).

The organic pollutant acts as an electron donor and $CO_2$ as an electron acceptor during anaerobic digestion. The organic pollutant acts as the electron donor during denitrification as shown in Table 2. Moreover, nitrate and nitrite act as electron acceptors (Dan et al., 2011; Nguyen et al., 2019).

Certain environmental factors, such as pH, temperature, oxygen content, and sunlight, are ideal for the metabolism of each variety of bacteria. The pH range of 6–7 is generally where most bacteria grow. The majority of bacteria thrive in temperatures between 28 and 30 °C. *Bacillus licheniformis*, *Sulfolobus solfataricus*, and *Thermomonas hydrothermalis* are examples of thermophilic bacterial species that can withstand extremely high temperatures (Mohammad et al., 2017). Wetland microbial activity is influenced by temperature, which impacts how well treatment procedures work. While high temperatures negatively impact microorganisms, cold temperatures slow down microbial activity. According to Maldonado and Quintana (2015), the continued presence of bacteria in wetlands is determined by a variety of factors, such as temperature variations, UV radiation, osmotic stress, nutrient availability, soil pH levels, electron acceptors, and chemical substances. According to Locatelli et al. (2013), besides the wetland environment, certain properties of microorganisms are also vital to their ability to survive; for instance, some bacteria form biofilms to shield themselves from external influences. Bacteria can also endure by forming spores (Maldonado & Quintana, 2015). Bacteria eventually produce endospores with an extremely robust shell that protects them after dying in harsh environments and from malnutrition (McKenney et al., 2013). When living conditions are extremely poor, endospores usually fall into dormancy or sleep; when conditions improve, the spore germinates and grows. (Santos et al., 2021). *Chlorella vulgaris* and *Scenedesmus obliquus* strains can extract selenium from river water through volatilization and accumulation and serve as two examples of algae capable of removing pollutants from artificial wetlands (Huang et al., 2013). Microorganisms and plants have the potential to boost selenium volatilization. Volatilization of selenium depends on nutrition (N and P). Prior studies have demonstrated that increases in P are more likely to cause selenium volatilization than increases in nitrogen (Gupta & Gupta, 2017).

**Table 2** Bacterial strains tangled in nitrification and denitrification in constructed wetlands

| Bacteria | Process | Source |
|---|---|---|
| *Pseudomonas flava WD-3* | Denitrification | Tang et al. (2015) |
| *Alcaligenes faecalis strain WT14* | Heterotrophic nitrification and aerobic denitrification | Chen et al. (2021) |
| *Albidiferax* | Nitrification | Wang et al. (2018) |
| *Candidatus Nitrosotenuis* | Nitrification | Wang et al. (2019) |
| *Nitrosomonas* | Nitrification | Wang et al. (2018) |
| *Nitrosopumilus* | Nitrification | Wang et al. (2018) |

## 6 Economical Assessment and Tasks in the Field

The construction costs for subsurface flow constructed wetlands (SSF CW) and surface flow constructed wetlands (SF CW) are categorized into various phases. The acquisition design, construction, and commissioning are phases, respectively. The first phase encompasses land acquisition, geotechnical investigations, and system design. The second phase involves site mobilization (including the establishment of a site office, fencing, and access roads); construction activities; supervision fees; and commissioning, which includes the trial period (Saeed et al., 2012). Particular care should be given to the expenses related to water accessibility and any supplementary land that may be needed. The total cost of constructed wetlands (CWs) mainly hinges on the design features and components, which can vary substantially from one system to another. Key design components encompass the selection of media, plant species, hydraulic infrastructure, pumps, and filtration systems. Choosing the right components for wastewater treatment is crucial to manage maintenance and operational costs efficiently. Cost-effective CW systems should incorporate minimal upkeep expenses and low operational costs. The expenses cover personnel needs, energy expenditures, upkeep of facility structures; plant renewal; cleaning; replacement of pumps and equipment; and quality control associated with maintenance, which includes sampling and analysis costs (Saeed et al., 2012). Some plant species release significantly more $CH_4$ into the atmosphere than others do. Examples of plants such as *Zizania latifolia, J. effusus*, and *P. australis* are known to produce substantial amounts of $CH_4$ when grown in constructed wetlands. For instance, *Z. latifolia, J. effusus*, and *P. australis* are known to generate significant amounts of $CH_4$ when cultivated in constructed wetlands. In contrast, species such as *Phalaris arundinacea, Cyperus papyrus*, and *T. latifolia* emit lower levels of $CH_4$ than the previously mentioned species. The plant community's composition and intensity also affect $CH_4$ emissions. Using biogas, which includes methane ($CH_2$), as an energy source is a potential strategy for reducing greenhouse gas emissions from artificial wetlands (Avellán & Gremillion, 2019). Constructed wetlands can potentially supply of the demand for unverifiable and harmful solid fuels, by utilizing established small-scale biogas technologies that can act as a community-based energy source. Using this approach can alleviate pressure on ecosystems and decrease dependence on synthetic fertilizers (Avellán & Gremillion, 2019).

## 7 Conclusion and Future Directions

Constructed wetlands serve as a simple and environmentally friendly treatment approach that is crucial to the global ecosystem. Both native and migratory wildlife are provided with habitats, thereby maintaining a balance between land development and the preservation of green areas (Zhi & Ji, 2012). The scientific community and engineering associations have been focusing on the potential uses of constructed wetlands in agricultural settings. These applications include enhancing soil fertility, establishing buffers to protect agricultural areas from storm runoff, and providing shade to mitigate wind effects. Furthermore, CWs can serve as effective strategies for climate change mitigation and adaptation by lessening the negative impacts associated with greenhouse gas emissions. Plants in CWs play a vital role in carbon cycling, as carbon is a primary component of greenhouse gases. Additionally, certain microorganisms can be employed in carbon fixation processes. While CWs have the capacity to sequester substantial amounts of carbon, they may also emit carbon dioxide ($CO_2$), methane ($CH_4$), and nitrous oxide (NO) when disturbed. Biodiversity considerations in CWs are critical, as they support a variety of microorganisms, avian species, and plant life, creating suitable habitats for numerous organisms. The vegetation and microbial communities within CWs provide a conducive environment for a diverse range of species, including air-breathing invertebrates such as water beetles, bugs, and snails, which contribute to ecological cycling. CWs represent a biological technology that, with optimal maintenance, effectively treats wastewater under ambient conditions. Current research regarding the accumulation and retention of microplastics in plants is still limited, although there is significant potential to deepen our understanding of the interactions between micro- and nanoplastics and aquatic macrophytes. Studies have shown that integrated vertical flow CW systems can remove 83–89% of microplastics, while surface flow constructed wetlands achieve a removal efficiency of 27–60%. Furthermore, CW systems have demonstrated effectiveness in removing persistent organic pollutants (POPs), including polychlorinated biphenyls (PCBs) and polycyclic aromatic hydrocarbons (PAHs), with phytoremediation capacities ranging from 66 to 100%. However, several limitations warrant consideration, including the seasonal temperature fluctuations and the specific approval ways that need to be clarified to accurately conclude the contribution of each uptake mechanism to the overall pollutant removal process.

## References

Adewuyi, G. O., & Olowu, R. A. (2012). *Assessment of oil and grease, total petroleum hydrocarbons and some heavy metals in surface and groundwater within the vicinity of NNPC oil depot in Apata, Ibadan metropolis, Nigeria.*

Al-Baldawi, I. A., Sheikh Abdullah, S. R., Anuar, N., Suja, F., & Idris, M. (2013). Performance assessment of pilot horizontal sub-surface flow constructed wetlands for removal of diesel from wastewater by *Scirpus grossus*. *Water Science and Technology, 68*(10), 2271–2278.

Almuktar, S. A., Abed, S. N., & Scholz, M. (2018). Wetlands for wastewater treatment and subsequent recycling of treated effluent: A review. *Environmental Science and Pollution Research, 25*, 23595–23623.

Aslam, M. M., Malik, M., Baig, M. A., Qazi, I. A., & Iqbal, J. (2007). Treatment performances of compost-based and gravel-based vertical flow wetlands operated identically for refinery wastewater treatment in Pakistan. *Ecological Engineering, 30*(1), 34–42.

Avellán, T., & Gremillion, P. (2019). Constructed wetlands for resource recovery in developing countries. *Renewable and Sustainable Energy Reviews, 99*, 42–57.

Ávila, C., Pedescoll, A., Matamoros, V., Bayona, J. M., & García, J. (2010). Capacity of a horizontal subsurface flow constructed wetland system for the removal of emerging pollutants: An injection experiment. *Chemosphere, 81*(9), 1137–1142.

Azubuike, C. C., Chikere, C. B., & Okpokwasili, G. C. (2016). Bioremediation techniques–classification based on site of application: Principles, advantages, limitations and prospects. *World Journal of Microbiology and Biotechnology, 32*, 1–18.

Bakhshoodeh, R., Alavi, N., Majlesi, M., & Paydary, P. (2017). Compost leachate treatment by a pilot-scale subsurface horizontal flow constructed wetland. *Ecological Engineering, 105*, 7–14.

Barik, D. (Ed.). (2018). *Energy from toxic organic waste for heat and power generation*. Woodhead Publishing.

Brovelli, A., Carranza-Diaz, O., Rossi, L., & Barry, D. A. (2011). Design methodology accounting for the effects of porous medium heterogeneity on hydraulic residence time and biodegradation in horizontal subsurface flow constructed wetlands. *Ecological Engineering, 37*(5), 758–770.

Bruch, I., Fritsche, J., Bänninger, D., Alewell, U., Sendelov, M., Hürlimann, H., Hasselbach, R., & Alewell, C. (2011). Improving the treatment efficiency of constructed wetlands with zeolite-containing filter sands. *Bioresource Technology, 102*(2), 937–941.

Carvalho, P. N., Basto, M. C. P., Almeida, C. M. R., & Brix, H. (2014). A review of plant–pharmaceutical interactions: From uptake and effects in crop plants to phytoremediation in constructed wetlands. *Environmental Science and Pollution Research, 21*, 11729–11763.

Chen, Y., Li, T., Hu, H., Ao, H., Xiong, X., Shi, H., & Wu, C. (2021). Transport and fate of microplastics in constructed wetlands: A microcosm study. *Journal of Hazardous Materials, 415*, Article 125615.

Cronk, J. K. (1996). Constructed wetlands to treat wastewater from dairy and swine operations: A review. *Agriculture, Ecosystems & Environment, 58*(2–3), 97–114.

Dan, T. H., Chiem, N. H., & Brix, H. (2011). Treatment of high-strength wastewater in tropical constructed wetlands planted with Sesbania sesban: Horizontal subsurface flow versus vertical downflow. *Ecological Engineering, 37*(5), 711–720.

Das, A. P., & Mishra, S. (2009). Hexavalent chromium [Cr (VI)]: Yellow water pollution and its remediation. *Sarovar Saurabh ENVIS Newsl Wetl Ecosyst, 5*(2).

Duis, K., & Coors, A. (2016). Microplastics in the aquatic and terrestrial environment: Sources (with a specific focus on personal care products), fate and effects. *Environmental Sciences Europe, 28*(1), 2.

Fernandez-Fernandez, M. I., Vega, P. M. D. L., Jaramillo-Morán, M. A., & Garrido, M. (2020). Hybrid constructed wetland to improve organic matter and nutrient removal. *Water, 12*(7), 2023.

Gikas, G. D., Pérez-Villanueva, M., Tsioras, M., Alexoudis, C., Pérez-Rojas, G., Masís-Mora, M., Lizano-Fallas, V., Rodríguez-Rodríguez, C. E., Vryzas, Z., & Tsihrintzis, V. A. (2018). Low-cost approaches for the removal of terbuthylazine from agricultural wastewater: Constructed wetlands and biopurification system. *Chemical Engineering Journal, 335*, 647–656.

Gogoi, A., Mazumder, P., Tyagi, V. K., Chaminda, G. T., An, A. K., & Kumar, M. (2018). Occurrence and fate of emerging contaminants in water environment: A review. *Groundwater for Sustainable Development, 6*, 169–180.

Guittonny-Philippe, A., Masotti, V., Combroux, I., Malleret, L., Boudenne, J. L., Petit, M. E., Monnier, Y., Coulomb, B., Viglione, J., & Laffont-Schwob, I. (2015). Proposal of a new ecotoxicity evaluation tool based on morphological responses of five helophytes to mixtures of pollutants: The Helophyte Development Index. *Ecological Engineering, 77*, 180–188.

Gupta, M., & Gupta, S. (2017). An overview of selenium uptake, metabolism, and toxicity in plants. *Frontiers in Plant Science, 7*, 2074.

Haberl, H., Wackernagel, M., Krausmann, F., Erb, K. H., & Monfreda, C. (2004). Ecological footprints and human appropriation of net primary production: A comparison. *Land Use Policy, 21*(3), 279–288.

Hadad, H. R., Maine, M. A., & Bonetto, C. A. (2006). Macrophyte growth in a pilot-scale constructed wetland for industrial wastewater treatment. *Chemosphere, 63*(10), 1744–1753.

Hassan, I., Chowdhury, S. R., Prihartato, P. K., & Razzak, S. A. (2021). Wastewater treatment using constructed wetland: Current trends and future potential. *Processes, 9*(11), 1917.

Helcoski, R., Yonkos, L. T., Sanchez, A., & Baldwin, A. H. (2020). Wetland soil microplastics are negatively related to vegetation cover and stem density. *Environmental Pollution, 256*, Article 113391.

Herath, I., & Vithanage, M. (2015). Phytoremediation in constructed wetlands. In *Phytoremediation: Management of environmental contaminants* (Vol. 2, pp. 243–263).

Hua, S. C. (2003). The use of constructed wetlands for wastewater treatment. *Wetlands International—Asia Pacific*, Selangor, Malaysia.

Huang, J. C., Suárez, M. C., Yang, S. I., Lin, Z. Q., & Terry, N. (2013). Development of a constructed wetland water treatment system for selenium removal: Incorporation of an algal treatment component. *Environmental Science & Technology, 47*(18), 10518–10525.

Ignatowicz, K., Puchlik, M., & Łozowicki, J. (2020). Removal of pesticides from wastewater by the use of constructed wetlands. *Journal of Ecological Engineering, 21*(1), 219–223.

Jain, M., Majumder, A., Ghosal, P. S., & Gupta, A. K. (2020). A review on treatment of petroleum refinery and petrochemical plant wastewater: A special emphasis on constructed wetlands. *Journal of Environmental Management, 272*, Article 111057.

Kumar, M., Pandit, S., Patel, V., Khanna, N., Nag, M., Lahiri, D., Ray, R. R., Das, A., & Das, D. (2023). Maximization of energy recovery from starch processing wastewater by thermophilic dark fermentation coupled with microbial fuel cell technology. *Geomicrobiology Journal*, 1–12. https://doi.org/10.1080/01490451.2023.2209555

Kumar, R., Sharma, P., & Bandyopadhyay, S. (2021). Evidence of microplastics in wetlands: Extraction and quantification in freshwater and coastal ecosystems. *Journal of Water Process Engineering, 40*, Article 101966.

Li, R., Yu, L., Chai, M., Wu, H., & Zhu, X. (2020). The distribution, characteristics and ecological risks of microplastics in the mangroves of Southern China. *Science of the Total Environment, 708*, Article 135025.

Lim, P. E., Tay, M. G., Mak, K. Y., & Mohamed, N. (2003). The effect of heavy metals on nitrogen and oxygen demand removal in constructed wetlands. *Science of the Total Environment, 301*(1–3), 13–21.

Liu, M., Li, X., He, Y., & Li, H. (2020). Aquatic toxicity of heavy metal-containing wastewater effluent treated using vertical flow constructed wetlands. *Science of the Total Environment, 727*, Article 138616.

Locatelli, A., Spor, A., Jolivet, C., Piveteau, P., & Hartmann, A. (2013). Biotic and abiotic soil properties influence survival of Listeria monocytogenes in soil. *PLoS ONE, 8*(10), Article e75969.

Luederitz, V., Eckert, E., Lange-Weber, M., Lange, A., & Gersberg, R. M. (2001). Nutrient removal efficiency and resource economics of vertical flow and horizontal flow constructed wetlands. *Ecological Engineering, 18*(2), 157–171.

Maldonado, L. A., & Quintana, E. T. (2015). Unexpected properties of micromonosporae from marine origin. *Advances in Microbiology, 5*(6), 452–456.

Marín-Muñiz, J. L., Hernández, M. E., Gallegos-Pérez, M. P., & Amaya-Tejeda, S. I. (2020). Plant growth and pollutant removal from wastewater in domiciliary constructed wetland microcosms with monoculture and polyculture of tropical ornamental plants. *Ecological Engineering, 147*, Article 105658.

Mateos-Cárdenas, A., van Pelt, F. N., O'Halloran, J., & Jansen, M. A. (2021). Adsorption, uptake and toxicity of micro-and nanoplastics: Effects on terrestrial plants and aquatic macrophytes. *Environmental Pollution, 284*, Article 117183.

McKenney, P. T., Driks, A., & Eichenberger, P. (2013). The Bacillus subtilis endospore: Assembly and functions of the multilayered coat. *Nature Reviews Microbiology, 11*(1), 33–44.

Mishra, S., & Das, A. P. (2021). Current treatment technologies for removal of microplastic and microfiber pollutants from wastewater. In *Wastewater treatment: Cutting-edge molecular tools, techniques and applied aspects* (pp. 237–251). https://doi.org/10.1016/B978-0-12-821881-5.00011-8

Mishra, S., & Das, A. P. (2022). Treatment of the wastewater polluted with synthetic microfiber released from washing machine. In B. B. Das, H. Hettiarachchi, P. K. Sahu, & S. Nanda (Eds.), *Lecture Notes in Civil Engineering* (Vol. 207, pp. 109–117). Springer Science and Business Media Deutschland GmbH. https://doi.org/10.1007/978-981-16-7509-6_9/COVER

Mishra, S., & Das, A. P. (2023). Microbial remediation of synthetic microfiber contaminated wastewater. In M. P. Shah (Ed.), *Microbial technologies in industrial wastewater treatment* (pp. 337–350). Springer. https://doi.org/10.1007/978-981-99-2435-6_17

Mishra, S., Dash, D., & Das, A. P. (2022a). Detection, characterization and possible biofragmentation of synthetic microfibers released from domestic laundering wastewater as an emerging source of marine pollution. *Marine Pollution Bulletin, 185*, Article 114254. https://doi.org/10.1016/J.MARPOLBUL.2022.114254

Mishra, S., Dash, D., & Das, A. P. (2024a). Isolation and molecular characterization of polyester-degrading bacterial strains from municipal wastewater and their application in microplastic degradation in contaminated wastewater. *Geomicrobiology Journal*. https://doi.org/10.1080/01490451.2024.2376101

Mishra, S., Dash, D., & Das, A. P. (2024b). Microbial degradation of polyester microfibers using indigenously isolated bacterial strain *Exiguobacterium* Sp. *CLEAN—Soil, Air, Water, 52*(12), e202300343. https://doi.org/10.1002/CLEN.202300343

Mishra, S., Singh, R. P., Rout, P. K., & Das, A. P. (2022). Membrane bioreactor (MBR) as an advanced wastewater treatment technology for removal of synthetic microplastics. In *Development in wastewater treatment research and processes* (pp. 45–60). https://doi.org/10.1016/B978-0-323-85583-9.00022-3

Mohammad, B. T., Al Daghistani, H. I., Jaouani, A., Abdel-Latif, S., & Kennes, C. (2017). Isolation and characterization of thermophilic bacteria from Jordanian hot springs: *Bacillus licheniformis* and *Thermomonas hydrothermalis* isolates as potential producers of thermostable enzymes. *International Journal of Microbiology, 2017*(1), 6943952.

Moral Pajares, E., Gallego Valero, L., & Román Sánchez, I. M. (2019). Cost of urban wastewater treatment and ecotaxes: Evidence from municipalities in southern Europe. *Water, 11*(3), 423.

Mustapha, H. (2018). *Treatment of petroleum refinery wastewater with constructed wetlands*. CRC Press.

Nan, X., Lavrnić, S., & Toscano, A. (2020). Potential of constructed wetland treatment systems for agricultural wastewater reuse under the EU framework. *Journal of Environmental Management, 275*, Article 111219.

Nawaz, H., Asghar, N., Ali, A., Ameer, A., Shehzad, M. A., Nawaz, F., Mehmood, A., Iqbal, M. S., Iqbal, U., Kaleem, M., Khan, A., & Ahmad, K. S. (2023). Optimization of citric acid and EDTA levels under Ni stress using rapeseed *Brassica napus* L. for Phytoremediation. *Water, Air, and Soil Pollution, 234*(4), 1–18. https://doi.org/10.1007/S11270-023-06229-4/METRICS

Nguyen, P. M., Afzal, M., Ullah, I., Shahid, N., Baqar, M., & Arslan, M. (2019). Removal of pharmaceuticals and personal care products using constructed wetlands: Effective plant-bacteria synergism may enhance degradation efficiency. *Environmental Science and Pollution Research, 26*, 21109–21126.

Omondi, D. O., & Navalia, A. C. (2020). Constructed wetlands in wastewater treatment and challenges of emerging resistant genes filtration and reloading. In *Inland waters-dynamics and ecology*. IntechOpen.

Papaevangelou, V. A., Gikas, G. D., Tsihrintzis, V. A., Antonopoulou, M., & Konstantinou, I. K. (2016). Removal of endocrine disrupting chemicals in HSF and VF pilot-scale constructed wetlands. *Chemical Engineering Journal, 294*, 146–156.

Papaevangelou, V. A., Gikas, G. D., Vryzas, Z., & Tsihrintzis, V. A. (2017). Treatment of agricultural equipment rinsing water containing a fungicide in pilot-scale horizontal subsurface flow constructed wetlands. *Ecological Engineering, 101*, 193–200.

Parlakidis, P., Mavropoulos, T., Vryzas, Z., & Gikas, G. D. (2021). Fluopyram removal from agricultural equipment rinsing water using HSF pilot-scale constructed wetlands. In *Environmental science and pollution research* (pp. 1–13).

Pat-Espadas, A. M., Loredo Portales, R., Amabilis-Sosa, L. E., Gómez, G., & Vidal, G. (2018). Review of constructed wetlands for acid mine drainage treatment. *Water, 10*(11), 1685.

Polińska, W., Kotowska, U., Kiejza, D., & Karpińska, J. (2021). Insights into the use of phytoremediation processes for the removal of organic micropollutants from water and wastewater; A review. *Water, 13*(15), 2065.

Priyadarsini, S., & Das, A. P. (2025). Lithium bioleaching: A review on microbial-assisted sustainable technology for lithium bio-circularity. *Journal of Water Process Engineering, 69*, Article 106744. https://doi.org/10.1016/j.jwpe.2024.106744

Qasaimeh, A., AlSharie, H., & Masoud, T. (2015). A review on constructed wetlands components and heavy metal removal from wastewater. *Journal of Environmental Protection, 6*(07), 710.

Saeed, T., & Sun, G. (2012). A review on nitrogen and organics removal mechanisms in subsurface flow constructed wetlands: Dependency on environmental parameters, operating conditions and supporting media. *Journal of Environmental Management, 112*, 429–448.

Saien, J., & Shahrezaei, F. (2012). Organic pollutants removal from petroleum refinery wastewater with nanotitania photocatalyst and UV light emission. *International Journal of Photoenergy, 2012*(1), Article 703074.

Santos, M., Melo, V. F., Monte Serrat, B., Bonfleur, E., Araujo, E. M., & Cherobim, V. F. (2021). Hybrid technologies for remediation of highly Pb contaminated soil: Sewage sludge application and phytoremediation. *International Journal of Phytoremediation, 23*(3), 328–335.

Schmidt, I., Sliekers, O., Schmid, M., Cirpus, I., Strous, M., Bock, E., Kuenen, J. G., & Jetten, M. S. (2002). Aerobic and anaerobic ammonia oxidizing bacteria–competitors or natural partners? *FEMS Microbiology Ecology, 39*(3), 175–181.

Stefanakis, A., Akratos, C. S., & Tsihrintzis, V. A. (2014). VFCW types. In *Vertical flow constructed wetlands* (pp 27–38).

Stefanakis, A. I. (2018). *Introduction to constructed wetland technology* (pp. 1–21). John Wiley & Sons, Ltd.

Stefanakis, A. I. (2020). Constructed wetlands for sustainable wastewater treatment in hot and arid climates: Opportunities, challenges and case studies in the Middle East. *Water, 12*(6), 1665.

Tang, M., Zhang, F., Yao, S., Liu, Y., & Chen, J. (2015). Application of Pseudomonas flava WD-3 for sewage treatment in constructed wetland in winter. *Environmental Technology, 36*(9), 1205–1211.

Tee, H. C., Lim, P. E., Seng, C. E., & Nawi, M. A. M. (2012). Newly developed baffled subsurface-flow constructed wetland for the enhancement of nitrogen removal. *Bioresource Technology, 104*, 235–242.

Thorslund, J., Jarsjo, J., Jaramillo, F., Jawitz, J. W., Manzoni, S., Basu, N. B., Chalov, S. R., Cohen, M. J., Creed, I. F., Goldenberg, R., & Hylin, A. (2017). Wetlands as large-scale nature-based solutions: Status and challenges for research, engineering and management. *Ecological Engineering, 108*, 489–497.

Tony, M. A., Purcell, P. J., & Zhao, Y. (2012). Oil refinery wastewater treatment using physicochemical, Fenton and Photo-Fenton oxidation processes. *Journal of Environmental Science and Health, Part A, 47*(3), 435–440.

Truu, J., Truu, M., Espenberg, M., Nõlvak, H., & Juhanson, J. (2015). Phytoremediation and plant-assisted bioremediation in soil and treatment wetlands: A review. *The Open Biotechnology Journal, 9*(1).

Verlicchi, P., & Zambello, E. (2014). How efficient are constructed wetlands in removing pharmaceuticals from untreated and treated urban wastewaters? A review. *Science of the Total Environment, 470*, 1281–1306.

Vymazal, J. (2011). Constructed wetlands for wastewater treatment: Five decades of experience. *Environmental Science & Technology, 45*(1), 61–69.

Vymazal, J., Zhao, Y., & Mander, Ü. (2021). Recent research challenges in constructed wetlands for wastewater treatment: A review. *Ecological Engineering, 169*, Article 106318.
Wang, M., Zhang, D., Dong, J., & Tan, S. K. (2018). Application of constructed wetlands for treating agricultural runoff and agro-industrial wastewater: A review. *Hydrobiologia, 805*, 1–31.
Wang, Q., Hernández-Crespo, C., Santoni, M., Van Hulle, S., & Rousseau, D. P. (2020). Horizontal subsurface flow constructed wetlands as tertiary treatment: Can they be an efficient barrier for microplastics pollution? *Science of the Total Environment, 721*, Article 137785.
Wojciechowska, E. (2013). Removal of persistent organic pollutants from landfill leachates treated in three constructed wetland systems. *Water Science and Technology, 68*(5), 1164–1172.
Yeh, T. Y. (2008). Removal of metals in constructed wetlands. *Practice Periodical of Hazardous, Toxic, and Radioactive Waste Management, 12*(2), 96–101.
Yin, L., Wen, X., Huang, D., Zeng, G., Deng, R., Liu, R., Zhou, Z., Tao, J., Xiao, R., & Pan, H. (2021). Microplastics retention by reeds in freshwater environment. *Science of the Total Environment, 790*, Article 148200.
Zhang, C., Yu, Z. G., Zeng, G. M., Jiang, M., Yang, Z. Z., Cui, F., Zhu, M. Y., Shen, L. Q., & Hu, L. (2014). Effects of sediment geochemical properties on heavy metal bioavailability. *Environment International, 73*, 270–281.
Zhang, D. Q., Hua, T., Gersberg, R. M., Zhu, J., Ng, W. J., & Tan, S. K. (2013). Carbamazepine and naproxen: Fate in wetland mesocosms planted with *Scirpus validus*. *Chemosphere, 91*(1), 14–21.
Zhi, W., & Ji, G. (2012). Constructed wetlands, 1991–2011: A review of research development, current trends, and future directions. *Science of the Total Environment, 441*, 19–27.
Zhi, W., Yuan, L., Ji, G., & He, C. (2015). Enhanced long-term nitrogen removal and its quantitative molecular mechanism in tidal flow constructed wetlands. *Environmental Science & Technology, 49*(7), 4575–4583.

MIX
Papier aus verantwortungsvollen Quellen
Paper from responsible sources
FSC® C105338

If you have any concerns about our products,
you can contact us on
**ProductSafety@springernature.com**

In case Publisher is established outside the EU,
the EU authorized representative is:
**Springer Nature Customer Service Center GmbH**
**Europaplatz 3, 69115 Heidelberg, Germany**

Printed by Libri Plureos GmbH
in Hamburg, Germany